Sensitivity of Mangrove Ecosystem to Changing Climate

Abhijit Mitra

Sensitivity of Mangrove Ecosystem to Changing Climate

Springer

Abhijit Mitra
Department of Marine Science
University of Calcutta
Kolkata, West Bengal, India

ISBN 978-81-322-1508-0   ISBN 978-81-322-1509-7 (eBook)
DOI 10.1007/978-81-322-1509-7
Springer New Delhi Heidelberg New York Dordrecht London

Library of Congress Control Number: 2013945852

Printed on acid-free paper

Springer is part of Springer Science+Business Media (www.springer.com)

Summary

There are no differences of opinion in saying that the climate of the planet Earth is changing with the passage of time. Although several factors beyond our control (like variations in the Earth's orbital characteristics, volcanic eruptions, variations in solar energy output and plate tectonics) do play a major role in this change, the influence of human intervention cannot be ignored. The rapid industrialization and urbanization in the recent *era* has hiked up the concentration of greenhouse gases in the atmosphere, and as the successive events, the temperature has increased, the seawater has expanded, the agricultural productivity has gone down, the existing socio-economic profile has altered, the vector-borne diseases have proliferated and the biodiversity of the world oceans and estuaries has changed. Question arises whether these chain reactions are irreversible in nature. Question also arises regarding the rate of alteration (acceleration or retardation) of environment around us, and we are uncertain about these answers. All the nations of the world have converged on a common platform to fight this invisible demon with irregular and uncertain behaviour. There is no point other than adaptation to minimize the impact of climate change. Adaptation here refers to the ability to respond and adjust to actual or potential impacts of changing conditions of the climate in ways that moderate harm or takes advantage of any positive opportunities that the climate may afford. It is therefore a mode of coping with the surrounding environment through development and alteration/modification of some attitude, habit, habitat, etc. Even changes in the policy level or planning processes or decision-making processes are also the part of adaptation.

This manuscript critically focuses on all the events of climate change starting from causes to effects on the matrix of mangrove ecosystem. Sundarbans being the biggest mangrove chunk of the globe has been used as the test bed to distinguish between human factors (effects of dam construction, barrage discharge, industries, etc.) and natural causes of environmental change, which often creates a *noise* while explaining the issue of climate change. The emphasis has been given on Indian Sundarbans because the subcontinent has been identified as one among 27 countries which are most vulnerable to the impact of global-warming-related accelerated sea-level rise (UNEP 1989). Available literatures state that the sea level is rising at the rate of 3.14 mm/year in Indian Sundarbans, which is greater than the average sea-level rise in rest part of the globe. The Indian Sundarbans sustains 34 true mangrove species, which not only act as the line of defence against sea-level rise, cyclones and tidal surges, but their role in carbon sequestration and minimization of carbon dioxide at the regional level has been critically analyzed in the present book.

Apart from ecological approach in reducing carbon dioxide level of the atmosphere, the existing policies and laws have also been discussed in the matrix of Indian subcontinent. In India, there are no laws, policies, insurance or funds to protect, compensate and rehabilitate the vulnerable communities related to climate change or weather extremes. This creates a situation where the properties and life along the coastline are exposed to extreme risks. Building of cyclone shelters and tidal surge warning system along with a well-orchestrated disaster management plan can ameliorate the adversity of impact to a certain extent. The authors of the present book feel the necessity of introducing some kind of climate insurance policies (like health insurance and car insurance) to cover the risk of climate-change-related disasters.

The book has its own individuality not because of the lucidness of language and presentation of relevant case studies but for bringing in frontline the long-term data (30 years) of this biggest mangrove chunk, which have just started to see the light of publication. Criticism and debates are the hallmarks of any climate-change-related books, as a 'noisy zone', always lies in such subjects, where it is difficult to segregate the effects of climate change from man-made intervention. In this connection, the Farakka dam issue has been discussed very clearly with its major role in regulating the salinity at the confluence of the Hooghly–Bhagirathi system and Bay of Bengal.

Finally the untapped marine and estuarine living resources (like seaweeds and edible oyster) have also been highlighted with the aim to develop alternative livelihood schemes as steps towards adapting to a new situation of rising salinity.

Contents

Acknowledgements

The author, from the innermost core of mind and heart, acknowledges the following personalities for their direct and indirect contributions:

- *Professor Amalesh Choudhury*: The author developed the base on mangrove ecology from Professor Choudhury not only in the capacity of a student but also as a disciple. He taught the author to work in the rigorous environment of Sundarbans with sincerity, dedication and tireless effort to reveal the ground reality of the ecosystem.
- *Dr. Atanu Kumar Raha*: The author got inspiration from Dr. Raha when he was the PCCF and Head of Forest Force, Govt. of West Bengal, during 2007–2012. Dr. Raha with excellent expertise in remote sensing and GIS technology helped the author to develop map on phytopigment variations in Sundarban ecosystem. Few maps in the book have been taken from the Ph.D. thesis of Dr. Raha.
- *Mr. Goutam Roychowdhury*: The author and his team members greatly acknowledge the infrastructural facility offered by Techno India University, Salt Lake Campus, while preparing the manuscript. The author also received inspiration for touching the sky from Sri Roychowdhury. The statistical software used to reflect the spatio-temporal variation in explaining the biotic community structure in mangrove ecosystem was offered under the guidance of Sri Roychowdhury.
- *Dr. Kakoli Banerjee*: The author gratefully acknowledges the field data generated by Dr. Banerjee (in the field of phytoplankton) without which the chapter on plankton could not be completed. Dr. Banerjee also critically scanned the manuscript and provided constructive inputs for the betterment of the manuscript.
- *Dr. Sufia Zaman*: While writing the manuscript the entire chapterization and corrections were critically done by Dr. Zaman. In addition the data on fish community and mangrove biomass were collected by Dr. Zaman during her field work in Indian Sundarbans. The author acknowledges her effort to bring the data of Indian mangrove ecosystem to the view of the readers.
- *Dr. Subhro Bikash Bhattacharyya*: The author is indebted to Dr. Bhattacharyya for his tireless sampling from the Sundarban Mangrove Forest, which helped the author to carry out the scientific analysis of soil, water and plankton community from time to time.

Apart from this the author cannot forget the contributions of his scholars like Dr. Ananda Gupta, Dr. Amitabha Aich, Dr. Kiran Lal Das, Dr. Aftab Alam, Ms. Rajrupa Ghosh, Ms. Kasturi Sengupta, Mr. Kunal Mondal,

Mr. Subhasmita Sinha, Ms. Mahua Roychowdhury, Mr. Saumya Kanti Ray and Mr. Saurav Sett for their inputs and reference collections.

The author also acknowledges the support team of field like Mr. Bidhan Mondal, Mr. Deb Kumar Samanta and Mr. Sanjoy Halder.

Finally the author expresses his gratefulness to his wife Shampa, daughter Ankita and mother Manjulika whose inspirations and encouragements act as booster to complete the manuscript.

About the Author

Abhijit Mitra, Associate Professor and former Head, Department of Marine Science, University of Calcutta (India), has been active in the sphere of Oceanography since 1985. He is now the Adjunct Professor of Techno India University, Salt Lake Campus, Kolkata. He obtained his Ph.D. as NET qualified scholar in 1994. Since then, he joined Calcutta Port Trust, WWF (World Wide Fund) and Ministry of Environment & Forests, Govt. of India, in various capacities to carry out research programmes on environmental science, biodiversity conservation, climate change and carbon sequestration. He has to his credit about **240** scientific publications in various national and international journals and **20** books of postgraduate standards. Dr. Mitra is presently the member of several committees like PACON International, IUCN and SIOS and has successfully completed about **15** projects on biodiversity loss in fishery sector, coastal pollution, alternative livelihood, climate change and carbon sequestration. Dr. Mitra also visited as faculty member and invited speakers in several foreign universities of Singapore, Kenya and the USA. Presently his domain of expertise includes environmental science, mangrove ecology, sustainable aquaculture, climate change and carbon sequestration. Dr. Mitra successfully supervised **17** Ph.D. scholars in the branches like Marine Science, Chemistry, Environmental Science and Zoology.

About the Book

The word 'mangrove' is being used in the Oxford English Dictionary since 1613, while the Americans, the Spanish and the Portuguese use the word 'mangle' which is interpreted from the Haytian Arawak language for the trees and shrubs of the genus *Rhizophora*. Later, the word was modified as 'mangrove' and also included other tree genera growing in the intertidal zone and exposed to brackish water. At present the term 'mangrove' or 'mangrove forest' or 'mangrove ecosystem' is commonly accepted and has become the synonym of 'tidal forest'. Scientists theorize that the earliest mangrove species originated in the Indo-Malayan region. This theory is supported by the fact that there are more mangrove species present in this region than anywhere else in the world. Due to many species, unique floating propagules and seeds, early mangroves spread westward, by ocean currents, to India and East Africa, and eastward to the Americas, arriving in Central and South America during the upper Cretaceous period and lower Miocene epoch (66 and 23 million years ago). This may explain why the mangroves of the Americas contain fewer and similar colonizing species, while those of Asia, India and East Africa contain a fuller range of mangrove species.

Mangroves are basically salt-tolerant forest ecosystems found mainly in tropical and subtropical intertidal regions. Till about 1960s, mangroves were largely viewed as 'economically unproductive areas' and were therefore destroyed for reclaiming land for various economic and commercial activities. Gradually, with the passage of time, the economic and ecological benefits of mangroves have become visible and their importance is now well appreciated.

It is now accepted that these rich ecosystems provide a wide range of ecological and economic products and services and also support a variety of other coastal and marine ecosystems, which again provide several economic and ecological benefits. The full value of mangroves is, however, still not recognized in most economies, as most of these benefits have not yet entered the market (e.g. carbon credit from mangrove forest and soil). Consequently, mangroves have been highly undervalued and neglected ecosystems, resulting in their heavy depletion and degradation in most parts of the world. Recently researchers have initiated compiling their value in monetary terms in some parts of the world. Most of this valuation is for the developed countries, and even when it is for the developing countries, the nature of benefits and costs estimated are very much different from the same in India, Indonesia, Sri Lanka, Pakistan or Bangladesh where mangroves play an extremely important role in the life and livelihood of coastal population. It is therefore a need to compute monetary value of mangroves in developing countries to understand its importance in the economic matrix. Today, mangroves are observed in about 30 countries in tropical subtropical regions covering an area of about 99,300 km². However, during the past 50 years, over 50 % of the mangrove cover has been lost, mainly because of the increased pressure of human activities like shrimp farming and agriculture, forestry, salt extraction, urban development, tourist development and infrastructure. Also, dam on rivers and contamination of seawaters caused by heavy metals, oil spills, pesticides and other products, etc. have been found to be responsible for the decline of mangroves. Shrimp farming is directly linked to the loss of mangrove in tropical countries. About 2,000 km² in Vietnam, 350 km² in India and 90 km² in Bangladesh have been lost to shrimp farming.

Climate change and subsequent sea-level rise is also a major threat to mangrove ecosystems of the world. Researchers are of the opinion that mangroves may be affected by climate change-related increases in temperature and sea-level rise. Although the temperature effect on growth and species diversity is not known, sea-level rise may pose a serious threat to these ecosystems. In Bangladesh, for instance, there is a threat to mangrove species in the three distinct ecological zones (hypersaline, mid-saline and hyposaline regions) that make up the Sundarbans – the largest continuous mangrove area in the world. If the saline waterfront moves further inland, *Heritiera fomes* (the dominant species in the landward freshwater zone of Bangladesh and locally referred to as Sundari) could be threatened. Species in the other two ecological zones *Nypa fruticans* and *Sonneratia apetala* also could suffer. These changes could result severe adverse impacts on the socio-economic profile of the coastal zone. A large section of the population – who are directly employed in the industries that use raw materials from the mangrove forests (e.g. wood-cutting; collection of thatching materials, honey, beeswax and shells; fishing) – may lose their sources of income. Sea-level rise also may threaten a wide range of mammals, birds, amphibians, reptiles and crustaceans living in the mangrove forests.

Some ecologists believe that mangrove communities are more likely to survive the effects of sea-level rise in macrotidal, sediment-rich environments – such as northern Australia, where strong tidal currents redistribute sediment – than in microtidal, sediment-starved environments like those in many small islands (e.g. in the Caribbean). Most small islands fall within the latter classification; therefore, they are expected to suffer reductions in the geographical distribution of mangroves. Furthermore, where the rate of shoreline recession increases, mangrove stands are expected to become compressed and suffer reductions in species diversity in the face of rising sea levels. The present book addresses all these important issues in separate chapters with some interesting case studies whose data may serve as a pathfinder for future researches in the sphere of the influence of climate change on mangrove ecosystem. According to me, what seems to be needed, at the least, is a massive programme of research and development to broaden the choices of alternative livelihood and a firm national and international commitment to mangal conservation instead of few sporadic afforestation programmes of mangrove species that are often taken up by the government sectors and private organizations. The role of mangroves in the sector of bioremediation is a unique feather in the crown of this coastal and brackish-water vegetation that may be taken up by the coastal industries in order to maintain the health of ambient environment.

That the developing nations cannot afford to be without more brackish-water-related livelihood options is the general essence of this book. But policy questions such as those mentioned in Chap. 2 (in the form of conservation policies) do not receive much attention here. Instead, this book seeks to discover and to assess the vulnerability of climate change on mangrove flora and fauna, their role in carbon sequestration and some interesting case studies by some groups of dedicated researchers that may serve as the basis of future climate-related policies.

The Author

Climate Change: A Threat of the *Era*

1

It is only in the most recent, and brief, period of his tenure that man has developed in sufficient numbers, and acquired enough power, to become one of the most potentially dangerous organisms that the planet has ever hosted.

John McHale

1.1 Climate Change: An Overview

Weather changes all the time. It is highly dynamic in nature. The average pattern of weather, called *climate*, usually remains uniform for centuries if it is left to itself. However, the Earth is not being left alone. People are taking multidimensional actions that are gradually changing the morphology, physiology and anatomy of the planet Earth and its climate in large scale. The single human activity that is most likely to have a large impact on the climate is the burning of 'fossil fuels' such as coal, oil and natural gas. These fuels contain carbon. Burning them liberates carbon dioxide gas in the atmosphere. Since the early 1800s, when people started burning large amounts of coal and oil, the amount of carbon dioxide in the Earth's atmosphere has increased by nearly 30 %, and average global temperature appears to have risen between 1 and 2 °F. This increment of temperature is keenly related to the basic property of the gas. Carbon dioxide gas traps solar heat in the atmosphere, partly in the same way as glass traps solar heat in a sunroom or a greenhouse. For this reason, carbon dioxide is sometimes called a 'greenhouse gas'. As more carbon dioxide is added to the atmosphere, solar heat faces more

trouble in getting out. The result is that, if everything else remains unchanged, the average temperature of the atmosphere would increase. With increased rate of industrialization and urbanization, the demand for fossil-fuel-based energy has hiked up. As people burn more fossil fuels for energy, they add more carbon dioxide to the atmosphere. This creates a blanket of carbon dioxide over the Earth's surface, which allows only the short waves of the sun to penetrate the Earth's atmosphere, but prevents the long-wave radiations (emitted from the Earth's surface) to get out. If this activity continues for a long period of time, the average temperature of the atmosphere will almost certainly rise. This is commonly referred to as *global warming*. Global warming is thus the increase in the average temperature of the Earth's near-surface air and oceans in recent decades and its projected continuation. The term 'global warming' is a subset of the universal set climate change, which also encompasses another subset called 'global cooling'. The United Nations Framework Convention on Climate Change (UNFCCC) uses the term 'climate change' for human-induced changes and 'climate variability' for other changes. Climate change is therefore any long-term significant change in the

A. Mitra, *Sensitivity of Mangrove Ecosystem to Changing Climate*,
DOI 10.1007/978-81-322-1509-7_1, © Springer India 2013

'average weather' that a given region experiences and involves changes in the variability or average state of the atmosphere over durations ranging from decades to millions of years. The roots of these changes can be related to several dynamic processes on Earth, external forces including variations in sunlight intensity and more recently by human activities.

Climate change poses various effects on physical, chemical and biological components of the planet Earth. If global warming occurs, every day or every place will not be warmer uniformly, but on average, most places will be warmer. This will cause changes in the amount and pattern of rain and snow, in the length of growing seasons, in the frequency and severity of storms and in sea level. Grasslands, forests, oceans and other eco-systems and their flora and fauna in the natural environment will all be affected. Acidification of seawater is also associated with climate change, which again has serious adverse effects on coral reef ecosystem, molluscan community and several organisms with calcareous components.

Adverse impact of climate change will also be witnessed on the economic profile of an area by increasing the number of environmental refugees. Changing climate will displace millions of the poor people from coastal zone due to increased frequency of cyclones and natural hazards, sea-level rise or alteration in the crop production volume and pattern. Small island states and low coastal areas and deltas such as southern Bangladesh are most at risk. In many cases, those displaced will have few opportuni-ties to re-establish their lives, except in urban areas, where livelihood opportunities are limited without the skills, capital and contacts needed to cope up with urban life. Even where people are not physically displaced, rising seas will reduce the reservoir of natural capital in ecosystems such as coastal fisheries, mangroves and wetlands that are essential to sustain the livelihood patterns of poor communities. The dangers of salinization of drinking water supplies and agricultural lands (due to ingression of seas) are also important threats to the socio-economic condition of the people.

1.2   Causes of Climate Change

The general state of the Earth's climate is a function of the amount of energy stored by the climate system. More specifically, it can be stated that the Earth's climate is regulated by the balance between the amount of energy the Earth receives from the Sun, in the form of light and ultraviolet radiation, and the amount of energy the Earth releases back to space, in the form of infrared heat energy. The basic causes of climate change involve any process that can alter this global energy balance. Scientists call this 'climate forcing'. Climate forcing 'forces' or induces the climate to change, although the acceleration or pace of the process is highly variable.

There are many climate forcing processes, but broadly speaking, they can be classified into internal and external types (Fig. 1.1). *External processes* operate outside the planet Earth and include changes in the global energy balance due to extraterrestrial factors like variations in the Earth's orbit around the Sun and changes in the amount of energy received from the Sun. *Internal processes* operate from within the Earth's climate system and include changes in the global energy balance due to changes in ocean circulation or changes in the composition of the atmosphere. Other climate forcing processes include the impacts of large volcanic eruptions and collisions with comets or meteorites. Luckily, the Earth is not hit by large comets or meteorites very often, perhaps every 20–30 million years or so, and therefore their associated climate changes occur rarely throughout Earth's history. However, other causes of climate change influence the Earth on much shorter time scales, with changes sometimes occurring within a single generation. Indeed, our present oscillation of the composition of atmosphere due to emission of greenhouse gases may be causing the global climate to change with an increased trend of atmospheric temperature. This man-made climate change associated with increasing trend of atmo-spheric temperature is popularly known as global warming.

Fig. 1.1 External and
internal processes affecting
Earth's climate

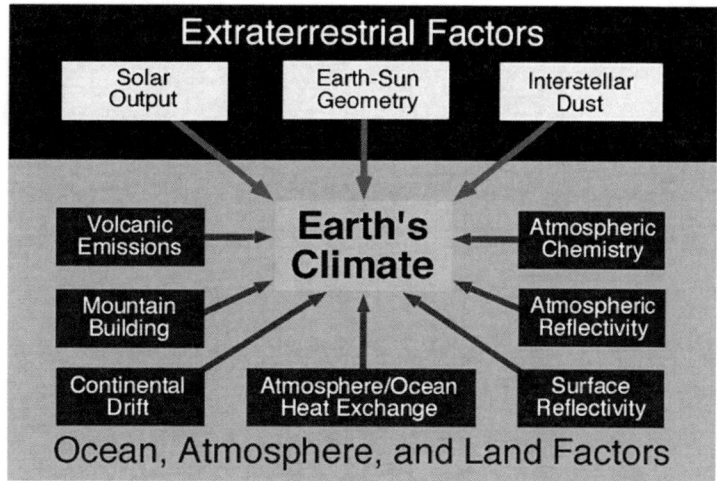

Fig. 1.1 External and internal processes affecting Earth's climate

For convenience of the readers, the causes of climate change may be divided into two broad domains: natural and man-made.

1.2.1 Natural Factors

The work of climatologists have found evidences to suggest that only a limited number of factors are primarily responsible for most of the past episodes of climate change on the Earth. These factors include:

- Variations in the Earth's orbital characteristics
- Atmospheric carbon dioxide variations
- Volcanic eruptions
- Variations in solar output
- Plate tectonics

1.2.1.1 Variations in the Earth's Orbital Characteristics

The Milankovitch theory suggests that normal cyclical variations in three of the Earth's orbital characteristics are probably responsible for some past climatic change. The basic idea behind this theory assumes that over time these three cyclic events result in the variation of the amount of solar radiation that is received on the surface of the planet Earth.

The *first cyclical variation*, known as *eccentricity*, controls the shape of the Earth's orbit around the Sun. The orbit gradually changes from being elliptical to being nearly circular and then back to elliptical in a period of about 100,000 years. The greater the eccentricity of the orbit (i.e. the more elliptical it is), the greater is the variation in solar energy received at the top of the atmosphere between the Earth's closest (perihelion) and farthest (aphelion) approach to the Sun. Currently, the Earth is experiencing a period of low eccentricity. The difference in the Earth's distance from the Sun between perihelion and aphelion (which is only about 3 %) is responsible for approximately a 7 % variation in the amount of solar energy received at the top of the atmosphere. When the difference in this distance is at its maximum (9 %), the difference in solar energy received is about 20 %.

The *second cyclical* variation results from the fact that, as the Earth rotates on its polar axis, it wobbles like a spinning top changing the orbital timing of the equinoxes and solstices (Fig. 1.2). This effect is known as the *precession of the equinox*. The precession of the equinox has a cycle of approximately 26,000 years. According to illustration (a), the Earth is closer to the Sun in January (perihelion) and farther away in July (aphelion) at the present time. Because of precession, the reverse will be true in 13,000 years and the Earth will then be closer to the Sun in July (illustration b). This means, of

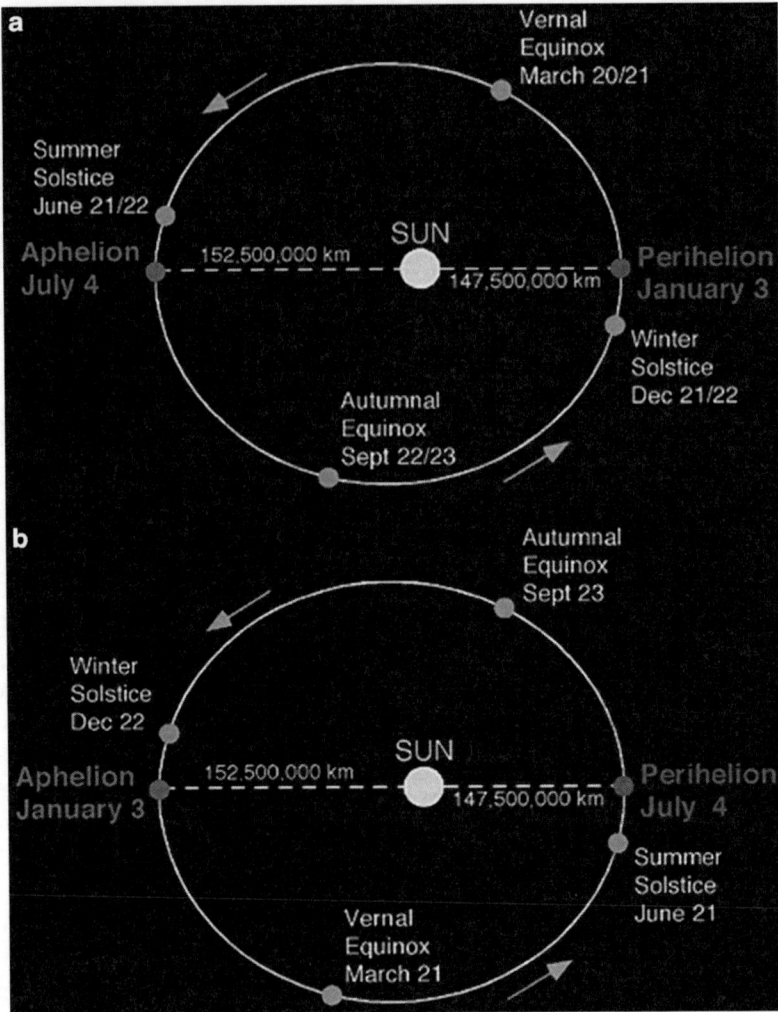

Fig. 1.2 Modification of the timing of aphelion and perihelion over time. (**a**) Today. (**b**) 13,000 years into the future

course, that if everything else remains constant, 13,000 years from now, seasonal variations in the Northern Hemisphere should be greater than at present (colder winters and warmer summers) because of the closer proximity of the Earth to the Sun.

The *third cyclical variation* is related to the changes in the *tilt* (obliquity) of the Earth's axis of rotation over a 41,000-year period. During the 41,000 year cycle, the tilt can deviate from approximately 22.5–24.5°. At the present time, the tilt of the Earth's axis is 23.5°. When the tilt is small, there is less climatic variation between the summer and winter seasons in the middle and high latitudes. Winters tend to be milder and

summers cooler. Warmer winters allow for more snow to fall in the high latitude regions. When the atmosphere is warmer, it has a greater ability to hold water vapour, and therefore more snow is produced at areas of frontal or orographic uplift. Cooler summers cause snow and ice to accumulate on the Earth's surface because less of this frozen water is melted. Thus, the net effect of a smaller tilt would be more extensive formation of glaciers in the polar latitudes.

Periods of a larger tilt result in greater seasonal climatic variation in the middle and high latitudes. At these times, winters tend to be colder and summers warmer. Colder winters produce more snow because of lower atmospheric temperatures.

As a result, more snow and ice accumulates on the ground surface. Moreover, the warmer summers produced by the larger tilt provide additional energy to melt and evaporate the snow that fall and accumulate during the winter months. In conclusion, glaciers in the polar regions should be generally receding, with other contributing factors constant, during this part of the obliquity cycle.

Computer models and historical evidences suggest that the Milankovitch cycles exert their greatest cooling and warming influence when the troughs and crests of all three cycles coincide with each other.

1.2.1.2 Atmospheric Carbon Dioxide Variations

Studies on long-term climate change have revealed a connection between the concentrations of carbon dioxide in the atmosphere and mean global temperature. Carbon dioxide is one of the most important gases responsible for the greenhouse effect. Certain atmospheric gases, like carbon dioxide, water vapour and methane, are able to alter the energy balance of the Earth by their property to absorb long-wave radiation emitted from the Earth's surface. The net result of this process is the rise of Earth's temperature. Without the greenhouse effect, the average global temperature of the Earth would be around $-18\,°C$ rather than the present $15\,°C$.

Researchers of the 1970s CLIMAP project documented strong evidence in deep-ocean sediments regarding the variations in the Earth's global temperature during the past several hundred thousand years of the Earth's history. Other subsequent studies have confirmed these findings and have discovered that these temperature variations were closely correlated to the concentration of carbon dioxide in the atmosphere and variations in solar radiation received by the planet as controlled by the Milankovitch cycles. Measurements indicated that atmospheric carbon dioxide levels were about 30 % lower during colder glacial periods. It was also theorized that the oceans were a major store of carbon dioxide and that they controlled the movement of this gas to and from the atmosphere. The amount of carbon

dioxide that can be held in oceans is a function of temperature. Carbon dioxide is released from the oceans when global temperatures become warmer and diffuses into the ocean when temperatures are cooler. Initial alterations in global temperature were triggered by changes in received solar radiation by the Earth through the Milankovitch cycles. The increase in carbon dioxide then amplified the global warming by enhancing the greenhouse effect.

Over the past three centuries, the concentration of carbon dioxide has been increasing in the Earth's atmosphere because of human influences (Fig. 1.3). Human activities like the burning of fossil fuels, conversion of natural prairie to farmland, intense industrialization, urbanization and deforestation have caused the release of carbon dioxide into the atmosphere. From the early 1700s, carbon dioxide has increased from 280 to 380 ppm in 2005. Many scientists believe that higher concentrations of carbon dioxide in the atmosphere will accelerate the greenhouse effect, making the planet warmer. Scientists also believe that the present *era* is already experiencing global warming due to an enhancement of the greenhouse gases. Most computer climate models suggest that the globe will warm up by 1.5–4.5 °C if carbon dioxide reaches the predicted level of 600 ppm by the year 2050. The direct correlation between the concentrations of carbon dioxide and temperature of the Earth's atmosphere is shown in Fig. 1.4.

1.2.1.3 Volcanic Eruptions

Climatologists have observed a connection between large explosive volcanic eruptions and short-term climatic changes (Fig. 1.5) after considering long-term data of these two variables. For example, one of the coldest years in the last two centuries occurred the year following the Tambora volcanic eruption in 1815. A number of regions across the planet Earth witnessed such lowering of atmospheric temperature. Several other major volcanic events also showed a pattern of cooler global temperatures lasting 1–3 years after their eruption.

At first, scientists thought that the dust emitted into the atmosphere from large volcanic eruptions

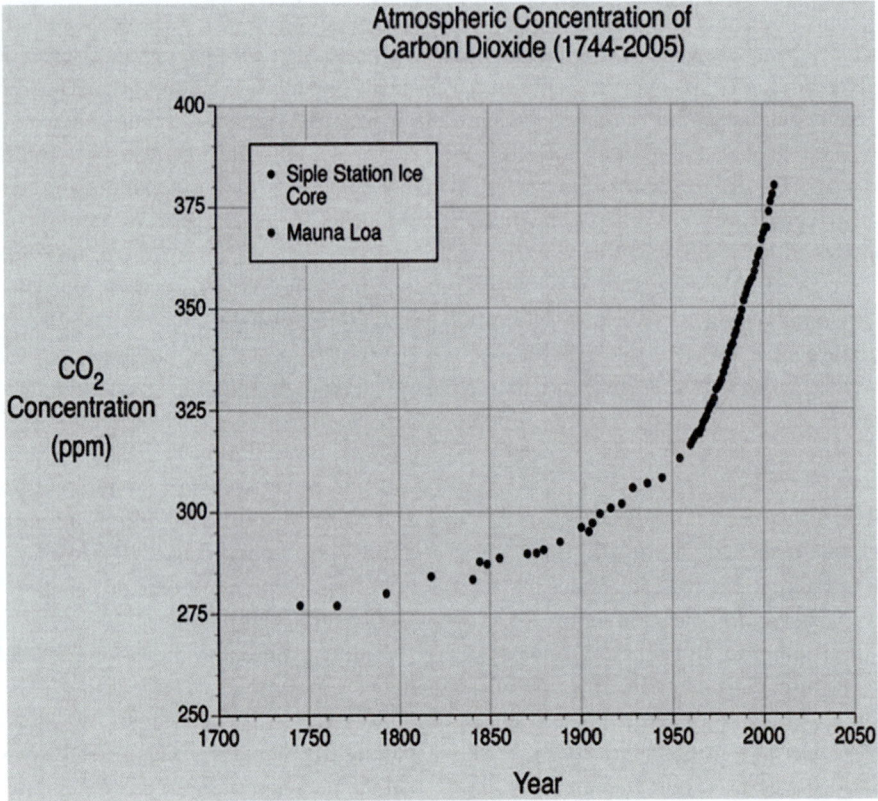

Fig. 1.3 The following graph illustrates the rise in atmospheric carbon dioxide from 1744 to 2005. Noteworthy is the exponential increase in carbon dioxide's concentration in the atmosphere (*Source*: Pidwirny (2006), http://www.physicalgeography.net/fundamentals/7y.html)

Fig. 1.4 Global average temperature and carbon dioxide concentrations during 1880–2004

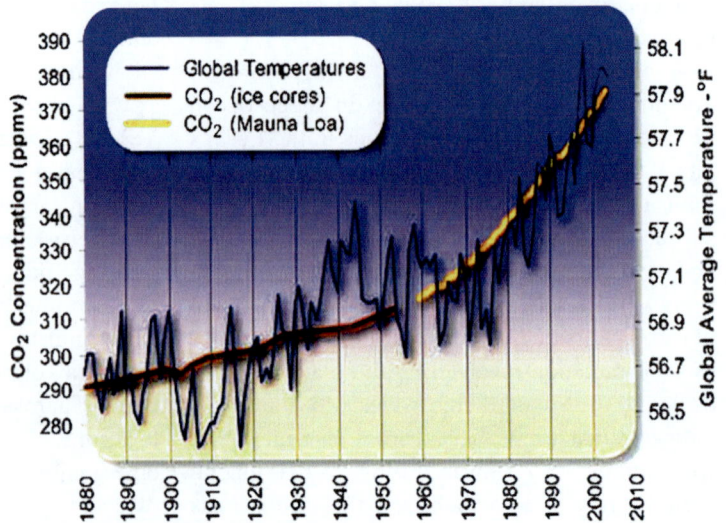

Fig. 1.5 Explosive volcanic eruptions have been shown to have a short-term cooling effect on the atmosphere if they eject large quantities of sulphur dioxide into the stratosphere. The present figure shows the eruption of Mount St. Helens on 18 May 1980 which had a local effect on climate because of ash, reducing the reception of solar radiation on the Earth's surface. Mount St. Helens had very minimal global effect on the climate because the eruption occurred at an oblique angle putting little sulphur dioxide into the stratosphere (*Source*: U.S. Geological Survey; photograph by Austin Post)

was responsible for the cooling by partially blocking the transmission of solar radiation to the Earth's surface. However, measurements indicate that most of the dust thrown in the atmosphere returned to the Earth's surface within six months. Recent stratospheric data suggests that large explosive volcanic eruptions also eject huge amounts of sulphur dioxide gas which remains in the atmosphere for as long as 3 years. Atmospheric chemists have determined that the ejected sulphur dioxide gas reacts with water vapour commonly found in the stratosphere to form a dense optically bright haze layer that reduces the atmospheric transmission of some of the Sun's incoming radiation.

In the last century, two significant climate modifying eruptions have occurred. El Chichon in Mexico erupted in April 1982, and Mount Pinatubo went off in the Philippines during June 1991 (Fig. 1.6). Of these two volcanic events, Mount Pinatubo had a greater effect on the Earth's climate and ejected about 20 million tonnes of sulphur dioxide into the stratosphere (Fig. 1.7). Researchers believe that the Pinatubo eruption was primarily responsible for the 0.8 °C drop in global average air temperature in 1992. The global climatic effects of the eruption of Mount Pinatubo are believed to have peaked in late 1993. Satellite data confirmed the interrelationship between the Mount Pinatubo eruption and reduction in global temperature in 1992 and 1993. The satellite data further indicated that the sulphur dioxide plume from the eruption caused a several per cent increase in the amount of sunlight reflected by the Earth's atmosphere back to space causing the surface of the planet to cool.

1.2.1.4 Variations in Solar Output

Until recently, many scientists thought that the Sun's output of radiation only varied by a fraction of a per cent over many years. However, measurements made by satellites equipped with radiometers in the 1980s and 1990s suggested that the Sun's energy output may be more variable than was once estimated. Measurements made during the early 1980s showed a decrease of 0.1 % in the total amount of solar energy, reaching the Earth over just an 18-month time period. If this trend were to extend over several decades, it could influence global climate. Numerical climatic models predict that a change in solar output of only 1 % per century would alter the Earth's average temperature between 0.5 and 1.0 °C.

Scientists have attempted to establish a linkage between *sunspots* and climatic change. Sunspots are huge magnetic storms that are seen as dark (cooler) areas on the Sun's surface (Fig. 1.8).

Sunspots have magnetic fields of strength up to 3,000 times as great as the average magnetic field of either the Sun or the Earth. Astronomers believe that the cause of sunspots is attributed to this fact. According to a standard explanation, the

Fig. 1.6 Ash column generated by the eruption of Mount Pinatubo on 12 June 1991. The strongest eruption of Mount Pinatubo occurred 3 days later on 15 June 1991 (*Source*: US Geological Survey)

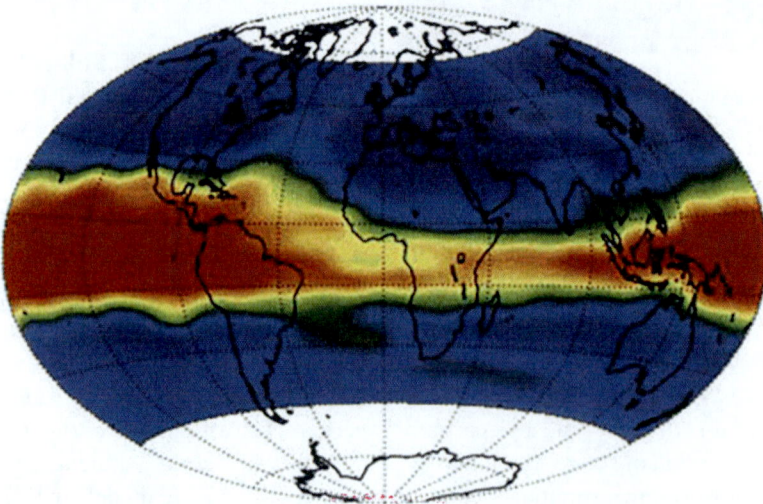

Fig. 1.7 Satellite image showing the distribution of Mount Pinatubo's sulphur dioxide and dust aerosol plume (*red* and *yellow* areas) between 14 June and 26 July 1991 (*Source*: SAGE II Satellite Project – NASA)

strong magnetic fields of the Sun have the shape of tubes just below the solar surface at the beginning of the sunspot cycle. These tubes lie perpendicular to the Sun's equator. The Sun rotates faster at its equator than at its poles, and so the tubes are stretched out in the east–west direction. Kinks then develop in the magnetic tubes and push through the solar surface. A pair of sunspots appears wherever a kink penetrates, because the kink both leaves and re-enters the surface. The number and size of sunspots show cyclical patterns, reaching a maximum about every 11, 90 and 180 years. The decrease in solar energy observed in the early 1980s corresponds to a period of maximum sunspot activity based on the 11-year cycle. In addition, measurements made

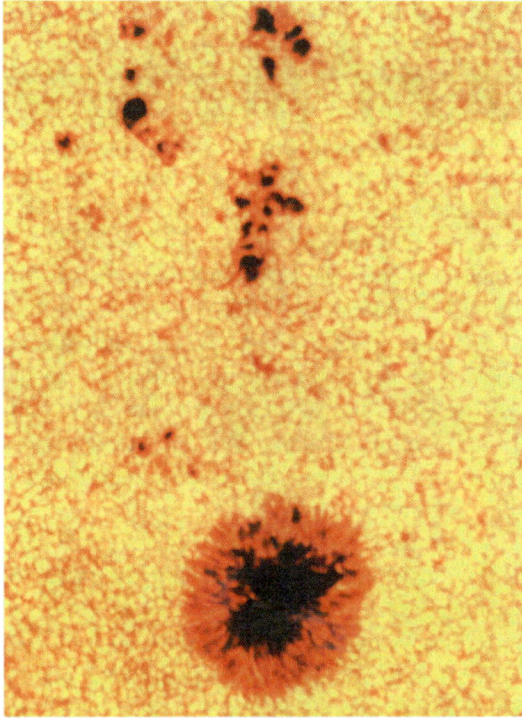

National Solar Observatory/National Optical Astronomy Observatory

Fig. 1.8 Sunspots on the Sun's surface

with a solar telescope from 1976 to 1980 showed that during this period, as the number and size of sunspots increased, the Sun's surface cooled by about 6 °C. Apparently, the sunspots prevented some of the Sun's energy from leaving its surface. However, these findings tend to contradict observations made on longer time scales. Observations of the Sun during the middle of the *Little Ice Age* (1650–1750) indicated that very little sunspot activity was occurring on the Sun's surface. The *Little Ice Age* was a time of a much cooler global climate and some scientists correlate this occurrence with a reduction in solar activity over a period of 90 or 180 years. Measurements have shown that these 90- and 180-year cycles influence the amplitude of the 11-year sunspot cycle. It is hypothesized that during times of low amplitude, like the Maunder Minimum, the Sun's output of radiation is reduced. Observations by astronomers during this period (1645–1715) revealed very little sunspot activity occurring on the Sun.

During periods of maximum sunspot activity, the Sun's magnetic field is strong. When sunspot activity is low, the Sun's magnetic field weakens. The magnetic field of the Sun also reverses every 22 years, during a sunspot minimum. Some researchers believe that the periodic droughts on the Great Plains of the United States are in some way correlated with this 22-year cycle.

1.2.1.5 Plate Tectonics

The phenomenon of plate tectonics plays a major role in the event of climate change. On the longer time scale, plate tectonics can reorient the position of continents, shape oceans, build and tear down mountains and generally serve to provide the matrix upon which climate exists. More recently, plate motions have been implicated in the intensification of the present ice age when, approximately three million years ago, the North and South American plates collided to form the Isthmus of Panama and shut off direct mixing between the Atlantic and Pacific Oceans. The movement of the plate and subsequent subduction also regulate the condition of the climate. Slow subduction is associated with narrow mid-oceanic ridge and low volcanic activity. Under this condition, there is maximum exposure of marine limestone to weathering, which draws more carbon dioxide from the atmosphere to participate in the reactions resulting in the cooling effect. The opposite phenomenon of warming is linked to faster subduction process.

1.2.2   Human Influences on Climate Change

Anthropogenic factors are human activities that change the environment and influence climate. The emission of carbon dioxide due to burning of fossil fuels and the increase of greenhouse gas concentrations due to rapid urbanization, industrialization and expansion of unplanned tourism are unquestionable human influences on climate change through the event of habitat destruction or modification. However, in some cases, the chain of causality is direct and unambiguous (e.g. by the effects of irrigation on temperature and humidity), while in others, it is less clear. Various hypotheses for human-induced climate change have been debated for many years.

According to Church and Gregory (2001), the amount of anthropogenic change in land water storage systems cannot be estimated with much confidence. A number of anthropogenic factors can contribute to sea-level rise. First, natural groundwater systems typically are in a condition of dynamic equilibrium where, over long time periods, recharge and discharge are in a balance. When the rate of groundwater pumping greatly exceeds the rate of recharge, as is often the case in arid and even semiarid regions, water is removed permanently from storage. The water that is lost from groundwater storage eventually reaches the ocean through the atmosphere or surface flow, resulting in sea-level rise. Second, wetland contains standing water, soil moisture and water in plants equivalent to water roughly 1 m deep. Hence, wetland destruction (for urbanization, industrialization and tourism, etc.) contributes to sea-level rise. Over time scales shorter than a few years, diversion of surface waters for irrigation in the internally draining basins of arid regions results in increased evaporation. The water lost from the basin hydrologic system eventually reaches the ocean. Third, forests store water in plant tissue both above and below ground. When a forest is removed, transpiration is eliminated so that run-off is favoured in the hydrologic budget.

It has been observed by several researchers that impoundment of water behind dams removes water from the ocean and lowers sea level. Dams have led to a sea-level drop over the past few decades from −0.5 to −0.7 mm/year (Chao 1994; Sahagian et al. 1994). Infiltration from dams and irrigation may raise the water table by storing more water. Gornitz (2001) estimated a range of −0.33 to −0.27 mm/year sea-level change equivalent held by dams (not counting additional potential storage due to subsurface infiltration).

It is very difficult to provide accurate estimates of the net anthropogenic contribution to climate change due to lack of worldwide information on individual factor, although the effect caused by dams is possibly better known than other effects.

According to Sahagian (2000), the sum of the anthropogenic effects could be of the order of 0.05 mm/year sea-level rise over the past 50 years, with an uncertainty several times as large.

In conclusion it can be advocated that the land-related contribution to sea-level change has not led to a reduction in the uncertainty compared to the IPCC Technical Assessment Report (2001), which estimated the rather wide ranges of −1.1 to +0.4 mm/year for 1910–1990 and −1.9 to +1.0 mm/year for 1990. However, indirect evidence by considering other contributions to the sea-level budget suggests that the land contribution either is small (<0.5 mm/year) or is compensated by unaccounted underestimated contributions.

Today the climate-change-related researches unanimously suggest that the biggest factor of present concern is the increase in CO_2 levels due to emissions from fossil fuel combustion, followed by aerosols (particulate matter in the atmosphere), which exert a cooling effect. Cement manufacture plays a vital role in the domain of human-induced climate change. Other factors, including land use, ozone depletion, animal husbandry, agriculture and deforestation, also exert considerable influence on the climatic profile of the planet Earth. These factors are discussed here in brief.

1.2.2.1 Fossil Fuels

Fossil fuels constitute the backbone of modern civilization. Beginning with the industrial revolution in the 1850s and accelerating ever since, the human consumption of fossil fuels has elevated CO_2 levels from a concentration of ~280 ppm to more than 380 ppm today (an increase of 35.7 %). These increases are projected to reach more than 560 ppm (an increase of 100 %) before the end of the twenty-first century. It is known that carbon dioxide levels are substantially higher now than at any time in the last 750,000 years. Along with rising methane levels, these changes are anticipated to cause an increase of atmospheric temperature within 1.4–5.6 °C between 1990 and 2100.

1.2.2.2 Aerosols

Anthropogenic aerosols, particularly sulphate aerosols from fossil fuel combustion, exert a cooling influence. This, together with natural variability, is believed to account for the relative 'plateau' in the graph of twentieth-century temperatures in the middle of the century.

1.2.2.3 Cement Manufacture

Cement manufacturing is one of the largest causes of human-induced carbon dioxide emissions. The gas is produced when calcium carbonate ($CaCO_3$) is heated to produce the calcium oxide (CaO, also called *quicklime*), which is the main ingredient for cement ($CaSiO_3$) production. The reaction steps in this manufacture are:

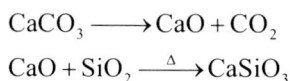

$$CaCO_3 \longrightarrow CaO + CO_2$$
$$CaO + SiO_2 \xrightarrow{\Delta} CaSiO_3$$

CO_2 produced in the first step has considerable contribution to global warming. While fossil fuel combustion and deforestation each produce significantly more carbon dioxide in the Earth's atmosphere, cement production alone is responsible for approximately 2.5 % of total worldwide emissions from industrial sources.

1.2.2.4 Land Use

Prior to widespread fossil fuel use, humanity's largest effect on local climate is likely to have resulted from alteration in land-use pattern. Irrigation, deforestation and agriculture fundamentally change the environment. These activities alter the amount of water going into and out of a given location. They also may change the local albedo by influencing the ground cover and altering the amount of sunlight that is absorbed. For example, there is evidence to suggest that the climate of Greece and other Mediterranean countries was permanently changed by widespread deforestation between 700 BC and 1 AD (the wood being used for ship building, construction and fuel), with the result that the modern climate in the region is significantly hotter and drier and the species of trees that were used for ship building in the ancient world are no longer be found in the area.

A controversial hypothesis by William Ruddiman called the early anthropocene hypothesis suggests that the rise of agriculture and the accompanying deforestation led to the increases in carbon dioxide and methane during the period 5,000–8,000 years ago. These increases, which reversed previous declines, may have been responsible for delaying the onset of the next glacial period, according to Ruddiman's overdue-glaciation hypothesis.

In modern times, a 2007 Jet Propulsion Laboratory study found that the average temperature of California has risen about 2 °F degrees over the past 50 years, with a much higher increase in urban areas. The change was attributed mostly to extensive human development of the landscape.

1.2.2.5 Livestock

According to a 2006 United Nations report, livestock is responsible for 18 % of the world's greenhouse gas emissions as measured in CO_2 equivalents. This however includes land usage change, meaning deforestation in order to create grazing land. In the Amazon Rainforest, 70 % of deforestation was done solely to make way for grazing land. This is the major factor in the UNFAO Report (2006), which was the first agricultural report to include land usage change into the radiative forcing of livestock. In addition to CO_2 emissions, livestock produces 65 % of human-induced nitrous oxide (which has 296 times the global warming potential of CO_2) and 37 % of human-induced methane (which has 23 times the global warming potential of CO_2).

Interplay of Factors

Balance is the art of nature, and the natural components always try to maintain a stable or ground state. If a certain forcing or inducing factor (e.g. solar variation) acts to change the climate, then there are mechanisms that act to amplify or reduce the effects. These are called positive and negative feedbacks, respectively. Researches have forwarded the view that the climate system is generally stable with respect to these feedbacks: positive feedbacks do not 'run away'. Part of the reason for this is the existence of a powerful negative feedback between temperature and emitted radiation: radiation increases as the fourth power of absolute temperature.

A number of important positive feedbacks prevail in the domain of climate change. The glacial and interglacial cycles of the present ice age provide a most relevant example. It is believed that orbital variations provide the timing for the growth and retreat of ice sheets. However, the ice sheets themselves reflect sunlight back into space and hence promote cooling and their own growth, known as the *ice-albedo feedback*. Further, falling sea levels and expanding ice decrease plant growth and indirectly lead to declines in carbon dioxide and methane. This leads to further cooling. Conversely, rising temperatures caused, for example, by anthropogenic emissions of greenhouse gases could lead to decreased snow and ice cover, revealing darker ground underneath, and consequently result in more absorption of sunlight. Water vapour, methane and carbon dioxide can act as significant positive feedbacks – their levels rising in response to a warming trend, thereby accelerating that trend. Water vapour acts strictly as a feedback (excepting small amounts in the stratosphere), unlike the other major greenhouse gases, which can also act as forcing.

More complex feedbacks involve the possibility of altered water currents within the oceans or air currents within the atmosphere. A significant concern is that melting glacial ice from Greenland may interfere in changing the thermohaline circulation of water in the North Atlantic, affecting the Gulf Stream, which brings warmer water to replace sinking colder water, which would affect the distribution of heat to Europe and the east coast of the United States.

Other potential feedbacks are not well understood and may either inhibit or promote warming. For example, it is unclear whether rising temperatures promote or inhibit vegetative growth, which could in turn draw down either more or less carbon dioxide. Similarly, increasing temperatures may lead to either more or less cloud cover. Since cloud cover has a strong cooling effect, any change to the abundance of clouds has high probability to affect the climate at local level and even at regional level.

1.3   Effects of Climate Change

Global warming is a phase of climate change, which is associated with the increasing trend in the average temperature of the Earth's near-surface air and oceans in recent decades and its projected continuation. Global average air temperature near the Earth's surface rose 0.74 ± 0.18 °C (1.3 ± 0.32 °F) during the past century. The Intergovernmental Panel on Climate Change (IPCC) concludes, 'most of the observed increase in globally averaged temperatures since the mid-twentieth century is very likely due to the observed increase in anthropogenic greenhouse gas concentrations (http://www.ipcc.ch/SPM2feb07.pdf)', which leads to warming of the surface and lower atmosphere by accelerating the greenhouse effect. Natural phenomena such as solar variation combined with volcanoes may probably have a small warming effect from pre-industrial times to 1950 but a small cooling effect since 1950.

There is still much uncertainty over the degree by which temperatures will actually rise and what the effects of these rises will be. What is not in doubt is that the Earth's temperature has increased by 0.5 °C over the past century and that recent years have been among the hottest on record. The likely effects of global warming have serious implications on the environment. A 20-cm rise in sea level by 2030 is expected to result from glacial melting and from the thermal expansion of the oceans as water temperatures rise. This may inundate some coastal regions and increase the risk of flooding in many areas. In case of Indian subcontinent population, density often exceeds 1,000 persons per km^2 within low elevation coastal zone (LECZ), which has placed India as one of the most vulnerable nations to climate-change-induced sea-level rise (Fig. 1.9).

Internationally, populations throughout much of China, Egypt, Denmark and Southeast Asia are most at risk – millions could lose their homes and livelihoods. In the UK, southeast England is most vulnerable to inundation, although floods, such as that which occurred in Towyn,

India

Pakistan

China

DELHI

Nepal

Bangladesh

CALCUTTA

MUMBAI

MADRAS

0   210   420 km   N

Sri
Lanka

Population Density within and outside of a 10 meter
low elevation coastal zone (LECZ), 2000

Persons per sq km	<25	25-100	100-250	250-500	500-1,000	>1,000
within LECZ						
outside LECZ						

largest urban areas

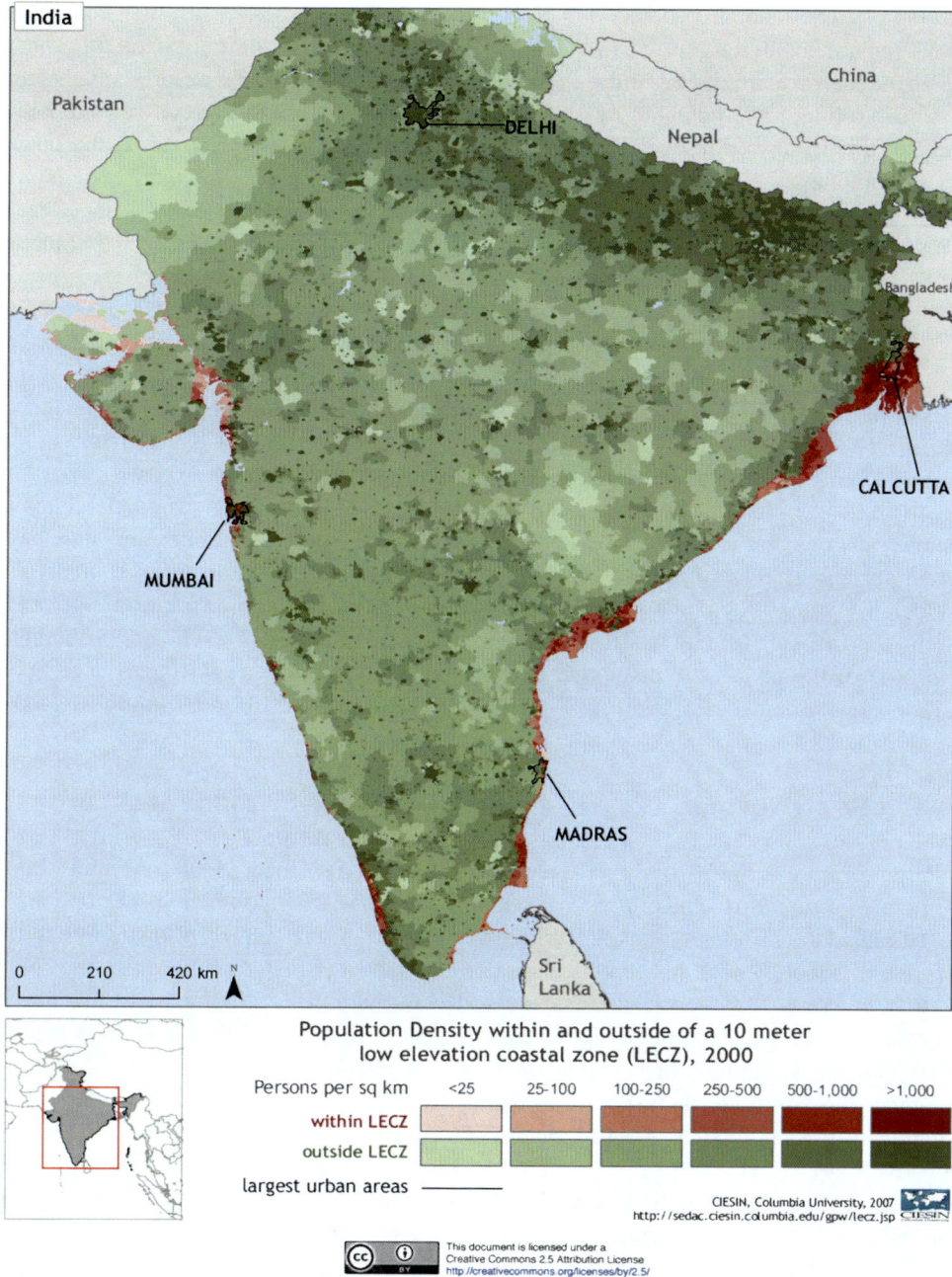

Fig. 1.9 Population density within a 10-m low elevation coastal zone (LECZ) often exceeds 1,000 persons per km^2

mid-Wales, in 1990, are likely to increase in both frequency and severity throughout the country. Agriculture and forestry and crop failure and famine will ensue throughout marginally productive regions of the world. The effects of climate change can be felt hardest in a country like India.

The fact that India has a large rural population who are dependent on the cycles of the climatic seasons for their agricultural activities, fisherman who work in the rivers and seas, farmers who need the seasonal monsoons and a large and varied indigenous population who live in harsh climatic regions of mountains, desert and river delta make India specially susceptible to a changing climate. Fishermen are among those hardest hit by climate change. The potential of a rise in sea levels due to global warming would wipe out existing shoreline, contaminate freshwater systems with salt water, diminish the minimum environment flow required for fishes in river and dry up floodplain wetlands, a rich habitat for fishes. The services that aquatic ecosystem provide to humans such as fisheries, flood protection, recreation and biodiversity conservation, which if converted in monetary terms would be in millions of rupees. A recent global assessment of the status of aquatic ecosystems showed that their capacity to provide goods and services appears to be drastically degraded.

Few important effects of global warming and associated climate change on weather, ecosystem, biodiversity, human health and economic profile of the nation are discussed here in detail.

1.3.1   Effects on Weather

A rise in temperature is likely to trigger the rate of precipitation (http://dpa.aapg.org/gac/papers/climate_change.cfm and http://www.agu.org/fora/eos/pdfs/2006EO360008.pdf), but the effects on storms (both in terms of frequency and intensity) are less clear. Extratropical storms partly depend on the temperature gradient, which is predicted to weaken in the Northern Hemisphere as the polar region warms more than the rest of the hemisphere.

Controversies also exist in the domain of correlation between sea-surface temperature and Atlantic basin hurricanes.

1.3.1.1 Increased Evaporation
Over the course of the twentieth century, evaporation rates have reduced worldwide (http://www.greenhouse.gov.au/impacts/overview/pubs/overview4.pdf); this is thought by many researchers to be explained by *global dimming*. As the climate grows warmer and the causes of global dimming are reduced, the rate of evaporation increases due to warmer oceans. Because the components of the planet Earth – its atmosphere, lithosphere and hydrosphere – comprise a closed system, this will cause heavier rainfall, more run-off and more erosion. Many scientists think that increased evaporation could result in more extreme weather as global warming progresses.

It has been predicted that each 1 % increase in annual precipitation would enhance the cost of catastrophic storms by 2.8 %. The Association of British Insurers (ABI) has stated that limiting carbon emissions would avoid 80 % of the projected additional annual cost of tropical cyclones by the 2080s. The cost is also increasing partly because of building in exposed areas such as coasts and floodplains. The ABI claims that reduction of the vulnerability to some inevitable impacts of climate change, for example, through more resilient buildings and improved flood defences, could also result in considerable cost savings in the long term (http://www.abi.org.uk/Display/File/Child/552/Financial_Risks-of_Climate.html).

1.3.1.2 Destabilization of Local Climates
The first recorded South Atlantic hurricane 'Catarina/Katrina', which hit Brazil in March 2004, is an example of climatic destabilization related to climate change. In the Northern Hemisphere, the southern part of the Arctic region (home to 4,000,000 people) has experienced a temperature rise of 1–3 °C (1.8–5.4 °F) over the last 50 years. Canada, Alaska and Russia are experiencing initial melting of permafrost. This may disrupt ecosystems, and by accelerated bacterial activity in the soil, these areas can become carbon sources instead of carbon sinks (http://www.grida.no/climate/ipcc_tar/wg1/295.htm). A study conducted on changes to eastern Siberia's permafrost suggests that it is gradually disappearing in the southern regions, leading to the loss of nearly 11 % of Siberia's 11,000 lakes since 1971 (http://www.gurdian.co.uk/international/story/0,1503170,00.html). At the same time, western Siberia is at the initial stage where melting permafrost is creating new lakes, which will eventually start disappearing as in the east.

Furthermore, permafrost melting will also cause methane release from melting permafrost peat bogs, thus posing a positive feedback on the overall global warming phenomenon.

1.3.2 Effects on Ocean

1.3.2.1 Sea-Level Rise

The increase in global temperature triggers the process of expansion of ocean water volume and additional input of waters from sources like glaciers, which were locked on the land. This causes the level of seawater to rise from its normal position. The sea level has elevated more than 120 m since the peak of the last ice age about 18,000 years ago. The bulk of that occurred before 6,000 years ago. From 3,000 years ago to the start of the nineteenth century, sea level was almost constant, rising at 0.1–0.2 mm/year. Since 1900, the level has risen at 1–2 mm/year (http://www.grida.no/climate/ipcc_tar/wg1/295.htm), and since 1992, satellite altimetry from TOPEX/Poseidon indicates a rate of about 3 mm/year (http://www.grida.no/climate/ipcc_tar/wg1/295.htm). An increase of 1.5–4.5 °C temperature is estimated to lead to an increase of 15–95 cm of the sea level (IPCC 2001).

The three primary contributing factors to sea-level rise often cited are (1) ocean thermal expansion, (2) glacial melt from Greenland and Antarctica (plus a smaller contribution from other ice sheets) and (3) the change in terrestrial storage. Until recently, ocean thermal expansion was expected to be the dominating factor behind the rise in sea level. However, new data sets on rates of deglaciation in Greenland and Antarctica suggest significant contribution of glacial melt, especially due to the uncertainty surrounding the dynamics of outlet glaciers (Hansen et al. 2005; Hanna et al. 2005; Howat et al. 2007; Krabill et al. 2004; Meier et al. 2007; Rignot and Kanagaratnam 2006; Stroeve et al. 2007; Velicogna and Wahr 2006). Although there is continued evidence of ice sheet growth in the Eastern regions of Antarctica (Davis et al. 2005), when coupled with the measured losses in the West, and including Greenland, the evidence

Table 1.1 Estimates of the various contributions to the budget of global mean sea-level change for 1961–2003 and 1993–2003 compared with the observed rate of rise

Source	1961–2003	1993–2003
Thermal expansion	0.42 ± 0.12	1.6 ± 0.5
Glaciers and ice caps	0.50 ± 0.18	0.77 ± 0.22
Greenland ice sheet	0.05 ± 0.12	0.21 ± 0.07
Antarctic ice sheet	0.14 ± 0.41	0.21 ± 0.35
Sum	1.1 ± 0.5	2.8 ± 0.7
Observed	1.8 ± 0.5	3.1 ± 0.7
Difference (observed-sum)	0.7 ± 0.7	0.3 ± 1.0

Sources: Willis et al. (2004), Antonov et al. (2005), Ishii et al. (2006), Lombard et al. (2006)

Remark: Ice sheet mass loss of 100 Gt/year is equivalent to 0.28 mm/year of sea-level rise. A GIA correction has been applied to observations from the tide gauges and altimetry. For the sum, the error has been calculated as the square root of the sum of squared errors of the contributions. The thermostatic sea-level changes are for the 0–3,000 m layer of the ocean

appears to point in the direction of increased sea-level rise. The implications of these findings for sea-level rise could be dramatic. The Greenland and Antarctic ice sheets contain enough water to raise the sea level by almost 70 m (Hansen et al. 2005), so even small changes in the their volume would have a significant effect.

The magnitude of contribution of freshwater to global sea-level rise is highlighted in Table 1.1.

Paleoclimatic information also indicated that the warmth of the last half century is unusual in at least the past 1,300 years. About 125,000 years ago, polar regions were relatively warmer than at present. Reductions in polar ice volume led to 4–6 m of sea-level rise (IPCC 2007). As new researches on the impacts of warming are coming in the light of publications, more confidence is developing with respect to the ranges of impact, but current evidence leads to an alarming possibility that a threshold triggering many metres of sea-level rise could be crossed well before the end of this century (Hansen 2006; Overpeck et al. 2006).

Several regional and local level case studies have been cited in favour of sea-level rise due to global warming in different parts of the planet Earth. However, many case studies are not directly related to increase in temperature or expansion of ocean water mass; rather other factors like sea bed rise or tectonic reasons are also standing in ques with significant proofs behind sea-level rise (Raha et al. 2012).

In context to India, *The Independent* (a news agency) reported in December 2006 that the first island claimed by rising sea levels caused by global warming in Indian Sundarbans was Lohachara Island. However, several scientists have contradicted the correlation between erosion of Lohachara Island and global warming and forwarded tectonic cause as one of the important reasons behind the loss of landmass of Lohachara Island. The Indian Sundarbans at the apex of the Bay of Bengal (between 21°13′ and 22°40′ N latitude and 88°03′ to 89°07′E longitude) is basically a deltaic complex of approximately 426,300 ha formed by the depositional activities of the Ganges and the Brahmaputra. A group of islands and a dense network of rivers, canals and creeks comprise the area. It has been suggested that in the past, the tidal swamps of the Sundarbans extended landward to the base of the Rajmahal hills and filled up in the later part of the Tertiary period (Fergusson 1863). A number of geomorphological and resultant hydrological changes have contributed to the present location and condition of the Sundarbans. Tectonic movements in the late Tertiary period in northwestern Punjab changed the course of the Indo-Brahma River. Some workers consider this phenomenon as the cause behind the beginning of the southeastern flow of the River Ganges (Wadia 1961). Geologists believe that there was a general southeastern slope of the Bengal Basin during the Tertiary period. The Ganges started flowing down this slope and created a new delta through 1,000 years of time by silt deposition. Gradually the western part of Bengal below the Rajmahal hills rose in level and became suitable for human settlement. Later evidence suggests that the Bengal Basin tilted eastward during the twelfth century because of neo-tectonic movement (Morgan and McIntire 1959). The tectonic movements of the basin in different geological periods are one of the important causes for the loss of landmass in some islands.

The frequent subsidence of some part of Sundarbans, causing the complete disappearance of forest chunk and settlement areas, is the effect of 'Swatch of no ground' phenomenon. This zone also known as *Ganga Trough* is situated south of the Raimangal–Malancha estuary between 21°00′ and 21°22′ latitude in the Bay of Bengal region (Chaudhuri and Choudhury 1994) and has a comparatively flat floor with walls of about 12° inclination. According to Oldham (1893), 'there is a singularly deep area in the sea outside the middle of the delta, which is marked in the chart as *Swatch of no ground*. Here the soundings are from 5 to 10 fathoms all round that change almost abruptly to 200 and even to 300 fathoms'. All the estuaries in the Sundarbans delta (Hugli, Muriganga, Saptamukhi, Thakuran, Matla, Bidyadhari, Gosaba and Harinbhanga) are believed to be influenced by Swatch of no ground, and the unstable condition of Sundarbans is due to the existence of this submarine zone (Chaudhuri and Choudhury 1994). These discussions lead to conclude that there are a lot of uncertainties to relate the erosion of islands and global-warming-induced sea-level rise as geological settings, geographic features and tectonic movements often generate signals of noise in such direct cause–effect relationship.

Sea-level rise has considerable impact on the economic profile of the developing countries. A research analysis conducted by Dasgupta et al. (2009) for the 84 developing countries is summarized in Table 1.2. The analysis indicates that approximately 0.3 % (194,000 km^2) of the territory of the 84 developing countries would be impacted by a 1-m sea-level rise. This would increase to 1.2 % in a 5-m sea-level rise scenario. Though this remains relatively small in percentage terms, approximately 56 million people (or 1.28 % of the population) in these countries would be impacted under a 1-m sea-level rise scenario. This would increase to 89 million people for 2-m sea-level rise (2.03 %) and 245 million people (5.57 %) for 5-m sea-level rise. The impact of sea-level rise on GDP is slightly larger than the impact on population, because GDP per capita is generally above average for coastal populations and cities. Wetland would experience

Table 1.2 Impacts of sea-level rise across at the global level

	1 m	2 m	3 m	4 m	5 m
Land area (total = 63,332,530 km²)					
Impacted area	194,309	305,036	449,428	608,239	768,804
% of total area	0.31	0.48	0.71	0.96	1.21
Population (total = 4,414,030.000)					
Impacted population	56,344,110	89,640,441	133,049,836	183,467,312	245,904,401
% of total population	1.28	2.03	3.01	4.16	5.57
GDP (total = 16,890.948 million USD)					
Impacted GDP (USD)	219,181	357,401	541,744	789,569	1,022,349
% of total GDP	1.30	2.12	3.21	4.67	6.05
Urban areas (total = 1,434,712 km²)					
Impacted area	14,646	23,497	35,794	50,742	67,140
% of total area	1.02	1.64	2.49	3.54	4.68
Agricultural land (total = 17,975,807 km²)					
Impacted area	70,671	124,247	196,834	285,172	377,930
% of total area	0.39	0.69	1.09	1.59	2.10
Wetlands area (total = 4,744,149 km²)					
Impacted area	88,224	140,355	205,697	283,009	347,400
% of total area	1.86	2.96	4.34	5.97	7.32

Source: Dasgupta et al. (2009)

significant impact even with a 1-m sea-level rise. Up to 7.3 % of wetlands in the 84 countries would be impacted by a 5-m sea-level rise. However, these impacts are not uniformly distributed across the regions and countries of the developing world.

Impacts were also estimated for individual countries/territories. Table 1.3 summarizes the results for each indicator by presenting the top ten impact cases (as a percentage of their national values). For this purpose, the 1-m sea-level rise scenario was used. In terms of population impacted, the top ten countries/territories worldwide are Vietnam, A.R. Egypt, Mauritania, Suriname, Guyana, French Guiana (Fr), Tunisia, United Arab Emirates, the Bahamas and Benin. Around 10 % of the populations of Vietnam and the A.R. of Egypt would be displaced with a 1-m sea-level rise for land area. The Bahamas is by far the most impacted country, with close to 12 % of its area inundated. Ten per cent of the urban areas of Vietnam and Guyana would be inundated by a 1-m sea-level rise, and the A.R. of Egypt's agricultural land would experience the largest percentage impact: 13 % would be submerged. The research indicates that areas inundated by a 1-m sea-level

rise would account for 10 % of GDP in Vietnam and more than 5 % in Mauritania, A.R. Egypt, Suriname and Benin. Finally, nearly 28 % of the wetlands in Vietnam, Jamaica and Belize would be inundated by a 1-m sea-level rise. For all of the indicators used in this research, Vietnam ranks among the top five most impacted countries, with the A.R. of Egypt, Suriname and the Bahamas consistently ranking among the highest.

1.3.2.2 Ocean Warming

The rise of atmospheric temperature in recent *era* has also been reflected through increase of aquatic temperature. The temperature of the Antarctic Southern Ocean rose by 0.17 °C (0.31 °F) between the 1950s and the 1980s, nearly twice the rate for the world's oceans as a whole. The temperature rise in the Gulf of Mexico is another prominent case study from the United States. As hurricanes cross the warm Loop Current coming up from South America, they can gain great strength in under a day (as did Hurricane Katrina and Hurricane Rita in 2005), with water above 85 °F seemingly promoting Category 5 storms. Hurricane season ends in November as the waters cool.

Table 1.3 Top ten most impacted countries with a 1-m sea-level rise (percentage impact in *parenthesis*)

Rank	Land area	Population	GDP	Urban areas	Agricultural land	Wetlands
1	The Bahamas (11.57)	Vietnam (10.79)	Vietnam (10.21)	Vietnam (10.74)	A.R. Egypt (13.09)	Vietnam (28.67)
2	Vietnam (5.17)	A.R. Egypt (9.28)	Mauritania (9.35)	Guyana (10.02)	Vietnam (7.14)	Jamaica (28.16)
3	Qatar (2.70)	Mauritania (7.95)	A.R. Egypt (6.44)	French Guiana (Fr) (7.76)	Suriname (5.60)	Belize (27.76)
4	Belize (1.90)	Suriname (7.00)	Suriname (6.35)	Mauritania (7.50)	The Bahamas (4.49)	Qatar (21.75)
5	Puerto Rico (1.64)	Guyana (6.30)	Benin (5.64)	A.R. Egypt (5.52)	Argentina (3.19)	The Bahamas (17.75)
6	Cuba (1.59)	French Guiana (Fr) (5.42)	The Bahamas (4.74)	Libya (5.39)	Jamaica (2.82)	Libya (15.83)
7	Taiwan, China (1.59)	Tunisia (4.89)	Guyana (4.64)	United Arab Emirates (4.80)	Mexico (1.60)	Uruguay (15.14)
8	The Gambia (1.33)	United Arab Emirates (4.59)	French Guiana (Fr) (3.02)	Tunisia (4.50)	Myanmar (1.48)	Mexico (14.85)
9	Jamaica (1.27)	The Bahamas (4.56)	Tunisia (2.93)	Suriname (4.20)	Guyana (1.16)	Benin (13.78)
10	Bangladesh (1.12)	Benin (3.93)	Ecuador (2.66)	The Bahamas (3.99)	Taiwan, China (1.05)	Taiwan, China (11.70)

Source: Dasgupta et al. (2009)

1.3.2.3 Acidification

The world's oceans soak up much of the carbon dioxide produced by living organisms, either as dissolved gas or in the skeletons of tiny marine creatures that fall to the bottom to become chalk or limestone. Oceans currently absorb about one tonne of CO_2 per person per year. It is estimated that the oceans have absorbed around half of all CO_2 generated by human activities since 1800 (120,000,000,000 tonnes or 120 petagrams of carbon).

In the aquatic phase, carbon dioxide becomes a weak carbonic acid resulting in the lowering of pH value. It has been observed that the increase in the greenhouse gas since the industrial revolution has already lowered the average pH of seawater by 0.1 units. Predicted emissions could lower it further about 0.5 unit by 2100, to a level not seen for millions of years. There are concerns that lowering of aquatic pH could have a detrimental effect on corals (16 % of the world's coral reefs have died from bleaching caused by warm water in 1998, which coincidentally was the warmest year ever recorded) and other marine organisms (particularly the molluscan community) due to dissolution of their calcium carbonate shells. Increased acidity may also directly affect the growth and reproduction of fish as well as the plankton on which they rely for food.

1.3.2.4 Shutdown of Thermohaline Circulation

There is some speculation that global warming could, via a shutdown or slowdown of the thermohaline circulation, trigger localized cooling in the North Atlantic and lead to cooling, or lesser warming, in that region. This would affect in particular areas like Scandinavia and Britain that are warmed by the North Atlantic drift. More significantly, it could lead to an oceanic anoxic event.

The chances of this near-term collapse of the circulation are unclear; there is some evidence for the short-term stability of the Gulf Stream and possible weakening of the North Atlantic drift. However, whether the degree of weakening will be sufficient to shutdown the circulation is under debate. There is no evidence for cooling in northern Europe or nearby seas.

1.3.2.5 Effects on Coral Reef

Coral reefs are colourful underwater taxonomically diverse units that vibrate with a wide spectrum of life forms and act as a natural protective barrier for coastal regions. Global warming affects the coral reefs in several ways. Studies show that warmer-than-normal sea temperatures are contributing to coral bleaching, which has been on the rise in recent years, weakening and killing reefs in many parts of the world. Coral bleaching occurs when the coral is stressed and expels the zooxanthellae. Without them, the coral loses its colour and becomes white. Sometimes the coral ingests new zooxanthellae, but if it fails, the coral will die. Corals thrive in water with temperature ranging between 66 and 86 °F. If the temperature changes 2 °F or 3 °F degrees outside this range, the coral faces a great stress. The trend of rising ocean water temperature in recent years due to anthropogenic-induced global warming phenomenon is thus a potential threat to the coral reef community. The warmer the water, the more the ocean becomes unfavourable for the survival and growth of corals, which eventually leads to coral bleaching.

Sea-level rise caused by melting sea ice and thermal expansion of the oceans could also cause problems for some reefs by making them too deep to receive adequate sunlight, which is another factor important for survival of corals.

Higher levels of carbon dioxide in the atmosphere are changing the chemistry of the oceans, making it more difficult for corals to build up calcium carbonate skeletons. The ocean absorbs almost a third of atmospheric carbon dioxide. This was, at one time, seen as positive due to the diminishing effect it has on global warming; however, high levels of carbon dioxide in the ocean lower the pH or increase the acidity, which is harmful to the growth and survival of corals. The pH has already dropped a tenth of a unit and might drop up to a third of unit in the coming 100 years. The source of the majority of these carbon dioxide emissions is industrial exhaust, such as cars and factories. This affects corals because their skeletons are made of calcium carbonate. When the atmospheric carbon dioxide is absorbed and mixed with seawater, it forms carbonic acid. This substance complicates the building of body parts by marine animals, such as coral. Corals' growth rate is slowed, and the shells of marine creatures gradually become thin and finally dissolve. Several observations state that in one year alone, 16 % of the world's coral reefs were wiped out. However, there are other views also. In 1997 an experiment was initiated with samples of the coral *Acropora grandis* that were sampled from the hot water outlet of a nuclear power plant near Nanwan Bay, Taiwan. In 1998, the year the power plant began full operation, the coral samples were completely bleached within 2 days of exposure to a temperature of 33 °C. Two years later, however, a different picture was observed. It was found that samples taken from the same area did not even start bleaching until 6 days after exposure to 33 °C temperatures.

1.3.3   Effects on Human Health

The effects of climate change will have an enormously negative impact on human health worldwide. Increasing temperatures, drought and flooding will cause increases in starvation, disease and deaths in human populations. One of the most devastating effects of climate change on human health will be the increase in the outbreaks of diseases, such as malaria, dengue fever, encephalitis, rift valley fever, Hantavirus pulmonary syndrome, West Nile virus, Lyme diseases, cholera and respiratory illness caused by increasing episode of fire.

The increase in outbreaks of mosquito-borne diseases in developing countries is one of the most dangerous effects associated with climate change. Malaria is one of the most prevalent and deadly of these diseases, and it is currently plaguing many developing countries, causing between one and three million deaths per year. The most dangerous form of malarial parasite is *Plasmodium falciparum* and is a widely abundant parasite in Africa. Recent evidences suggest that this form of malaria is also becoming increasingly abundant in Venezuela and Sri Lanka. Outbreaks of malaria caused by flooding and abnormally wet conditions can also be linked to drought conditions.

For example, flooding in Mozambique in February and March of 2000 caused a malaria epidemic. In Brazil, abnormal droughts in the 1980s and 1990s caused crop failure and famine. Under these circumstances, people were forced to migrate to other areas of the Amazon where they subsequently contracted malaria and brought it back to their home towns when the drought subsided. With increasing extreme weather events associated with climate change, it can be expected that more droughts and periods of abnormally wet conditions will result in more cases of malaria throughout the world and, more specifically, in developing countries. This conclusion can also be drawn for other mosquito-vectored illness, such as dengue fever.

People are also experiencing increase in respiratory illnesses due to the changing climate. For instance, many people suffer from asthma, and in developing countries, asthma can be a very debilitating illness. It has been shown that increase in CO_2 causes plants to photosynthesize at a faster rate, producing more pollen. Warming temperatures have also been causing plants to release pollen earlier each year, resulting in longer growing seasons. This phenomenon has negative effects for those who suffer from asthma. Researchers also documented that droughts cause exponential increase in dust in countries like Africa and the Caribbean island, which further agitates respiratory ailments. The risks of forest fires also accelerate under drought condition as seen in Southeast Asia and the Amazon.

Heat waves are very dangerous to human health and are likely to become more frequent as climate change progresses. For instance, during the summer of 2003, Europe experienced what was considered the hottest summer since at least AD 1500. An event such as this, if it occurred in the United States, 'could cause thousands of excess deaths in the inner cities and could precipitate extensive blackouts' (*Harvard Medical School Bulletin*, 2005; page 53). Heat waves cause death among the early, ill and very young age groups. Increased humidity associated with heat waves also has a negative effect on human health, causing discomfort, respiratory ailments and further exacerbating the problem of increasing insect populations leading to instances of insect-vectored diseases.

1.3.4 Effects on Economic Profile

There exists considerable interrelationship between climate change and economic profile of an area. The range of vulnerabilities that poor people face in different parts of the world encompasses all aspects of life that are related to climate change in some way or other. There are many ways to evaluate the relationship between climate change and vulnerability on the matrix of economic profile, but the 2001 IPCC Working Group II report on Impacts, Adaptation and Vulnerability gives insights that may serve as a starting point of the subject. Alteration of crop production both with respect to volume and pattern, salinization of coastal land, gradual vanishing of economically important fisheries and loss of mangroves are direct hit to the economic condition of the coastal people. Secondary impacts will likely include increases in food prices, frequency of fish diseases and greater problems associated with local services such as water supply and sanitation that affect the poor community.

Few case studies from the mangrove-dominated deltaic complex of Indian Sundarbans revealed huge economic loss due to outbreak of viral diseases in the sector of tiger prawn culture. The frequency of the disease increases during the hot summer months (premonsoon season), when the water temperature of the culture pond exceeds ~34 °C. Warming of the Earth climate will accelerate the disease problems in the sphere of shrimp culture. This will affect a large population of coastal zone in India particularly in the maritime states of Andhra Pradesh, Orissa and West Bengal, where shrimp culture is the dominant livelihood of a considerable fraction of the local population.

The changing climate patterns, and especially the increased frequency and/or severity of *extreme events*, will increase vulnerability to natural disasters, both slower-onset ones such as droughts and rapid-onset disasters such as floods and cyclones. These will affect many areas, but semiarid areas (prone to droughts) and coastal and deltaic regions (prone to floods and storms) are particularly vulnerable. The incidence of AILA (a severe tropical cyclone) on 25 May 2009 in the Gangetic delta region is a relevant example in this context. AILA was formed in the central Bay of Bengal as the net output of several factors. Around 20 May 2009, a monsoon initiated at Andaman. The moisture-laden south westerlies accelerated the moisture content in the winds of the Bay of Bengal. The wind speed was also variable in the northern and southern Bay of Bengal. In southern Bay of Bengal, the wind speed in the lower troposphere was around 37 km/h, whereas in northern Bay of Bengal, it was about 9 km/h. These variations led to the curling of winds, which is known as positive relative vorticity. Basically an area of depression developed in the Bay of Bengal on 20 May 2009 which transformed into a cyclone on May 23 and hit the deltaic complex of Indian Sundarbans on May 25, destroying the lives and properties of millions of island dwellers. A preliminary IMD report said that the cyclone retained its intensity for about 15 h after it hit the landmasses as it was close to the Bay of Bengal. It lays centred over the Gangetic delta for quite some time, ascertaining the availability of moisture. This is peculiar nevertheless because premonsoon storms rarely hit the maritime state of West Bengal (Fig. 1.10).

The effect of AILA was severe, resulting in the death of people, damages to properties and alterations in the physico-chemical properties of the soil and water. Embankments were destroyed, agricultural fields and freshwater ponds lost their productivity due to intrusion of saline water and many people became homeless. Seventy island dwellers died in the Indian Sundarbans region. Over 8,000 people could not be traced, and about a million became homeless in India and Bangladesh. The embankments, which act as the line of defence of the villages, were completely smashed by tidal surges. Out of 3,500 km long stretch of embankment, 400 km was completely broken and another 565 became highly vulnerable. Two thousand cattle died during Aila and 200

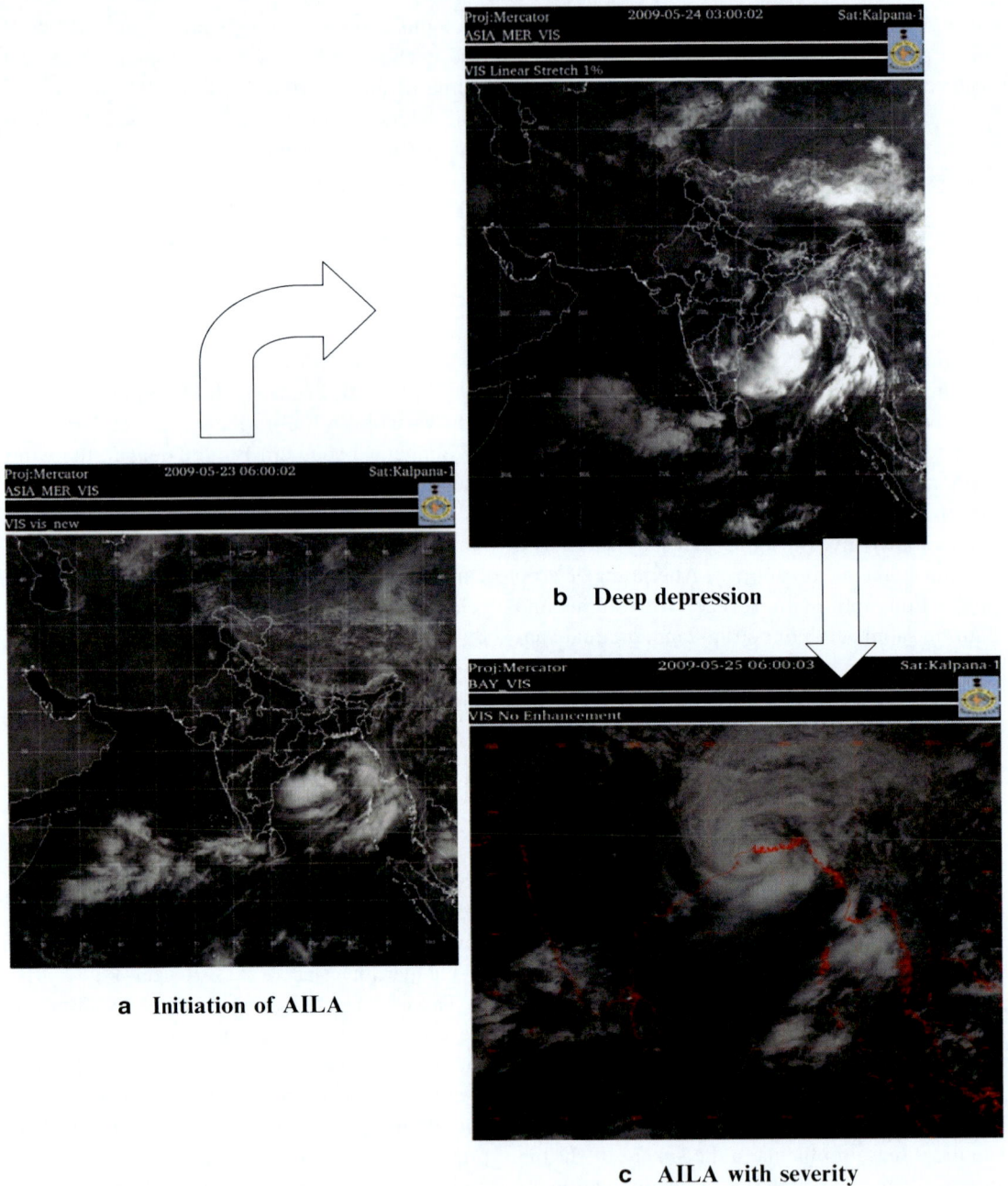

b **Deep depression**

a **Initiation of AILA**

c **AILA with severity**

Fig. 1.10 (**a**) Initiation of AILA. (**b**) Deep depression. (**c**) AILA with severity (*Source*: Regional Specialized Meteorological Center – Tropical Cyclone, New Delhi)

died because they drank saline water. Even honey makers' business was completely destroyed. One hundred and fifty boxes of bees were destroyed. The agricultural fields were submerged by sea water, and all standing crops were destroyed. The freshwater ponds were transformed into brackish water bodies, and several fish species of saline water were recovered from these ponds. The entire spectrum of damages is summarized in Table 1.4.

Dangers of erosion, landslides and flash floods will also increase, particularly in many hilly and mountainous areas.

Changing climate patterns and more extreme events will also have negative impacts on several new livelihood activities (such as tourism) that will limit diversification of opportunities which, combined with damage to infrastructure and other types of physical capital, will affect the wider range of vulnerabilities (such as limited access to markets), the poor face. The weak social and political capital of many underdeveloped nations, along with extremely limited access to financial capital, mean that the poor communities are least likely to be protected by investments in infrastructure or disaster mitigation and relief systems. This will ultimately disrupt the total economic spectrum of the poor class particularly of those living below the poverty line.

The predicted adverse impact on human health associated with climate change will affect the poor in particular throughout the developing countries. These risks are primarily associated with waterborne (such as dysentery or cholera) and vector-borne (such as malaria) diseases as well as heat stress morbidity and mortality. The health impacts pose a double jeopardy for poor people's livelihoods: the contribution of key productive members of the household is lost and the cost of health care is expensive and time consuming. Such risks will be widespread with the passage of time, but the dearth of medical care systems in many more remote, poorer areas of Africa and Asia in particular will accelerate the vulnerability and magnitude of the risks.

The deterioration of the availability or quality of water supplies in many areas (again due to wider resource stresses that climate change will exacerbate) will significantly increase many of these health risks, while poorer nutritional states caused by declining food security will make a large percentage of poor people more vulnerable to the effects of diseases when they do strike.

Few significant changes associated with climate change are listed in Table 1.5.

Table 1.4 Damage spectrum of AILA

Affected	West Bengal	Sundarbans (Indian part)
Villages	28,359	4,249
Population	670,000	254,000
Human lives lost	137	70
Livestock lost	71,196	70,811
Crop area affected (ha)	290,000	125,000

Table 1.5 Twentieth-century changes in the Earth's atmosphere, climate and biophysical system

Indicator characteristic	Observed change
Concentration indicator	
Atmospheric concentration of CO_2	280 ppm for the period 1,000–1,750 to 368 ppm in year 2000 (~31.4 % increase)
Atmospheric concentration of CH_4	700 ppb for the period 1,000–1,750 to 1,750 ppb in year 2000 (151 ± 25 % increase)
Atmospheric concentration of N_2O	270 ppb for the period 1,000–1,750 to 316 ppb in year 2000 (17 ± 5 % increase)
Tropospheric concentration of O_3	Increased by 35 ± 15 % from the years 1750 to 2000; varies with region
Stratospheric concentration of O_3	Decreased over the years 1970–2000; varies with altitude and latitude
Atmospheric concentration of HFCs, PFCs and SF_6	Increased globally over the last 50 years
Weather indicators	
Global mean surface temperature	Increased by 0.6–0.2 °C over the twentieth century; land areas warmed more than the oceans (very likely)
Northern Hemispheric surface temperature	Increased over the twentieth century; greater than during any other century in the last 1,000 years; the 1990s was the warmest decade of the millennium (likely)

(continued)

Table 1.5   (continued)

Indicator characteristic	Observed change
Diurnal surface temperature range	Decreased over the years 1950–2000 over land: night-time minimum temperatures increased at twice the rate of day time
Hot days/heat index	Maximum temperature (likely)
Cold/frost days	Increased (likely)
Continental precipitation	Decreased for nearly all land areas during the twentieth century (very likely). Increased by 5–10 % over the twentieth century in the Northern Hemisphere (very likely) although decreased in some region (e.g. north and west Africa and parts of the Mediterranean)
Heavy precipitation	Increased at mid- and high northern latitudes (likely)
Frequency and severity of droughts	Increased summer drying and associated incidence of drought in a few areas (likely)
	In some regions, such as parts of Asia and Africa, the frequency and intensity of droughts have been observed to increase in recent decades
Biological and physical indicators	
Global mean sea level	Increased at an average annual rate of 1 to 2 mm during the twentieth century
Duration of ice cover of rivers and lakes	Decreased by about 2 weeks over the twentieth century in mid- and high latitudes of the Northern Hemisphere (very likely)
Arctic sea ice extent and thickness	Thinned by 40 % in recent decades in late summer to early autumn (likely) and decreased in extent by 10–15 % since the 1950s in spring and summer
Nonpolar glaciers	Widespread retreat during the twentieth century
Snow cover	Decreased in area by 10 % since global observation become available from satellites in the 1960s (very likely)
Permafrost	Thawed, warmed and degraded in parts of the polar, subpolar and mountainous regions
El Nino events	Became more frequent, persistent and intense during the last 20–30 years compared to the previous 100 years
Growing season	Lengthened by about 1–4 days per decade during last 40 years in the Northern Hemisphere, especially at higher latitudes
Plant and animal ranges	Shifted poleward and up in elevation for plants, insect, birds and fish
Coral reef bleaching	Increased frequency, especially during *El Nino* events
Economic indicators	
Weather-related economic losses	Global inflation; adjusted losses increased over the last 40 years. Part of the observed upward trend is linked to climatic factors

1.3.5   Challenges to IPCC Reports

1.3.5.1 Carbon Dioxide Emission Estimation

The IPCC reports have been contradicted by several researchers on the ground of overestimation of emission and underestimation of sink capacity of carbon dioxide. According to Peter Dietze (1997), '... the IPCC burns about 2,300 Gt C for scenario S750, though the available fossil reserves are 720 Gt conventional or 1,000 Gt including unconventional (Houghton et al. 1995).

The IPCC reference scenario IS92a burns about 1,500 Gt C until 2,100. The IPCC's concentration rises up to 600 ppm carbon dioxide, which is far above 500 ppm that could be reached at maximum if we assume all conventional fossil fuel reserves are burnt and 40 % of the emission remains in the air for a long time...'.

1.3.5.2 Underestimation of Carbon Sinks

The IPCC has underestimated the magnitude of different carbon sinks to scale the net increase of

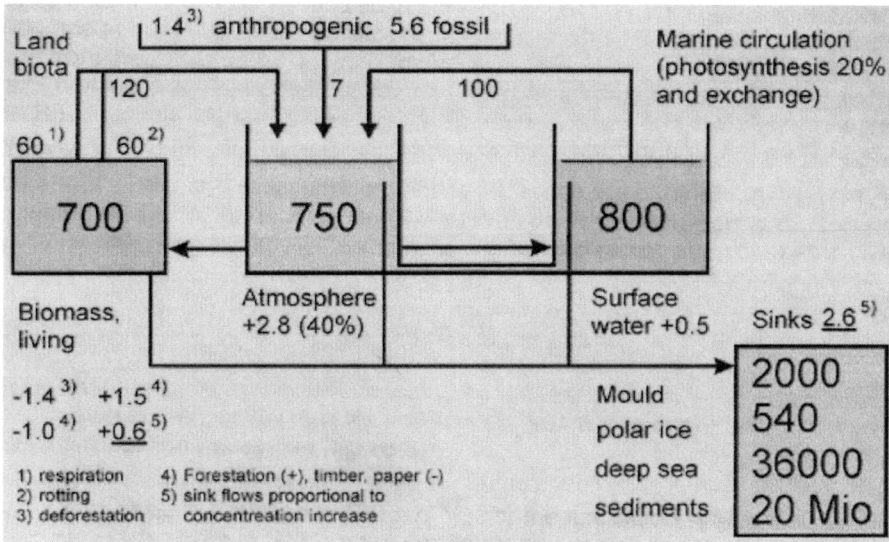

Fig. 1.11   Coupled Waterbox model with proportional sink flows (reservoirs in Gt C, flows in Gt C/year)

greenhouse gas in the Earth's atmosphere. It is known that oceans contain 50 times more carbon than the atmosphere and may take up to nearly six times more carbon dioxide at equilibrium and the photosynthesis of land floral community may increase up to 18 Gt C/year for a concentration doubling. At present the oceans are still mostly on a pre-industrial level.

A new global carbon cycle model with a realistic CO_2 e-fold lifetime of 55 years (half lifetime: 38 years) reveals that the temperature will increase by ~0.3 °C only if the present global CO_2 emission is kept constant until 2,100. In IPCC scenario, it is assumed that far more fossil fuels would be burnt than is physically recoverable. Using an eddy diffusion ocean model, the IPCC has grossly underestimated the future oceanic CO_2 uptake. Hardly coping with biomass response and taking a double to treble, temperature sensitivity have led to an IPCC error factor (Dietze 1997).

The Coupled Waterbox Model: An Approach to Contradict IPCC Model

To develop a new global carbon cycle model, the Waterbox model was extended in the form of coupled Waterbox model (Fig. 1.11). In this figure, the three boxes represent the land biota (700 Gt C),

the atmosphere (750 Gt C) and the mixed ocean layer (800 Gt C), which is closely coupled with the atmosphere through the process of precipitation and gas diffusion, exchanging about 100 Gt C/year with the atmosphere. The model depicts that net photosynthesis of land biota amounts to about 60 Gt C/year and marine photosynthesis is roughly around 20 Gt C/year. In high latitudes, the icy cold salt water takes up large amounts of CO_2. This is taken into the deep sea that mixes via the conveyor belt into all oceans. The central link is the Atlantic Circumpolar Current.

In the figure, the lower box represents all sinks of CO_2, which are detritus, polar ice, deep sea and sediments including shells and corals. These are to be allocated to carbon-fertilized biomass, solubility in the mixed layer, polar water and ice as well as decreasing degassing of upwelling pre-industrial water against the increasing atmospheric concentration. The upwelling water is degassing in tropical latitudes with a time delay of 400–1,000 years.

The deep sea is still mostly in a pre-industrial CO_2 level condition, which means the sink flows will be, in rough approximation, proportional to the atmospheric difference to the CO_2 ocean bulk equilibrium. Any greenhouse science statement that the surface water limits the CO_2 uptake, thus becoming independent of concentration level (or even reducing with increasing concentration as in IPCC's HILDA model), cannot be verified. So far, a vivid deep-water formation has been observed.

Ocean and biota uptake are controlled by the atmospheric CO_2 concentration without 'knowing' how much CO_2 is generated through human emission. However, the HILDA model (Seigenthaler and Joos 1992; Enting et al. 1994) exactly splits the excess of this emission into the oceanic and biotic compartments. So in spite of increasing concentration, the ocean and biomass uptake are decreasing in proportion to the emission in IPCC's stabilization scenario S550. This behaviour is absolutely implausible. After the concentration has doubled, the IPCC ocean returns to normal state, taking less than 1.8 Gt/year. The IPCC biomass returns to about zero net uptake even though photosynthesis probably increases by 30 %, i.e. 18 Gt more. The sink flows at 550 ppm are supposed to be about 12 Gt C/year, which is six times more than assumed by the IPCC. This expanded magnitude of sink for CO_2 has been underestimated in the IPCC approach.

The reason for IPCC's very small stabilization fluxes is the Oeschger/Siegenthaler eddy diffusion ocean model. To support a flux, the diffusion needs a concentration gradient from the mixed layer down to the deep ocean. Because increasing back pressure builds up, even a constant flux needs a permanently increasing concentration in the air. So to avoid a climate disaster, future emissions have to be reduced considerably with this model – even using a high vertical eddy diffusion coefficient of 7,685 m^2/year – chosen about double the measurement and 3×10^5 times the original value for diffusion. However, in reality, the CO_2 uptake of the ocean requires an eddy transport and deep-water formation model. Here an uptake flux can be maintained rather at a constant difference to the ocean bulk concentration.

1.3.5.3 Glacier Melting

The IPCC report was also strongly criticized on the issue of glacier melting in the Himalayas in context to a deadline year for the disappearance of the glacier. The Himalaya has a large concentration of glaciers and permanent snowfields. During winter, most of the high-altitude regions experience snowfall, and snow cover plays an important role in the ecology of the region. Melting from seasonal snow cover during summer forms an important source of many rivers originating in the Higher Himalaya. Therefore, understanding of snow accumulation and ablation is important for utilization of the Himalayan water resource and fate of the rivers originating from the Himalayan glaciers. Snowpack ablation is highly sensitive to climatic variations. Increase in atmospheric temperature can enhance energy exchange between the atmosphere and snowpack. This can increase snow melting. Investigations suggest that climate of the Earth has constantly changed in the course of time, during the past ten million years or so. During this time, the Earth has experienced alternate cycles of warm and cold periods. The difference in global mean temperature between the Last Glacial Maximum and the present warm period is about 5 °C (Kulkarni et al. 2002). The rate of climate change, however, accelerated probably in the twentieth century due to rapid, unplanned and intense industrialization. Large emissions of CO_2, other trace gases and aerosols have changed the composition of the atmosphere. This is changing the global radiation budget of the Earth–atmosphere system, which is subsequently elevating the temperature of the planet Earth. Obviously, this will have a profound impact on snow accumulation and ablation rate in the Himalaya, as snow and glaciers are sensitive to thermal energy in and around them.

An in-depth analysis on the effect of climate change on Himalayan glaciers was reported by Kulkarni et al. (2002) considering two important basins: Beas and Baspa basins in Himachal Pradesh (Fig. 1.12).

The Beas basin was studied up to Manali and Baspa up to Sangla. These basins are located in two different regions of the Himalaya. In addition, the Baspa basin is highly glacierized and located in the higher altitude range. Therefore, winter run-off of the Baspa basin is mostly

Fig. 1.12 Location of the study area in Himachal Pradesh (North India)

contributed by snow melt rather than rainfall, and it can be used as an indicator for long-term changes in snow cover. Investigations carried out in these two Himalayan glacier basins suggest that warmer winter is causing snowmelt in the higher altitude zone. Melting and retreat of snow cover was even observed in months of December and January (coldest months of the year) at an altitude of 5,400 m. This observation was originally made in the winter of 1998–1999, and the same trend was observed in 1999–2000 and in the winter of 2000–2001. Melting and retreat of snow in the middle of winter is an unusual observation. Available temperature data at Manali suggest that this unusual observation is due to high winter time temperature. The long-term trend of number of degree days for December at Manali from 1977 to 2001 indicates substantially increasing trend. Five-year running average degree days are increased from 6.2 to 8.4 between 1983 and 2001. This is also possibly linked with overall increase in average global temperature. Average stream run-off of Baspa River in December from 1966 to 1992 has gone up by almost 75 %. This is a substantial change in the stream run-off. Steady rise in the stream run-off of Baspa River from 1980 matches with the average global temperature rise from 1980 onwards and also with increasing trend at Manali. These observations suggest that global warming has started affecting snowmelt and stream run-off in the Himalaya.

However, the statement of total vanishing of some Himalayan glaciers within a stipulated period of time has put the IPCC report under severe criticism. The IPCC warned that climate change may likely to melt most of the Himalayan glaciers by 2035 – an idea considered ludicrous by most glaciologists.

Jointly awarded the Nobel Prize in 2007, the IPCC's report is relied upon by most governments and scientists as the best guide to the effects and projections of climate change. However, a sentence in the report that claimed there was a 'very high' chance of Himalayan glaciers melting entirely by 2035 has been questioned and widely criticized. The UN's top climate officials have launched a review of the 2007 report by the Intergovernmental Panel on Climate Change, following claims that a section of the report contained exaggerated and unreliable information about the rate of glacial melting in the Himalayas. Critics point out that Hasnain, of all people, should have known the claim that the Himalayan glaciers could melt by 2035 was bogus because he was meant to be a leading glaciologist specializing in the Himalayas. Considering this critical background of the Himalayan glacier melting, the Chair, Vice-Chairs and the Co-chairs of the IPCC regret the poor application of well-established IPCC procedures in this instance and issued the following statement on 20 January 2010.

IPCC Statement on the Melting of Himalayan Glaciers

The Synthesis Report, the concluding document of the Fourth Assessment Report of the Intergovernmental Panel on Climate Change (page 49), stated: '… Climate change is expected to exacerbate current stresses on water resources from population growth and economic and land-use change, including urbanization. On a regional scale, mountain snow pack, glaciers and small ice caps play a crucial role in freshwater availability. Widespread mass losses from glaciers and reductions in snow cover over recent decades are projected to accelerate throughout the twenty-first century, reducing water availability, hydropower potential, and changing seasonality of flows in regions supplied by melt-water from major mountain ranges (e.g., Hindu-Kush, Himalaya, Andes), where more than one-sixth of the world population currently lives…'.

This conclusion is robust, appropriate and entirely consistent with the underlying science and the broader IPCC assessment. However, a paragraph (IPCC Fourth Assessment Report of Working Group II – second paragraph in Section 10.6.2) on page 938 refers to poorly substantiated estimates of rate of recession and date for the disappearance of Himalayan glaciers. In drafting the paragraph in question, the clear and well-established standards of evidence, required by the IPCC procedures, were not applied properly.

1.3.5.4 Global Climate Models

The IPCC report shows confidence in the ability of general circulation models (GCMs) to simulate future climate and attribute observed climate change to anthropogenic emissions of greenhouse gases. The forecasts in the Fourth Assessment Report were not the outcome of validated scientific procedures. They are the opinions of scientists transformed by mathematics and obscured by complex writing. Today's state-of-the-art climate models fail to accurately simulate the physics of earth's radiative energy balance, resulting in uncertainties 'as large as, or larger than, the doubled CO_2 forcing'.

A long list of major model imperfections prevents models from properly modelling cloud formation and cloud–radiation interactions, resulting in large differences between model predictions and observations. Computer models have failed to simulate even the correct sign of observed precipitation anomalies, such as the summer monsoon rainfall over the Indian region.

1.3.5.5 Feedback Factors and Radiative Forcing

The IPCC reports have totally underestimated the cooling effect of aerosols. Studies have found that their radiative effect is comparable to or larger than the temperature forcing caused by all the increase in greenhouse gas concentrations recorded since pre-industrial times.

Higher temperatures are known to increase emissions of dimethyl sulphide (DMS) from the world's oceans, which increases the albedo of marine stratus clouds. This phenomenon triggers the cooling effect. Iodocompounds – created by marine algae – function as cloud condensation nuclei, which help create new clouds that reflect more incoming solar radiation back to space and thereby cool the planet.

1.3.5.6 Temperature Records: A Grey Area

Highly accurate satellite data sets, adjusted for orbit drift and other factors, show a much more modest warming trend in the last two decades of the twentieth century and a dramatic decline in the warming trend in the first decade of the twenty-first century. Temperature records in Greenland and other Arctic areas reveal that temperatures reached a maximum around 1930 and have decreased in recent decades. Longer-term studies depict oscillatory cooling since the Climatic Optimum of the mid-Holocene (~9,000–5,000 years BP), when it was perhaps 2.5 °C warmer than it is now.

1.3.5.7 Extreme Weather Events

The IPCC says, 'it is likely that future tropical cyclones (typhoons and hurricanes) will become more intense, with larger peak wind speeds and more heavy precipitation associated with ongoing increase of tropical sea-surface temperatures'. But despite the claim of 'unprecedented' warming of the twentieth century, there has been no increase in the intensity or frequency of tropical cyclones globally or in any of the specific oceans.

Singh et al. (2000, 2001) analyzed 122 years of tropical cyclone data from the North Indian Ocean over the period 1877–1998. Since this was the period of time during which the planet recovered from the global chill of the Little Ice Age, it is logical to assume that their findings can throw light on the frequency and nature of cyclone due to warming of the planet as claimed by the IPCC experts. However, Singh et al. (2000, 2001) found that on an annual basis, there was a slight decrease in tropical cyclone frequency, such that the North Indian Ocean, on average, experienced about one less hurricane per year at the end of the 122-year record in 1998 than it did at its start in 1877. In addition, based on data from the Bay of Bengal, they found that tropical cyclone numbers dropped during the months of most severe cyclone formation (November and May), when the El Nino–Southern Oscillation was in a warm phase.

1.3.5.8 Extinction of Species

According to IPCC Report, 'new evidence suggests that climate-driven extinctions and range retractions are already widespread' and the 'projected impacts on biodiversity are significant and of key relevance, since global losses in biodiversity are irreversible (very high confidence)'. These claims are not supported by scientific research.

The world's species have proven to be remarkably resilient to climate change. Most wild species are at least one million years old, which means they have all been through hundreds of climate cycles involving temperature changes on par with or greater than those experienced in the twentieth century.

The four known causes of extinctions are huge asteroids striking the planet, human hunting, human agriculture and the introduction of alien species (e.g. lamprey eels in the Great Lakes and pigs in Hawaii). None of these causes is connected with either global temperatures or atmospheric carbon dioxide concentrations. It has been recorded that most populations of polar bears are growing, not shrinking, and the biggest influence on polar populations is not temperature but hunting by humans, which historically has taken a large toll on polar bear populations.

The gradual decrease of the horseshoe crab population in the mangrove-dominated Indian Sundarbans is another example of insignificant correlation between climate change and species. The pharmaceutical value of horseshoe crab is immense. All the horseshoe crab species exhibit high sensitivity to bacterial endotoxin. The cell lysates obtained from the blue blood of these species are widely used for estimating the contamination of bacterial endotoxin. The reagents manufactured from the cell lysates, namely, Carcinoscorpius amoebocyte lysate (CAL), Tachypleus amoebocyte lysate (TAL) and Limulus amoebocyte lysate (LAL), are extremely sensitive and are used for the rapid and accurate assay of gram-negative bacteria even if they are present in a very minute quantity.

The horseshoe crabs species Carcinoscorpius rotundicauda are regular visitors of Sundarbans particularly during the premonsoon season (March to June) when the aquatic salinity reaches its peak. During this period, they are found in mating pairs in the mangrove creeks and mudflats. However, with the vague belief of the ability of this crustacean to cure arthritis, they are often killed by the local people by boiling the entire animal body, and the extract is sold in the district markets as medicine to cure arthritis. This activity has severely reduced the population of the species, which could otherwise be the source of CAL and TAL for pharmaceutical industries.

1.3.5.9 Coral Calcification

The IPCC reports claimed a positive correlation between global warming and coral bleaching. A study of historical calcification rates determined from coral cores retrieved from 35 sites on the Great Barrier Reef, Lough and Barnes (1997) observed a statistically significant correlation between coral calcification rate and local water temperature, such that a 1 °C increase in mean annual water temperature increased mean annual coral calcification rate by about 3.5 %. However, they also pointed 'declines in calcification in Porites on the Great Barrier Reef over recent decades'.

Fact Below the Carpet
The power of the Sun is natural, but the supreme power of the industry houses is acquired. These houses fund the research organizations, NGOs, government departments and even individual initiative to touch the goal, which is often shaped by them. Changes in the policies to control and mitigate climate change may pose an

(continued)

(continued)

adverse impact on their products. The car companies may suffer as personal vehicle usage may be reduced or better efficiency of their products has to be installed. Architecture and construction companies will have to divert more funds from their profit in the R&D sector for 'green' buildings which may become a norm in the coming days. The oil companies may face competition with the houses striving hard for nonconventional sources of energy, and above all greening of the environment, landscaping, carbon trading, etc. will be the burden of the present-day industrial houses. Very few houses are going to accept such radical change, and therefore a strong anti-climate voice has been raised from different corners of the society often triggered by these big industry houses. The disappearance of Himalayan glaciers by 2035 is definitely an overstatement, but it should be considered as line of caution rather than an open challenge as the freshwater supply (the basic raw material for all industries) is a function of glacier volume. The mega industry houses must be sensitized with the fact that attacking a paragraph of a huge multidimensional report is basically an indirect attack on their own raw materials in the planet Earth which is definitely finite.

Important References

Antonov JI, Levitus S, Boyer TP (2005) Steric variability of the world ocean, 1955–2003. Geophys Res Lett 32(12):L12602. doi:10.1029/2005GL023112

Chao B (1994) Man-made lakes and global sea level. Nature 370:258

Chaudhuri AB, Choudhury A (1994) Mangroves of the Sundarbans, vol I, India. IUCN – The World Conservation Union, Bangkok

Church JA, Gregory JM (2001) Chapter 11: Changes in sea level. In: Climate change 2001: the scientific basis. Contribution of working group I to the third assessment report of the Intergovernmental Panel on Climate Change. Cambridge University Press, New York.

Dasgupta S, Laplante B, Meisner C, Wheeler D, Yan J (2009) The impact of sea level rise on developing countries: a comparative analysis. Clim Chang 93:379–388

Davis CH, Li Y, McConnell JR, Frey MM, Hannah E (2005) Snowfall-driven growth in East Antarctica ice sheet mitigates recent sea-level rise. Science 308(5730): 1898–1901

Dietze P (1997) Little warming with new global carbon cycle model. ESEF vol II. http://www.esef.org. Accessed on 20 Feb 2010

Enting I, Wigley T, Heimann M (1994) Future emissions and concentrations of carbon dioxide: key ocean/atmosphere/land analyses, CSIRO Technical Paper No. 31. CSIRO Australia, 1994 (Electronic Edition). http://www.physicalgeography.net/fundamental.7y.html. Accessed on 03 July 2010

Fergusson J (1863) Recent changes in the delta of the Ganges. Quat J Geol Soc 19:321–354

Gornitz V (2001) Impoundment, groundwater mining, and other hydrologic transformations: impacts on global sea level rise. In: Douglas BC, Kearney MS, Leatherman SP (eds) Sea level rise: history and consequences. Academic, San Diego, pp 97–119

Hanna E, Huybrechts P, Janssens I, Cappelen J, Steffen K, Stephens A (2005) Runoff and mass balance of the Greenland ice sheet: 1958–2003. J Geophys Res 110:D13108

Hansen J (2006) Can we still avoid dangerous human-made climate change?. In: Presentation on December 6, 2005 to the American Geophysical Union in San Francisco. http://www.columbia.edu/~jeh1/newschool_text_and_sides.pdf

Hansen J, Nazarenko L, Ruedy R, Sato M, Willis J, Del Genio A, Koch D, Lacis A, Lo K, Menon S, Novakov T, Perlwitz J, Russell G, Schmidt G, Tausnev N (2005) Earth's energy imbalance: confirmation and implications. Science 308:1431–1435

Howat I, Joughin I, Scambos T (2007) Rapid changes in ice discharge from Greenland outlet glaciers. Science 315:1559–1561

Houghton JT, Meira Filho LG, Callander BA, Harris N, Kattenberg A (1995) In: Maskell K (ed) Climate change 1995: the science of climate change. Contribution of Working Group I to the second assessment report of the Intergovernmental Panel on Climate Change. Cambridge University Press, Cambridge, ISBN 0-521-56433-6

IPCC (2001) The scientific basis. Contribution of working group I to the third assessment report of the intergovernmental panel on climate change (IPCC). Cambridge University Press, Cambridge

IPCC (2007) Climate change 2007: the physical science basis. Summary for policymakers. IPCC Secretariat, Geneva

Ishii M, Kimoto M, Sakamoto K, Iwasaki SI (2006) Steric sea level changes estimated from historical ocean subsurface temperature and salinity analyses. J Oceanogr 62(2):155–170

Krabill W, Hanna E, Huybrechts P, Abdalati W, Cappelen J, Csatho B, Frederick E, Manizade S, Martin C, Sonntag J, Swift R, Thomas R, Yunge J

(2004) Greenland ice sheet: increased coastal thinning. Geophys Res Lett 31:L24402

Kulkarni AV, Mathur P, Rathore BP, Alex S, Thakur N, Kumar M (2002) Effect of global warming on snow ablation pattern in the Himalaya. Curr Sci 83(2):120–123

Lombard A, Cazenave A, Traon PYL, Guinehut S, Cecile C (2006) Perspectives on present-day sea level change: a tribute to Christial le Provost. Ocean Dyn 56(5–6): 445–451

Lough JM, Barnes DJ (1997) Several centuries of variation in skeletal extension, density and calcification in massive *Porites* colonies from the Great Barrier Reef: a proxy for seawater temperature and a background of variability against which to identify unnatural change. J Exp Mar Biol Ecol 211:29–67

Meier M, Dyurgerov M, Rick U, O'Neel S, Preffer W, Anderson R, Anderson S, Glazovsky A (2007) Glaciers dominate eustatic sea level rise in the 21st century. Science 317:1064–1067

Morgan JP, McIntire WG (1959) Quaternary geology of the Bengal Basin, East Pakistan and Burma. Bull Geol Soc Am 70:319–342

Oldham RD (1893) A manual of geology of India. Government of India, Calcutta

Overpeck J, Otto- Bliesner B, Miller G, Muhs D, Alley R, Kichl J (2006) Paleoclimatic evidence for future ice sheet instability and rapid sea level rise. Science 311:1064–1067

Pidwirny M (2006) Causes of climate change. In Fundamentals of physical geography, 2nd edn.

Raha A, Das S, Banerjee K, Abhijit M (2012) Climate change impacts on Indian Sunderbans: a time series analysis (1924–2008). Biodivers Conserv 21(5): 1289–1307. doi:10.1007/s10531-012-0260-z

UNFAO Report (2006) Livestock – a major threat to Environment: remedies urgently needed. Reported by Christopher Matthews, Media Relations, FAO, Rome

Rignot E, Kanagaratnam P (2006) Changes in the velocity structure of the Greenland ice sheet. Science 311:986–990

Sahagian DL (2000) Global physical effects of anthropogenic hydrological alterations: sea level and water redistribution. Global Planet Change 25:39–48

Sahagian DL, Schwartz FW, Jacobs DK (1994) Direct anthropogenic contributions to sea level rise in the twentieth century. Nature 367:54–56

Seigenthaler U, Joos F (1992) Use of a simple model for studying oceanic tracer distributions and the global carbon cycle. Tellus B 44:186–207

Singh OP, Ali Khan TM, Rahman S (2000) Changes in the frequency of tropical cyclones over the North Indian Ocean. Meteorol Atmos Phys 75:11–20

Singh OP, Ali Khan TM, Rahman S (2001) Has the frequency of intense tropical cyclones increased in the North Indian Ocean? Curr Sci 80:575–580

Stroeve J, Holland M, Meier W, Scambos T, Serreze M (2007) Arctic sea ice decline: faster than forecast. Geophys Res Lett 34:L09501

Velicogna I, Wahr J (2006) Measurements of time variable gravity show mass loss in Antarctica. Science 311:1754–1756

Wadia DN (1961) Geology of India. M.C. Muller & Co Ltd, London

Willis JK, Roemmich D, Cornuelle B (2004) International variability in upper-ocean heat content, temperature and thermosteric expansion on global scales. J Geophys Res 109:C12036. doi:10.1029/200 3JC002260

Internet References

http://www.physicalgeography.net/fundamentals/7y. html

http://www.grida.no/climate/ipcc_tar/wg1/295.htm

http://www.greenhouse.gov.au/impacts/overview/pubs/ overview4.pdf

http://www.gurdian.co.uk/international/story/0,1503170,00. html

http://dpa.aapg.org/gac/papers/climate_change.cfm

http://www.agu.org/fora/eos/pdfs/2006EO360008.pdf

http://www.abi.org.uk/Display/File/Child/552/Financial_ Risks-of_Climate.html

http://www.ipcc.ch/SPM2feb07.pdf

Website of US Geological Survey

Mangroves: A Unique Gift of Nature

2

Mother Earth has created several gifts in the ecosystem vaults, which need to be priced properly. A free gift invites destruction, degradation, spoilage and conflict.

The Author

2.1 Mangroves: An Overview

Mangroves are salt-tolerant forest ecosystems found mainly in the tropical and subtropical intertidal regions of the world. They encompass swamps, forestland within, and the surrounding water bodies. It is a matter of great surprise that mangrove floral species can thrive luxuriantly in saline habitat (which is basically physiologically dry in nature) through orientation of their morphological, anatomical and physiological systems. Thus, this vegetation is the most efficiently adapted biotic community in response to climate-change-induced sea-level rise.

The term 'mangrove' has originated from the Portuguese word *Mangue*, which means the community, and the English word *Grove*, which means trees or bushes. According to Mepham and Mepham (1984), the term has been inconsistent and confusing in the past. Mangroves are basically evergreen sclerophyllous, broad-leaved trees with aerial root like pneumatophore or stilt root and viviparously germinated seedlings (UNESCO 1973). They grow along protected sedimentary shores especially in tidal lagoons, embayment and estuaries (MacNae 1968). They also can grow far inland, but never isolated from the sea. These emergent, evergreen canopies are found along the sedimentary shores of both tropical and subtropical regions in association with intertidal flora and fauna commonly known as mangrove ecosystem, and the community of these mangroves was coined by MacNae (1968) as *Mangal*. Lear and Turner (1977) expressed the word 'mangrove' of coastal ecosystem in a holistic manner, including its common habitat or inhabiting fauna. The term 'mangrove' also denotes both the ecological group of flowering halophytic shrubs and trees of up to 30 m high belonging to several unrelated families and the complete community or association of plants which fringe sheltered tropical shores. About 60–75 % of tropical coastline is fringed with mangroves (Reimold and Queen 1974). Duke (1992) defined mangroves as '… A tree, shrub, palm and ground fern, generally exceeding one half meter in height and which normally grows above mean sea level in the intertidal zone of marine coastal environments or estuarine margins …'. This definition is acceptable except that ground ferns should be considered as mangrove associates rather than true mangroves. The term 'mangrove' often refers to both the plants and the forest community. To avoid confusion, MacNae (1968) stated that 'mangrove' should refer to halophytic plant species, while the entire forest community including micro- and macro-organisms should be considered as 'mangal'. The mangal is therefore a broad domain encompassing the entire biotic community comprising of individual plant species, associated microbes (like bacteria and fungi) and animals. The mangal and its associated abiotic factors constitute an ideal *mangrove ecosystem*, which is a unique ecosystem of the planet Earth.

The mangrove forests are highly productive ecosystems with productivity about 20 times more than the average oceanic production (Gouda and Panigrahy 1996). Moreover, it is a 'detritus-based' ecosystem unlike other coastal ecosystems, which are usually 'plankton-based'. The detritus supplied by the ecosystem saturates the ambient water with nutrients, which triggers the growth and development of planktonic community in the water bodies on which the fishery resource is also dependent. The greatest concentration of mangrove species is observed usually at the mouth of tidal creeks and rivers where salt and freshwater mix in ideal proportion and floodwaters deposit plenty of material to build up the banks. This unique coastal ecosystem of the world sustains a rich spectrum of floral and faunal community in and around its vicinity. The mangroves enrich the coastal waters with nutrients, yield commercial forest products, protect coastlines and support coastal fisheries (Kathiresan and Bingham 2001). Generally, the mangrove vegetations are well adapted to extreme conditions of salinity, tides, winds and temperature, although they show a preference for freshwater. There are no floral groups in the plant kingdom, which possess such well-organized and highly developed morphological, biological and physiological as well as ecological adaptations to extreme environmental conditions.

Mangrove plants tolerate salinity of the soil and water through three basic processes as listed here:

1. *Salt excretion*: Mangrove plants take saline water as such through roots. However, in the tissues of some species of mangroves, only water molecules and essential salts are retained. Excess salts are excreted through *salt glands* that are present in the leaves. The salt-excreting species of the mangrove community like *Avicennia alba*, *Aegiceras corniculatum*,

Acanthus ilicifolius and *Aegialitis rotundifolia* regulate their internal salt levels through foliar glands. In salt secreting (excreting) mangroves, the NaCl concentration of xylem sap is relatively high, about one-tenth of the concentration of salt in seawater. So, the salt-excreting species allow more salt into the xylem than do the non-excretors but still exclude about 90 % of the salts (Scholander et al. 1962; Azocar et al. 1992). Salt is only partially excluded at the roots. The absorbed salt is primarily excreted metabolically via specialized salt glands in the leaves. The salt in solution can crystallize by evaporation and either can be blown away or washed off. Since, in salt-excreting mangroves, superfluous salts are excreted by guttation through special salt glands, all these salt-excreting halophytes are often referred to as *crinohalophytes*.

It is interesting to note that salt excretion is an active process, as evidenced by ATPase activity in the plasmalemma of the excretory cells (Drennan and Pammenter 1982). The process is probably regulated by leaf hypodermal cells, which may store salt as well as water (Balsamo and Thomson 1995).

2. *Salt exclusion*: In some of the mangrove plants, the roots possess an *ultrafiltration* mechanism called *reverse osmosis* by which water and salts in the seawater are separated in the root zone itself and only water is taken inside and the salts are rejected. Many mangrove species can exclude 90 % of salt in the ambient seawater or estuarine system (http://www.epa.qld.gov.au/nature_conservation/habitats/mangroves_andwetlands/man groves). *Rhizophora mucronata*, *Ceriops decandra*, *Bruguiera gymnorrhiza*, *Kandelia candel*, etc. are few salt excluders of the mangrove community. Scholander (1968) demonstrated experimentally that the salt separation process in mangroves occurs at or near the root surface. This is mediated by physical processes alone, since it is not inhibited by poisons or high temperature, which may cause an inhibitory effect on metabolic process. In the root area, the physical mechanism for salt separation involves ultrafiltration which occurs either at

the root surface (epiblema) or at the root endodermis. However, the latter region might be the most preferable site (Tomlinson 1986) because the ultimate absorbing roots in most of the mangroves lack root hairs (e.g. capillary rootlets of *Rhizophora* sp.). This indicates that the absorbing area of mangroves is reduced in comparison to non-mangrove plants.

3. *Salt accumulation*: In this type of mangrove plants, the species possess neither salt glands nor ultrafiltration system, but they have the capacity to accumulate a large amount of salts in their leaves. This imparts succulence to their leaves. *Sonneratia apetala, Lumnitzera racemosa, Excoecaria agallocha, Sesuvium portulacastrum, Suaeda maritima* and *Suaeda nudiflora* are included in this category. Leaf succulence in mangroves has a simple explanation in terms of salt balance. The osmotic potential of the leaf cells of mangroves is high (Scholander et al. 1964) which is essential if mangroves are to draw water from the sea with its high negative water potential. However, Scholander (1968) noted that the salt concentration of mangrove leaves remains constant and independent of age. Measurement of salt content in xylem sap demonstrates incomplete salt exclusion at the roots. But, mangroves accumulate salt, and so, this accumulation is partly compensated by salt glands, mainly in the less efficient salt excluders. Since salt concentration is constant and independent of leaf age, salt must accumulate by an increase in the volume of the leaf cells inducing succulence. The leaf succulence in mangroves may therefore be accepted as a part of their adaptation in an environment that provides ample water at the expense of some compensation for high aquatic salinity.

Studies on salt tolerance in *Aegiceras corniculatum* and *Sesuvium portulacastrum* generated few interesting findings: (i) NaCl salinity has a considerable effect on the degree of succulence. With the increase of NaCl salinity in the ambient media, the mass and volume of the leaves increase due to increment in the water content. (ii) The effect of NaCl and Na_2SO_4 is more pronounced in *Sesuvium* than *Aegiceros*. (iii) In *Sesuvium* sp., the effect of chloride salinity is more prominent than that of sulphate salinity. (iv) NaCl is the most effective salt in promoting succulence (Van Eijk 1939). Succulence is due to expansion of the cell wall leading to increase in the size of cells. (v) Accumulation of NaCl is more in *Sesuvium* than *Aegiceros* due to their difference in the mode of salt regulation. (vi) Chlorophyll content decreases sharply at high concentration of NaCl in both the plant species. (vii) High concentration of Na_2SO_4 stimulates the synthesis of chlorophyll in *Aegiceras corniculatum* but inhibits the same in *Sesuvium* sp.

The mangroves not only stabilize the shoreline and act as a bulwark against the encroachment by the sea, but they also act as the abode of several species of fin fish, nursery of a wide range of finfish and shellfish juveniles and biopurifying matrix of wastes generated as a result of industrialization and urbanization. In mangrove ecosystem, different kinds of unrelated fauna and flora get themselves adapted to thrive under the influence of tidal inundation and brackish water. This ecosystem is thus a zone of adaptive convergence, which is a critical issue in the sphere evolutionary biology.

Mangroves are circumtropical in distribution, and this forest community occupies approximately 75 % of the total tropical coastline. Northern extension of this coastline occurs in Japan (31°22′ N) and Bermuda (32°20′ N), whereas southern extensions are in New Zealand (38°03′ S), Australia (38°45′ S) and on the east coast of South Africa (32°59′ S). Globally, mangroves are distributed in 112 countries and territories. It is interesting to note that mangrove plants are not native to the Hawaiian Islands – six species have been introduced there since the year 1900. The mangrove diversity is more in Southeast Asian countries (Fig. 2.1). The region holds nearly 75 % of the world's mangrove species with the highest species diversity found in Indonesia with 45 species, followed by Malaysia (36 species) and Thailand (35 species). India is no less in terms of the number of mangrove

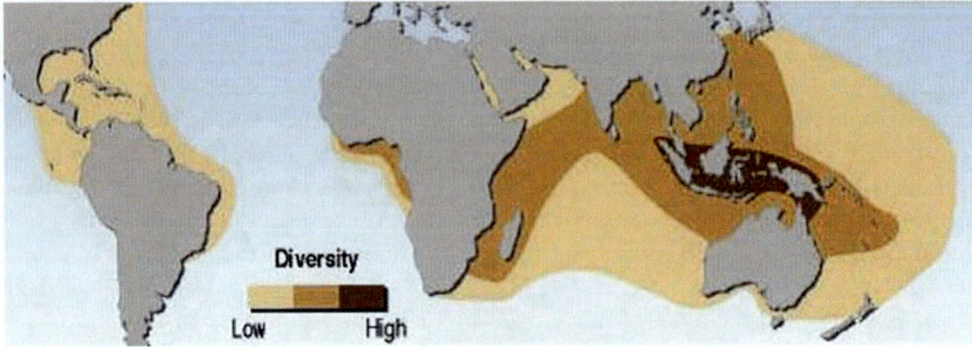

Fig. 2.1 Global distribution of mangrove diversity (After UNEP 2002)

Table 2.1 Estimates of mangrove areas (ha)

Reference	Reference year[a]	No. of countries included	Estimated total area
FAO/UNEP (1981)	1980	51	15,642,673
Saenger et al. (1983)	1983	65	16,221,000
FAO (1994)	1980–1985	56	16,500,000
Groombridge (1992)	1992	87	19,847,861
ITTO/ISME[b] (1993)	1993	54	12,429,115
Fisher and Spalding (1993)	1993	91	19,881,800
Spalding et al. (1997)	1997	112	18,100,077
Aizpuru et al. (2000)	2000	112[c]	17,075,600
FAO (2003)	2003	112	14,653,000

[a]Except for FAO/UNEP (1981) and Aizpuru et al. (2000), the reference year is the year of the publications in which the estimate is cited, not the weighted average of all the national area estimates
[b]Combined figure from three publications by Clough (1993), Diop (1993) and Lacerda (1993)
[c]New estimates were provided for 21 countries, and for the remaining countries, the study relied on Spalding et al. (1997)

species (34 species of true mangroves) and hence is considered as one of the mega-biodiversity countries in the world.

The total global coverage of mangroves has been variously estimated as 14–15 million hectares (Schwamborn and Saint-Paul 1996), 10 million hectares (Bunt 1992) and 24 million hectares (Twilley et al. 1992). Spalding (1997) gave a recent estimation of global mangrove coverage at around 18 million hectares with 41.4 % in South and Southeast Asia and additional 23.5 % in Indonesia. The most recent estimates suggest that mangroves presently occupy about 14,653,000 ha of tropical and subtropical coastline (Wilkie and Fortuna 2003) (Table 2.1).

A list of mangrove-dominated countries in and around Indian Ocean is shown in Table 2.2.

In the Indian Ocean region, the mangroves are found in a variety of coastal settings, ranging

Table 2.2 Mangrove-rich countries in the Indian Ocean region

Country	Area (km[2])
Indonesia	42,500
Myanmar	6,950
Malaysia	6,410
India	4,871
NW Australia	4,513
Bangladesh	4,500
Madagascar	4,200
Mozambique	4,000
Pakistan	2,600
Thailand	1,900

from arid areas through estuaries, lagoons and deltas to coastal fringes. The functional types of mangroves in the Indian Ocean region are:

1. Overwashed mangrove forests – small mangrove islands, frequently overwashed by the tides

2. Fringing mangrove forests – found along the waterways influenced by daily tides
3. Basin mangrove forests – stunted mangroves, located in the interior of swamps
4. Hammock mangrove forests – similar to basin type, but existing in more elevated sites
5. Scrub mangrove forests – dwarf stands of mangroves, existing on flat coastal fringes

The sheltered coasts support a luxuriant growth of mangroves and a higher biodiversity, and this is because of the favourable conditions, such as muddy sediment, frequent water exchange, high rainfall and high humidity, prevailing in the areas. The best examples are mangroves of Sundarbans (India and Bangladesh), Malaysia and Indonesia. In contrast, the arid regions of Arabian Gulf countries, Pakistan and Gujarat (India), where the sediment is sandy, highly saline and poor in nutrients have only dwarf mangrove stands.

2.1.1 Mangroves: A Multivalued Ecosystem

Mangroves have immense ecological value. They protect and stabilize the coastal zone, fertilize the coastal waters with nutrients, yield commercial forest products, support coastal fisheries and provide a surprising genetic reservoir that are the sources of several bioactive substances and extracts having high medicinal values. Thus, this unique vegetation of the globe provides various direct and indirect benefits to the stakeholders. The direct uses focus mainly on the timber, fire wood and honey production. The indirect uses may be related to litter and detritus contribution by this ecosystem due to which a food web is generated and large spectrum of finfish and shellfish juveniles are attracted and nourished. Mangrove forests are among the world's most productive sites, producing organic carbon well in excess of the ecosystem requirements and thus contributing significantly to the global carbon cycle. The carbon sequestration property of mangrove flora has added another feather to the crown of this vegetation particularly after the Clean Development Mechanism (CDM) concept has been introduced. The mangrove ecosystem forms

Table 2.3 Some traditional uses of true mangrove species

Mangrove species	Uses
Aegiceras corniculatum	Bark used as fish poison. Also contains tannin
Avicennia alba	Used as fodder and fuel
Bruguiera caryophylloides	Wood used for firewood and timber. Bark used for tannin production
Bruguiera gymnorrhiza	Wood used for house posts. Bark used as tannins. Also provides fuel
Bruguiera parviflora	Source of fuel. Leaves and barks contain tannin
Bruguiera sexangula	Timber used in house building
Ceriops tagal	Used for keel of boats and house posts. Provides good fuel charcoal. Bark is rich in tannin, used for dyeing fishing nets
Rhizophora mucronata	Bark is used for tannin and cattle fodder
Sonneratia apetala	Wood used in house building, packing cases and yield excellent fuel
Xylocarpus granatum	Yields gum, resin which are used in local medicines. Bark contains tannin

Source: Chaudhuri and Choudhury (1994)

the backbone of coastal economy in certain pockets of the globe for its various benefits to coastal population. The multiple ecological, economic and aesthetic benefits offered by this luxuriant ecosystem are pointed here in brief.

1. The mangrove vegetations and their associates are economically very important for their products like timber, firewood, honey, wax, alcohol, tannins (Table 2.3) and even extracts having surprising medicinal properties.
2. Mangroves are thought to possess the ability to control coastal water quality. The complexity of the mangrove forest habitat increases the residence time, which assists in the assimilation of inorganic nutrients and traps suspended particulate matter. The mangroves also function as flood control barrier and binder of sediment particles (http://www.fao.org/gpa/sediments/habitat.htm). This feature of mangroves is very important in context to climate-change-induced sea-level rise.
3. The ecosystem forms an ideal ecological asset because the production of leaf litter and

detritus matter from mangrove plants fulfils the nutritional requirements of prawn juveniles, adult shrimps, molluscs and fishes of high economic value. It is for this reason mangrove ecosystem is recognized as the world's most potential nursery.

There are three main factors to justify the nursery role of the mangrove habitat. They are:

- High levels of water turbidity, which increases the survival rate of larvae through reduced perception distance of predators.
- Tidal mixing, nutrient trapping and freshwater inflow that result in considerable primary productivity and provide base of a food web from zooplankton to post larval fishes and juveniles.
- The physical and structural complexity of the habitat itself provides a variety of niches favourable to survival of juveniles. Niche segregation often leads to minimization of inter- and intraspecific competition resulting in a high survival rate of finfish and shellfish juveniles.

4. The vibrating mangrove ecosystem provides nutritional inputs to adjacent shallow channels and bay system that constitute the primary habitat of a large number of aquatic species, algae of commercial importance, seaweeds, phytoplankton, etc.

5. Mangroves and mangrove habitats contribute significantly to the global carbon cycle. Twilley et al. (1992) estimated the total global mangrove biomass to be approximately 8.70 gigatons dry weight (which is equivalent to 4.00 gigatons of carbon). Accurate biomass estimation however needs the measurement of the volumes of individual trees. Thus, mangroves are vital carbon sink in the coastal ecosystem. Mangrove destruction can release large quantities of stored carbon and exacerbate global warming trends, while mangrove rehabilitation will increase sequestering of carbon (Kauppi et al. 2001; Ramsar 2001; Chmura et al. 2003). A case study on carbon storage by a widely distributed mangrove tree species *Avicennia marina* is presented at the end of this chapter as Annexure 2A.1. The study

clearly demonstrates the adverse impact of saline water intrusion (because of sea-level rise) on the biomass and carbon sequestration potential of mangroves.

6. Mangroves trap debris and silt, leading to the stabilization of the near shore environment (http://www.reefrelief.org/Documents/mangrove.html).

7. The highly specialized mangrove ecosystem also acts as the protector of the coastal landmass from storm surges, tropical cyclone, high winds, tidal bores, seawater seepage and intrusion. Large numbers of references are available in context to tsunami of 26 December 2004, suggesting that mangroves both dissipated the force of the tsunami and caught the debris washed up by it and thus helped to reduce damage (IUCN 2005). In several cases, mangroves were also instrumental in saving lives by preventing people caught in the backwash of the wave from being pulled out to sea. However, as with coral reefs, subsequent studies showed that the benefit of mangrove protection was rather variable. In India, bathymetry and coastal profile are most important in determining the impact, but less erosion was observed in the Andamans behind mangroves than where there were no mangroves (Department of Ocean Development 2005). A survey of 24 lagoons and estuaries along the southwest, south and southeast coasts of Sri Lanka which suffered the greatest damage showed that there was little destruction to the coast where good quality mangrove communities occurred, and the mangroves themselves were not badly harmed. However, forests dominated by less typical mangrove species (i.e. those that had been degraded in the past and were no longer dominated by genera such as *Sonneratia* or *Rhizophora*) were damaged (Dahdouh-Guebas et al. 2005). It therefore seems that the 'quality' of the mangrove forest contributes in large measure to its buffering capacity, in addition to its size and the extent of regrowth, if it had previously been cleared. Tree density is also an important factor related to the defensive property of mangroves: one study indicated

that a 100 m-wide belt of mangroves, with trees at a density of 30 per 100 m², would be sufficient to reduce the flow pressure from a tsunami by as much as 90 % (Hiraishi and Harada 2003).

8. Bioaccumulation of heavy metals by certain mangrove species reveals the most surprising feature about these plants as they can act as bio-purifier or bio-filter. Few species of mangroves are highly efficient in detecting or assessing the change of ambient environment. The concentration of heavy metal pollutants in different parts of mangrove plants may act as a pathfinder for water quality monitoring programme (Mitra et al. 2004).

9. Mangroves filter groundwater and storm water run-off that often contain harmful pesticides, herbicides and hydrocarbons. Mangroves also recharge underground water table by collecting rainwater and slowly releasing it into the underground reservoir.

10. Mangrove prop roots protect and offer habitat for mammals, amphibians, reptiles, countless unique plants, juvenile fish and invertebrates that filter water such as sponges, barnacles, oysters, mussels, crabs and shrimps.

11. Mangroves are ideal nesting grounds for many water birds such as the great white heron, reddish egrets, roseate spoonbills, white-crowned pigeons, cuckoos and frigate birds. The excretory materials of these birds (rich in phosphorus) serve as the fertilizers of the adjacent water bodies on which the primary production of the aquatic phase depends.

12. Mangrove forests are the housing complexes for bees, birds, mammals and reptiles from which honey, wax, food, etc. are obtained.

13. The molluscan species in the mangrove ecosystem (like oysters and gastropods) are the sources of lime.

14. Mangrove leaves are used as fodder and green manure. The cyanobacterial strains present on the forest floor of mangrove ecosystem are important sources of biofertilizer.

15. Extracts from mangrove and mangrove-dependent species have proven activity against some animal and plant pathogens. Moreover, mangrove extracts kill larvae of the mosquitoes, e.g. a pyrethrin-like compound in stilt roots of *Rhizophora apiculata* shows strong mosquito larvicidal activity (Thangam 1990).

16. Bioactive compound (ecteinascidin) extracted from the mangrove ascidian *Ecteinascidia turbinata* has shown strong in vivo activity against a variety of cancer cells.

17. Phenols and flavonoids in mangroves leaves serve as UV-screening compounds. Hence, mangroves can tolerate solar UV radiation and create a UV-free, under-canopy environment (Moorthy 1995).

18. Bark of *Ceriops* sp. is an excellent source of tannin, and a decoction of it is used to stop haemorrhage and as an application to malignant ulcers. Flowers of this plant are a rich source of honey and bee wax.

19. Mangrove ecosystem affords recreation to hunters, fishermen, bird watchers, photographers and others who treasure natural areas. However, the intrusive actions of noisy jet skis, campers and others, which disturb nesting and breeding areas, chop down mangroves and otherwise damage this fragile environment, threatening its existence. The recent trend of expanding shrimp culture activity at the expense of mangroves is another major threat to mangrove biodiversity.

20. Mangroves can adapt to sea-level rise if it occurs slowly enough (Ellison and Stoddart 1991), if adequate expansion space exists and if the ambient environmental conditions are congenial for their survival and growth. They have special aerial roots, supporting roots and buttresses to live in muddy, shifting and saline conditions. Mangroves may adapt to changes in sea level by growing upward in place or by expanding landward or seaward. This property of mangroves is known as *resilience*. Mangrove vegetation can expand their range despite sea-level rise if the rate of sediment accretion is sufficient to keep up with sea-level rise. However, their ability to migrate landward or seaward is also determined by local conditions, such as infrastructure (e.g. roads, agricultural fields, dikes, urbanization, seawalls and shipping channels) and topography (e.g. steep slopes). If inland migration or growth cannot occur fast enough to account

for the rise in sea level, then mangroves will become progressively smaller with each successive generation and may perish (UNEP 1994). The mangrove vegetation band width is thus directly proportional to their migration rate. A report published by McLeod and Salm (2006) identified important mangrove resilience factors related to site selection in context to climate-change-induced sea-level rise. These are listed here:

(a) *Factors that allow for peat building to keep up with sea-level rise*
- Association with drainage systems including permanent rivers and creeks that provide freshwater and sediment
- Sediment-rich macrotidal environments to facilitate sediment redistribution and accretion
- Actively prograding coast and delta
- Natural features (bays, barrier islands, beaches, sandbars, reefs) that reduce wave erosion and storm surges

(b) *Factors that allow for landward migration*
- Mangroves backed by low-lying retreat areas (e.g. salt flats, marshes, coastal plains) which may provide suitable habitat for colonization and landward movement of mangroves as sea level rises
- Mangroves in remote areas and distant from human settlements and agriculture, aquaculture and salt production developments
- Mangroves in areas where abandoned alternate land use provides opportunities for restoration, for example, flooded villages, tsunami-prone land and unproductive ponds

(c) *Factors that enhance sediment distribution and propagule dispersal*
- Unencumbered tidal creeks and areas with a large tidal range to improve flushing reduce ponding and stagnation and enhance sediment distribution and propagule dispersal
- Areas with a large tidal range may be better able to adjust to increases in sea level due to stress tolerance

- Permanent strong currents to redistribute sediment and maintain open channels

(d) *Factors that indicate survival over time*
- Diverse species assemblage and clear zonation over range of elevation (intertidal to dry land)
- Range in size from new recruits to maximum size class (location and species dependent)
- Tidal creeks and channel banks consolidated by continuous dense mangrove forest (which will keep these channels open)
- Healthy mangrove systems in areas which have been exposed to large increases in sea level due to climate-change-induced sea-level rise and tectonic subsidence

(e) *Factors that indicate strong recovery potential*
- Access to healthy supply of propagules, either internally or from adjacent mangrove areas
- Strong mangrove recruitment indicated by the presence, variety and abundance of established mangrove propagules
- Close proximity and connectivity to neighbouring stands of healthy mangroves
- Access to sediment and freshwater
- Limited anthropogenic stress
- Unimpeded or easily restorable hydrological regime
- Effective management regime in place such as the control of usual threats like dredging and filling, conversion to aquaculture ponds and construction of dams, roads and dikes that disrupt hydrological regime
- Integrated Coastal Management Plan or Protected Area Management Plan implemented

The decision tree related to site selection in context to climate-change-induced sea-level rise is presented in flowchart No. A.

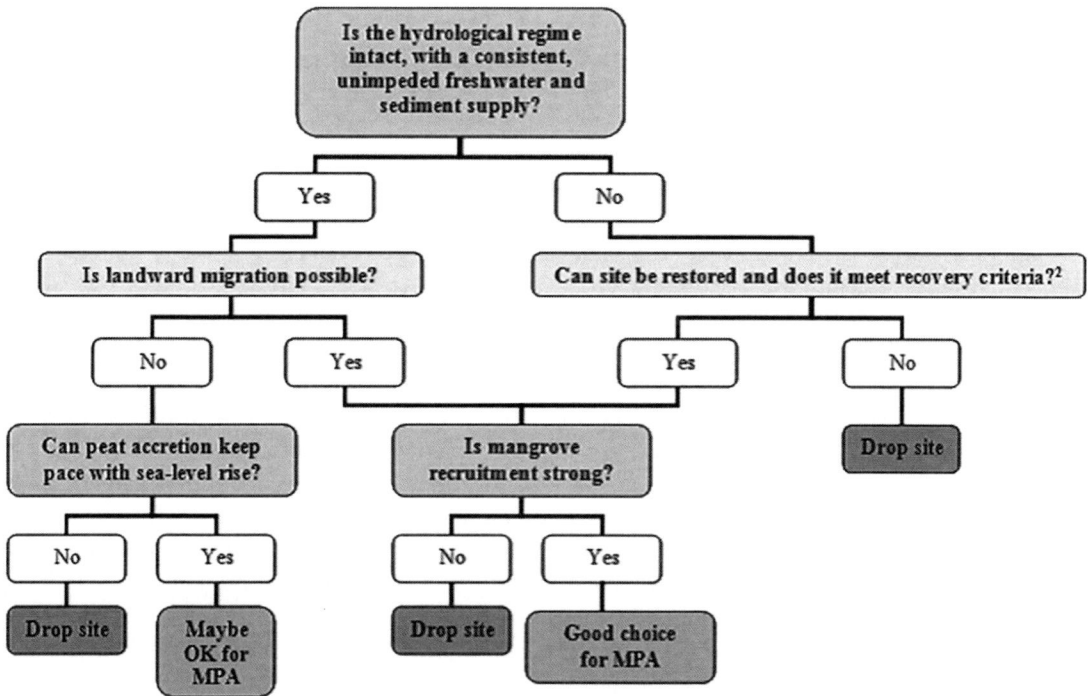

Flow chart No. A. Decision tree for mangrove recruitment/restoration in context to climate change induced sea level rise

21. The mangrove soil acts as unique reservoir of carbon. The carbon budget in the intertidal mudflat is, however, regulated by biological and physical factors. For interested readers, a technical report on mangrove soil carbon is presented as Annexure 2A.2 at the end of this chapter.

22. Mangroves, by way of fertilizing the adjacent estuarine and coastal water bodies, promote the growth of phytoplanktons that are unique sink of carbon dioxide. The hypersaline condition, however, limits the carbon sequestering potential of phytoplankton community by way of shrinking their volume. A detailed technical report on the role of salinity in regulating stored carbon in phytoplankton species is highlighted as Annexure 2A.3.

2.2 Threats to Mangrove Ecosystem

The mangrove ecosystems are gradually disappearing especially in the Indian Ocean region (Kathiresan and Rajendran 2005). The universal causes of the destruction are shrimp culture, woodchip and pulp industry, urban development and human settlements and domestic uses for timber, firewood and fodder. In dry areas, grazing by buffaloes, sheep, goats and camel can also lead to destruction of mangroves. In areas such as Gulf countries, oil pollution is often harmful for survival of mangroves. A natural cause that results in large-scale destruction of mangroves is cyclones in Bay of Bengal that gets aggravated with human interference (Blasco et al. 1994). Diversion of river water leading to hypersalinity in the downstream areas is a serious problem in the Indus delta of Pakistan and Cauvery and Sundarban deltas of India. Due to reduction in freshwater inputs, the freshwater-loving mangrove species like *Nypa fruticans* and *Heritiera fomes* become reduced in population density. In several coastal and estuarine sectors (e.g. in the upper stretch of Indian Sundarbans), because of low salinity of tidal water, tiger prawn hatcheries cannot be developed. In these areas, prawn seeds are collected by local villagers to meet the needs of shrimp farms. During this activity, finfish and shellfish seed resources of mangroves are largely destroyed. To cite an example, in Sundarban, about 40,000 people harvest about 540 million seeds of tiger prawn (*Penaeus monodon*) every year, and in this process, about 10.6 billion seeds of other fishes and shrimps are killed, which may have serious impact on fish diversity and fisheries resources (Chaudhuri and Choudhury 1994).

During the last two centuries, the highly productive mangrove ecosystems have been destroyed or degraded very rapidly. Although mangrove ecosystems have tremendous value for coastal communities and associated species, they are being destroyed at alarming rates. Over the last 50 years, about one-third of the world's mangrove forests have been lost (Alongi 2002). Human threats to mangroves include the overexploitation of forest resources by local communities; conversion into large-scale development such as agriculture, forestry, salt extraction, urban development and infrastructure; and diversion of freshwater for irrigation (UNEP 1994). The greatest human threat to mangroves is the establishment of shrimp aquaculture ponds. Because mangroves are often viewed as wastelands, many developing countries are replacing these forests with agricultural land and/or shrimp aquaculture production (Franks and Falconer 1999). Shrimp aquaculture accounts for the loss of 20–50 % of mangroves worldwide (Primavera 1997). Projections suggest that mangroves in developing countries are likely to decline another 25 % by 2025 (Ong and Khoon 2003). In some key countries like Indonesia, which has the world's largest intact mangroves, the projected rate of loss is even higher with 90 % loss in some provinces like Java and Sumatra (Bengen and Dutton 2003). In addition to these anthropogenic threats, mangroves are also threatened by the impact of global climate change. Global climate change and concomitant effects, such as changes in temperature and CO_2, altered precipitation patterns, storminess and eustatic sea-level rise as observed over recent decades, are primarily due to anthropogenic activities. Most of the observed warming over the last 50 years is attributed to an increase in greenhouse gas concentrations in the atmosphere (Houghton et al. 2001).

The alteration of hydrological parameters particularly salinity, pH, soil composition, etc. has degraded the mangrove ecosystem and posed threat to the survival and growth of few important mangrove species. The gradual vanishing of Sundari (*Heritiera fomes*) from the Indian Sundarbans sector (particularly from the central sector around the tide-fed Matla River) is a prominent example in this context. Over a period of 27 years, the salinity has shown an increasing trend (Fig. 2.2) in the central Indian Sundarbans.

Sundari (*Heritiera fomes*), being a freshwater-loving mangrove floral species, could not withstand this rising salinity of the ambient water and gradually vanished from the region. The authors of this book conducted an extensive comparative research on the impact of salinity on the growth of *Heritiera fomes* thriving in the Gangetic delta complex. The relatively higher growth rate of above-ground biomass of *Heritiera fomes* in the western sector of the deltaic complex (2.90 ± 0.26; range 2.65–3.3 t/ha during 2004, 3.56 ± 0.28; range 2.92–3.87 t/ha during 2005, 3.98 ± 0.29; range 3.15–4.08 t/ha during 2006, 4.32 ± 0.31; range 3.34–4.75 t/ha during 2007, 4.86 ± 0.37; range 3.92–5.45 t/ha during 2008 and 5.71 ± 0.43; range 4.62–6.38 t/ha during 2009) in comparison to that of the central sector (0.52 ± 0.09; range 0.3–0.77 t/ha during 2004, 0.53 ± 0.10; range 0.32–0.65 t/ha during 2005, 0.55 ± 0.13; range 0.34–0.70 t/ha during 2006, 0.59 ± 0.10; range 0.37–0.80 t/ha during 2007, 0.59 ± 0.19; range 0.39–0.83 t/ha in 2008 and 0.61 ± 0.17; range 0.42–0.93 t/ha during 2009) confirms the freshening and salinification of the western and central sectors, respectively, in the framework of lower Gangetic delta region. There are numerous studies on mangroves in terms of wood production, forest conservation, and ecosystem management (Putz and Chan 1986; Tamai et al. 1986; Komiyama et al. 1987; Clough and Scott 1989; McKee 1995; Ong et al. 1995; Chaudhuri and Choudhury 1994; Mitra and Pal 2002). This study is the first attempt from Indian subcontinent showing the impact of dilution of the estuarine system on the biomass of *Heritiera fomes*. Mangroves like other halophytes also decrease their water and osmotic potentials to maintain turgor pressure at higher salinity (Naidoo 1987; Khan et al. 1999; 2000a, b). There is a great deal of variation in the level of salinity required for

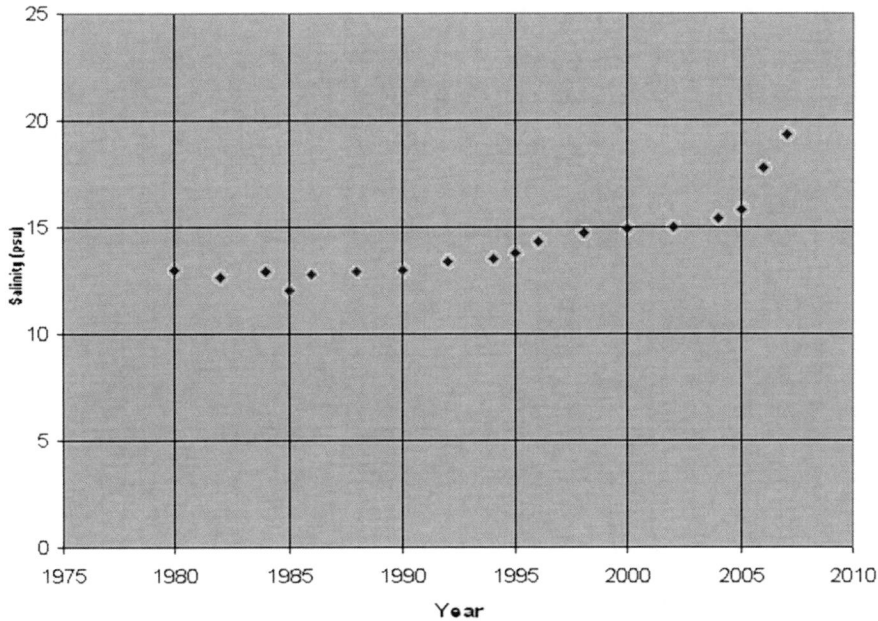

Fig. 2.2 Increasing salinity trend in the central part of Indian Sundarbans (*Data Sources*: A long term multi-disciplinary Research approach and Report on Mangrove ecosystem of Sundarbans (1980–1986 compilation) published by Department of Marine Science, University of Calcutta; Chakraborty and Choudhury (1985), Mitra et al. (1987, 1992, 2001, 2004, 2009), Mitra and Choudhury (1994), Saha et al. (1999), Banerjee et al. (2002, 2003), ADB – Asian Development Bank (2003))

optimal growth which varies from 10 to 50 % seawater (Downton 1982; Clough 1984; Naidoo 1987; Lin and Sternberg 1992, 1995; Karim and Karim 1993; Ball and Pidsley 1995; Smith and Snedaker 1995), and a decline in their overall growth is observed with a further increase in salinity. Similarly, decreased stomatal conductance, lower water potential and accumulation of inorganic ions are the result from extreme saline environments for most of the plants (Ball and Farquhar 1984; Naidoo 1987). *Heritiera fomes* is absolutely a freshwater-loving species, and hence, its growth has been greatly reduced in Central Indian Sundarbans over a period of time. The relatively higher growth of the above-ground biomass of the species in the western sector in comparison to that of the central sector (457.69 % higher in 2004, 571.70 % higher in 2005, 623.64 % higher in 2006, 632.20 % higher in 2007, 723.73 % higher in 2008 and 836.07 % higher in 2009) indicates the adverse impact of high salinity on *Heritiera fomes* due to difference in geophysical set-up (Fig. 2.3).

The study clearly demonstrates the response of certain mangrove species to changing salinity that may serve as indicator for climate-change-related studies. Similar works in Bangladesh revealed adverse impact of salinity on *Heritiera fomes* (Hoque et al. 2006). The species was detected as a rare species in strong salinity zones, whereas its presence was observed in the moderate and the low salinity zones of the Sundarbans forest of Bangladesh. Death of the species was also reported due to top dying disease which was frequently observed beside the river or the canals where inundation by the saline water is much and has water logging problem. Rahman et al. (1994) reported that top dying symptom was seen in areas where most of the pneumatophores have been buried partially and exposed to high salinity. The basic moral of the present study is the adverse role played by high aquatic salinity on certain mangrove species like *Heritiera fomes* and *Nypa fruticans* that may become very prominent in the near future due to climate-change-induced sea-level

Fig. 2.3 Above-ground biomass of *Heritiera fomes* in Central and Western Indian Sundarbans

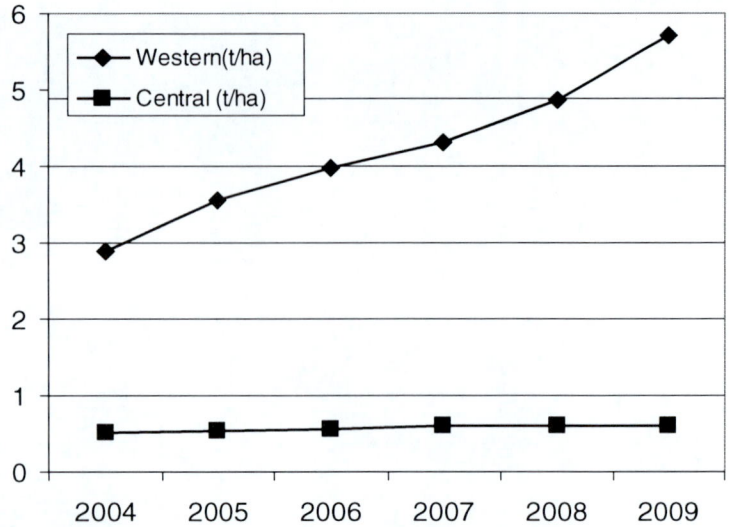

Fig. 2.4 *Heritiera fomes*

Fig. 2.5 *Nypa fruticans*

as indicator of sea-level rise and subsequent intrusion of seawater in the brackish water system.

Apart from fluctuation of hydrological parameters, mangroves are also degraded due to human factors. Due to extensive magnitude of human activity starting from agriculture to industry, the mangroves are degraded in various ways. Although the exact damage caused by pollution is difficult to assess, the mangroves are threatened by various kinds of pollutants that reach the coastal environment, namely, silt in the form of terrestrial and alluvial sediment, crude oil and petroleum derivatives, sewage, pesticides and solid and industrial wastes

rise and subsequent salinification of the deltaic or coastal soil. The freshwater-loving mangrove species (Figs. 2.4 and 2.5) can thus act

including toxic chemicals. Apart from this, logging and land development including agriculture, terrestrial and seabed excavations, insecticides and pesticides used in agriculture and their run-off, deliberate and operational discharges from ships, oil spills due to accident and industrial outflows also add to the pollution in the mangrove areas.

In context to Indian Sundarbans, sustaining about 34 true mangrove species, mangrove degradation is attributed in most cases to mushroom growth of shrimp farms along the tidal creeks and estuarine inlets. Ecologists have given their attention regarding the importance of mangrove and mangrove ecosystems only during the last two decades. Today it has been established that natural factors along with direct and indirect interference have largely changed the biological composition, ecosystem function and extent, productivity, regeneration and succession patterns within mangrove ecosystems.

The use of natural resources by man changes according to his needs. In turn, the needs of human populations vary locally and seasonally and may change also as a consequence of unpredictable episodic events and long-term climatic rhythms. Anthropogenic use and overexploitation of the mangrove forests have in many instances caused drastic changes of various extents in the coastal zone (Table 2.4). Major changes vary from total elimination of mangroves with consequent local edaphic, erosion or accretion processes to decreased productivity in terms of timber, firewood, charcoal and woodchip production, dependent fisheries and many others including the depletion of carbon sinks and its little known implication to global carbon budget and climate change (Gattuso et al. 1998). Total felling and overexploitation of mangrove forests are the historically recorded root causes of the desertification that has taken place in many coastal areas, as, for instance, along the coasts bordering the Arabian Gulf or the northern coast of the island of Socotra in the Indian Ocean.

The basic template of mangrove ecosystem conservation does not lie within the narrow band of afforestation of selective mangrove species, nor it is related to enforcement of laws, but community participation and generation of alternative source of income are some important strategies for conserving this fragile ecosystem. Conservation of mangrove ecosystem should be the priority to impart long-term sustainability to the entire matrix as all the species of the ecosystem are intra- and interconnected. In context to large-scale threats like sea-level rise, few important strategies have been identified that collectively hold promise to increase the viability of mangroves by enhancing their resilience. These are as follows:

1. Apply risk-spreading strategies to address the uncertainties of climate change.
2. Identify and protect critical areas that are naturally positioned to survive climate change.
3. Manage human stresses on mangroves.
4. Establish greenbelts and buffer zones to allow for mangrove migration in response to sea-level rise and to reduce impacts from adjacent land-use practices.
5. Restore degraded areas that have demonstrated resistance or resilience to climate change.
6. Understand and preserve connectivity between mangroves and sources of freshwater and sediment and between mangroves and their associated habitats like coral reefs and seagrasses.
7. Establish baseline data and monitor the response of mangroves to climate change.
8. Implement adaptive strategies to compensate for changes in species ranges and environmental conditions.
9. Develop alternative livelihoods for mangrove-dependent communities as a means to reduce mangrove destruction.
10. Build partnerships with a variety of stakeholders to generate the necessary finances and support to respond to the impacts of climate change.

Table 2.4 World mangrove loss and degradation approximate assessment in 2000 (in km²)

Country	Approx. total area 1990[a]	Approx. total area 2000	Closed-dense mangrove (forest or thickets)	Open-degraded or very degraded mangroves	Clear felling conversion to Aquaculture	Agriculture	Freshwater diversion, damming	Salt industry	Urban development	Oil spills	Other pollutions	Excessive siltation	Climate change[c]	Sea-level change
Australia	11,500	11,500	9,200 (80 %)	2,300 (20 %)	[d]				x		x			x
Bangladesh	6,000	6,300	4,095 (65 %)	2,205 (35 %)	xx		xxx					x	?	x
Brazil	13,800	10,150	6,090 (60 %)	4,060 (40 %)		xx			xx	xx	xx	?	?	?
Cameroon	2,500	2,400	1,680 (70 %)	720 (30 %)		xx			x	x	x		?	?
Columbia	4,000	3,600	2,520 (70 %)	1,080 (30 %)	xxx	xxx	xxx		xx		x	xx		?
Cuba	5,300	5,500	4,565 (83 %)	935 (17 %)					x		x			?
Ecuador	1,900	1,600	1,040 (65 %)	560 (35 %)	xxxx			xx	xx		x			
Guinea	3,100	2,900	2,320 (80 %)	580 (20 %)		xx			xx					?
Guinea Bissau	2,400	2,400	1,560 (65 %)	840 (35 %)		xxx			xx		x			?
India	6,700	6,700	3,015 (45 %)	3,685 (55 %)		xxx		xxx	xxx	x	xx	xx	?	x
Indonesia	42,500	40,000	20,000 (50 %)	20,000 (50 %)	xxx	xxx			xxx	xx	xx	xx		?
Madagascar	3,500	3,300	1,980 (60 %)	1,320 (40 %)	xxx	x			x		xx	xx		?
Malaysia	6,400	6,400	5,120 (80 %)	1,280 (20 %)	xx	x			x		x			
Mexico	5,300	5,000	3,000 (60 %)	2,000 (40 %)	xx	xx			xx	xx	xx			?
Myanmar	7,500	6,900	2,280 (33 %)	4,620 (67 %)	xxx	xxx			xx		x	xx		?
Nigeria	10,500	10,500	2,625 (25 %)	7,875 (75 %)		xxx			xxx	xxxx	xx	xx		?
Papua N. Guinea	4,100	4,100	3,690 (90 %)	410 (10 %)	xx	xx			x					?
Senegal	2,000	1,800	900 (50 %)	900 (50 %)		xxx	xxx	xx	xx		xx	xx	?	?
Philippines	1,400	1,300	780 (60 %)	520 (40 %)	xxxx									?
Venezuela	2,700	2,500	1,500 (60 %)	1,000 (40 %)		xx	xx		xx			xx	?	?
Vietnam	2,500	2,500	1,125 (45 %)	1,375 (55 %)	xxxx	xxx	xxx		xx		xx	xx		?
Total of 21 countries	145,600	137,350	79,085 (58 %)	58,265 (42 %)										
World Total	181,077	170,756	99,038	71,718										

Source: Laboratoire d'Ecologie Terrestre, Toulouse

Logs or conversion: 8.250 km² in the decade (5.7 %) or 825 km²/year. This result is based on the main 21 countries having mangroves representing about 80 % of the global total

World's mangrove loss during the decade has been about 10,321 km² or about 1,030 km²/year

[a] These figures should be used with caution; the year 1990 is only a reference; in many cases, essentially in SE Asia, the latest available figures for the decade 1980–1990 are those published by UNESCO/UNDP 1986

[b] Scale of impacts: xxxx, extreme; xxx, severe; xx, medium intensity; x intensity; ? Possibly, to be confirmed

[c] For instance:

  Higher frequency of low rainfall years in West Africa or Coastal Peru and Ecuador

  Higher frequency or intensity of cyclonic storm

[d] Blanks: No data or not observed from space

2.3 Conservation of Mangroves: Case Study of Indian Sundarbans

IUCN in 1980 defined conservation as '… the management of human use of the biosphere (i.e., all living things) so that it may yield the greatest sustainable benefit to present generations while maintaining its potential to meet the need and aspiration of future generations…'. Maintaining biodiversity is thus a central tenet of nature and wildlife conservation. To conserve marine life, information on the resource (the habitats, communities and species) needs to be collected and arranged in a structured way in order to identify the nature conservation importance of sites and to manage areas to conserve their important features (Hiscock 1997). It is to be noted that the conservation policy is not uniform throughout the globe; it varies from place to place depending on the resource base, pattern of exploitation and nature and magnitude of threat, e.g. the pattern of exploitation in Arctic or Antarctic zone is totally different from the tropical seas or mangrove regions. To make it clearer, the example of regional resource base can be cited. The coral is a biological resource scattered around the archipelagos of Andaman and Nicobar Islands of Indian subcontinent, which is gradually eroding due to temperature fluctuation or changes in physico-chemical properties of the seawater. However, in Indian Sundarbans region, fish germplasm constitutes an important resource, which is under threat due to wild harvest of prawn seeds, illegal fishing practice, etc. Thus, threats are always site specific in nature and intricately related to the nature of resource, their exploitation pattern and socio-economic profile of the area. On this basis, it has been suggested by Hiscock (1997) that there is a need for five main areas of information prior to conservation. These are:

1. Physical and biological resource data
2. Physical and chemical environment-related information
3. Information on the structure of biotic communities and on the key elements in their functioning
4. Information on natural variability
5. Information on effects of human activities

All these information will require a systematic approach to its application to conservation and management through:

1. Classification systems for resource data
2. Criteria to identify the nature conservation importance of an area including those to identify a comprehensive series of marine protected areas

The mangrove ecosystem of Indian Sundarbans is one of the most biologically productive, taxonomically diverse and aesthetically celebrated gene pool of the country sustaining 34 species of true mangroves along with a variety of associate species (Mitra and Pal 2002). This deltaic lobe is the cradle of several species of finfish, nursery of different variety of shellfish and reservoir of various nonliving resources of marine origin and encompasses a wide range of riverine, estuarine, coastal and marine habitats. On one hand, it exhibits enormous diversity based on its genesis, geographical location, hydrological regimes and substrate factors, and on the other hand, it also exhibits rare endemic genetic material, which demands preservation and proper sustainable utilization for the benefit of mankind. The biodiversity of this unique ecosystem has not yet been comprehensively assessed for several reasons. In India, due to lack of trained taxonomists and low emphasis on brackish water wetland micro-biodiversity studies, the exact role of micropelagic community is still not clearly known. Meiobenthos communities are seldom identified to the genus level and rarely into species. In general, very little information is available on microorganisms. A very important aspect of biodiversity is species-level genetic diversity, which has not received much attention. Both conservation and management of this unique ecosystem therefore demand a holistic approach and are very much dependent on the sciences and skills of geology, pedology, climatology, hydrology, botany, ecology, silviculture, forest technology and economics.

The landscape of Indian Sundarbans has changed remarkably due to large-scale human intervention since the beginning of the last century. As a result of overexploitation, demographic pressure, loss of habitat and change of ecological condition, several of the earlier reported species have become extinct or are in a very threatened or degraded state.

Large-scale clearing of the Sundarbans mangroves in the last 200 years for settlement of human population, aquaculture, etc. has invited a situation of great stress to the pristine mangrove ecosystem in this part of the Indian subcontinent. Hence, effective conservatory measures are essential to protect this gene pool. The integrated management of mangrove resources depends not only on an understanding of the ecological and silvicultural parameters for forest management but also on the biological role that the primary production from the forest plays in sustaining the food web of the adjacent aquatic subsystem (secondary production). An understanding of the role of key stone species in maintaining the equilibrium of this particular ecosystem is also very essential.

Being a mega-diversity country, India accounts for 7–8 % of the total recorded species of the world though occupying only 2.4 % of the land area. The Convention on Biological Diversity (CBD), to which India is a contracting party, deals with various aspects, like conservation of natural resources, their sustainable use and fair and equitable share of benefits. The convention has also identified forests, coastal and marine ecosystems, grasslands, wetlands and deserts as priority ecosystem for biodiversity conservation. Based on this convention, several strategies have been adopted to conserve the vibrating mangrove ecosystem of Indian Sundarbans. The important ones are listed here:

- Encouragement towards alternative livelihood scheme
- Afforestation
- Introduction of silviculture system
- Ecological rehabilitation of the Sundarbans under the Integrated Wasteland Project
- Captive breeding programmes
- Conservation policies

Each of the above strategies is discussed in detail.

2.3.1 Encouragement Towards Alternative Livelihood Scheme

Among 4.2 million people in the Sundarbans Biosphere Reserve area of Indian part, 50 % survive below poverty level (BPL). Improving the economic conditions of these people through promotion of sustainable utilization of natural resources is an optimistic and valid step towards ensuring the long-term sustainability of the adjacent mangrove ecosystem. Government departments and several nongovernmental organizations have undertaken important programmes on alternative livelihood in the deltaic complex to wean away people from random exploitation of mangrove resources.

2.3.1.1 Environment Education and Awareness Generation

Teacher-training workshop is used as one of the tools of conservation for generating environment awareness and imparting knowledge of endemic biotic resources. Teachers of the islands are exposed to modern pedagogic techniques to make environment education interesting rather than routine. As a part of awareness generation programme, a number of nature clubs have been opened by the nongovernmental organizations. The main purpose of such programmes is to imbibe the conservation ethics among the students through their education system. Finally the student community is motivated to spread the light of conservation in different tiers of the society.

2.3.1.2 Vocational Training for Prawn Seed Collectors

Most of the women in Sundarbans are involved in collection of tiger prawn (*Penaeus monodon*) seeds in the estuarine water. This creates a major loss in population of finfish and shellfish juveniles, which may cause an adverse impact on the pelagic and demersal fish stock of the aquatic subsystem in the near future. To divert the prawn seed collectors from such destructive practice, vocational trainings on several aspects are given, which include training on tailoring, net making

Table 2.5 Total coliform load (five test tube method) in water samples collected from the sampling stations during April 2007

Sl. No.	Stations	MPN/100 ml in LST	MPN/100 ml in BGL
1	Namkhana	>1,600	>1,600
2	Frasergaunge	1,600	1,600
3	Sajnekhali	550	350

and handicrafts. Apart from these, trainings on kroiler rearing, piggery, campbell duckery, etc. are also conducted on a regular basis. Kroiler (a breed of hen) rearing has involved large number of villagers in eastern Indian Sundarbans as the economic returns are quite high and had enthused youth of the village to start small-scale business venture dealing only with kroiler poultry. Presently pig farming has been taken up by a group of beneficiaries on an experimental basis. Fish feed preparation using seaweed-based protein is also taught to some group of beneficiaries through which they can boost up fish production in their household ponds and brackish water bodies. The Department of Science and Technology, Government of India, also completed successfully a pilot project on fish feed preparation using the red seaweed *Catenella repens* (Sanction No.SSD/SS/046/2006; date: 05.07.2007). Researchers from University of Calcutta documented the biochemical composition of the seaweed species (Banerjee et al. 2009) as baseline data of the livelihood project.

2.3.1.3   Canal Excavation

Excavation of canals is done to stock freshwater or to harvest the rainwater so that agricultural activity may be carried out throughout the year in this saline belt of the country. Government departments like Forest Department and Fishery Department are regularly involving considerable village population in canal excavation. The canal excavation has brought about a radical change in the sphere of agriculture through a shift from mono-cropping system to multiple-cropping system. Today, the people of the islands are growing chilli, tomato, potato, brinjal, cauliflower, cabbages, etc. along with the main crop paddy. Many beneficiaries have also initiated freshwater prawn culture (*Macrobrachium rosenbergii*) in these

canals. It is expected that this constructive activity will defray a sizeable chunk of the population from penetrating into the adjacent forest zone.

2.3.1.4   Oyster Culture

In Indian Sundarbans, edible oyster species like *Saccostrea cucullata* and *Crassostrea madrasensis* are common. These species have high export value and are also good source of calcium, which can be used in poultry and cattle feed industry. The ability to produce, transport and market healthy disease-free shellfish is crucial to the success of the Indian oyster industry. Hence, microbial load in the oyster tissue and ambient media was monitored by Mitra et al. (2006) to evaluate the quality of the product and the suitability of the environment to initiate the culture (Tables 2.5, 2.6, 2.7, and 2.8).

The culture of edible oyster can open up a new avenue in the livelihood sector of the Sundarban people, but due to drastic fall of aquatic salinity in the monsoon, the species may face high mortality. Turbidity of the aquatic phase is another hindrance to oyster culture in the lower Gangetic region. Construction of sheltered closed system with provisions of brine and sedimentation tank may be effective to initiate oyster culture in the area.

2.3.1.5   Crab Fattening

The Fisheries Department, Govt. of West Bengal, and few local NGOs working in Indian Sundarbans have taken up this venture. *Scylla serrata* is the most common crab species in this deltaic lobe with high market value and demand. Megalops of this species collected from the wild are stocked in brackish water ponds and fed voraciously with trash fishes to increase the fat content of the species within 120 days. The male crabs usually attain a weight of 300–350 g, while

Table 2.6 Total coliform load (three test tube method) in sediment samples collected from the sampling stations during April 2007

Sl. No.	Stations	MPN/g in LST	MPN/g in BGL
1	Namkhana	30	30
2	Frasergaunge	24	24
3	Sajnekhali	9.5	9.5

Table 2.7 Total coliform load (three test tube method) in edible oyster samples collected from the sampling stations during April 2007

Sl. No.	Stations	MPN/g in LST	MPN/g in BGL
1	Namkhana	>110	>110
2	Frasergaunge	110	110
3	Sajnekhali	46	45

Table 2.8 Total faecal coliform load (five test tube method) in water samples collected from the sampling stations during April 2007

Sl. No.	Stations	MPN/100 ml in EC
1	Namkhana	40
2	Frasergaunge	50
3	Sajnekhali	4

the female crabs attain a weight of 180–250 g, which fetch good price in the market.

2.3.1.6 Apiculture

Several cooperatives have initiated apiculture in Indian Sundarbans including the Forest Department, Govt. of West Bengal. Schemes involving the harvest of honey from wild *Apis indica* by placing artificial beehives inside the forest have also been adopted. Although this species cannot be tamed, the yield has been about 20–25 kg per box over a period of 2 months during trials in comparison to yields generated from *Apis dorsata*, which is usually 3–4 kg per box. An annual harvest of 65,000 kg of honey from the *Apis indica* beehives is expected.

2.3.1.7 Pisciculture

Considering the water resources in and around Indian Sundarbans and the seed resources of several commercially important fish species in the brackish water system, several private industries initiated polyculture in islands of central Sundarbans during March 2008. Such programmes

aimed the blending of biotechnology with pisciculture practice. The species of polyculture like tiger prawn (*Penaeus monodon*), parse (*Liza parsia*) and bhangone (*Liza tade*) was fed with specially formulated feed having seaweeds and other plant extracts. The feed also improved the water quality and controlled the disease problem of tiger prawn to a great extent. This programme of sustainable polyculture through formulated fish feed created an impulse among the local people, and an overwhelming response was observed in terms of the number of beneficiaries. The Department of Science and Technology, Govt. of India, also funded similar type of projects through WOSB scheme during 2009–2012 with the aim to boost up sustainable pisciculture and put momentum to conservation process of mangroves (Fig. 2.6).

2.3.2 Afforestation

There are enormous degraded areas in Sundarbans forests, which may be categorized as follows:
1. Degraded areas devoid of any vegetation (Category 1)
2. Degraded areas under scrubby growth of *Ceriops decandra* (Category 2)
3. Degraded areas under scrubby growth of *Excoecaria agallocha* (Category 3)

Category 1 is characterized by saucer-shaped areas with no vegetation. Floodwater remains in the depressions from March to December. Salt accumulates on the topsoil and no plantation can be undertaken. Salinity can be reduced by improved drainage of these areas, and the soil may be improved through the use of gypsum, sulphur and natural grasses.

The area under *Category 2* is covered with a scrubby growth of *Ceriops decandra* and requires silvicultural improvement through planting of upper storey species. Some experiments have been carried out where strips are cleared and upper storey species have been planted.

Category 3 has hard soil and, apart from a scrubby growth of *Excoecaria agallocha*, no seed naturally transported or artificially dibbled germinates in this area.

Fig. 2.6 Harvested freshwater prawn (*Macrobrachium rosenbergii*) – an output of conservation-oriented DST, Govt. of India funded project

- A mangrove rehabilitation project recently launched envisages the afforestation of embankments with mangrove species over 3,300 ha and protection of degraded mangrove forestland covering 25,000 ha, through Forest Protection Committees (FPCs). More recently, NGOs have also taken active role in afforestation programme involving the local population.

2.3.3 Introduction of Silviculture System

The silviculture systems are designed to harvest the mature crop while allowing regeneration of the forests. Trees above a certain fixed diameter are harvested as long as these do not create a permanent gap in the canopy.

Until 1931, the system of exploitation was based upon the issue of permits from the revenue stations along the forest border. Following the introduction of *Curtis's plan* in 1931, a system of ranges, working circles, felling cycles and coupes were introduced. Natural regeneration was regarded as healthy, and therefore, a system of selection felling on cycles varying from 16 to 40 years was prescribed.

Choudhuri's plan was a modified version of Curtis's plan, which made the exploitation and laying out of coupes easier but did not put adequate stress on the silviculture of mangrove species. In the late 1950s, the assumption still remained that the mangroves were capable of regenerating themselves with considerable recovery rate and in sufficient numbers to justify the selection method of felling.

The *Forest Working Plan for 24 Parganas Forest Division*, following the partition of India in 1947, adopted a felling cycle of 20 years. The rules under this plan, for the most part, exist till the present day. According to this plan, the exploitable diameter at breast height (i.e. 130 cm from the ground level) of different mangrove plants is mentioned in Table 2.9.

Fishing was allowed free in tidal water, provided that the fishing boats are registered with the Forest Directorate on payment of a registration fee and royalty for dry fuel wood to be consumed during each fishing trip. However, no fishing was allowed within the core area of the Sundarbans Tiger Project. Permits were only issued for entry inside the reserved forest for exploitation of fuel, timber and other forest products. The number of persons and quantity of production to be allocated

Table 2.9 Exploitable diameter at breast height of different mangrove plants

Mangrove plants	Exploitable diameter at breast height (cm)
Avicennia sp.	12.5
Bruguiera sp.	15.0
Excoecaria agallocha	10.0
Heritiera fomes	7.5
Xylocarpus granatum	12.5
Xylocarpus mekongensis	15.0
Avicennia sp.	22.5–40.0
Sonneratia apetala	23.5–45.5

Table 2.10 Zonation within the Project Tiger area

Zone	Area (km²)
Revised core area	1,699.50
Buffer area	885.27
Wildlife sanctuary area	362.33
Available exploitable area	496.78
Erosion buffer	26.16

Source: Chaudhuri and Choudhury (1994)

to each permit holder was fixed by the Divisional Forest Officer (DFO). For this, an essential prerequisite was to register a boat, and only boats with carrying capacities ranging from 50 to 400 quintals were registered. A published schedule of prices for different species of timber, fuel wood and other forest products was used for the assessment of the value of the harvested products.

From 1981 to 1982, regular coupes were laid out in the buffer zone. The first Management Plan of Tiger Project prescribed a felling cycle of 20 years. Felling was controlled by area, where a selection-cum-improvement system was followed. Permits were granted for the collection of honey, wax, *Nypa* palm and minor forest products, although the collection of *Nypa* palm was prohibited in 1978. Fishing was permitted in the buffer zone of the project area to permit holders who had registered boats. These permit holders generally apply to get permission for fuel wood collection as well.

In the core area of Tiger Project, no harvesting has been allowed since 1984. Felling has also been prohibited in the wildlife sanctuaries, following their gazettement. The remaining areas utilized for silviculture are managed on a selection-cum-improvement felling cycle. Selective felling of trees for fuel wood and timber over the non-sanctuary area in the buffer zone each year is conducted in a felling series over 1/20th of the area. Plantations are raised in the mangrove blocks mainly with the species like *Heritiera fomes, Xylocarpus granatum* and *Xylocarpus mekongensis*. Dry fuel wood permits are also issued.

Previously the total area of reserved forest was estimated at 2,585 km² and the core area of the tiger reserve as 1,330 km². Based upon satellite imagery, the size of the core area has been revised as 1,699 km² (Table 2.10).

The actual area available for timber harvest is therefore 587.35 km² where 29.26 km² is exploited annually on a 20-year felling cycle. This area was further divided into fair weather and rough weather coupes to account for potentially hazardous working conditions, such as strong winds. Following the policy to restrict felling in the Sundarbans, the area covered by these coupes was systematically reduced and the present coupe area is 800 ha, to be worked annually in fair weather only. Hence, the annual felling area in the Project Tiger area, calculated for a 20-year felling period, amounts to 1/20th of the total available exploitable area or 24.84 km². The present felling regulations in the felling area are:

1. Fair and rough weather coupes are to be marked from the north.
2. The exploitable diameter for *Excoecaria* trees is 25 cm.
3. *Heritiera fomes, Sonneratia apetala* and *Xylocarpus mekongensis* are not to be felled.
4. Dry fuel wood may be collected by villagers for local consumption, only from coupes already worked over.

In the 24 Parganas district, outside the Project Tiger area, a new scheme has been drawn for the 10-year period 1992–1993 to 2001–2002. Under this scheme, the annual coupe will be maintained at 1,150 ha, under four felling series. The felling cycle has been fixed for 15 years. Felling of *Heritiera fomes, Sonneratia* sp., *Xylocarpus granatum* and

Xylocarpus mekongensis has been restricted. The rough weather-felling coupe has been closed in 1989–1990.

2.3.4 Ecological Rehabilitation of the Sundarbans Under the Integrated Wasteland Project

The project envisages a total economic uplift of local communities presently living below the poverty line in remote areas of Sundarbans. The project is being implemented through the Sundarbans Biosphere Reserve (SBR), and the scheme has been designed to combine a full spectrum of activities comprising afforestation, conservation of fragile areas, development of pasture, soil conservation, minor irrigation, cottage industries and other socio-economic and ecological components.

2.3.5 Captive Breeding Programmes

2.3.5.1 Crocodile

The estuarine crocodile (*Crocodylus porosus*) is one of the largest saline water reptiles, which was indiscriminately killed for the purpose of making luxury goods from its skin. The level of poaching became so severe that the population subsequently declined, making the species endangered.

On 1 April 1975, the Government of India, with the assistance of FAO and UNDP, requisitioned the services of an FAO expert and initiated crocodile farming in India. The saltwater crocodile scheme was initiated in the Sundarbans in March 1976 at Bhagabatpur on Lothian Island, and efforts were focused on reducing the high mortality rate of the egg and newly hatched stages. Today, the saltwater crocodile scheme in the Sundarbans is one of the principal crocodile breeding centres in India. By 1990, more than 197 crocodiles had been released into the Sundarbans (Chaudhuri and Choudhury 1994). This is a success indicator of the crocodile breeding programme.

2.3.5.2 Olive Ridley Turtles

Marine turtles are protected under Schedule 1 of the Wildlife Protection Act. The olive ridley turtle (*Lepidochelys olivacea*), which extends its migration to the Sundarbans deltaic region, is an endangered species. Previously there was a period when millions of olive ridley eggs were harvested immediately after the breeding season. The poaching of mating pairs was a lucrative business during the breeding season (January and February) when thousands of animals were brought to the shore at Digha, West Bengal, for selling in the Calcutta market. This reduced their population to a great extent, and, therefore, attempts were made to rear and rehabilitate this species at the Bhagabatpur Crocodile Breeding-cum-Rearing Centre and at the Saptamukhi Hatchery by the Department of Forests, Government of West Bengal. The programme showed encouraging results when out of 117 hatchlings that emerged from artificial nests at Bhagabatpur, 99 healthy hatchlings were released into the ambient water and 18 were segregated for further research work (Banerjee 1985).

2.3.5.3 Horseshoe Crabs

Horseshoe crabs, chelicerate arthropods belonging to the class Merostomata, are considered to be the oldest 'living fossils'. They have been considered as Schedule IV animal by the Ministry of Environment and Forests (Govt. of India) in 2008 under Wildlife Protection Act (1976). There are only four species of the horseshoe crab reported in the world, among which *Carcinoscorpius rotundicauda* is plentiful in the deltaic Sundarbans region. Although *Tachypleus gigas* is also reported from the deltaic Sundarbans, the frequency of occurrence is very low (3–5 %) among the total horseshoe crab population. The species is destroyed at random during the fish catch in the neritic zone and wild harvests of tiger prawn seeds. With the unfounded belief that they can cure arthritis, the extracts of different organs of the species are sold in Calcutta markets. These activities caused a decline in the population of the species in recent years in the northeast coast of the Bay of Bengal. The species have been found to be a potential source of a bioactive substance, Carcinoscorpius amoebocyte lysate (CAL), from its blue blood (Chatterji and Abidi 1993). This reagent is highly sensitive and useful for the rapid and accurate assay of gram-negative bacteria, even if present in a very minute quantity, up to the level of 10^{-10} g. Hence, the CAL has been proved to be a valuable diagnostic reagent in the detection of endotoxins in several pharmaceutical products, especially injectables. Given the immense biomedical value of the species, a project entitled 'Application of biotechnology and molecular biology in the resource exploration and sustainable management of Sundarbans mangroves' was sanctioned to the S. D. Marine Biological Research Institute at Sagar Island by the Department of Biotechnology under the Ministry of Science and Technology, Government of India, on 26 December 1997. This aimed to develop culture technology for *Carcinoscorpius rotundicauda* and *Tachypleus gigas* suited to the Sundarbans environment. The project made a thorough study on the optimum range of various physico-chemical variables (such as water temperature, salinity, pH, nitrate, phosphate, silicate, dissolved oxygen, dissolved copper, organic carbon of sediment) in relation to the fertilization and condition value of the three different weight groups of horseshoe crabs. The development and standardization of culture technology of this species along with cell line culture may pave the way for manufacturing Carcinoscorpius amoebocyte lysate (CAL) and Tachypleus amoebocyte lysate (TAL) on commercial basis from the amoebocytes of these species without damaging their natural population.

2.3.6 Conservation Policies

A number of acts and policies have been adopted to protect India's biological diversity and its biotic and its associated abiotic components, which were also applicable for the protection of the Sundarbans mangrove ecosystem. There are three main acts for the protection of the environment. These are the *Wildlife Protection Act* (1972), the *Forest Conservation Act* (1980) and the *Environment Protection Act* (1986), with their

various amendments (Government of India 1992a). The implementation of *Coastal Regulation Zone* (*CRZ*) is another bold approach towards conservation of natural resources in the marine, coastal and estuarine fronts of the country through restriction of certain activities.

The National Conservation Strategy and Policy Statement on Environment and Development (Government of India 1992b) provide directives for the integration and internalization of environmental issues in the policies and programmes of various sectors. The National Forest Policy (1988) complements the Forest Conservation Act (1980) by increasing the participation of local people in forest conservation activities.

The semi-intensive shrimp culture farms that deteriorated the coastal ecosystem by way of discharging (untreated) wastes have also been restricted by the Supreme Court of India on the basis of the Environment Protection Act (1986). On 11 December 1996, the Court ordered the coastal states with aquaculture farms other than traditional and improved traditional ones, operating within 500 m of the HTL in the coastal zone to demolish them by March, 1997. The Court also directed that no shrimp farms could be set up within 500 m of the HTL in the zone. In accordance with the principles of 'sustainable development' and 'polluter pays', the Court also directed the government to set up a body under clause (3) of Section 3 of the Environment Protection Act 1986. The said body would issue permits to the farmers of traditional aquaculture to adopt improved traditional farming systems. The Court also ruled that the owners of the closed aquaculture units would be liable to pay 6-year wages as compensation to the workers as well as for eco-restoration of the affected areas by way of creating an Environment Protection Fund. There was an excellent response to this notification and now only traditional and improved traditional shrimp farms exist in the northeast coast of the Bay of Bengal.

India has several legislative acts that are relevant to various uses of the oceans and coastal zones under which the Coastal Regulation Zone (CRZ) was notified in 1991 through the Ministry of Environment and Forests. Under Section 3(1)

Table 2.11 Structure and activities (permitted and restricted) in CRZ

Category	Area	Important conservation-oriented activities
CRZ – I	All the ecologically sensitive and important areas, namely, the Sajnekhali, Lothian and Halliday sanctuaries; reserve forests, wildlife habitat and mangrove forests in the Sundarbans; New Moore and Sandhead islands; important breeding sites of finfishes, shellfishes and other species; important hotspots from the point of biodiversity; the lower long sand area for its aesthetic beauty and biological diversity; historical and heritage sites at Sagar Island that fall within the high tide level (HTL) and low tide level (LTL)	Some eco-friendly communication facilities, construction of schools and hospitals will be allowed
		Aquaculture will not be permitted in mangrove-rich area
		Seed hatchery will be encouraged where by 5 % will be released in the creek
		A marine park will be established in the southern part of the Sagar Island
		Culture of algae and edible oyster will be encouraged
CRZ – II	Areas that have already been developed and close to the shoreline. Such areas include those within the municipal limits (e.g. Haldia along the Hugli river) and other legally designated areas which have drainage facilities, approach roads, water supply, etc. The Haldia town and eastern part of Digha fall within this zone	Shore protection measures will be adopted
		Dredging activities will be regulated
		EIA will be compulsory for any projects
		Regular monitoring of pollution will be carried out
CRZ – III	Areas that are yet to be developed and include Shankarpur and western Digha	A special body of the State Government will select aquaculture sites on the basis of recommendation
CRZ – IV	Coastal stretches in the Andaman and Nicobar, Lakshadweep and small islands	No new construction of buildings shall be permitted within 200 m of the HTL
		The buildings between 200 and 500 m from the HTL shall not have more than 2 floors (ground floor and first floor), the total covered area on all floors shall not be more than 50 % of the plot size and the total height of construction shall not exceed 9 m
		The design and construction of buildings shall be consistent with the surrounding landscape and local architectural style
		Dredging and underwater blasting in and around coral formations shall not be permitted

Source: Mitra and Pal (2002); examples of West Bengal coast have been cited except for CRZ IV
Remark: The distance from the High Tide Line *shall* apply *to both sides* in the case of rivers, creeks and back waters and may be modified on a case by case basis for reasons to be recorded while preparing the Coastal Zone Management Plans. However, this distance shall not be less than *50** (**This provision has been struck down by the Supreme Court of India*) 100 m or the width of the creek, river or backwater whichever is less

and (2) of the Environment Protection Act (1986) and Rule 5(3) (D) of the Environment Protection Rule (1986), the CRZ covers the coastal stretches of seas, bays, estuaries, creeks, rivers and backwaters which are influenced by tidal actions in the landward side up to 500 m from the High Tide Line (HTL) and the land between the Low Tide Line (LTL) and the HTL. In general, the HTL is estimated to be the line up to which the highest tide reaches during the spring time. In relation to the maritime states of the Indian sub-continent, the CRZ has been classified into four categories with distinct gradation of activities for a more scientific management of the ecosystem (Table 2.11).

The entire discussion pinpoints that resource base of any ecosystem may be claimed as sustainable, when it can be carried out over long term at an acceptable level of biological and economic productivity, without leading to ecological changes that reduce options for future generations. A sustainable resource base, therefore, has three major components – ecological, social and economic. Considering the salinity-based ecological zonation, social structure and weak economic level of Indian Sundarbans, a proper guideline needs to be developed not by considering the deltaic lobe as a whole, but by emphasizing on the segmental approach, since every segment of Indian Sundarbans is significantly different from the other in terms of environmental parameters, resource base, threats and social and economic structure. Community participation and alternative livelihood scheme implementations are mere experimental tools of conservation, which may have both positive and negative outputs. Hence, such experiment needs to be conducted in selective pockets of this deltaic lobe because failure of the experiment may produce severe risk on the gene bank.

Fact Below the Carpet

Conservation is never possible when man is hungry. Out of hunger he destroys nature, intrudes into the forests, exploits the seas and demolishes the mountains for minerals. Hunger has several dimensions. The affluent society becomes hungry when the dinner is not served in right time or the breakfast could not be consumed because of the morning meeting with the boss. However, for majority of the people on Earth, there is a different meaning for hunger. To them hunger is the result of starvation, the result of natural disaster that has snatched their only patch for growing crops. The essence of conservation cannot reach this section of the society unless they get minimum food for satisfying their beloved children. The island dwellers of Sundarbans are the representatives of this class of the society. In this mysterious land of tides and tigers, law becomes lawless, resource base becomes baseless and policy cries in the voice of non-implementation under the pressure of this poorer section of society.

Important References

Aizpuru M, Achard F, Blasco F (2000) Global assessment of cover change of the mangrove forests using satellite imagery at medium to high resolution. EEC research project no. 15017-1999-05 FIEF ISP FR – Joint Research Centre, Ispra

Alongi DM (2002) Present state and future of the world's mangrove forests. Environ Conserv 29(3):331–349

ADB – Asian Development Bank (2003) TA No. 3784 – IND, Interim report, vol I

Azocar A, Rada F, Orozco A (1992) Water relations and gas exchange in two mangrove species with contrasting mechanisms of salt regulation. Ecotropicos 5:11–19

Ball MC, Farquhar GD (1984) Photosynthesis and stomatal responses of two mangrove species, *Aegiceras corniculatum* and *Avicennia marina* to long term salinity and humidity conditions. Plant Physiol 74:1–6

Ball MC, Pidsley SM (1995) Growth responses to salinity in relation to distribution to two mangrove species, *Sonneratia alba* and *Sonneratia lanceolata* in Northern Australia. Funct Ecol 9:77–85

Balsamo RA, Thomson WW (1995) Salt effects on membranes of the hypodermis and mesophyll cells of *Avicennia germinans* (Avicenniaceae): a freeze-fracture study. Am J Bot 82:435–440

Banerjee R (1985) Captive rearing of Olive Ridley Turtle. Hamadryad: 12–14

Banerjee K, Mitra A, Bhattacharyya DP, Choudhury A (2002a) Role of nutrients on phytoplankton diversity in the North-east coast of Bay of Bengal. Ecol Ethol Aquat Bot 6:102–109

Banerjee K, Mitra A, Bhattacharyya DP (2003) Phytopigment level of the aquatic sub-system of Indian Sundarbans at the apex of Bay of Bengal. Sea Explorers 6:39–46

Banerjee K, Ghosh R, Homechaudhuri S, Mitra A (2009) Seasonal variation in the biochemical composition of red seaweed (*Catenella repens*) from Gangetic delta, northeast coast of India. J Earth Syst Sci 118(5):1–10, Springer Verlag

Bengen DG, Dutton IM (2003) Interactions between mangroves and fisheries in Indonesia. In: Northcote TG, Hartman GF (eds) Fishes and forestry – worldwide watershed interactions and management. Blackwell Scientific, Oxford, pp 632–653

Blasco F, Janodet E, Bellan MF (1994) Natural hazards and mangroves in the Bay of Bengal. J Coast Res 12:277–288

Bunt JS (1992) Introduction. In: Robertson AI, Alongi DM (eds) Tropical mangrove ecosystem. American Geophysical Union, Washington, DC, pp 1–6

Chakraborty SK, Choudhury A (1985) Distribution of fiddler crabs in Sundarbans mangrove estuarine complex, India. In: Proceedings of national symposium on biology. Utilization and conservation of mangroves. Shivaji University, Kohlapur, Maharashtra, India, pp 467–472

Chatterji A, Abidi SAH (1993) The Indian horseshoe crab – a living fossil. J Ind Ocean Stud 1(1):43–48

Chaudhuri AB, Choudhury A (1994) Mangroves of the Sundarbans, India, vol I. Published by IUCN, Bangkok

Chmura GL, Anisfeld SC, Cahoon DR, Lynch JC (2003) Global carbon sequestration in tidal, saline wetland soils. Global Biogeochem Cycles 17(4):111–121

Clough BF (1984) Growth and salt balance of the mangroves *Avicennia marina* (Forssk) Vierh and *Rhizophora stylosa* (Griff) in relation to salinity. Aust J Plant Physiol 11:419–430

Clough BF (1993) Constraints on the growth, propagation and utilization of mangroves in arid regions. In: Lieth H, Al Masoom A (eds) Towards the rational use of high salinity tolerant plants, vol I. Kluwer Academic Publishers, Amsterdam, pp 341–352

Clough BF, Scott K (1989) Allometric relationship for estimating above ground biomass in six mangrove species. For Ecol Manage 27:117–127

Dahdouh-Guebas F, Jayatisse LP, Di Nitto D, Bosire JO, Lo Seen D, Koedam N (2005) How effective were mangroves as a defense against the recent tsunami? Curr Biol 15(12):R443–R447

Department of Ocean Development (2005) Preliminary assessment of impact of tsunami in selected coastal areas of India. Department of Ocean Development, Integrated Coastal Marine Area Management Project Directorate, Chennai

Diop ES (1993) Conservation and sustainable utilization of mangrove forests in Latin America and Africa regions. Part II – Africa, pp 245–261. Mangrove ecosystems technical reports, vol 3. ITTO/ISME project PD 114/90. Okinawa, ISME, 262 pp

Downton WJS (1982) Growth and osmotic relations of the mangrove *Avicennia marina*, as influenced by salinity. Aust J Plant Physiol 9:519–528

Drennan P, Pammenter NW (1982) Physiology of salt secretion in the mangrove *Avicennia marina* (Forsk.) Vierh. New Phytol 91:1000–1005

Duke NC (1992) Mangrove floristics and biogeography. In: Robertson AI, Alongi DM (eds) Tropical mangrove ecosystems. American Geophysical Union, Washington, DC, pp 63–100

Ellison JC, Stoddart DR (1991) Mangrove ecosystem collapse during predicted sea-level rise: Holocene analogues and implications. J Coastal Res 7:151–165

FAO (1994) Mangrove forest management guidelines. FAO, Rome, p 319

FAO (2003) Status and trends in mangrove area extent worldwide. In: Wilkie ML, Fortuna S (eds) Forest resources assessment working paper no. 63. Forest Resources Division, FAO, Rome

Fisher P, Spalding MD (1993) Protected areas with mangrove habitat. Draft Report World Conservation Centre, Cambridge, 60pp

Franks T, Falconer R (1999) Developing procedures for the sustainable use of mangrove systems. Agric Water Manage 40:59–64

Gattusso JP, Frankignoulle M, Wollast R (1998) Carbon and carbonate metabolism in coastal aquatic ecosystems. Annu Rev Ecol Evol Syst 29:405–434

Gouda R, Panigrahy RC (1996) The mangroves of Orissa. J Ind Ocean Stud 3(3):228–237

Government of India (1992a) National policy statement for abatement and pollution. Ministry of Environment and Forests, New Delhi

Government of India (1992b) National conservation strategy and policy statement on environment and development. Ministry of Environment and Forests, New Delhi

Groombridge B (1992) Global biodiversity: status of the earth's living resources. WCMC/The National History Museum/IUCN/UNEP/WWF/WRI, Chapman and Hall, London, 594pp

Hiraishi T, Harada K (2003) Greenbelt tsunami prevention in South-Pacific region. Rep Port Airport Res Inst 42(2):1–23

Hiscock K (1997) Conserving biodiversity in North-East Atlantic marine ecosystems. In: Ormond RFG, Gage JD, Angel MV (eds) Marine biodiversity: patterns and processes. Cambridge University Press, Cambridge, 415pp

Hoque MA, Sarkar MSKA, Khan SAKU, Moral MAH, Khurram AKM (2006) Present status of salinity rise in Sundarbans area and its effect on Sundari (*Heritiera fomes*) species. Res J Agric Biol Sci 2:115–121

Houghton J, Ding Y, Griggs D, Noguer M, Van der Linden P, Dai X, Maskell K, Johnson CA (eds) (2001) Climate change 2001: the scientific basis. Published for the Intergovernmental Panel on Climate Change. Cambridge University Press, Cambridge/New York, 881pp

ITTO/ISME (1993) Project PD 71/89. Rev. 1 (F). International Society for Mangrove Ecosystems (ISME), Okinawa

IUCN (2005) Early observations of Tsunami effects on mangroves and coastal forests. Statement from the IUCN forest conservation programme. http://www.iucn.org/info_and_news/press/TsunamiForest. 7 Jan 2005

Karim J, Karim A (1993) Effect of salinity on the growth of some mangrove plants in Bangladesh. In: Lieth H, Al Masoom A (eds) Towards the rational use of high salinity tolerant plants, vol 1. Kluwer, Dordrecht, pp 193–216

Kathiresan K, Bingham BL (2001) Biology of mangroves and mangrove ecosystems. Adv Mar Biol 40:81–251

Kathiresan K, Rajendran N (2005) Mangrove ecosystems of the Indian Ocean region. Ind J Mar Sci 34(1):104–113

Kauppi P, Sedjo R, Apps M, Cerri C, Fujimori T, Janzen H, Krankina O, Makundi W, Marland G, Masera O, Nabuurs G, Razali W, Ravindranath N (2001) Chapter 4: Technological and economic potential of options to enhance, maintain, and manage biological carbon reservoirs and geo-engineering. In: Intergovernmental Panel on Climate Change. Climate Change 2001: mitigation. A report of working group III of the Intergovernmental Panel on Climate Change, Geneva

Khan NY, Saeed T, Al-Ghadban AN, Beg MD, Jacob P, Al-Dosari AM, Al-Shemmari H, Al-Mutairi M, Al-Obaid T, Al-Matrook K (1999) Assessment of sediment quality in Kuwait's territorial waters. Phase I: Kuwait Bay. Kuwait Institute for Scientific Research. Kuwait report no. KISR 5651, Kuwait

Khan MA, Ungar IA, Showalter AM (2000a) Effects of salinity on growth, water relations and ion accumulation of the subtropical perennial halophyte, *Atriplex griffithii* var. stocks ii. Ann Bot 85:225–232

Khan MA, Ungar IA, Showalter AM (2000b) The effects of salinity on the growth, water status and ion content of a leaf succulent perennial halophyte, *Suaeda fruticosa* (L.) Forssk. J Arid Environ 45:73–84

Komiyama A, Ogino K, Aksomkoae S, Sabhasri S (1987) Root biomass of a mangrove forest in southern Thailand 1. Estimation by the trench method and the zonal structure of root biomass. J Trop Ecol 3:97–108

Lacerda LD (1993) Conservation and sustainable utilization of mangrove forests in Latin America and Africa regions. Part I – Latin America. Mangrove ecosystems technical reports ITTO/ISME project PD 114/90 (F), vol 2, Okinawa, 272pp

Lear R, Turner T (1977) Mangrove of Australia. University of Queensland Press, St. Lucia

Lin G, Sternberg L (1992) Effects of growth form, salinity, nutrient, and sulfide on photosynthesis, carbon isotope discrimination and growth of red mangrove (*Rhizophora mangle* L.). Aust J Plant Physiol 19:509–517

Lin G, Sternberg L (1995) Variation in propagule mass and its effect on carbon assimilation and seedlings growth of red mangrove (*Rhizophora mangle* L.) in Florida, USA. J Trop Ecol 11:109–119

MacNae, W. (1968). A general account of the fauna and flora of mangrove swamps and forests in the Indo-West Pacific region. Adv Mar Biol 6:73–270

McKee KL (1995) Interspecific variation in growth, biomass partitioning, and defensive characteristics of neotropical mangrove seedlings: response to light and nutrient availability. Am J Bot 82:299–307

McLeod E, Salm RV (2006) Managing mangroves for resilience to climate change. IUCN, Gland, 64pp

Mepham RH, Mepham JS (1984) The flora of tidal forests – a rationalization of the use of the term 'mangrove'. South Afr J Bot 51:75–99

Mitra A, Choudhury A (1994) Dissolved trace metals in surface waters around Sagar Island. Ind J Eco Biol 6(2):135–139

Mitra A, Pal S (2002) Components of mangrove ecosystem. In: Banerjee S, Tampal F (eds) The oscillating mangrove ecosystem and the Indian Sundarbans. Published by WWF-India, WBSO, Kolkata

Mitra A, Ghosh PB, Choudhury A (1987) A marine bivalve *Crassostrea cucullata* can be used as an indicator species of marine pollution. In: Proceedings of national seminar on estuarine management, Trivandrum, 1987, pp 177–180

Mitra A, Choudhury A, Zamaddar YA (1992) Effects of heavy metals on benthic molluscan communities in Hooghly estuary. Proc Zool Soc 45:481–496

Mitra A, Banerjee K, Pal S, Majumdar S, Mahapatra B, Halder KC, Das KL, Choudhury A, Bhattacharyya DP (2001) Seasonal variations of phytopigments in the Northwestern Bay of Bengal. Res J Chem Environ 3:27–35

Mitra A, Banerjee K, Bhattacharyya DP (2004) Impact of pollution on mangroves. In: The other face of mangroves. Published by Department of Environment, Govt. of West Bengal, Kolkata, pp 69–86

Mitra A, Banerjee K, Samanta S, Jana HK, Basu S (2006) Edible oyster culture in Indian Sundarbans. In: Mitra A, Banerjee K (eds) Training manual on non-classical use of mangrove resources of Indian Sundarbans for alternative livelihood programmes, 1st edn. Unit I, WWF-India, Canning Field Office, West Bengal, Kolkata, India, pp 11–28

Mitra A, Gangopadhyay A, Dube A, Andre CKS, Banerjee K (2009) Observed changes in water mass properties in the Indian Sundarbans (Northwestern Bay of Bengal) during 1980–2007. Curr Sci 97(10):1445–1452

Moorthy P (1995) Effects of UV-B radiation on mangrove environment: physiological responses of *Rhizophora apiculata* Blume. Ph.D. thesis, Annamalai University, Chidambaram, 130pp

Naidoo G (1987) Effects of salinity and nitrogen on growth and plant water relations in the mangrove *Avicennia marina* (Forssk.) Vierh. New Phytol 107:317–326

Ong JE, Khoon GW (2003) Carbon fixation in mangrove ecosystem and carbon credits. Theme B from the East Asian Sea Congress: essential cross-sectoral processes and approaches to achieving sustainable development conference bulletin. Issue 2, 11 July 2012

Ong JE, Gong WK, Clough BF (1995) Structure and productivity of a 20-year old stand of *Rhizophora apiculata* BL mangrove forest. J Biogeogr 55:417–424

Primavera JH (1997) Fish predation on mangrove-associated penaeids. The role of structures and substrate. J Exp Mar Biol Ecol 215:205–216

Putz FE, Chan HT (1986) Tree growth, dynamics, and productivity in a mature mangrove forest in Malaysia. For Ecol Manage 17:211–230

Rahman MA, Shah MS, Murtaza MG, Martin MA (1994) Degeneration of Bangladesh's Sundarbans mangrove: a management issue. Commonwealth Forestry Association. Int For Rev 6(2):123–135

Ramsar S (2001) Wetland values and functions. Gland. Tasmanian Government, Department of Primary industries, water and Environment (ISBN 0724662545, 9780724662548)

Reimold RJ, Queen WH (eds) (1974) Ecology of halophytes. Academic Press, Inc., New York, 605pp

Saenger P, Hegerl EJ, Davie JDS (1983) Global status of mangrove ecosystems. Commission on ecology papers no. 3. IUCN, Gland, 88pp

Saha SB, Mitra A, Bhattacharyya SB, Choudhury A (1999) Heavy metal pollution in Jagannath canal, an important tidal water body of the north Sundarbans aquatic ecosystem of West Bengal. Ind J Environ Prot 19(11):801–804

Scholander PF (1968) How mangroves desalinate sea water. J Plant Physiol 21:251–261

Scholander PF, Hammel HT, Hemmingsen EA, Garey W (1962) Salt balance in mangroves. Plant Physiol 37:722–729

Scholander PF, Hammel HT, Hemmingsen EA, Bradstreet ED (1964) Hydrostatic pressure and osmotic potential in leaves of mangroves and some other plants. Proc Nat Acad Sci USA 52:119–125

Schwamborn R, Saint-Paul U (1996) Mangroves – forgotten forests? Nat Res Dev 43(44):13–36

Smith SM, Snedaker SC (1995) Salinity responses in two populations of viviparous Rhizophora mangle L. seedlings. Biotropica 27:435–440

Spalding, M. (1997). The global distribution and status of mangrove ecosystems. International newsletter of coastal management – inter-coast network, special edition 1, Coastal Resources Center, University of Rhode Island, Narragansett, pp 20–21

Spalding MD, Blasco F, Field CD (eds) (1997) World mangrove atlas. International Society for Mangrove Ecosystems, Okinawa, p 178

Tamai S, Nakasuga T, Tabuchi R, Ogino K (1986) Standing biomass of mangrove forests in Southern Thailand. J Jpn Forest Soc 68:384–388

Thangam TS (1990) Studies on marine plants for mosquito control. Ph.D. thesis, Annamalai University, Chidambaram, 68pp

Tomlinson PB (1986) The botany of mangroves. Cambridge University Press, London

Twilley RR, Chen R, Hargis T (1992) Carbon sinks in mangroves and their implication to carbon budget of tropical ecosystems. Water Air Soil Pollut 64:265–288

UNEP (2002) Global distribution of coral, mangrove and sea grass diversity. Retrieved 26 May 2007 from: http://maps,grida.no/go/graphic/distributionofcoral-mangrove-and-seagrass-diversity

UNEP United Nations Environment Programme (1994) Assessment and monitoring of climatic change impacts on mangrove ecosystems. UNEP regional seas reports and studies. Report No. 154. UNEP, Nairobi

UNEP-FAO (1981) Tropical forest resources assessment project, forest resources of tropical Asia, FAO, UNEP, Rome, 475pp

UNESCO (1973) International classification and mapping of vegetation. United Nations Education, Scientific and Cultural Organisation, Paris

Van Eijk M (1939) Analyse der Wirkung des NaCl auf die Entwicklung Sukhulenz and Transpiration bei Salicornia herbacea, Sowie untersuchungen uber den Einfluss der Salzauf Nahme auf die Wurzelatmung bei Aster tripolium. Recueil des Travaux. Botaniques Neerlandais 36:559–657

Wilkie ML, Fortuna S (2003) Status and trends in mangrove area extent worldwide, vol 63, Forest resources assessment working paper. Food and Agriculture Organization of the United Nations, Forest Resources Division, Rome

Internet References

http://www.reefrelief.org/Documents/mangrove.html

http://www.epa.qld.gov.au/nature_conservation/habitats/mangroves_andwetlands/man groves

http://www.fao.org/gpa/sediments/habitat.htm

Annexure 2A.1: Spatial Variation of Stored Carbon in *Avicennia alba* of Indian Sundarbans

Kasturi Sengupta, Mahua Roy Chowdhury,
G. Roychowdhury, Atanu Kumar Raha,
Sufia Zaman and Abhijit Mitra

Abstract

We evaluated stored carbon in the above-ground biomass (AGB), below-ground biomass (BGB) and total biomass (TB) of 12-year-old trees of *Avicennia alba* in the western, central and eastern sectors of Indian Sundarbans during premonsoon (June) of 2012. We also analyzed the soil organic carbon (SOC) simultaneously in these sectors with the aim to find the interrelationships between AGB, BGB, TB and SOC, which exhibited significant spatial variations ($p < 0.05$). In all the three sectors, significant positive correlations ($p < 0.01$) were observed between the mangrove carbon (both above-ground carbon or AGC and below-ground root carbon or BGC) and SOC indicating considerable contributions of stem, branch, leaf and roots of *A. alba* to SOC.

Keywords

Above-ground biomass • Below-ground biomass • Total biomass • Soil organic carbon • Indian Sundarbans

K. Sengupta • M.R. Chowdhury • S. Zaman
Department of Marine Science, University
of Calcutta, 35, B.C Road, Kolkata,
West Bengal 700 019, India

G. Roychowdhury
Chancellor, Techno India University, Salt Lake Campus,
Kolkata, West Bengal 700 091, India

A.K. Raha
Department of Environment & Forest Science Techno
India University, Salt Lake Campus,
Kolkata, West Bengal 700 091, India

A. Mitra (✉)
Department of Oceanography, Techno India
University, Salt Lake Campus, Kolkata,
West Bengal 700 091, India
e-mail: abhijit_mitra@hotmail.com

2A.1.1 Introduction

Forests form a major component of the carbon reserves in the world's ecosystems (Whittaker and Likens 1975) and greatly influence the lives of other organisms as well as human societies. Tropical forests are an important compartment in the global carbon cycle and represent 30–40 % of the terrestrial net primary production (Clark et al. 2001). Although the area covered by mangrove forest represents only a small fraction of the tropical forests, their position at the terrestrial–ocean interface and potential exchange of nutrients with coastal water suggest that these forests make a unique contribution to carbon biogeochemistry in coastal oceans (Twilley et al. 1992). Mangroves are a taxonomically diverse group of salt-tolerant, mainly arboreal, flowering plants that grow primarily in tropical and subtropical regions (Ellison and Stoddart 1991). Estimates of mangrove area vary from several million hectares (ha) to 15 million ha worldwide (FAO 1981). The most recent estimates suggest that mangroves presently occupy about 14,653,000 ha of tropical and subtropical coastline (Wilkie and Fortuna 2003). The coastal zone (<200 m depth), covering ~7 % of the ocean surface (Gattuso et al. 1998), has an important role in the oceanic carbon cycle, and various estimates indicate that the majority of mineralization and burial of organic carbon, as well as carbonate production and accumulation, take place in the coastal ocean (Gattuso et al. 1998; Mackenzie et al. 2004). The 'outwelling' hypothesis, first proposed for mangroves by Odum (1968) and Odum and Heald (1972), suggests that a large fraction of the organic matter produced by mangrove trees is exported to the coastal ocean, where it forms the basis of a detritus food chain and thereby supports coastal fisheries. Despite the large number of case studies dealing with various aspects of organic matter cycling in mangrove systems (Kristensen et al. 2008), there is very limited consensus on the carbon sequestering potential of mangroves.

The present study aims to establish a baseline databank on the carbon storage by above-ground and below-ground structures of a dominant and most common mangrove species *A. alba* in the Indian Sundarbans along with soil organic carbon (SOC) in order to critically address two important issues:

1. Whether spatial difference in salinity causes variation in carbon storage capacity of the mangrove tree
2. Whether any interrelationships exist between mangrove biomass, mangrove carbon and soil carbon

2A.1.2   Material and Methods

2A.1.2.1   Study Site Description

The mighty River Ganga emerges from a glacier at Gangotri, about 7,010 m above mean sea level

in the Himalayas, and flows down to the Bay of Bengal covering a distance of 2,525 km. At the apex of Bay of Bengal, a delta has been formed which is recognized as one of the most diversified and productive ecosystems of the tropics and is referred to as the Indian Sundarbans. The deltaic complex has an area of 9,630 km^2 and houses about 102 islands (Mitra 2000). Eighteen sampling sites were selected each in the western, central and eastern sectors of Indian Sundarbans (Fig. 2A.1.1). The western sector of the deltaic lobe receives the snowmelt water of mighty Himalayan glaciers after being regulated through several barrages on the way. The central sector, on the other hand, is fully deprived from such supply due to heavy siltation and clogging of the Bidyadhari channel since the fifteenth century (Chaudhuri and Choudhury 1994). The eastern sector of Indian Sundarbans is adjacent to the Bangladesh Sundarbans (which comprises 62 % of the total Sundarbans) and receives the freshwater from the River Raimangal and also from the Padma–Meghna–Brahmaputra river system of Bangladesh Sundarbans through several creeks and inlets. Samplings in these sectors were carried out in low tide period during June 2012.

In each site, selected forest patches were ~12 years old. Fifteen sample plots (10 m × 10 m) were established (in the river bank) through random sampling in the various qualitatively classified biomass levels for each site. The population density of *A. alba* was evaluated to estimate the magnitude of stored carbon by the species in a particular site.

AGB of individual trees of the species in each plot was estimated, and the average values of 15 plots from each site were finally converted into biomass (t/ha) in the study area.

AGB is the sum total of stem, branch and leaves of the tree. Hence, the biomass of the vegetative parts was estimated separately.

2A.1.2.2   Above-Ground Stem Biomass (AGSB) Estimation

The stem volume for each tree of the species in every plot was estimated using the Newton's formula (Husch et al. 1982) as per the expression:

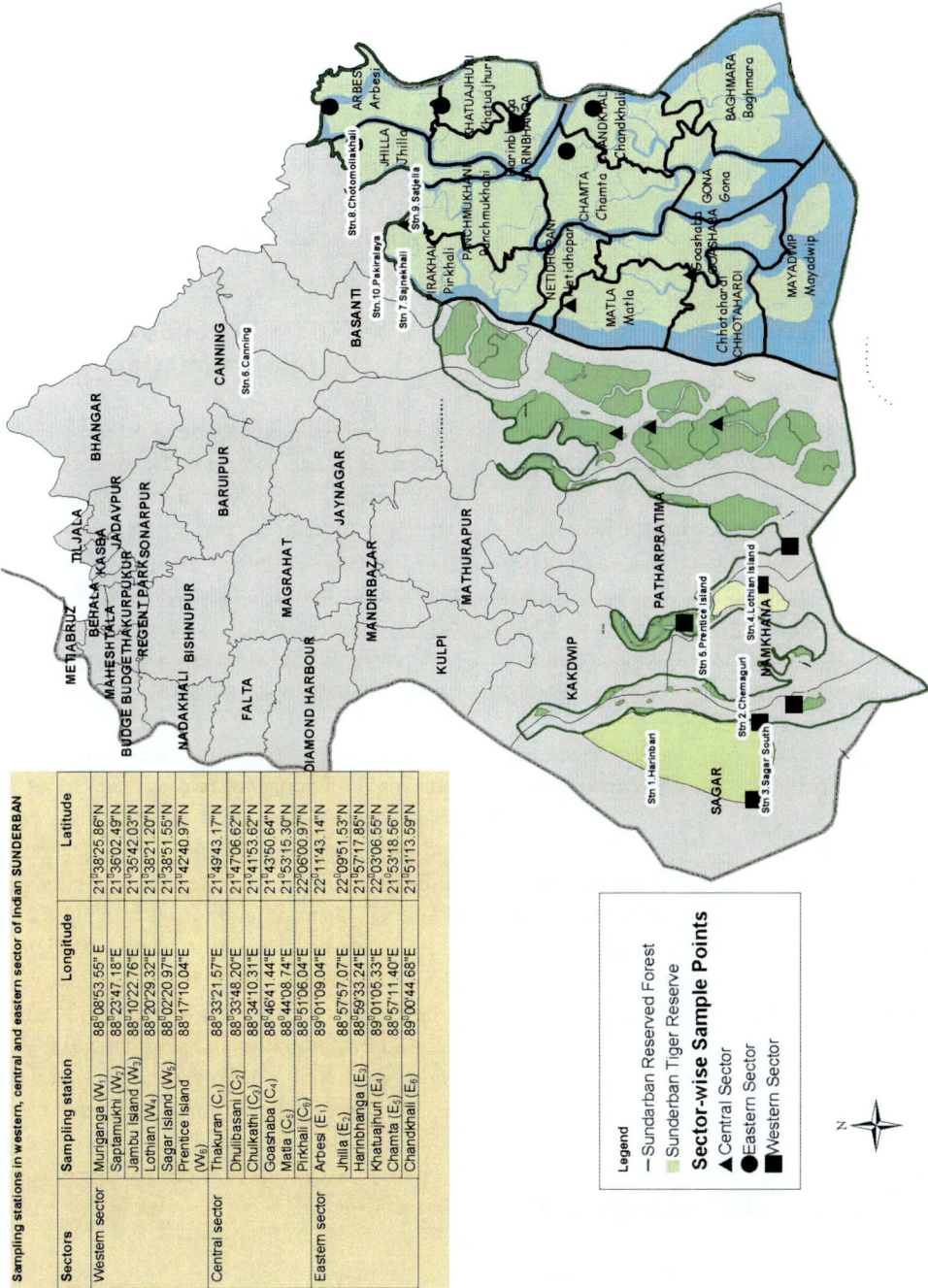

Sectors	Sampling station	Longitude	Latitude
Western sector	Muriganga (W_1)	88°08'53.55" E	21°38'25.86"N
	Saptamukhi (W_2)	88°23'47.18"E	21°36'02.49"N
	Jambu Island (W_3)	88°10'22.76"E	21°35'42.03"N
	Lothian (W_4)	88°20'29.32"E	21°38'21.20"N
	Sagar Island (W_5)	88°02'20.97"E	21°38'51.55"N
	Prentice Island (W_6)	88°17'10.04"E	21°42'40.97"N
Central sector	Thakuran (C_1)	88°33'21.57"E	21°49'43.17"N
	Dhulibasani (C_2)	88°33'48.20"E	21°47'06.62"N
	Chulkathi (C_3)	88°34'10.31"E	21°41'53.62"N
	Goashaba (C_4)	88°46'41.44"E	21°43'50.64"N
	Matla (C_5)	88°44'08.74"E	21°53'16.30"N
	Pirkhali (C_6)	88°51'06.04"E	22°06'00.97"N
Eastern sector	Arbesi (E_1)	89°01'09.04"E	22°11'43.14"N
	Jhilla (E_2)	88°57'57.07"E	22°09'51.53"N
	Harinbhanga (E_3)	88°59'33.24"E	21°57'17.85"N
	Khatuajhuri (E_4)	89°01'05.33"E	22°03'06.55"N
	Chamta (E_5)	88°57'11.40"E	21°53'18.56"N
	Chandkhali (E_6)	89°00'44.68"E	21°51'13.59"N

Sampling stations in western, central and eastern sector of Indian SUNDERBAN

Legend
— Sundarban Reserved Forest
  Sundarban Tiger Reserve
Sector-wise Sample Points
▲ Central Sector
● Eastern Sector
■ Western Sector

Fig. 2A.1.1 Location of sampling stations of Indian Sundarbans

$$V = \frac{h}{6\left(A_b + 4A_m + A_t\right)}$$

where V is the volume (in m^3), h the height measured with laser beam (BOSCH DLE 70 Professional model) and A_b, A_m and A_t are the areas of the selected tree at base, middle and top, respectively. Specific gravity (G) of the wood was estimated taking the stem cores, which was further converted into stem biomass (B_S) as per the expression $B_S = GV$. The stem biomass of individual tree was finally multiplied with the number of trees of the species in 15 selected plots for every site in the western, central and eastern sectors of Indian Sundarbans.

2A.1.2.3 Above-Ground Branch Biomass (AGBB) Estimation

The total number of branches irrespective of size was counted on each of the sample trees. These branches were categorized on the basis of basal diameter into three groups, namely, <6, 6–10 and >10 cm. Fresh weight of two branches from each size group was recorded separately using the equation of Chidumaya (1990).

Total branch biomass (dry weight) of individual tree was determined as per the expression:

$$B_{db} = n_1 \mathrm{bw}_1 + n_2 \mathrm{bw}_2 + n_3 \mathrm{bw}_3 = \Sigma n_i \mathrm{bw}_i$$

where B_{db} is the dry branch biomass per tree, n_i the number of branches in the ith branch group, bw_i the average weight of branches in the ith group and $i = 1, 2, 3, \ldots, n$ are the branch groups. The branch biomass of individual tree was finally multiplied with the number of trees of the species in all the 15 plots for each site.

2A.1.2.4 Above-Ground Leaf Biomass (AGLB) Estimation

Leaves from nine branches (three of each size group) of individual trees were plucked, weighed and oven-dried separately to a constant weight at 80 ± 5 °C. Three trees per plot were considered for estimation. The leaf biomass was then esti-

mated by multiplying the average biomass of the leaves per branch with the number of branches in a single tree and the average number of trees per plot as per the expression:

$$L_{db} = n_1 \mathrm{Lw}_1 N_1 + n_2 \mathrm{Lw}_2 N_2 + \cdots n_i \mathrm{Lw}_i N_i$$

where L_{db} is the dry leaf biomass of selected mangrove species per plot, $n_1 \ldots n_i$ are the number of branches of each tree of the species, $\mathrm{Lw}_1 \ldots \mathrm{Lw}_i$ are the average dry weight of leaves removed from the branches and $N_1 \ldots N_i$ are the number of trees of the species in the plots.

2A.1.2.5 Below-Ground Biomass (BGB) Estimation

An excavation method (Bledsoe et al. 1999) was used to estimate root biomass of the same trees that were selected for above-ground biomass (AGB) estimate. According to our observation, very few roots in our sampling plots were distributed deeper than 1 m in sediments. We also found canopy diameter of these trees was usually smaller than 2 m. Most roots of the selected species were distributed within the projected canopy zone. Therefore, for below-ground biomass (BGB, referring to root biomass in this study), we excavated all roots (of two trees/species) in 1 m depth within the radius of 1 m from the tree centre and then washed the roots. We excavated all the sediments within the sampling cylinder (2 m in diameter × 1 m in height) and washed them with a fine screen to collect all roots. The roots were sorted into four size classes: extreme fine roots (diameter < 0.2 cm), fine roots (diameter 0.2–0.5 cm), small roots (diameter 0.5–1.0 cm) and coarse roots (diameter > 1 cm). We did not separate live or dead roots. The roots after thorough washing were oven-dried to a constant weight at 80 ± 5 °C and biomass was estimated for each species.

2A.1.2.6 Carbon Estimation

Direct estimation of per cent carbon in the AGB was done by *Vario MACRO elementar* CHN analyzer, after grinding and random mixing the oven-dried stems, branches and leaves separately. For this, a portion of fresh sample of stem, branch and leaf from trees were oven-dried at 70 °C, randomly mixed separately and ground to pass through a 0.5 mm screen (1.0 mm screen for leaves). Carbon content of the oven-dried root system (BGB) was also estimated directly in the same instrument.

2A.1.2.7   Soil Organic Carbon (SOC) Estimation

Soil samples from the upper 5 cm were collected from all the 15 plots at each site and dried at 60 °C for 48 h. For analysis, visible plant particles were handpicked and removed from the soil. After sieving the soil through a 2-mm sieve, the samples of the bulk soil (50 g from each plot) were ground finely in a ball mill. The fine-dried sample was randomly mixed to get a representative picture of the study site. Modified version of Walkley and Black method (1934) was then followed to determine the organic carbon of the soil in %.

2A.1.2.8   Statistical Analysis

Data of the present study generated from 18 sampling sites were subject to sector-wise analysis of correlation coefficient (r) in order to evaluate the interrelationships between mangrove biomass, mangrove carbon and soil organic carbon. Analysis of variance (ANOVA) was performed to assess whether biomass, tree carbon and soil organic carbon content varied significantly between sectors; possibilities ($p < 0.01$) were considered statistically significant. All statistical calculations were performed with SPSS 9.0 for Windows.

2A.1.3   Results and Discussion

The recent thrust on global warming phenomenon has generated tremendous interest in the carbon-storing ability of mangroves. The carbon sequestration in this unique producer community is a function of biomass production capacity, which in turn depends upon interaction between edaphic, climate and topographic factors of an area. Hence, results obtained at one place may not be applicable to another. Therefore, region-based potential of different land types or substratum characteristics needs to be worked out. In the present study, spatial variations of stored carbon have been worked out separately for AGB, BGB and SOC, and statistical tools were employed to evaluate the significance of variation.

2A.1.3.1   Relative Abundance

The relative abundance of *A. alba* was highest in the eastern sector (45.73 %), followed by western sector (30.65 %) and central sector (23.62 %).The lowest relative abundance may be attributed to the hypersaline condition in the central sector, where the freshwater supply has been almost disconnected due to siltation of the Bidyadhari River since the late fifteenth century as well as upcoming of densely populated habitations and agricultural activities after reclamation of wetland in the upper catchment of the estuary.

2A.1.3.2   Above-Ground Biomass (AGB)

The AGB of the mangrove species was relatively higher in the sampling sites of the eastern sector compared to the western and central sectors.

In the present study, the AGB ranged from 50.75 t/ha (W_3) to 60.85 t/ha (W_6) in the western sector, 31.55 t/ha (C_5) to 58.75 t/ha (C_4) in the central sector and 48.25 t/ha (E_6) to 66.58 t/ha (E_1) in the eastern sector of the Indian Sundarbans. The AGB varied as per the order eastern sector (59.28 t/ha) > western sector (55.02 t/ha) > central sector (46.99 t/ha). ANOVA results also confirm significant spatial variations in AGB ($p < 0.05$) (Table 2A.1.1 and Fig. 2A.1.2).

Table 2A.1.1 Spatial variations of AGB, BGB, TB and AGC, BGC, TC of *Avicennia alba* and SOC of Indian Sundarbans

Variable	F_{cal}	F_{crit} ($p < 0.05$)
AGB		
Between sectors	4.193	4.102
Between stations	0.499	3.325
AGC		
Between sector	4.658	4.102
Between station	0.576	3.325
BGB		
Between sector	4.792	4.102
Between station	0.793	3.325
BGC		
Between sector	5.045	4.102
Between station	0.793	3.325
TB		
Between sector	4.412	4.102
Between station	0.961	3.325
TC		
Between sector	4.812	4.102
Between station	0.635	3.325
SOC		
Between sector	4.501	4.102
Between station	0.961	3.325

2A.1.3.3 Below-Ground Biomass (BGB)

The order of BGB in the study area is eastern sector (17.67 t/ha) > western sector (16.08 t/ha) > central sector (13.12 t/ha). In this study, the BGB ranged from 13.95 t/ha (W_3) to 17.85 t/ha (W_6) in the western sector, 8.27 t/ha (C_5) to 17.23 t/ha (C_4) in the central sector and 13.11 t/ha (E_6) to 20.64 t/ha (E_1) in the eastern sector in the Indian Sundarbans. These results are in accordance with the ANOVA that exhibit significant spatial variation in BGB ($p < 0.05$) (Table 2A.1.1 and Fig. 2A.1.2).

2A.1.3.4 Total Biomass

In this study the TB ranged from 64.70 t/ha (W_3) to 78.69 t/ha (W_6) in the western sector, 39.82 t/ha (C_5) to 75.98 t/ha (C_4) in the central sector and 61.36 t/ha (E_6) to 87.22 t/ha (E_1) in the eastern sector of Indian Sundarbans. The TB varied as per the order eastern sector (76.95 t/ha) > western sector

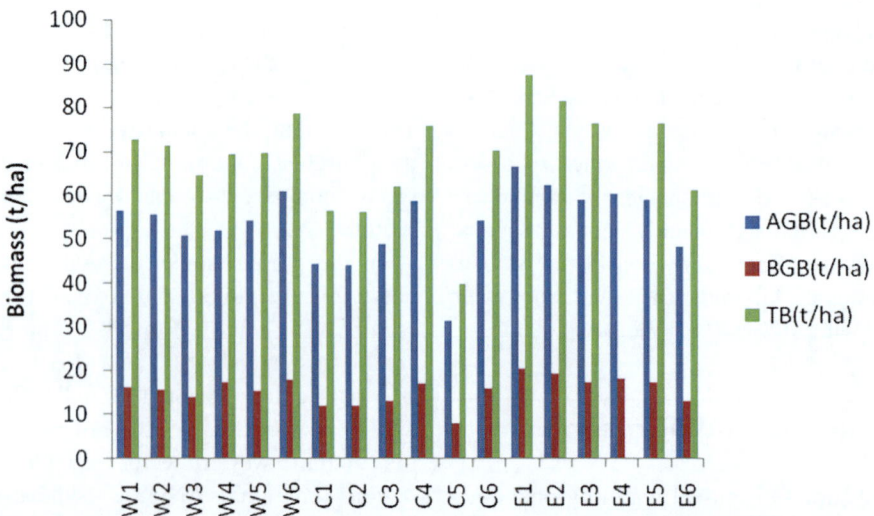

Fig. 2A.1.2 Spatial variations of biomass (t/ha) in western, central and eastern sectors of Indian Sundarbans

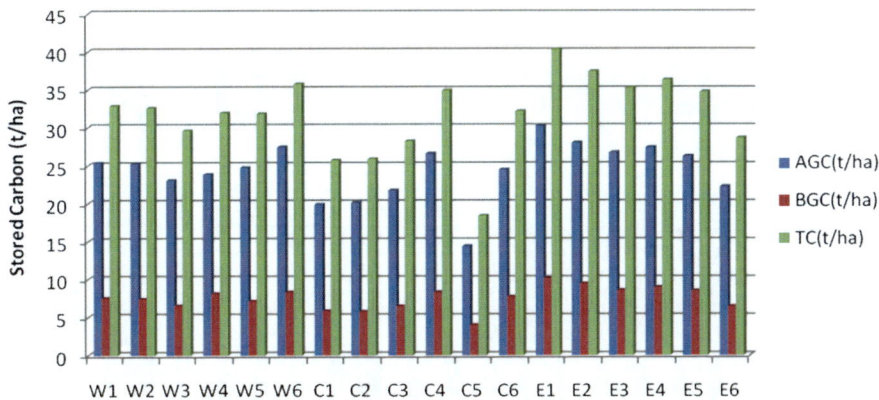

Fig. 2A.1.3 Spatial variations of stored carbon (t/ha) in biomass of species in western, central and eastern sectors of Indian Sundarbans

(71.10 t/ha)>central sector (60.11 t/ha). ANOVA results confirm significant spatial variation in TB of *A. alba* between western, central and eastern sectors ($p < 0.05$) (Table 2A.1.1 and Fig. 2A.1.2).

2A.1.3.5 Stored Carbon in *A. alba* Biomass

In this study, the stored carbon in AGB of the selected species ranged from 23.08 t/ha (W_3) to 27.45 t/ha (W_6) in the western sector, 14.43 t/ha (C_5) to 26.60 t/ha (C_4) in the central sector and 22.24 t/ha (E_6) to 30.22 t/ha (E_1) in the eastern sector. Stored carbon in below-ground biomass ranged from 6.53 t/ha (W_3) to 8.34 t/ha (W_6) in the western sector, 3.99 t/ha (C_5) to 8.34 t/ha (C_4) in the central sector and 6.42 t/ha (E_6) to 10.13 t/ha (E_1) in the eastern sector. The total stored carbon in *A. alba* thus followed the order eastern sector>western sector>central sector. ANOVA results indicate significant spatial differences in carbon content in AGB, BGB and TB (Table 2A.1.1 and Fig. 2A.1.3).

2A.1.3.6 Soil Organic Carbon

The order of SOC in the study area is eastern sector (1.41 %)>western sector (1.18 %)>central sector (1.05 %). The SOC ranged from 1.02 % (W_3 and W_4) to 1.61 % (W_6) in the western sector,

0.79 % (C_5) to 1.30 % (C_4) in the central sector and 1.15 % (E_6) to 1.66 % (E_1) in the eastern sector of Indian Sundarbans. Significant spatial variations ($p < 0.05$) were observed in SOC of deltaic Sundarbans (Table 2A.1.1 and Fig. 2A.1.4).

It is observed that AGB, BGB and TB exhibit significant spatial variations with highest value in the eastern sector followed by the western and central sectors. Similar trend is also observed in case of stored carbon. Two plausible reasons behind such spatial variations may be attributed to aquatic salinity and anthropogenic influence.

A long-term study conducted on salinity variation by several researchers in the Indian Sundarbans exhibits three distinct regimes in terms of salinity. The western sector is relatively low saline owing to freshwater discharge from the Farakka barrage. Ten-year surveys (1999–2008) on water discharge from Farakka dam revealed an average discharge of $(3.7 \pm 1.15) \times 10^3$ m³/s. Higher discharge values were observed during the monsoon with an average of $(3.81 \pm 1.23) \times 10^3$ m³/s, and the maximum of the order 4,524 m³/s during freshet (September). Considerably lower discharge values were recorded during premonsoon with an average of $(1.18 \pm 0.08) \times 10^3$ m³/s, and the minimum of the order 846 m³/s during May. During postmonsoon discharge, values were moderate with an average of $(1.98 \pm 0.97) \times 10^3$ m³/s.

The central sector, on contrary, exhibits hypersaline condition and an increasing trend in

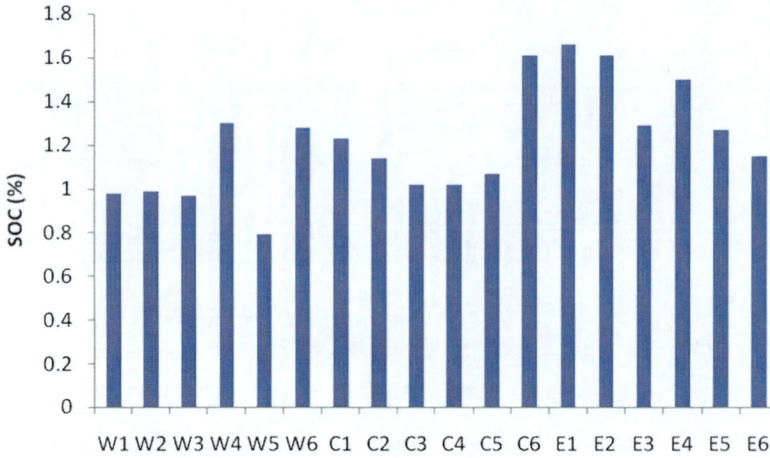

Fig. 2A.1.4 Spatial variations of SOC (%) in western, central and eastern sector of Indian Sundarbans

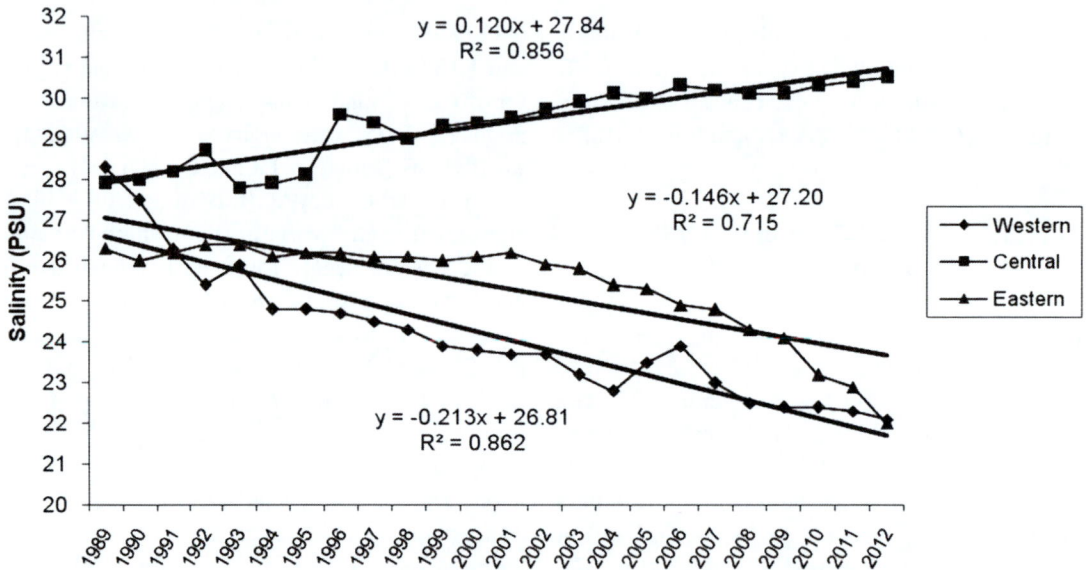

Fig. 2A.1.5 Spatio-temporal variations of surface water salinity (in psu) in Indian Sundarbans

salinity through time. This is due to complete siltation of the Bidyadhari channel since the fifteenth century that has almost cut off the freshwater supply from the upstream region (Chaudhuri and Choudhury 1994). The eastern sector of Indian Sundarbans adjacent to Bangladesh Sundarbans receives freshwater from several channels and creeks from the Padma–Meghna–Brahmaputra river system and their tributaries.

The mangroves are salt-tolerant species, but under hypersaline condition, they exhibit stunted growth (Mitra et al. 2004). The order of biomass and stored carbon in *A. alba* in the present study is basically the reflection of salinity, which is highest, and exhibits an increasing trend in the central sector of Indian Sundarbans compared to the other two sectors. In the western and eastern sectors, the salinity decreased by 21.91 and 16.35 %, respectively, over a period of 24 years, whereas in the central sector there has been a steady increase in the salinity by 9.32 % during the same period (Fig. 2A.1.5).

Table 2A.1.2 Interrelationships between biomass and SOC in western, central and eastern sectors of Indian Sundarbans

Combination	*r* value			*p* value		
	Western sector	Central sector	Eastern sector	Western sector	Central sector	Eastern sector
SOC × AGB	0.938989	0.940957	0.872343	<0.01	<0.01	<0.01
SOC × BGB	0.9651	0.633564	0.899707	<0.01	<0.01	<0.01
SOC × TB	0.946405	0.924799	0.880887	<0.01	<0.01	<0.01

Table 2A.1.3 Interrelationships between stored carbon content and SOC in western, central and eastern sectors of Indian Sundarbans

Combination	*r* value			*p* value		
	Western sector	Central sector	Eastern sector	Western sector	Central sector	Eastern sector
SOC × AGC	0.947859	0.937395	0.893225	<0.01	<0.01	<0.01
SOC × BGC	0.964988	0.617627	0.903088	<0.01	<0.01	<0.01
SOC × TC	0.953211	0.90812	0.897461	<0.01	<0.01	<0.01

The degree of anthropogenic stress is also less in the eastern Indian Sundarbans compared to western and central sectors. This is because of the location of eastern sector in the Reserve Forest area, where human entry is highly restricted. This causes more natural growth rate of mangroves in the eastern sector compared to the western and central sectors, where the mangroves are mostly cut down for timber, fuel, etc.

The SOC in the mangrove ecosystem is contributed by the vegetative and reproductive parts of the halophytes, although the contributions of riverine inputs pose a regulatory effect on the SOC budget of the mangrove soil (Banerjee et al. 2012). The present study reveals significant contribution of AGB and BGB of *A. alba* in the matrix of SOC (Table 2A.1.3). The stored carbon in AGB (24.96 % in the western sector, 21.22 % in the central sector and 26.82 % in the eastern sector) and BGB (7.51 % in the western sector, 6.35 % in the central sector and 8.67 % in the eastern sector) are the main players in regulating the SOC in the intertidal mudflats of Indian Sundarbans. The significantly high correlation coefficient values of SOC with AGB, AGC, BGB, BGC, TB and TC of *A. alba* support the dependency of SOC on mangrove biomass and carbon (Tables 2A.1.2 and 2A.1.3).

The overall discussion thus explains a congenial environment for the growth of *A. alba* in eastern Indian Sundarbans and also indicates the necessity of freshwater supply to accelerate the biomass and subsequently the stored carbon in *A. alba* thriving in the Indian Sundarbans region in the lower Gangetic delta complex.

2A.1.4   Conclusion

This study has demonstrated that dwarfing of the mangrove species *A. alba* in the central Indian Sundarbans is a complex phenomenon influenced by a variety of hydro-edaphic conditions. Lack of freshwater supply from the upper catchments coupled with tidal influence from the Bay of Bengal has increased the salinity of central Indian Sundarbans over a period of time. The conditions in the western and eastern Indian Sundarbans are comparatively congenial for the growth of the mangrove species due to freshwater supply from the upstream region. These findings tend to support several other researches that point towards the necessity of dilution of saline water for proper growth and survival of mangroves. Additional research is needed to test and further define these conclusions.

References of Annexure 2A.1

Banerjee K, Roy Chowdhury M, Sengupta K, Sett S, Mitra, A (2012) Influence of anthropogenic and natural factors on the mangrove soil of Indian Sundarban Wetland. Arch Environ Sci 6:80–91

Bledsoe CS, Fahey TJ, Day FP, Ruess RW (1999) Measurement of static root parameters—biomass, length, and distribution in the soil profile. In: Robertson GP, Coleman DC, Bledsoe CS, Sollins P (eds) Standard soil methods for long-term ecological research. Oxford University Press, New York

Chaudhuri AB, Choudhury A (1994) Mangroves of the Sundarbans. The World Conservation Union, Dhaka

Chidumaya EN (1990) Above ground woody biomass structure and productivity in a Zambezian woodland. Forest Ecol Manage 36:33–46

Clark DA, Brown S, Kiicklighter DW, Chambers JQ, Thomlinson JR, Ni J, Holland EA (2001) Measuring net primary production in forest: an evaluation and synthesis of existing field data. J Appl Ecol 11:371–384

Ellison JC, Stoddart DR (1991) Mangrove ecosystem collapse during predicted sea-level rise: Holocene analogues and implications. J Coastal Res 7:151–165

Gattuso JP, Frankignoulle M, Wollast R (1998) Carbon and carbonate metabolism in coastal ecosystems. Annu Rev Ecol Evol Syst 29:405–434

Husch B, Miller CJ, Beers TW (1982) Forest mensuration. Ronald Press, New York

Kristensen E, Bouillon S, Dittmar T, Marchand C (2008) Organic carbon dynamics in mangrove ecosystems: a review. Aquat Bot 89:201–219

Mackenzie FT, Lerman A, Andersson AJ (2004) Past and present of sediment of carbon biogeochemical cycling models. Biogeoscience 1:11–32

Mitra A (2000) Chapter 62: The Northeast coast of the Bay of Bengal and deltaic Sundarbans. In: Sheppard C (ed) Seas at the millennium – an environmental evaluation. Elsevier Science, Amsterdam, pp 143–157

Mitra A, Banerjee K, Bhattacharyya DP (2004) The other face of mangroves. Published by Department of Environment, Govt. of West Bengal, Kolkata

Odum EP (1968) A research challenge: evaluating the productivity of coastal and estuarine waters. In: Proceedings of the 2nd sea grant conference, University of Rhode Island Kingston, Rhode Island, 1968, pp 63–64

Odum WE, Heald EJ (1972) Trophic analysis of an estuarine mangrove community. Bull Mar Sci 22: 671–738

Twilley RR, Chen R, Hargis T (1992) Carbon sinks in mangroves and their implication to carbon budget of tropical ecosystems. Water Air Soil Pollut 64:265–288

UNEP-FAO (1981) Tropical forest resources assessment project, forest resources of tropical Asia, FAO, UNEP, Rome, 475pp

Walkley A, Black IA (1934) An examination of the Degtjareff method for determining soil organic matter and a proposed modification of the chromic acid titration method. Soil Sci 37:29–38

Whittaker RH, Likens GE (1975) The biosphere and man. In: Lieth H, Whittaker RH (eds) Primary productivity of the biosphere, vol 14, Ecological studies. Springer, Berlin, pp 305–328

Annexure 2A.2: Influence of Anthropogenic and Natural Factors on the Mangrove Soil of Indian Sundarbans Wetland

Kakoli Banerjee, Mahua Roy Chowdhury, Kasturi Sengupta, Saurov Sett, and Abhijit Mitra

Abstract

Soil organic carbon, pH and salinity were monitored in mangrove ecosystem of Indian Sundarbans in five successive years (2006–2010). Samplings were carried out at 14 stations in four different depths (0.01–0.10, 0.10–0.20, 0.20–0.30 and 0.30–0.40 m) during premonsoon period. High organic carbon load is observed in the stations of western Indian Sundarbans (mean = 1.02 wt%) which are near to the highly urbanized city of Kolkata. The central and eastern sectors under the protected forest area show comparatively less soil organic carbon (mean = 0.64 wt%). A unique spatial variability in soil salinity and pH was observed with lower values in the western and eastern sectors compared to central sector. Soil pH exhibited a lower value (7.47 ± 0.071) in reserve forest zone (central and eastern sectors) compared to western sector (7.57 ± 0.067). The soil salinity increased with depth, while organic carbon and pH decreased with depth in all the stations. The paper depicts the increase of soil organic carbon and pH due to anthropogenic activities in western Indian Sundarbans, which if continued may decrease the potential of Sundarban soil as carbon sink and make the soil highly saline. Hence, curbing of anthropogenic activities may keep the soil characteristics ecologically safe.

Keywords

Indian Sundarbans • Soil organic carbon • Soil pH • Soil salinity

K. Banerjee • M.R. Chowdhury • K. Sengupta • S. Sett
Department of Marine Science, University
of Calcutta, 35, B.C. Road, Kolkata,
West Bengal 700 019, India

A. Mitra (✉)
Department of Oceanography, Techno India University,
Salt Lake Campus, Kolkata, West Bengal 700 091, India
e-mail: abhijit_mitra@hotmail.com

2A.2.1 Introduction

In mangrove ecosystem, organic carbon usually originates via the riverine introduction of pollutants, including industrial and domestic wastes; agricultural, aquacultural and mining run-off; accidental spillages; and decomposition of debris from marine organisms. However, different factors may control the partitioning and also the bioavailability of the organic compounds within the benthic ecosystem. These factors include sediment characteristics, such as grain size distribution, mineral composition and organic content (Lambert 1967; Forstner 1977; Khalaf et al. 1981). Surface sediments may be resuspended and redistributed by the action of waves and currents (Cahoon and Reed 1995). As these phenomena trigger the process of erosion and accretion, therefore, the top most layers of the sediments contain recently deposited organic matter. Total organic carbon has a major influence on both chemical and biological processes that take place in sediments (Kamaruzzaman et al. 2010). The amount of organic carbon has a direct role in determining the redox potential and pH in sediment, thus regulating the behaviour of other chemical species such as metals (Eshleman and Hemond 1985; Kerekes et al. 1986). Natural processes and human activities have resulted in elevated content of total organic carbon in mangrove soils and adjacent estuaries and creeks. These include diverse input through fall, stream flow, inappropriate animal waste applications and disposals, forest clearance, agricultural practices and changes in land uses (Moore and Jackson 1989). Also mangrove litter fall and decomposition of organisms regulate the organic carbon budget in the intertidal mudflats (Mitra et al. 2004).

The mangrove ecosystem of Indian Sundarbans, at the apex of Bay of Bengal, covers an area of about 4,266.6 km². On the basis of satellite imagery, the Forest Survey of India (1999) estimated the area of Indian Sundarbans as 2,125 km², excluding the network of creeks and backwaters, which are the vital matrix of mangrove ecosystem. Mangrove communities often exhibit distinct patterns of species distribution (Chapman 1976; Lugo and Snedaker 1974; Macnae 1968; Tomlinson 1986) that contribute to the organic carbon level in the intertidal soil through decomposition of litter and organisms. Since the mangrove habitat is basically saline, several studies have attempted to correlate salinity with the standing crop of vegetation and productivity (Chen and Twilley 1998, 1999;

Lugo 1980; Mall et al. 1987; Ukpong 1991; Mitra et al. 2011a). Sundarbans shelters one of the most important mangrove communities of the world. A few published works deal with the community structure of this forest (Joshi and Ghose 2002; Matilal et al. 1986). However, very few reports are available on the organic carbon profile of mangrove soil (Mitra et al. 2004) that can reflect the status of this unique ecosystem in terms of natural (Mitra and Banerjee 2005) or anthropogenic influences (Mitra 1998, 2000; Mitra et al. 2011b). The aim of this paper is to determine what role the anthropogenic and natural factors have on mangrove soil and how the soil characteristics change over time.

2A.2.2   Materials and Methods

2A.2.2.1   The Study Area

The Indian Sundarbans (between 21°13′ N and 22°40′ N latitude and 88°03′ E and 89°07′ E longitude) is bordered by Bangladesh in the east, the Hugli River (a continuation of the River Ganga) in the west, the Dampier and Hodges line in the north and the Bay of Bengal in the south. The temperature is moderate due to its proximity to the Bay of Bengal in the south. Average annual maximum temperature is around 35 °C. The summer (premonsoon) extends from the middle of March to mid-June and the winter (postmonsoon) from mid-November to February. The monsoon usually sets in around the middle of June and lasts up to the middle of October. Rough weather with frequent cyclonic depressions occurs during mid-March to mid-September. Average annual rainfall is 1920 mm. Average humidity is about 82 % and is more or less uniform throughout the year. Thirty-four true mangrove species and some 62 mangrove associate species have been documented in Indian Sundarbans (Mitra 2000). The ecosystem is extremely prone to erosion, accretion, tidal surges and several natural disasters (Mitra and Banerjee 2005), which directly affect the topsoil of the intertidal mudflats encircling the islands. The average tidal amplitude is around 3.0 m. Some sea-facing islands experience high tidal amplitude (~5.0 m).

We conducted survey at 14 stations in the Indian Sundarbans region during premonsoon (May) from 2006 to 2010. Station selection was primarily based on anthropogenic activities, salinity and mangrove floral richness. The western Indian Sundarbans is a stressed zone (stations 1–7). On the contrary, stations 8–14 are within the reserve forest areas with luxuriant mangrove vegetation and diversity and have been considered as control zone in this study. The major activities influencing the nature of soil in the selected stations are highlighted in Table 2A.2.1.

2A.2.2.2   Sampling and Analysis

Table 2A.2.1 and Fig. 2A.2.1 represent the study site in which sampling plots of 10 m × 5 m were considered for each station. Care was taken to collect the samples within the same distance from the estuarine edge, tidal creeks and the same micro-topography. Under such conditions, spatial variability of external parameters such as tidal amplitude and frequency of inundation (Ovalle et al. 1990), inputs of material from the adjacent bay/estuary and soil granulometry and salinity (Lacerda et al. 1993) are minimal.

Ten cores were collected from the selected plots in each station by inserting PVC core of known volume into the soil to a maximum depth of 0.40 m during low tide condition. Each core was sliced into four equal parts, placed in aluminium foil and packed in ice for transport. In the laboratory, the collected samples were carefully sieved and homogenized to remove roots and other plant and animal debris prior to oven-drying to constant weight at 105 °C. Total organic carbon was analyzed by rapid dichromate oxidation method of Walkley and Black (1934).

Measurement of soil pH was done with fresh samples in the field with a Systronics pH Meter with glass–calomel electrode (sensitivity ±0.01)

and standardized with buffer 7.0 to avoid oxidation of iron pyrites (a common constituent of mangrove soils) to sulphuric acid (English et al. 1997). Soil salinity was determined in supernatant of 1:5 soil–water mixtures using a refractometer.

2A.2.3   Results

2A.2.3.1   Soil Organic Carbon

The soil organic carbon differs significantly between stations and years considering the mean values (0.99 ± 0.07) of all four depths (Fig. 2A.2.2). We observe relatively higher values of organic carbon in the stations of anthropogenically stressed western sector (Stn. 1–7) compared to those in the central (Stn. 8–11) and eastern sectors (Stn. 12–14) of Indian Sundarbans that encompass mainly the reserve forest with almost no human intrusion. The mean value of soil organic carbon in the western sector (stressed zone) is 1.02 wt%. In the central and eastern sectors (control zone), the value is 0.64 wt%. In all the selected stations, the soil organic carbon content decreases with depth (Fig. 2A.2.3). The gradual increase of organic carbon (composite figure of four depths) through years in all the stations clearly reflects the efficacy of Sundarban mangrove as potential sink of carbon. In western Indian Sundarbans, the rate of increase is 0.031 % per year, whereas in the stations of central and eastern sectors that are mostly within the reserve forest, the value is 0.026 % per year.

2A.2.3.2   Soil pH

Soil pH decreases significantly at those sites which fringe the salt-marsh grass (*Porteresia coarctata*) bed and sustain rich mangrove vegetation particularly in the reserved forest area (Stn. 8–14). The average soil pH in this zone (considering all depths and years) is 7.47 ± 0.071. In the western sector of Indian Sundarbans (Stn. 1–7), comparatively higher soil pH is observed (average pH value of all four depths considering

Table 2A.2.1 Major activities influencing the organic carbon pool in Indian Sundarbans

Station name and Stn. no. (as in map)	Geographical location		Major activity	Magnitude
	Longitude (°E)	Latitude (°N)		
Kachuberia (1)	88°08′04.43″	21°52′26.50″	Navigational channel	+++
			Passenger vessel jetties	+++
			Shrimp culture farms	+
			Marketplace	++
Harinbari (2)	88°04′52.98″	21°47′01.36″	Mangrove patches ($n=7$; AGMB = 89 t/ha)	++
			Unorganized fishing activities	+
Sagar South (3)	88°03′06.17″	21°38′54.37″	Pilgrims	+++
			Tourism	+++
			Navigational channel	+++
			Erosion (sea facing)	+++
			Mangrove patches ($n=11$; AGMB = 94 t/ha)	++
Chemaguri (4)	88°10′07.03″	21°39′58.15″	Fish-landing stations	+++
			Tourism	+++
			Mangrove patches ($n=17$; AGMB = 71 t/ha)	++
			Shrimp culture farms	++
Frazergaunge (5)	88°15′15.63″	21°33′11.84″	Shrimp culture farms	++
			Mangrove forest ($n=17$; AGMB = 124 t/ha)	+++
			Fish-landing stations	+
			Marketplace	++
Prentice Island (6)	88°17′10.04″	21°42′40.97″	Mangrove patches ($n=18$; AGMB = 108 t/ha)	+++
Lothian Island (7)	88°22′13.99″	21°39′01.58″	Mangrove forest (protected area; $n=31$; AGMB = 136 t/ha)	+++
			Navigational channel	+++
			Erosion	+++
			Wave action	+++
Sajnekhali (8)	88°46′10.08″	22°05′13.04″	Mangrove forest (protected area; $n=31$; AGMB = 136 t/ha)	+++
			Tourism	++
Amlamethi (9)	88°44′26.07″	22°03′54.02″	Mangrove forest ($n=17$; AGMB = 148 t/ha)	+++
			Tourism (occasional)	+
			Shrimp and prawn culture farms	+
Jharkhali (10)	88°41′47.25″	22°05′52.82″	Mangrove forest (protected; $n=13$; AGMB = 141 t/ha)	+++
Dobanki (11)	88°45′20.06″	21°59′24.04″	Accretion zone	++
			Mangrove forest ($n=16$; AGMB = 112 t/ha)	+++
Netidhopani (12)	88°44′39.4″	21°55′14.9″	Mangrove forest ($n=16$; AGMB = 112 t/ha)	+++

(continued)

Table 2A.2.1 (continued)

Station name and Stn. no. (as in map)	Geographical location		Major activity	Magnitude
	Longitude (°E)	Latitude (°N)		
Haldibari (13)	88°46′44.9″	21°43′01.4″	Accretion zone	++
			Mangrove forest (n = 16; AGMB = 112 t/ha)	+++
Burirdabri (14)	89°01′43.6″	22°04′39.2″	Mangrove forest (protected area; n = 21; AGMB = 136 t/ha)	+++

Data Sources: Chaudhuri and Choudhury (1994), Mitra et al. (2010) and Mitra et al. (2011)

+, ++ and +++ indicate low, medium and high magnitude, respectively, for the major activities in the selected stations; n and *AGMB* represent number of mangrove species and above-ground mangrove biomass (t/ha) of three dominant species, respectively

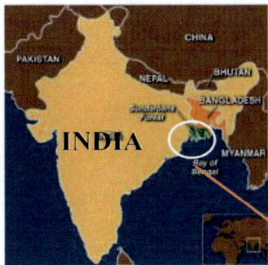

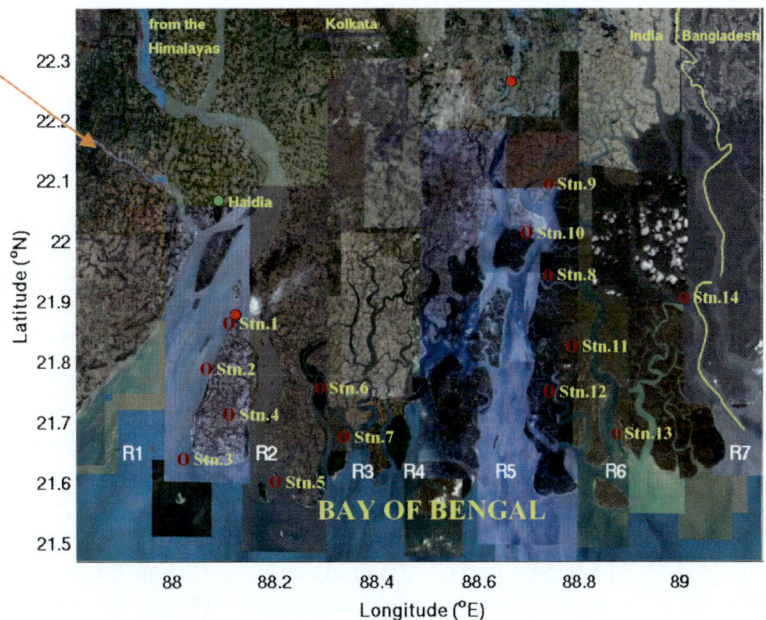

Fig. 2A.2.1 Map of the study region showing the sampling stations. R1–R7 are the seven rivers of Sundarbans starting from west to east, namely, Hugli, Mooriganga, Saptamukhi, Thakuran, Matla, Gosaba and Harinbhanga

5 successive years = 7.57 ± 0.067). We also observe a decreasing trend in soil pH with depth in all the stations (Fig. 2A.2.4). The uppermost layers are alkaline and slightly acidic soil is observed within the depth of 0.20–0.40 m mainly in the stations 8–14. Significant yearly variations of soil pH are observed in the present study area (Fig. 2A.2.5). In stations 1–7, the mean values (of all depths) are 7.67 ± 0.23, 7.60 ± 0.21, 7.55 ± 0.21, 7.53 ± 0.21 and 7.49 ± 0.19 in 2006, 2007, 2008, 2009 and 2010, respectively. In stations 8–14, the values are 7.57 ± 0.23, 7.51 ± 0.21, 7.43 ± 0.16,

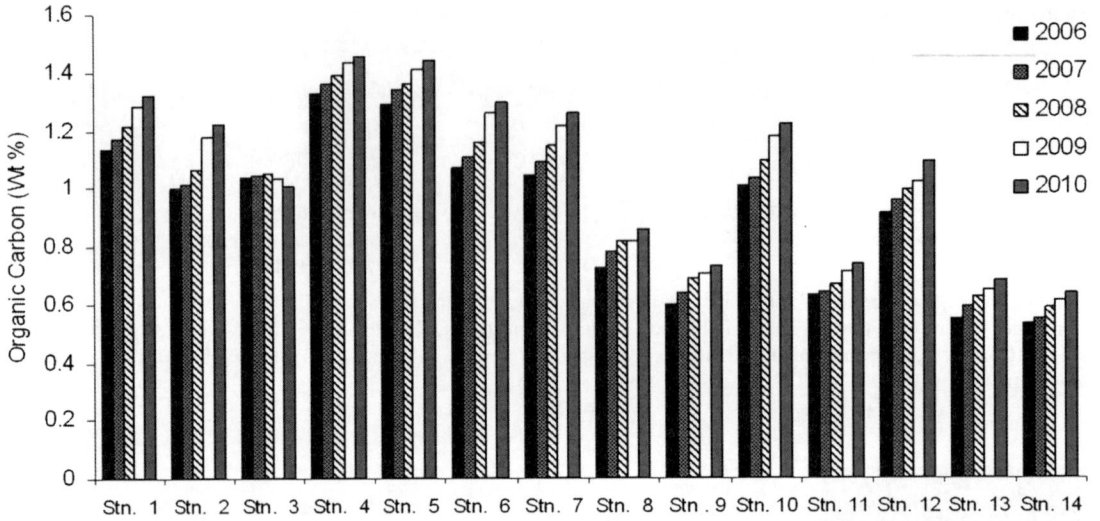

Fig. 2A.2.2 Spatial and temporal variations of soil organic carbon (wt%) during premonsoon 2006–2010 (composite value of the depths)

7.44 ± 0.21 and 7.40 ± 0.19 in 2006, 2007, 2008, 2009 and 2010, respectively.

2A.2.3.3 Soil Salinity

The soil salinity exhibits a unique spatial and temporal trend (Fig. 2A.2.6). In the western Indian Sundarbans (Stn. 1–7), the values are relatively lower (mean value of all depths and years = 9.75 psu), while in stations 8–11, adjacent to Matla River in the central sector of Indian Sundarbans, the values are relatively higher (mean value of all depths and years = 13.85 psu). Interestingly, in the eastern Indian Sundarbans encompassing stations 12–14 (adjacent to Bangladesh Sundarbans), the soil salinity again decreases significantly (mean value considering all depths and years = 6.98 psu). An apparent increase in soil salinity with depth is observed in all the stations (Fig. 2A.2.7). It is observed that the soil salinity exhibits a decreasing trend with years in stations 1–7 (10.92, 10.14, 9.80, 9.16 and 8.71 psu in 2006, 2007, 2008, 2009 and 2010, respectively) and stations 12–14 (7.70, 7.35, 7.04, 6.64 and 6.18 psu in 2006, 2007, 2008, 2009 and 2010, respectively), but the values increase in stations 8–11 (13.32, 13.52,

13.64, 14.26 and 14.49 psu in 2006, 2007, 2008, 2009 and 2010, respectively).

2A.2.4 Discussion

2A.2.4.1 Soil Organic Carbon

The significant variation ($p < 0.0001$) of soil organic carbon between anthropogenically stressed western zone and non-disturbed central and eastern zones may be attributed to a large extent by human activities, mangrove floral richness and physical factors like accretion and erosion. Anthropogenic activities like fish landing, tourism and unplanned urban development and shrimp farms contribute appreciable amount of organic load in stations like Kachuberia (Stn. 1) and Frazergaunge (Stn. 5). The presence of shrimp farms at Chemaguri (Stn. 6) along with a 12-year-old mangrove vegetation (17 species) may be attributed to highest organic carbon level in the soil core. The western Indian Sundarbans (encompassing stations 1–7) is under severe stress due to intense industrialization, rapid urbanization and unplanned tourism and aquaculture activities (Mitra et al. 1994; Mitra 1998) which contribute appreciable organic carbon in

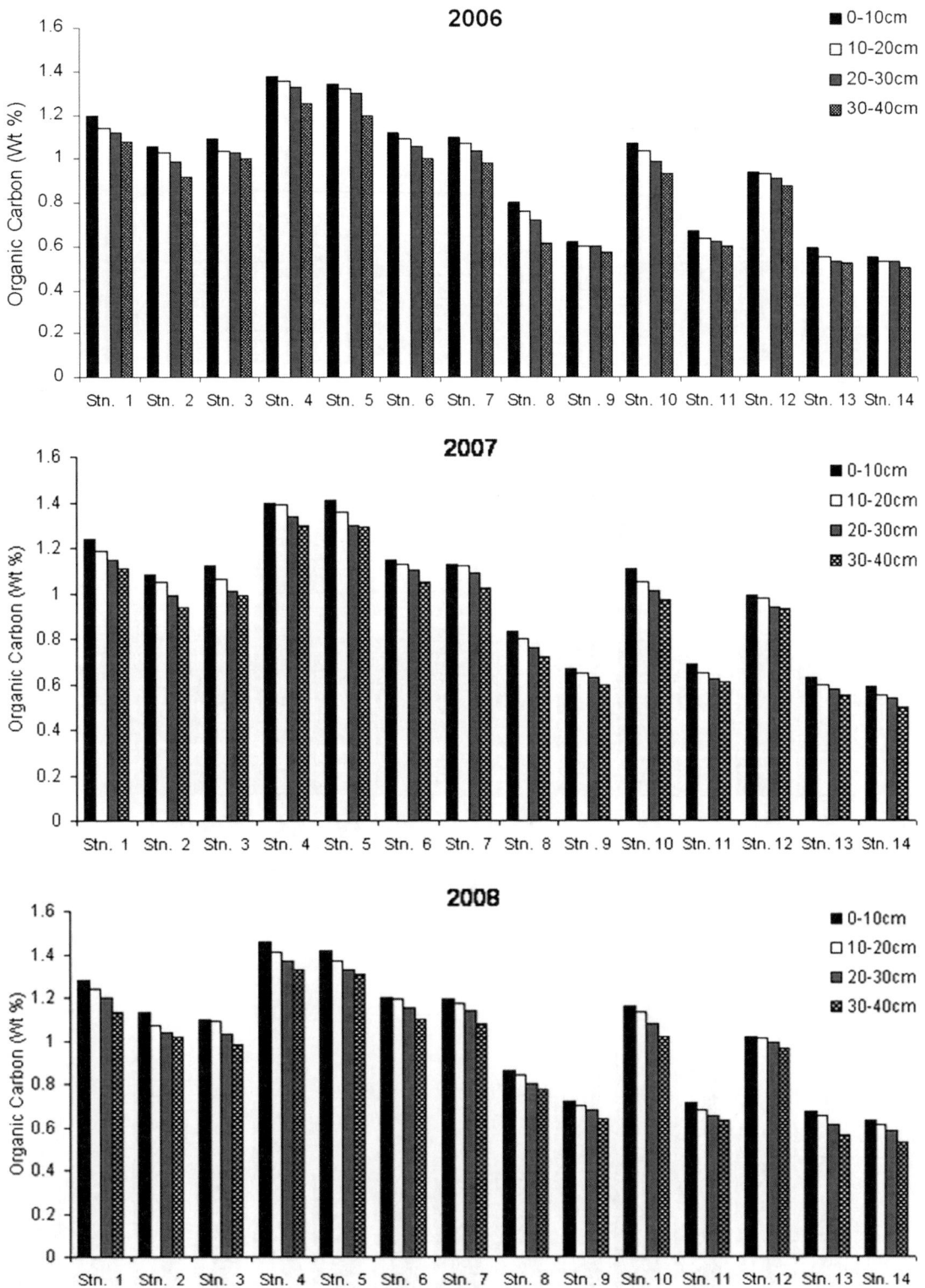

Fig. 2A.2.3 Yearly variations of soil organic carbon (wt%) with depth in the selected stations

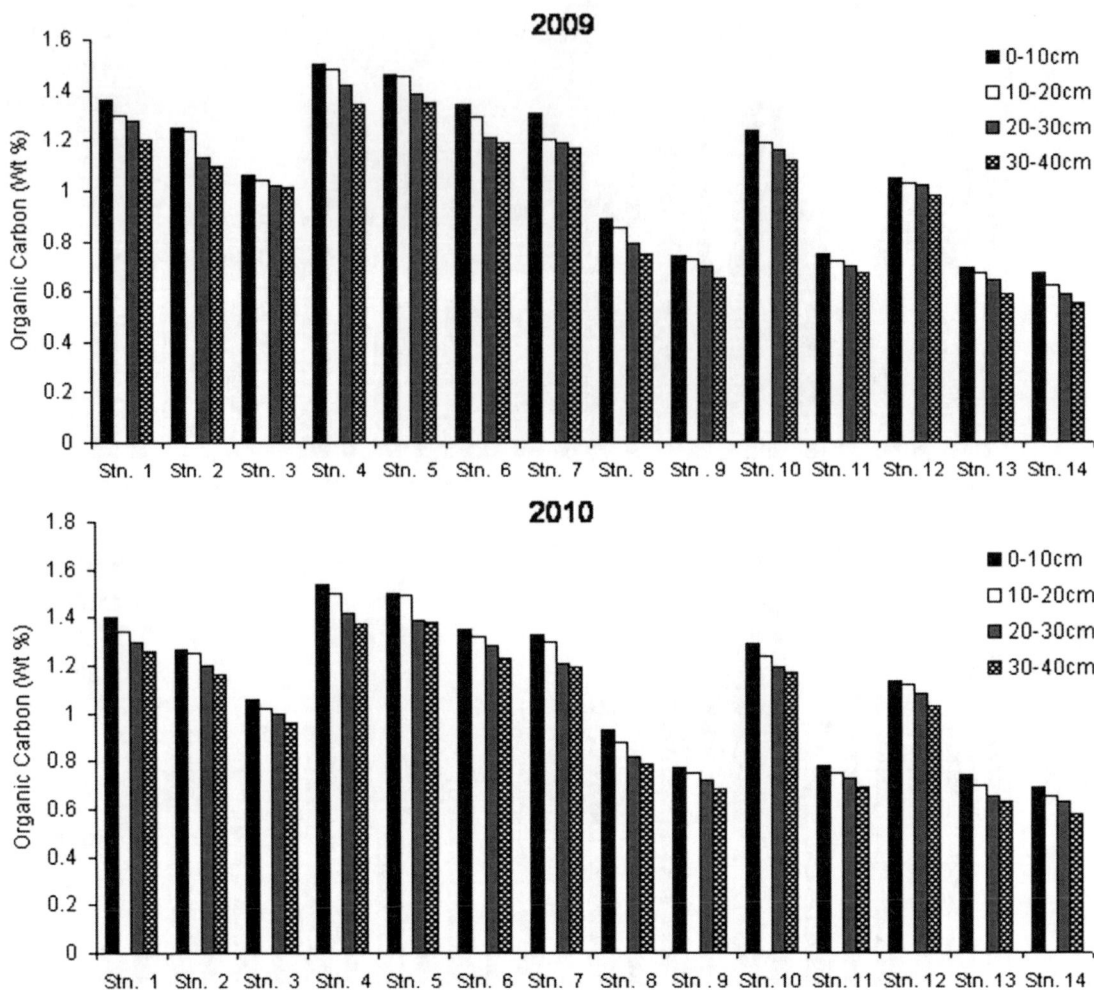

Fig. 2A.2.3 (continued)

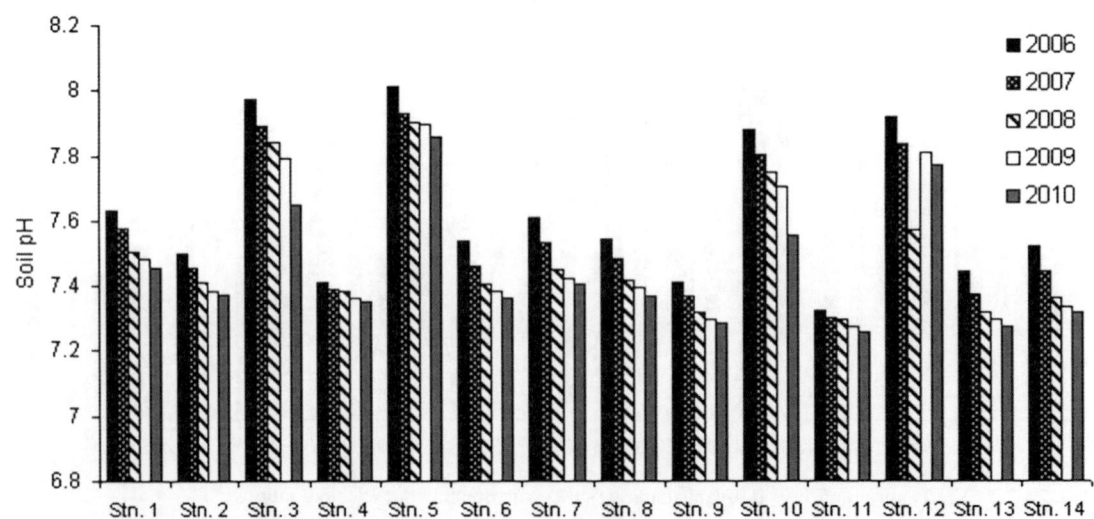

Fig. 2A.2.4 Spatial and temporal variations of soil pH during premonsoon 2006–2010 (composite value of the depths)

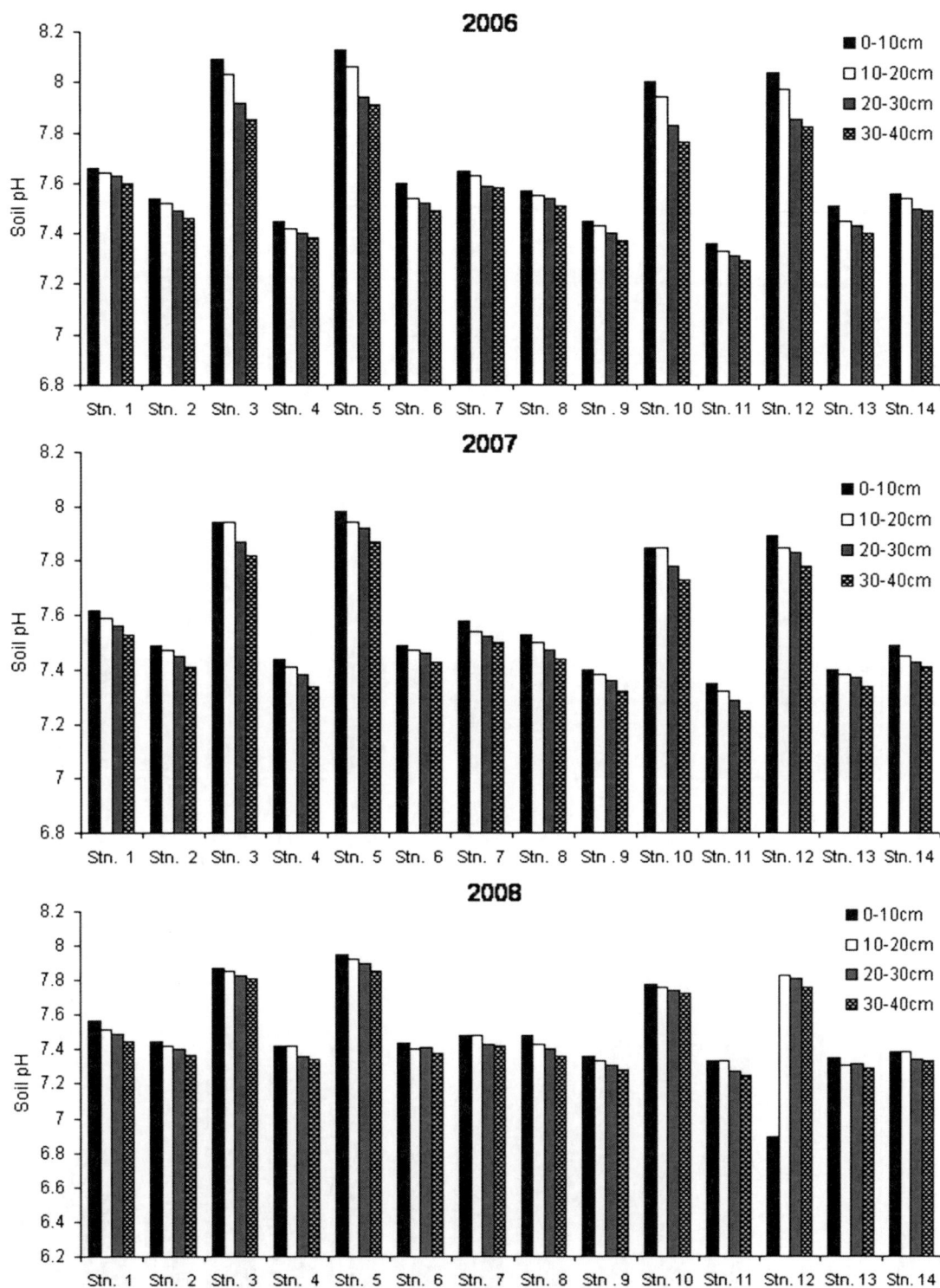

Fig. 2A.2.5 Temporal variations of soil pH with depth in the selected stations

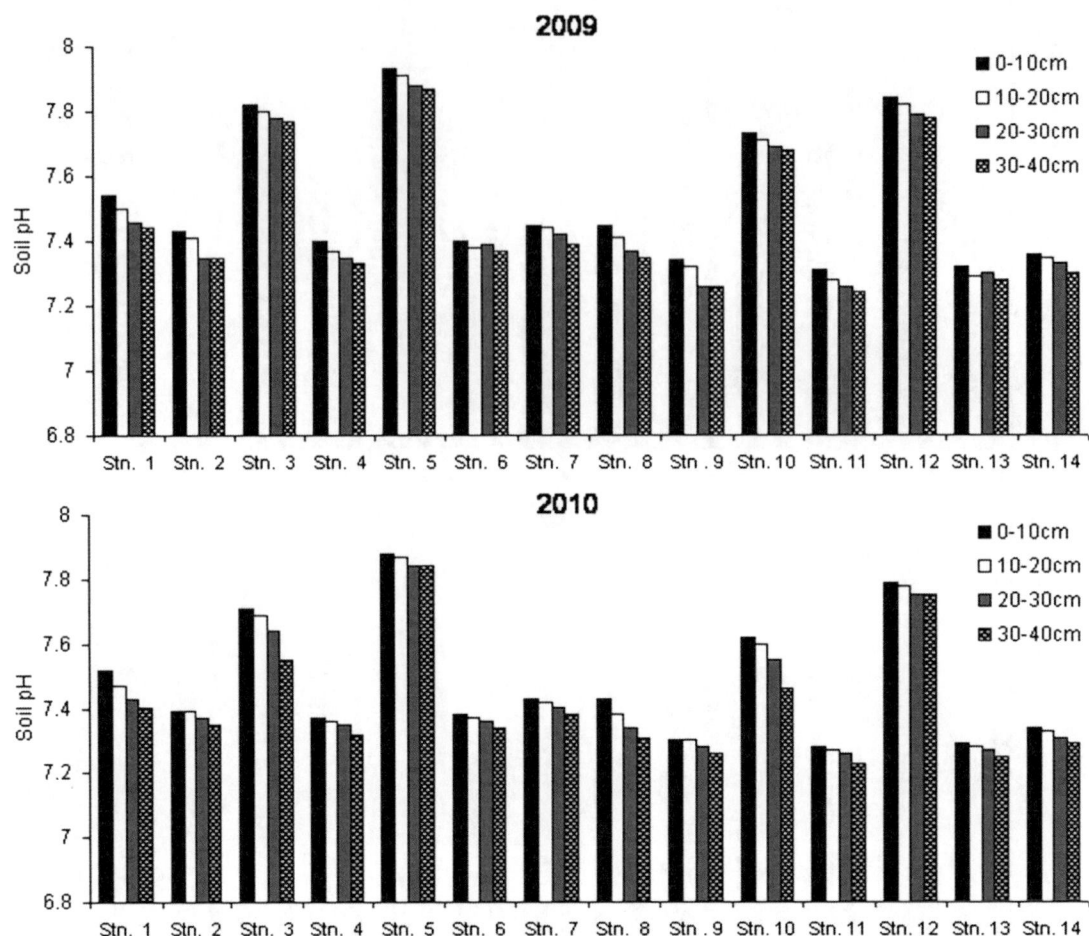

Fig. 2A.2.5 (continued)

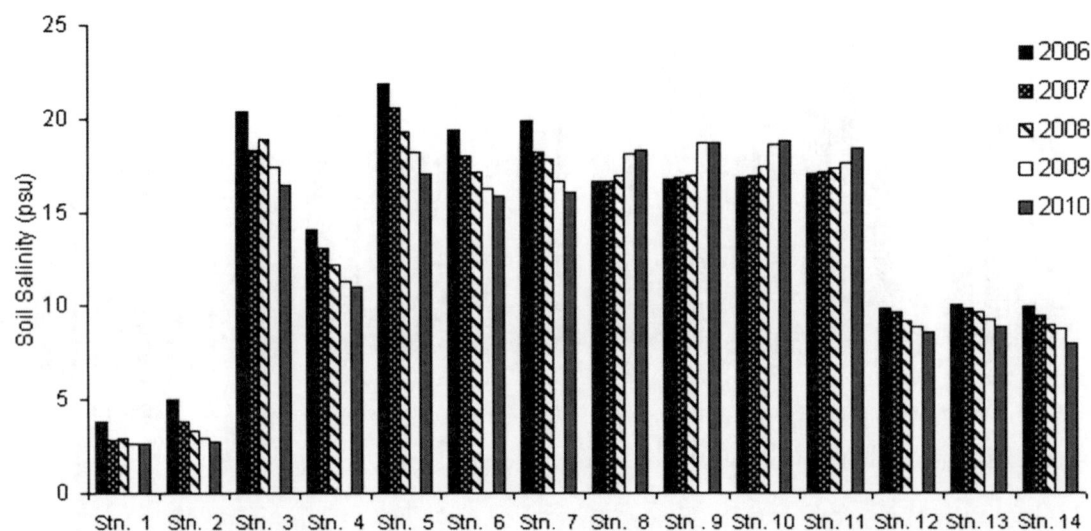

Fig. 2A.2.6 Spatial and temporal variations of soil salinity (psu) during premonsoon 2006–2010 (composite value of the depths)

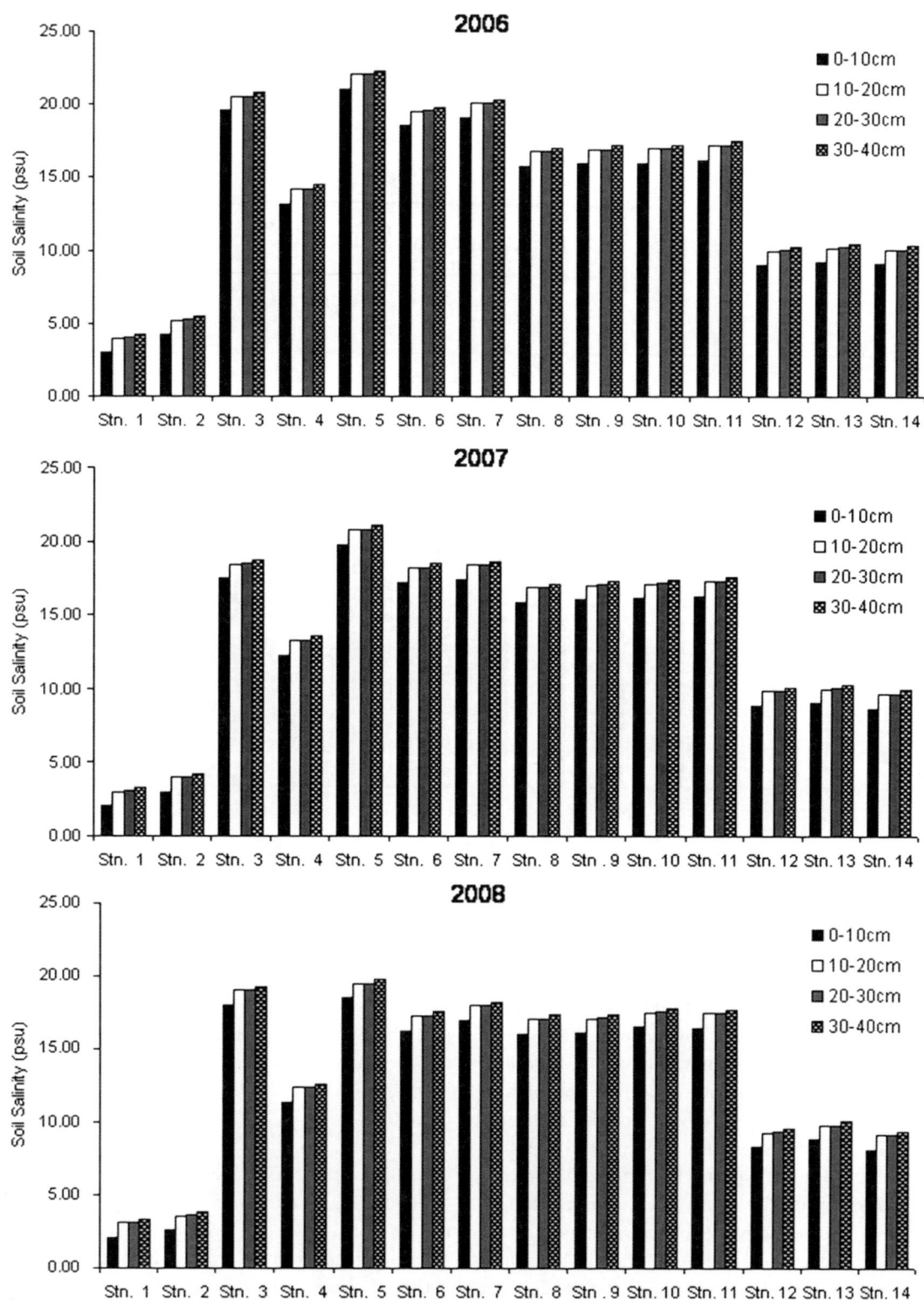

Fig. 2A.2.7 Yearly variations of soil salinity (psu) with depth in the selected stations

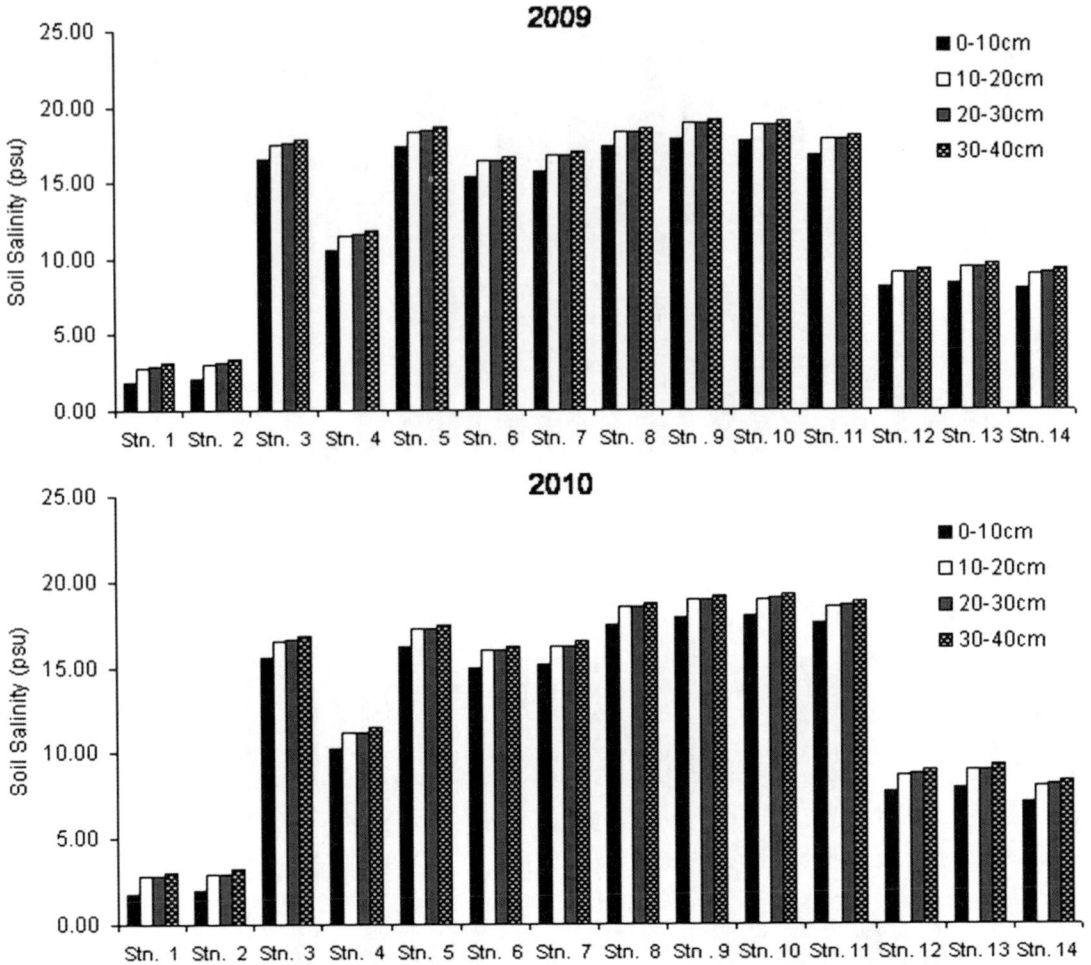

Fig. 2A.2.7 (continued)

the soil. The relatively low organic carbon at Sagar South (Stn. 3) is due to its location at seafront where wave action and tidal amplitude are maximum (range 3.0–5.0 m and mean = 3.5 m). Continuous erosion of this island may be the reason behind minimum retention of organic matter in the intertidal zone. The low organic carbon at stations 8–14 confirms the anthropogenic origin of organic load, which is almost absent in these stations (control zone). Being located within the reserve forest area, stations 8–14 receive complete legal protection, and hence, the major source of organic carbon in this zone is primarily the mangrove detritus. The variation of organic carbon in the Indian Sundarbans is thus regulated through an intricate interaction of biological, physical and anthropogenic activities (Table 2A.2.1).

The decrease in soil organic carbon with increased depth ($p < 0.0001$) is in accordance with the findings of Lacerda et al. (1995), where the organic carbon levels under *Rhizophora mangle* soil were 2.80, 2.70 and 2.70 % in the 0.01–0.05, 0.05–0.10 and 0.10–0.15 m depth, respectively. Similar trend was also observed under *Avicennia* soil (Lacerda et al. 1995). Report of decreasing mangrove soil organic carbon below 1 m is presented by Donato et al. (2011). The factor governing variation of below-ground carbon storage in mangrove soils is difficult to pinpoint (Bouillon et al. 2009; Alongi 2008) as it is not a simple function of measured flux rates but also integrates thousands of years of variable deposition, transformation and erosion dynamics associated with fluctuating sea levels and episodic disturbances (Chmura et al. 2003).

The present study is significant from the point that the area has not yet witnessed the light of documentation of soil carbon content although AGMB and carbon storage have been studied by several workers (Mitra et al. 2011a). Donato et al. (2011) quantified whole-ecosystem C storage in mangroves across a broad tract of the Indo-Pacific region, which includes the Bangladesh Sundarbans. The study however did not cover the lower Gangetic soil sustaining 38 % of the total Sundarbans in the Indian part. The present approach is thus an attempt to fill this gap area and establish a continuous 5-year baseline data of soil carbon in the mangrove-dominated Indian part of Sundarbans.

2A.2.4.2   Soil pH

The acidity of the soil influences the chemical transformation of most nutrients and their availability to plants. Most mangrove soils are well buffered, having a pH in the range of 6–7, but some have a pH as low as 5 (Clarke 1985). In this study, soil pH (7.47 ± 0.071) is lower in the reserved forest area (Stn. 8–14) that sustains rich mangrove and several associate floral species. The organic acids released from these vegetations may drive the pH of soil to lower value. In anthropogenically stressed western Indian Sundarbans, the soil pH is comparatively higher (7.57 ± 0.067). The spatial variation of soil pH is highly significant ($p < 0.0001$). The spatial variation of soil pH is highly significant ($p < 0.0001$).

A significant decrease in soil pH with depth at all locations ($p < 0.0001$) may indicate the production of organic acids and carbon dioxide by actively metabolizing mangrove roots. The surface soils are usually neutral to slightly acid in mangrove ecosystem due to the influence of alkaline estuary water (Clarke 1985), and in the present system, the value ranges from 7.90 to 8.30 depending on season (Mitra et al. 2011b). Soil pH in all the stations exhibit significant yearly variations ($p < 0.0001$), which may be attributed

to climatic factors that regulate the ambient aquatic pH through precipitation, run-off and biological phenomenon like mangrove litter fall and their subsequent decomposition in the soil of intertidal mudflats.

2A.2.4.3   Soil Salinity

Soil salinity reflects the geophysical features of the ecosystem. It is also an indicator of dilution caused by run-off, stream discharge, barrage discharge and other anthropogenic activities. The relatively low soil salinity in the stations at western sector (stations 1–7) is the effect of Farakka barrage discharge that release freshwater through the main Hugli channel. The Hugli estuary in the western Indian Sundarbans marked by the outer drainage of Ganga River system receives high volume of freshwater discharge all round the year. The annual freshwater discharge through the estuary accounts for 6,7200, 16,200 and 62,100 million ft^3 from the main channel of the River Ganga, Damodar and Roopnarayan covering an aggregate of about 11,900 km^2 of catchment area. The siltation of the Bidyadhari River since the late fifteenth century blocked the flow of freshwater in the central and eastern Indian Sundarbans. Interestingly stations 12, 13 and 14 in the eastern Indian Sundarbans exhibit low soil salinity profile and also a decreasing trend with time. This is due to proximity of these stations to Bangladesh Sundarbans that receive the maximum freshwater flow from the Himalayan glacier through the River Padma. The presence of numerous creeks and channels in the eastern most part of Indian Sundarbans may act as conveyer belt of freshwater from the Bangladesh part to the eastern Indian Sundarbans. The increase of soil salinity with depth ($p < 0.0001$) is the effect of percolation during tidal inundation of the intertidal mudflats (twice daily). The bottom layer is not washed away unlike the topsoil layer by daily tidal action which results in accumulation of salts in the bottom layer.

It is to be noted that increase of salinity in the stations adjacent to the Matla River (Stn. 8–11 in the central sector) may pose serious threat to certain mangrove species like *Heritiera fomes* (locally known as Sundari, from which the name

Sundarbans has originated). Symptoms of excess chloride include burning and firing of leaf tips or margins, bronzing, premature yellowing, abscission of leaves and, less frequently, chlorosis. Smaller leaves and slower growth also are typical. Symptoms of excess sodium also include necrotic areas on the tips, margins or interveinal areas. High salinity also results in the stunted growth of mangroves (Mitra et al. 2009, 2010, 2011a). This may have far-reaching impact on the aquatic subsystem of central Indian Sundarbans as mangrove litter and detritus, which are the primary sources of soil organic carbon, may reduce in quantum. This may eventually lead to poor productivity of the adjacent water bodies. Therefore, efforts should be made to develop better understanding of the problem so that appropriate management strategy could be adopted for improved and sustainable ecological management of the central sector of Indian Sundarbans with particular reference to siltation problem that has cut off the freshwater supply in the region.

2A.2.5 Conclusions

Few core findings are listed below:

1. The estimated mean soil organic carbon in the western Indian Sundarbans is 1.02 wt%, where the soil salinity is low due to dilution of the system with Farakka barrage discharge and run-off from highly urbanized townships and agricultural fields around the area. The average pH (7.57 ± 0.067) is relatively higher in this sector.

2. The stations in the central and eastern part of Indian Sundarbans are free from anthropogenic influences due to their locations within the reserve forest area. The luxuriant mangrove vegetation in these areas associated with salt-marsh grass (*Porteresia coarctata*) has caused low pH in the soils of intertidal mudflats (7.47 ± 0.071). The organic carbon is comparatively low (0.64 wt%) in the absence of any human activities like aquaculture, agriculture and sewage disposal.

3. It can be concluded from the soil organic carbon data that the carbon budget in the soil is mostly influenced by physical (waves, tides, erosion, accretion, etc.), biological (vegetation type and density) and anthropogenic (urbanization, barrage discharge and nature of livelihood) factors.

4. The gradual increase of soil organic carbon with time in the western Indian Sundarbans is a clear signature of anthropogenic role in regulating soil organic carbon in the present geographical locale.

References of Annexure 2A.2

Alongi DM (2008) Mangrove forests: resilience, protection from tsunamis, and responses to global climate change. Estuar Coast Shelf Sci 76:1–13

Bouillon S, Rivera-Monroy VH, Twilley RR, Kairo JG (2009) Mangroves. In: Dan Laffoley D, Grimsditch GD (eds) The management of natural coastal carbon sinks. IUCN, Gland

Cahoon DR, Reed DJ (1995) Relationship among marsh surface topography, hydroperoid and soil accretion in a deteriorating Louisiana Salt Marsh. J Coast Res 11(2):357–369

Chapman VJ (1976) Mangrove vegetation. J. Cramer, Vaduz, 447pp

Chaudhuri AB, Choudhury A (1994) Mangroves of the Sundarbans – India. IUCN – The World Conservation Union, Bangladesh

Chen R, Twilley RR (1998) A gap dynamic model of mangrove forest development along gradients of soil salinity and nutrient resources. J Ecol 86:37–52

Chen R, Twilley RR (1999) Patterns of mangrove forest structure and soil nutrient dynamics along the Shark River estuary, Florida. Estuaries 22:955–970

Chmura GL, Anisfeld SC, Cahoon DR, Lynch JC (2003) Global carbon sequestration in tidal, saline wetland soils. Glob Biogeochem Cycles 17:1111

Clarke PJ (1985) Nitrogen pools and soil characteristics of a temperate estuarine wetland in eastern Australia. Aquat Bot 23:275–290

Donato DC, Kauffman BJ, Murdiyarso D, Sofyan K, Melanie S, Markku K (2011) Mangroves amongst the most carbon-rich forests in the tropics. Nat Geosci 4:293–297

English S, Wilkinson C, Basker V (1997) Survey manual for tropical marine resources. Australian Institute of Marine Science, Townsville, pp 119–195

Eshleman KN, Hemond HF (1985) The role of soluble organics in acid base status of waters at Bickford Watershed Massachusetts. Water Resour Res 21:1503–1510

Forest Survey of India (1999) The state of forest report. Forest Survey of India, Ministry of Environment and Forests, Dehradun

Forstner U (1977) Metal concentrations in fresh water sediments, natural background and cultural effects. In: Golterman HL (ed) Interactions between sediments and freshwater. Junk, The Hague, pp 94–103

Joshi H, Ghose M (2002) Structural variability and biomass production of mangroves in Lothian Island of Sundarbans, India. In: Javed S, de Souza AG (eds) Research and management options for mangrove and saltmarsh ecosystems. ERWDA, Abu Dhabi, pp 46–158

Kamaruzzaman BY, Siti Waznanh A, Shahbuddin S, Jalal KCA, Ong MC (2010) Temporal variation of organic carbon during the premonsoon and postmonsoon season in Pahang River-estuary, Pahang, Malaysia. Orient J Chem 26(4):1309–1313

Kerekes J, Beauchamp S, Tordon R, Tremblay C, Pollock T (1986) Organic versus anthropogenic acidity in tributaries of the kejimkujik watersheds in western Nova Scotia. Water Air Soil Pollut 30:165–174

Khalaf F, Literathe P, Anderlini V (1981) Vanadium as a tracer of chronic oil pollution in the sediments of Kuwait. In: Proceedings of the 2nd international symposium on interaction between sediment and freshwater Ontario, 1981, pp 14–18

Lacerda LD, Carvalho CEV, Tanizaki KF, Ovalle ARC, Rezende CE (1993) The biogeochemistry and trace metals distribution of mangrove rhizospheres. Biotropica 25:251–256

Lacerda LD, Ittekkot V, Patchineelam SR (1995) Biochemistry of mangrove soil organic matter: a comparison between *Rhizophora* and *Avicennia* soils in South-eastern Brazil. Estuar Coast Shelf Sci 40:713–720

Lambert SM (1967) Functional relation between absorption in soil and chemical structure. J Agric Food Chem 16(2):340–343

Lugo AE (1980) Mangrove ecosystems: successional or steady state? Tropical succession. Biotropica Suppl 12:65–72

Lugo AE, Snedaker SC (1974) The ecology of mangroves. Ann Rev Ecol Syst 5:39–64

MacNae, W. (1968). A general account of the fauna and flora of mangrove swamps and forests in the Indo-West Pacific Region. Adv Mar Biol 6:73–270

Mall LP, Singh VP, Garge A, Pathak SM (1987) Ecological studies on mangrove forests of Ritchie's Archipelago in relation to substrata. Trop Ecol 28:182–197

Matilal S, Mukherjee BB, Chatterjee N, Gupta MD (1986) Studies on soil and vegetation of mangrove forests of Sundarbans. Ind J Mar Sci 15:181–184

Mitra A (1998) Status of coastal pollution in West Bengal with special reference to heavy metals. J Ind Ocn Stud 5(2):135–138

Mitra A (2000) Chapter 62: The Northeast coast of the Bay of Bengal and deltaic Sundarbans. In: Sheppard C (ed) Seas at the millennium – an environmental evaluation. Elsevier Science, Amsterdam, pp 143–157

Mitra A, Banerjee K (2005) Living resources of the sea: focus Indian Sundarbans. WWF-India, Sundarbans landscape project, Canning Field Office, West Bengal, Kolkata, India

Mitra A, Trivedi S, Choudhury A (1994) Inter-relationship between gross primary production and metal accumulation by *Crassostrea cucullata* in the Hooghly estuary. Pollut Res 13:391–394

Mitra A, Banerjee K, Bhattacharyya DP (2004) The other face of mangroves. Department of Environment, Govt. of West Bengal, Kolkata

Mitra A, Banerjee K, Sengupta K, Gangopadhyay A (2009). Pulse of climate change in Indian Sundarbans: a myth or realty. Natl Acad Sci Lett 32(1–2):1–7

Mitra A, Banerjee K, Sengupta K (2010) The affect of salinity on the mangrove growth in the lower Gangetic plain. J Coast Environ 1(1):71–82

Mitra A, Sengupta K, Banerjee K (2011a) Standing biomass and carbon storage of above-ground structures in dominant mangrove trees in the Sundarbans. For Ecol Manage 261(7):1325–1335

Mitra A, Mondal K, Banerjee K (2011b) Spatial and tidal variations of physico-chemical parameters in the lower Gangetic delta region, West Bengal, India. J Spatial Hydrol Am Spatial Hydrol Union 11(1):52–69

Moore TR, Jackson RJ (1989) Dynamics of dissolved organic carbon in forested and catchments, wetland, New Zealand, Larry River. Water Resour Res 5:1331–1339

Ovalle ARC, Rezende CE, Lacerda LD, Silva CAR (1990) Factors affecting the hydrochemistry of a mangrove tidal creek, Sepetiba Bay. Brazil Estuar Coast Shelf Sci 31:639–650

Ukpong IE (1991) The performance and distribution of species along soil salinity gradients of mangrove swamps in southeastern Nigeria. Vegetatin 95:63–70

Walkley A, Black IA (1934) An examination of the Degtjareff method for determining soil organic matter and a proposed modification of the chromic acid titration method. Soil Sci 37:29–38

Annexure 2A.3: Carbon Content in Phytoplankton Community of a Tropical Estuarine System

Abhijit Mitra, Subhro Bikash Bhattacharyya, Sufia Zaman, Kakoli Banerjee, Subhasmita Sinha, and Atanu Kumar Raha

Abstract

Seasonal variations of cell carbon content in diatoms, dinoflagellates, cyanobacteria and green algae were studied during 2011 in the western and central sectors of Indian Sundarbans, a mangrove-dominated tropical estuary in the lower Gangetic delta region. Twelve geometric shapes were assigned to 47 species documented from 12 stations to calculate the cell volume and carbon content. The carbon content of the species varied significantly with seasons and sites ($p < 0.01$). This variation may be attributed to unique seasonal variations of salinity and contrasting environmental conditions prevailing in the selected sectors of Indian Sundarbans. The western sector is relatively less saline and favourable for the growth of the phytoplankton species. The central sector is hypersaline on account of massive siltation that prevents the mixing of freshwater of the River Ganga with the River Matla. Such a hypersaline condition posed adverse effect on the cell volume and carbon content of the phytoplankton species. Among 47 phytoplankton species, cell volume and carbon content of 6 species exhibited significant negative relationships, and 2 species exhibited significant positive relationships with aquatic salinity. Our results imply that rising salinity in the central sector of Indian Sundarbans may be a threat to phytoplankton cell carbon reservoir by way of shrinking their volumes.

Keywords

Cell carbon • Indian Sundarbans • Phytoplankton • Seasonal variation

A. Mitra (✉)
Department of Oceanography, Techno India
University, Salt Lake Campus, Kolkata,
West Bengal 700 091, India
e-mail: abhijit_mitra@hotmail.com

S.B. Bhattacharyya • S. Zaman • K. Banerjee • S. Sinha
Department of Marine Science,
University of Calcutta, 35 B.C. Road,
Kolkata, West Bengal 700 019, India

A.K. Raha
Department of Environment & Forest Science,
Techno India University, Salt Lake Campus,
Kolkata, West Bengal 700 091, India

2A.3.1 Introduction

In the marine and estuarine ecosystems, free-floating microscopic photoautotrophic floral communities referred to as phytoplankton account for approximately half the production of organic matter on Earth. This community strongly influences climate processes (Boyce et al. 2010) and biogeochemical cycles (Sabine et al. 2004; Roemmich and McGowan 1995) particularly the carbon cycle (Boyce et al. 2010). Despite this far-reaching importance, empirical estimates of carbon content in this community remain limited. Biovolume and surface area calculations for phytoplankton cells are important for studying many related ecological parameters (Malone 1980; Sournia 1981; Chisholm 1992), such as biomass, growth, photosynthesis, respiration, assimilation, sinking and grazing. Mullin et al. (1966) determined cell carbon, cell volume and surface area for a variety of phytoplankton organisms and concluded that cell volume gave a better estimate of cell carbon than did surface area. Phytoplankton cell size varies greatly among different genera or even between different individuals. Sizes range from a few micrometres (or even less than 1 mm) to a few millimetres. Hence, there is a wide range of nine orders in magnitude for cell biovolume of phytoplankton. Several automated and semi-automatic methods for biovolume estimation have been described in the literature, such as the Coulter Counter (Hastings et al. 1962; Maloney et al. 1962; Boyd and Johnson 1995), the micrographic image analysis system (Gordon 1974; Krambeck et al. 1981; Estep et al. 1986), flow cytometry (Olson et al. 1985; Wood et al. 1985; Steen 1980; Cunningham and Buonnacorsi 1992) and holographic scanning technology (Brown et al. 1989). Indian Sundarbans, being a major tropical estuarine system and a World Heritage site at the apex of Bay of Bengal, has no baseline databank on cell volume and carbon content of phytoplankton although 106 species have been documented till date (Mitra et al. 2004).

In this study, the carbon content in diatoms, dinoflagellates, cyanobacteria and green algae was compared in two different physiographic regions of Indian Sundarbans (western and central sectors) having contrasting salinity levels.

2A.3.2 Methods

2A.3.2.1 Study Sites

Sundarbans delta is one of the dynamic mangrove-dominated estuarine deltas of the world (Banerjee et al. 2012), which is situated at the apex of Bay of Bengal. A major portion of this delta (62 %) lies in Bangladesh and the remaining 38 % is within the Indian subcontinent. In the Indian Sundarbans, approximately 2,069 km^2 of area is occupied by the tidal river system or estuaries, which finally end up in the Bay of Bengal. The seven main riverine estuaries from west to east are Hugli, Muriganga, Saptamukhi, Thakuran, Matla, Gosaba and Harinbhanga. The flow of the mighty Ganges River through the Hugli estuary in the western sector of the Indian Sundarbans, ending up in the Bay of Bengal, results in an ecological situation totally different from that of the central sector, where five major rivers have lost their upstream connections with the Ganges (Fig. 2A.3.1) due to heavy siltation and solid waste disposal from the adjacent cities and towns (Chakrabarti 1998; Mitra et al. 2009, 2011). Presently, the rivers in the western sector (Hugli and Muriganga) are connected to the Himalayan glaciers through the Ganges originating at the Gangotri Glacier, whereas the five central and eastern sector rivers like Saptamukhi, Thakuran, Matla, Gosaba and Harinbhanga are all tide fed. The tidal flow from the Bay of Bengal (mean salinity = 32 psu) in the south of this deltaic complex sustains these five tide-fed rivers.

We conducted seasonal survey at 12 stations in the Indian Sundarbans region (between 21°40′ N and 22°40′ N latitude and 88°03′ E and 89°07′ E longitude) during 2011 in May (premonsoon), September (monsoon) and December (postmon-

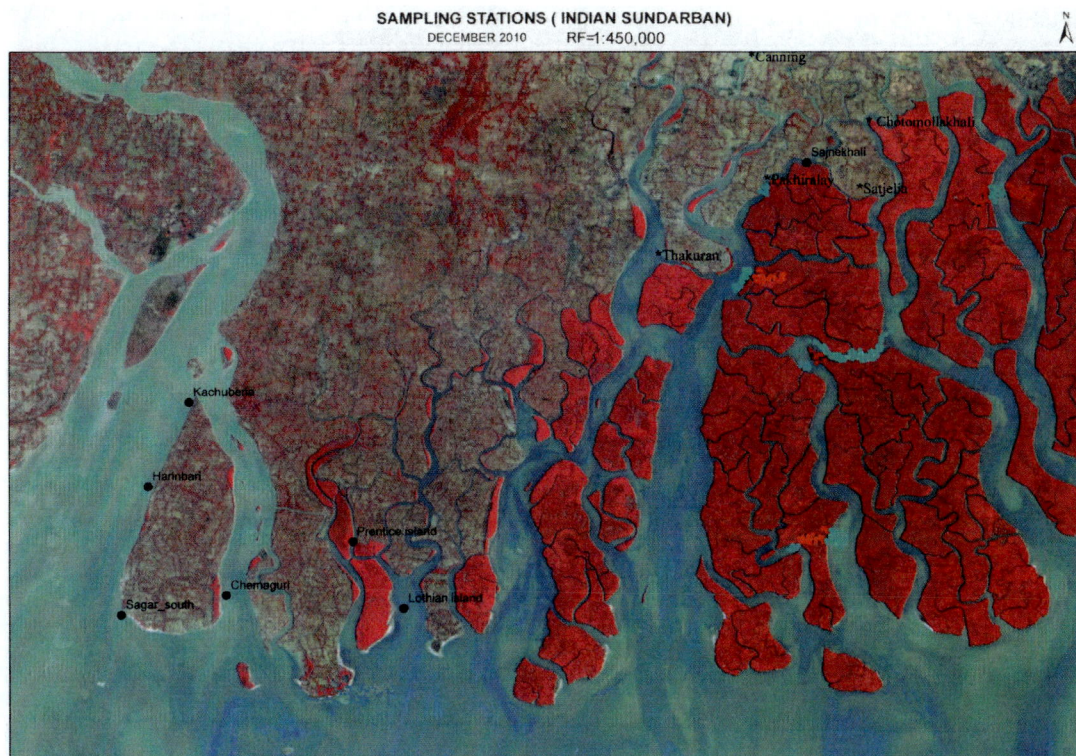

Fig. 2A.3.1 Map of Indian Sundarbans highlighting the sampling stations

Table 2A.3.1 Sampling stations with coordinates

Station name	Station code	Geographical location Longitude	Latitude
Kachuberia	Stn. 1	88°08′04.43″	21°52′26.50″
Harinbari	Stn. 2	88°04′52.98″	21°47′01.36″
Chemaguri	Stn. 3	88°10′07.03″	21°39′58.15″
Sagar South	Stn. 4	88°03′06.17″	21°38′54.37″
Lothian island	Stn. 5	88°22′13.99″	21°39′01.58″
Prentice island	Stn. 6	88°17′10.04″	21°42′40.97″
Canning	Stn. 7	88°41′16.20″	22°18′40.25″
Sajnekhali	Stn. 8	88°46′10.08″	22°05′13.04″
Choto mollakhali	Stn. 9	88°54′26.71″	22°10′40.00″
Satjelia	Stn. 10	88°52′49.51″	22°05′17.86″
Pakhiralaya	Stn. 11	88°48′29.00″	22°07′07.23″
Thakuran	Stn. 12	88°38′45.20″	21°35′33.10″

soon). Station selection was primarily based on aquatic salinity (Table 2A.3.1 and Fig. 2A.3.1). The discharge of Farakka barrage (the biggest barrage in the Gangetic plain) through Hugli channel has made the western sector of the study area (stations 1–6) relatively low saline (Mitra et al. 2009). On the contrary, stations 7–12 are high-saline zone due to complete blockage of the freshwater because of siltation of the Bidyadhari River (Chaudhuri and Choudhury 1994; Mitra et al. 2011) that used to contribute freshwater to the tidal rivers of central Indian Sundarbans in the fifteenth century (Chaudhuri and Choudhury 1994).

2A.3.2.2　Salinity

The surface water salinity in the selected stations was recorded during high tide condition by means of an optical refractometer (Atago, Japan) and cross-checked in laboratory using Mohr-Knudsen method. The correction factor was found out by titrating silver nitrate solution against standard seawater (IAPO Standard Seawater Service Charlottenlund, Slot Denmark, chlorinity = 19.376 psu). Our method was applied to estimate the salinity of standard seawater procured from NIO, and a standard deviation of 0.02 % was obtained for salinity. The average accuracy for salinity (for triplicate sampling) was ± 0.24 psu.

2A.3.2.3   Cell Volume

Net samples were collected during high tide condition (around 12.00 noon) with a conical nylon net bag (30 cm diameter) made of a 30 No. bolting silk from 12 selected stations and preserved in 4 % neutral formaldehyde. Phytoplankton samples were observed with a ZEISS research microscope coupled with an image analyzing system. Phytoplankton cell identifications were based on standard taxonomic keys (Verlencar and Desai 2004; Botes 2003). Linear dimensions of the phytoplankton species were measured on the basis of taxonomic information and shape code as provided by Sun and Liu (2003). For each species, the best fitting geometric shape and corresponding equation were used to calculate the cell volume.

2A.3.2.4   Cell Carbon

Standard mathematical expressions (specific for different phytoplankton group) were used to transform cell volume into cell carbon.

The cell volume of diatoms was converted into cell carbon as per the expression cell carbon (pg) = 0.288 [live cell volume $(\mu m^3)]^{0.811}$. For dinoflagellates, the expression cell carbon (pg) = 0.760 [live cell volume $(\mu m^3)]^{0.819}$ (Montagnes and Berges 1994; Menden-Deuer et al. 2001; Davidson et al. 2002) was used. For phytoplankton species other than dinoflagellates and diatoms, the expression $Log_{10}C = 0.76 \, Log_{10} V - 0.29$ (Mullin et al. 1966) was used to estimate carbon content (pg) per cell. The phytoplankton population (in $No.l^{-1}$) was enumerated simultaneously using Sedgewick Rafter Cell Counter as per the method of McAlice (McAlice 1971). This approach is appropriate for larger phytoplankton

species (>10–15 µ) having relatively higher population densities ($\geq 10^5$cells/l). The species-wise carbon content per unit volume of water was calculated by the product of population and mean cell carbon content of each species.

2A.3.2.5   Statistical Analysis

To explore the relationships between phytoplankton cell volume and cell carbon scatter plots, allometric equations and correlations were computed separately for diatoms, dinoflagellates, cyanobacteria and green algae (n = 225 for each group). To assess whether cell volume and carbon content varied significantly between sectors (western vs. central) and seasons, two-way ANOVA was performed. All statistical calculations were performed with SPSS 9.0 for Windows.

2A.3.3   Results

2A.3.3.1   Salinity

The stations in western and central sectors of Indian Sundarbans exhibited significant differences in aquatic salinity during the study period. In the western sector, salinity of surface water ranged from 1.89 psu (at station 1 during September 2011) to 27.99 psu (at station 6 during May 2011), and the average salinity was 14.98 ± 9.16 psu. In the central sector, the lowest salinity was recorded at station 7 (4.01 psu during September 2011) and the highest salinity was at station 10 (29.98 psu during May 2011) with an average value of 20.60 ± 8.82 psu. In both the sectors, the seasonal trend in salinity was premonsoon > monsoon > postmonsoon (Fig. 2A.3.2).

ANOVA computed for aquatic salinity (Table 2A.3.2) exhibits significant spatial (western vs. central Indian Sundarbans) and temporal (seasonal) variations ($p < 0.05$).

2A.3.3.2   Cell Volume

A total of 47 phytoplankton taxa were identified to species level, which exhibited 12 different

Fig. 2A.3.2 Spatial variations of salinity during 2010–2011 in selected stations

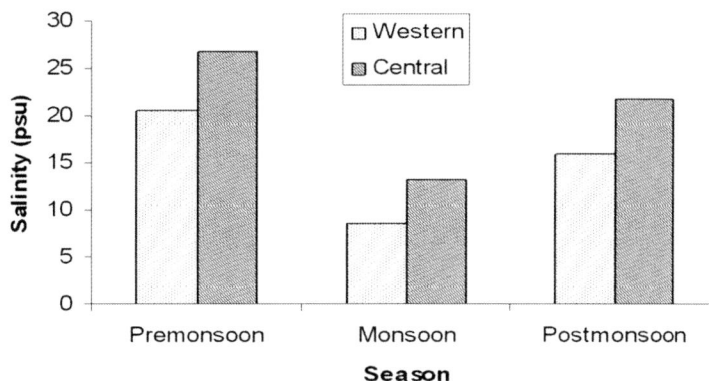

Table 2A.3.2 ANOVA (spatial and seasonal variations) of aquatic salinity, phytoplankton cell volume and phytoplankton cell carbon content

Variables		F_{cal}	F_{crit}
Salinity			
Between sectors		101.07	18.51
Between seasons		177.00	19
	Phytoplankton group		
Cell volume			
Between sectors	Diatom	4,899.41	18.51
	Dinoflagellates	59.78	18.51
	Cyanobacteria and green algae	467.54	18.51
Between seasons	Diatom	703.76	19
	Dinoflagellates	1,440.36	19
	Cyanobacteria and green algae	120.22	19
Cell carbon			
Between sectors	Diatom	4,811.32	18.51
	Dinoflagellates	59.71	18.51
	Cyanobacteria and green algae	66.27	18.51
Between seasons	Diatom	1,034.39	19
	Dinoflagellates	903.76	19
	Cyanobacteria and green algae	18.49	19

geometric shapes. The cell volume ranged from 86.89 μm^3 (*Hemidiscus hardmanninus*) to 271,405.63 μm^3 (*Planktoniella sol*) in the western sector and from 85.39 μm^3 (*Asterionella japonica*) to 227,153.21 μm^3 (*Planktoniella sol*) in the central sector (Table 2A.3.3).

We observed relatively higher cell volume in the hyposaline western Indian Sundarbans com-

pared to the hypersaline central sector. Almost all the phytoplankton species (except *Rhizosolenia alata*, *Ceratium trichoceros*, *Ceratium furca*, *Nitzschia closterium*, *Trichodesmium erythraea* and *Chlorella marina*) exhibited relatively higher cell volume in the western sector (Table 2A.3.3).

In most cases, the cell volume showed unique seasonal variations in both the sectors as per the order monsoon > postmonsoon > premonsoon (exceptional species: *Rhizosolenia crassipina*, *Chaetoceros peruvianus* and *C. compressus* in central Indian Sundarbans).

ANOVA results indicate significant seasonal and sectoral differences in cell volume for all the three groups of phytoplankton (Table 2A.3.2).

2A.3.3.3 Cell Carbon

The cell carbon content of the phytoplankton ranged from 10.76 pg (*Hemidiscus hardmanninus*) to 7,346.10 pg (*Planktoniella sol*) in the western sector and 10.61 pg (*Asterionella japonica*) to 6,358.68 pg (*Planktoniella sol*) in the central sector (Table 2A.3.4). The relatively higher cell carbon content in the western sector (with some exceptions like *Rhizosolenia alata*, *Ceratium trichoceros*, *Ceratium furca*, *Nitzschia closterium*, and *Trichodesmium erythraea*) compared to the central sector is due to higher cell volume of the phytoplankton in the estuaries of western Indian Sundarbans.

With few exceptional species like *Rhizosolenia crassipina*, *Chaetoceros peruvianus* and *C. compressus* in central Indian Sundarbans, all species

Table 2A.3.3 Seasonal variation of cell volume (in μm^3) of phytoplankton in Indian Sundarbans during 2010–2011

Sl. No.	Species	Western sector			Central sector		
		Premonsoon	Monsoon	Postmonsoon	Premonsoon	Monsoon	Postmonsoon
1.	*Coscinodiscus eccentricus*	10,637.39	14,968.71	12,477.54	8,952.38	12,632.06	10,132.49
2.	*Coscinodiscus jonesianus*	12,132.22	13,996.13	12,336.21	8,797.29	10,700.41	9,133.41
3.	*Coscinodiscus lineatus*	4,196.70	4,778.10	4,317.48	2,250.07	2,762.49	2,384.49
4.	*Coscinodiscus radiatus*	4,4451.19	45,918.19	45,240.19	31,950.14	35,252.09	33,519.31
5.	*Coscinodiscus gigas*	4,266.90	4,942.07	4,387.07	4,158.55	4,699.30	4,316.18
6.	*Coscinodiscus oculusiridis*	3,839.30	4,551.30	3,978.08	3,404.25	4,046.49	3,522.94
7.	*Planktoniella sol*	264,994.76	271,405.63	269,383.6	218,202.63	227,153.21	225,131.2
8.	*Cyclotella striata*	25,031.36	26,984.78	25,962.78	21,894.74	24,735.60	23,381.79
9.	*Thalassiosira subtilis*	770.75	1,327.55	966.22	651.18	1,209.82	867.82
10.	*Ceratium tripos*	357.90	580.75	442.07	324.78	533.65	413.65
11.	*Skeletonema costatum*	1,503.43	1,738.43	1,636.43	1,255.07	1,555.88	1,420.80
12.	*Paralia sulcata*	19,855.18	21,165.56	20,709.56	19,240.80	20,768.98	20,300.37
13.	*Rhizosolenia crassipina*	3,764.91	4,726.57	4,141.13	3,757.13	4,301.63	3,603.63
14.	*Rhizosolenia setigeara*	4,341.69	5,465.69	4,760.69	4,171.14	5,334.15	4724.94
15.	*Rhizosolenia alata*	3,726.28	4,370.98	4,098.28	3,764.90	4,590.08	4,152.08
16.	*Ceratium teres*	1,186.60	1,711.99	1,395.95	1,005.77	1,474.05	1,185.05
17.	*Ceratium trichoceros*	994.36	1,280.98	1,160.36	1,009.27	1,284.20	1,170.16
18.	*Bacteriastrum delicatulum*	39,595.75	41,915.00	40,684.01	36,521.99	39,088.83	37,792.22
19.	*Bacteriastrum varians*	38,028.52	42,328.52	41,028.52	37,024.92	42,018.74	40,750.11
20.	*Bacteriastrum comosum*	36,959.33	40,688.09	39,759.33	35,577.78	39,671.08	38,358.4
21.	*Chaetoceros didymus*	4,287.00	5,705.20	4,652.00	2,241.10	3,861.23	2,259.56
22.	*Chaetoceros peruvianus*	3,844.48	4,727.96	4,161.48	1,901.92	2,565.09	1,832.76
23.	*Chaetoceros compressus*	3,170.71	4,215.71	3,467.71	1,649.00	2,467.42	1,634.5

24.	*Ditylum sol*	1,678.78	2,048.78	1,863.78	1,644.05	2,018.14	1,838.47
25.	*Triceratium favus*	2,552.06	3,557.06	2,856.06	2,151.66	2,704.98	2,338.18
26.	*Triceratum reticulatum*	3,704.79	4,414.38	3,937.79	3,188.87	3,821.97	3,478.13
27.	*Biddulphia sinensis*	16,700.84	19,441.51	17,395.84	15,691.40	17,228.50	16,495.92
28.	*Biddulphia mobiliensis*	27,305.85	29,792.85	28,805.85	26,456.24	29,066.02	28,315.68
29.	*Hemidiscus hardmanninus*	86.89	264.89	112.89	85.80	242.27	108.37
30.	*Climacosphenia elongate*	8,439.61	9,428.61	8,975.61	8,316.51	9,318.79	8,852.51
31.	*Fragillaria oceanica*	33,704.85	30,691.85	32,711.80	33,211.39	30,309.34	31,155.48
32.	*Rhaphoneis amphiceros*	3,618.27	4,539.27	3,783.27	3,213.38	3,901.00	3,290.73
33.	*Thalassionema nitzschioides*	357.36	598.70	442.36	212.06	404.23	283.30
34.	*Thalassiothrix longissima*	2,079.97	2,613.97	2,279.97	1,981.33	2,507.89	2,152.19
35.	*Thalassiothrix fraunfeldii*	356.87	521.10	425.87	240.10	409.46	299.45
36.	*Asterionella japonica*	98.89	209.70	141.22	85.39	197.536	127.60
37.	*Ceratium extensum*	159.20	283.72	193.72	154.16	266.12	185.46
38.	*Gyrosigma balticum*	2,854.95	3,230.95	2,932.95	2,548.63	2,969.27	2,654.14
39.	*Pleurosigma normanii*	1,396.71	1,750.71	1,520.71	1,216.92	1,558.98	1,327.03
40.	*Pleurosigma elongatum*	49,151.35	55,262.35	53,251.35	41,648.81	48,377.66	45,702.66
41.	*Diploneis smithii*	2,404.30	2,924.06	2,674.95	2,228.71	2,731.72	2,480.27
42.	*Cymbella marina*	391.05	630.09	506.72	386.43	613.50	492.50
43.	*Nitzschia sigma*	404.62	694.05	537.62	318.33	629.39	449.50
44.	*Nitzschia closterium*	425.50	718.91	581.50	439.20	732.92	587.80
45.	*Ceratium furca*	442.74	724.22	586.74	475.34	745.29	613.20
46.	*Trichodesmium erythraea*	3,667.96	4,232.77	3,927.96	4,479.54	5,170.93	4,868.80
47.	*Chlorella marina*	568.54	342.54	546.54	585.46	358.90	561.50

Table 2A.3.4 Mean seasonal variation of phytoplankton cell carbon content (in pg) in estuarine water of Indian Sundarbans during 2010–2011

Sl. No.	Species	Cell carbon (in pg)					
		Western			Central		
		Premonsoon	Monsoon	Postmonsoon	Premonsoon	Monsoon	Postmonsoon
1.	Coscinodiscus eccentricus	531.08	700.59	604.44	461.76	610.50	510.54
2.	Coscinodiscus jonesianus	590.84	663.45	598.88	455.26	533.63	469.32
3.	Coscinodiscus lineatus	249.79	277.50	255.60	150.67	177.95	157.93
4.	Coscinodiscus radiatus	1,693.66	1,738.85	1,717.99	1,295.75	1,403.33	1347.12
5.	Coscinodiscus gigas	253.17	285.20	258.94	247.94	273.79	255.54
6.	Coscinodiscus oculusiridis	232.39	266.77	239.18	210.79	242.51	216.74
7.	Planktoniella sol	7,205.07	7,346.10	7,301.69	6,154.72	6,358.68	6312.74
8.	Cyclotella striata	1,063.07	1,129.87	1,095.04	953.69	1,052.88	1005.89
9.	Thalassiosira subtilis	63.19	98.22	75.90	55.12	91.09	69.58
10.	Ceratium tripos	93.83	139.48	111.55	86.66	130.15	105.64
11.	Skeletonema costatum	108.64	122.22	116.38	93.85	111.70	103.78
12.	Paralia sulcata	880.98	927.85	911.60	858.81	913.73	896.97
13.	Rhizosolenia crassipina	228.35	275.07	247.10	228.73	254.84	220.75
14.	Rhizosolenia setigeara	256.76	309.47	276.68	248.55	303.42	274.99
15.	Rhizosolenia alata	226.83	258.17	245.03	228.73	268.61	247.63
16.	Ceratium teres	250.42	338.10	286.06	218.70	299.10	250.15
17.	Ceratium trichoceros	216.67	266.61	245.88	219.33	267.17	247.57
18.	Bacteriastrum delicatulum	1,542.00	1,614.86	1,576.29	1,444.19	1,525.97	1484.79
19.	Bacteriastrum varians	1,492.32	1,627.77	1,587.10	1,460.29	1,618.09	1578.36
20.	Bacteriastrum comosum	1,458.19	1,576.41	1,547.17	1,413.83	1,544.38	1502.80
21.	Chaetoceros didymus	254.14	320.43	271.55	150.18	233.47	151.18
22.	Chaetoceros peruvianus	232.65	275.14	248.08	131.47	167.56	127.58
23.	Chaetoceros compressus	198.99	250.70	213.98	117.10	162.37	116.26

24.	*Ditylum sol*	118.81	139.64	129.33	116.82	137.95	127.89
25.	*Triceratium favus*	166.87	218.44	182.82	145.30	174.94	155.44
26.	*Triceratum reticulatum*	225.77	260.24	237.22	199.91	231.54	214.49
27.	*Biddulphia sinensis*	765.66	866.06	791.39	727.90	785.22	758.03
28.	*Biddulphia mobiliensis*	1140.76	1,224.33	1,191.33	1,111.89	1,200.05	1174.86
29.	*Hemidiscus hardmanninus*	10.76	27.58	13.30	10.65	24.72	12.87
30.	*Climacosphenia elongate*	440.19	481.58	462.73	434.98	477.03	457.58
31.	*Fragilaria oceanica*	1,353.16	1,109.63	1,217.93	1,437.08	1,137.37	1400.20
32.	*Rhaphoneis amphiceros*	221.48	266.20	229.64	201.16	235.42	205.08
33.	*Thalassionema nitzschioides*	33.88	51.49	40.28	22.19	37.44	28.06
34.	*Thalassiothrix longissima*	141.36	170.15	152.29	135.90	164.52	145.33
35.	*Thalassiothrix fraunfeldii*	33.84	46.00	39.06	24.54	37.83	29.36
36.	*Asterionella japonica*	11.95	21.99	15.96	10.61	20.95	14.70
37.	*Ceratium extensum*	48.33	77.58	56.76	47.07	73.61	54.77
38.	*Gyrosigma balticum*	182.76	202.05	186.79	166.69	188.67	172.26
39.	*Pleurosigma normanii*	102.35	122.92	109.66	91.53	111.89	98.19
40.	*Pleurosigma elongatum*	1,837.50	2,020.70	1,960.86	1,606.54	1,814.00	1732.29
41.	*Diploneis smithii*	158.99	186.34	173.36	149.50	176.34	163.05
42.	*Cymbella marina*	36.45	53.67	44.97	36.09	52.52	43.95
43.	*Nitzschia sigma*	37.47	58.04	47.19	30.85	53.62	40.80
44.	*Nitzschia closterium*	39.03	59.72	50.29	40.05	60.67	50.73
45.	*Ceratium furca*	111.69	167.13	140.66	118.38	171.01	145.83
46.	*Trichodesmium erythraea*	262.39	292.57	276.42	305.45	340.65	325.42
47.	*Chlorella marina*	63.62	47.94	60.15	65.06	52.40	61.38

in the study area exhibited a sharp seasonal trend in cell carbon content with highest value during monsoon followed by postmonsoon and premonsoon.

ANOVA results indicate significant differences in cell carbon content ($p < 0.01$) between seasons and sectors in all three groups of phytoplankton (Table 2A.3.2).

2A.3.4   Discussion

Aquatic salinity seems to be the key player in regulating the cell volumes and carbon content of phytoplankton in the present study area. The relatively lower salinity in the western sector of the Sundarban delta region (Indian part) may be attributed to Farakka barrage that releases freshwater on regular basis through Ganga–Bhagirathi–Hugli River system. The central sector, on the contrary, does not receive the riverine discharge due to massive siltation of the Bidyadhari River that has blocked the freshwater flow in the region (Mitra et al. 2009; 2011; Raha et al. 2012). Ten-year survey (1999–2008) on water discharge from Farakka barrage revealed an average discharge of $(3.1 \pm 1.2) \times 10^3$ m^3/s. Higher discharge values were observed during the monsoon with an average of $(2.9 \pm 1.2) \times 10^3$ m^3/s, and the maximum of the order 4,185 m^3/s during freshet (September). Considerably lower discharge values were recorded during premonsoon with an average of $(1.0 \pm 0.09) \times 10^3$ m^3/s, and the minimum of the order 820 m^3/s during May. During postmonsoon discharge, values were moderate with an average of $(1.9 \pm 0.95) \times 10^3$ m^3/s. The lower Gangetic deltaic lobe also experiences considerable rainfall (1,400 mm average rainfall) and surface run-off from the 60,000 km^2 catchment areas of Ganga–Bhagirathi–Hugli system and their tributaries. All these factors (dam discharge + precipitation + run-off) increase the dilution factor of the Hugli estuary in the western sector of Indian Sundarbans (Mitra et al. 2011). The central sector does not receive the freshwater input on account of siltation of the Bidyadhari River since the fifteenth century, and the stations in this sector (stations 7–12)

receive only the tidal waters from the Bay of Bengal. Such significant variations in salinity within the same deltaic lobe may be the probable reason for variation in phytoplankton cell volume. We observed significant negative relationships between aquatic salinity and cell volume of *Coscinodiscus eccentricus, C. jonesianus, C. lineatus, C. radiatus, C. oculusiridis* and *Planktoniella sol* (Table 2A.3.5).

It is possible that the salinity has a direct effect on cell morphogenesis of certain phytoplankton species. Hildebrandt et al. (2006) observed that the height of the centric diatom is reduced at increased salinity. Similar observations were also reported by other researchers (Hildebrand et al. 2006; Roubeix and Lancelot 2008) who observed a lower size of the species *Thalassiosira pseudonana* when grown at higher NaCl concentration. Many researchers put forward several views on the negative impact of salinity on phytoplankton cell volume. According to some researchers (Pickett-Heaps et al. 1990; Harold 2002), the elongation of diatom cells during the interphase preceding division is driven by turgor pressure, which makes the siliceous components of the cell walls slide apart. At increased salinity, freshwater diatom might not be able to produce the intracellular osmolarity needed to generate the same turgor pressure as at low salinity. Thus, if cell elongation is less efficient before each cell division, cell height might decrease faster in high-saline water (Roubeix and Lancelot 2008). This may be a possible cause for lowering of cell volume of the species *Coscinodiscus eccentricus, C. jonesianus, C. lineatus, C. radiatus, C. oculusiridis* and *Planktoniella sol* with the increase of aquatic salinity. However, two species *Fragilaria oceanica* and *Chlorella marina* exhibited significant positive relationships with aquatic salinity (Table 2A.3.5) and confirm the tolerance of the species to high salinity. Similar results were obtained while conducting the growth experiment on *Cyclotella meneghiniana* (Pickett-Heaps et al. 1990). According to a classification type of phytoplankton (Harold 2002), few species of phytoplankton are holoeuryhaline in nature that are able to grow from almost freshwater to marine conditions with a wide range of tolerance. Such

Table 2A.3.5 Inter-relationship between phytoplankton cell volume and salinity during 2010–2011

Species	Correlation coefficient (r) between phytoplankton cell volume and salinity			p value		
	Premonsoon	Monsoon	Postmonsoon	Premonsoon	Monsoon	Postmonsoon
Coscinodiscus eccentricus	−0.4425	−0.9088	−0.6294	<0.01	<0.01	<0.01
Coscinodiscus jonesianus	−0.5355	−0.5821	−0.5055	<0.01	<0.01	<0.01
Coscinodiscus lineatus	−0.6674	−0.6504	−0.6250	<0.01	<0.01	<0.01
Coscinodiscus radiatus	−0.6718	−0.5723	−0.6274	<0.01	<0.01	<0.01
Coscinodiscus gigas	0.0654	−0.1021	0.0790	IS	IS	IS
Coscinodiscus oculusiridis	−0.7708	−0.7077	−0.7550	<0.01	<0.01	<0.01
Planktoniella sol	−0.8030	−0.8463	−0.7965	<0.01	<0.01	<0.01
Cyclotella striata	−0.3647	−0.3334	−0.3611	IS	IS	IS
Thalassiosira subtilis	−0.2364	−0.2766	−0.2103	IS	IS	IS
Ceratium tripos	−0.1692	−0.1171	−0.1924	IS	IS	IS
Skeletonema costatum	−0.1591	−0.0072	−0.0927	IS	IS	IS
Paralia sulcata	−0.3471	−0.3547	−0.3450	IS	IS	IS
Rhizosolenia crassipina	−0.0707	−0.4175	−0.3827	IS	IS	IS
Rhizosolenia setigeara	−0.4142	−0.3765	−0.3360	IS	IS	IS
Rhizosolenia alata	−0.3186	0.0466	−0.2770	IS	IS	IS
Ceratium teres	−0.1247	−0.2556	−0.1973	IS	IS	IS
Ceratium trichoceros	−0.3453	−0.1612	−0.2761	IS	IS	IS
Bacteriastrum delicatulum	−0.2859	−0.3076	−0.2709	IS	IS	IS
Bacteriastrum varians	−0.1518	0.0903	0.1061	IS	IS	IS
Bacteriastrum comosum	−0.2296	−0.2015	−0.2596	IS	IS	IS
Chaetoceros didymus	−0.3033	−0.3837	−0.3691	IS	IS	IS
Chaetoceros peruvianus	−0.3026	−0.2977	−0.3872	IS	IS	IS
Chaetoceros compressus	−0.2859	−0.1558	−0.3836	IS	IS	IS
Ditylum sol	−0.0872	−0.0748	−0.0623	IS	IS	IS
Triceratium favus	−0.3937	−0.0605	−0.3720	IS	IS	IS
Triceratum reticulatum	−0.4089	−0.1785	−0.3317	IS	IS	IS
Biddulphia sinensis	−0.3512	−0.0733	−0.3595	IS	IS	IS
Biddulphia mobiliensis	−0.1966	−0.1630	−0.0486	IS	IS	IS
Hemidiscus hardmanninus	−0.3699	0.3786	−0.3962	IS	IS	IS
Climacosphenia elongate	−0.3814	−0.3488	−0.3487	IS	IS	IS
Fragilaria oceanica	0.6087	0.6917	0.6494	<0.01	<0.01	<0.01
Rhaphoneis amphiceros	−0.3125	0.0127	−0.4025	IS	IS	IS
Thalassionema nitzschioides	−0.3351	−0.1317	−0.3896	IS	IS	IS
Thalssiothrix longissima	0.1337	−0.2601	−0.3458	IS	IS	IS
Thalassiothrix fraunfeldii	−0.3426	−0.2797	−0.3882	IS	IS	IS
Asterionella japonica	−0.2804	−0.3653	−0.3392	IS	IS	IS
Ceratium extensum	0.0439	0.3595	−0.3795	IS	IS	IS
Gyrosigma balticum	−0.3085	−0.3851	−0.3732	IS	IS	IS
Pleurosigma normanii	−0.2107	−0.2793	−0.2186	IS	IS	IS
Pleurosigma elongatum	−0.2648	−0.2591	−0.2329	IS	IS	IS
Diploneis smithii	−0.0587	−0.3192	−0.2034	IS	IS	IS
Cymbella marina	0.0415	−0.3165	−0.3705	IS	IS	IS
Nitzschia sigma	−0.3280	−0.3674	−0.3703	IS	IS	IS
Nitzschia closterium	−0.3477	−0.3463	−0.3847	IS	IS	IS
Ceratium furca	−0.2220	−0.3118	−0.3300	IS	IS	IS
Trichodesmium erythraea	0.3508	0.4193	0.3436	IS	IS	IS
Chlorella marina	0.7043	0.7633	0.7201	<0.01	<0.01	<0.01

species may serve as ideal indicators of salinity through variation of their cell size and volume. The cell volume may, however, vary depending on the adaptive efficiency of the species through variation of turgor pressure and cell morphogenesis at varying salinity.

One of the interesting aspects in the present estuarine system is the dominancy of diatoms (85.10 %) due to which the group plays the key role in carbon storage. The relative abundances of dinoflagellates and other phytoplankton (like green algae and cyanobacteria) are 10.64 and 4.26 %, respectively. Converting the carbon values cited in into carbon dioxide equivalent, it is observed that in the western sector for diatoms, dinoflagellates, blue green algae and green algae, the mean carbon dioxide equivalents per cell are 2,467.39, 624.08, 1,017.05 and 259.00 pg, respectively. Similarly in the central sector, the mean values of carbon dioxide equivalents for diatoms, dinoflagellates, blue green algae and green algae are 2,211.43, 595.80, 1,188.49 and 243.25 pg, respectively. Considering the standing stock of phytoplankton community and species-wise cell carbon content, it is observed that carbon content per litre of estuarine water (Table 2A.3.6) is maximum in diatoms (4,069.07 pg in western sector and 4,997.22 pg in the central sector) followed by blue green algae and green algae (2,512.33 pg in the western sector and 3,288.23 pg in the central sector) and dinoflagellates (186.98 pg in western sector and 483.13 pg in the central sector). It is also interesting to note that the monsoon period is most congenial for storing carbon in the phytoplankton cell (as revealed from significant negative relationship between salinity and phytoplankton cell carbon in Table 2A.3.7) due to increase in volume that may be attributed to enrichment of the present tropical estuarine waters with nutrients from mangrove litter and run-off from adjacent landmasses and agricultural fields (Banerjee et al. 2002, 2003; Chowdhury et al. 2011). The highly significant

positive correlations between cell volume and carbon content of phytoplankton (Fig. 2A.3.3a, b, c, d, e, and f) confirm the contribution of cell volume to stored carbon in phytoplankton cell. The variations of allometric equations between the western and central Indian Sundarbans (Fig. 2A.3.1) emphasize the role of environmental variables (preferable salinity) in regulating the volume and carbon content in phytoplankton cell.

2A.3.5   Conclusion

The carbon sequestration in phytoplankton community is a direct function of cell volume that depends on the interaction between salinity, climate and several hydrological parameters. Hence, results obtained at one location may not be equally applicable to another location, and site-specific allometric models need to be developed to predict the stored carbon on the basis of cell volume.

The present study demonstrates that the cell volume and carbon storage capacity of phytoplankton species vary with spatial locations due to varying salinity, perhaps moderated by anthropogenic factors (like barrage discharge) and climatic factors (like monsoon). Effective freshwater flow (through artificial canalization or channelization from rainwater harvested pools) into the tropical mangrove system is important for increasing the carbon storage potential of phytoplankton.

Acknowledgements The authors are grateful to Department of Science and Technology, Govt. of India, for financial support [Project Sanction No. DST/IS-STAC/ CO2-SR-59/09 (Pt. II) dated 24.05.2011] and Directorate of Forest, Government of West Bengal, for providing infrastructural support during the tenure of the work. One of the authors (Sufia Zaman) is grateful to Department of Science and Technology Govt. of India for financial assistance (under Women Scientist Scheme-B Project, Sanction No. SSD/SS/028/2010/G; dated 28.09.2011).

Table 2A.3.6 Mean seasonal variation of phytoplankton carbon (in pg) $\times$ 10^5/l of estuarine water in Indian Sundarbans during 2010–2011

Sl. No.	Species	Cell carbon (in pg) $\times$ 10^5/l					
		Western			Eastern		
		Premonson	Monsoon	Postmonsoon	Premonsoon	Monsoon	Postmonsoon
1.	*Coscinodiscus eccentricus*	11,064.17	13,661.51	10,638.14	10,759.01	12,881.55	11,129.77
2.	*Coscinodiscus jonesianus*	9,118.631	7,596.503	6,827.232	7,967.05	8,778.214	7,884.576
3.	*Coscinodiscus lineatus*	3,422.123	3,260.625	3,118.32	2,370.541	2,429.018	2,368.95
4.	*Coscinodiscus radiatus*	30,147.15	26,256.64	23,536.46	26,260.53	26,172.1	29771.35
5.	*Coscinodiscus gigas*	2,793.309	2,752.18	2,459.93	2,933.957	1,930.22	3,296.466
6.	*Coscinodiscus oculusiridis*	1,975.315	1,667.313	1,243.736	1749.557	1,612.692	2,189.074
7.	*Planktoniella sol*	67,007.15	52,891.92	37,968.79	60,931.73	57,863.99	73,859.06
8.	*Cyclotella striata*	9,780.244	8,474.025	6,460.736	11,635.02	9,897.072	12,473.04
9.	*Thalassiosira subtilis*	787.7687	1213.017	774.18	793.728	924.5635	967.162
10.	*Ceratium tripos*	268.9793	251.064	278.875	378.4153	475.0475	485.944
11.	*Skeletonema costatum*	383.8613	281.106	162.932	469.25	363.025	529.278
12.	*Paralia sulcata*	4,845.39	4,917.605	3190.6	5,754.027	5,025.515	4,843.638
13.	*Rhizosolenia crassipina*	639.38	605.154	593.04	1,021.661	815.488	1,103.75
14.	*Rhizosolenia setigeara*	770.28	1,021.251	719.368	579.95	743.379	797.471
15.	*Rhizosolenia alata*	650.246	361.438	514.563	1,158.899	1,061.01	1,262.913
16.	*Ceratium teres*	250.42	0	0	641.52	269.19	625.375
17.	*Ceratium trichoceros*	570.5643	133.305	295.056	921.186	948.4535	1,015.037
18.	*Bacteriastrum delicatulum*	6,682	3,229.72	4,728.87	8,328.162	6,714.268	9,502.656
19.	*Bacteriastrum varians*	8,655.456	5,941.361	6,030.98	11,000.85	8,575.877	1,3573.9
20.	*Bacteriastrum comosum*	4,034.326	3,546.923	2,011.321	5,890.958	4,401.483	5,710.64
21.	*Chaetoceros didymus*	728.5347	384.516	190.085	725.87	712.0835	604.72
22.	*Chaetoceros peruvianus*	496.32	426.467	347.312	403.1747	259.718	280.676
23.	*Chaetoceros compressus*	431.145	513.935	449.358	308.3633	251.6735	290.65
24.	*Ditylum sol*	502.9623	586.488	413.856	767.118	634.57	869.652
25.	*Triceratium favus*	1,507.392	1.736.598	1,188.33	1,593.457	1,618.195	1,725.384
26.	*Triceratium reticulatum*	1,407.3	1,366.26	1,613.096	1,779.199	1,354.509	922.307
27.	*Biddulphia sinensis*	2,883.986	2,684.786	3,165.56	3,857.87	3,297.924	4,244.968
28.	*Biddulphia mobiliensis*	6,274.18	4,897.32	5,599.251	7,857.356	6,600.275	4,934.412
29.	*Hemidiscus hardmanninus*	36.22533	55.16	22.61	59.64	85.284	30.888

(continued)

Table 2A.3.6 (continued)

Sl. No.	Species	Cell carbon (in pg) × 10⁵/l					
		Western			Eastern		
		Premonson	Monsoon	Postmonsoon	Premonsoon	Monsoon	Postmonsoon
30.	*Climacosphenia elongate*	74.8323	149.2898	41.6457	239.239	126.413	137.274
31.	*Fragilaria oceanica*	6,089.22	5,726.039	5,104.548	8,468.173	5,965.086	4,900.7
32.	*Rhaphoneis amphiceros*	39.12813	33.275	4.5928	76.4408	36.4901	30.762
33.	*Thalassionema nitzschioides*	84.7	128.725	40.28	89.49967	95.472	50.508
34.	*Thalassiothrix longissima*	1,022.504	961.3475	639.618	1,082.67	929.538	944.645
35.	*Thalassiothrix fraunfeldii*	170.328	163.3	132.804	170.144	213.7395	184.968
36.	*Asterionella japonica*	11.55167	17.592	3.192	21.92733	23.045	30.87
37.	*Ceratium extensum*	90.216	58.185	56.76	175.728	125.137	191.695
38.	*Gyrosigma balticum*	724.948	545.535	523.012	850.119	669.7785	723.492
39.	*Pleurosigma normanii*	921.15	934.192	756.654	1,006.83	839.175	1,060.452
40.	*Pleurosigma elongatum*	4,838.75	1,919.665	1,764.774	7,604.289	3,718.7	7,275.618
41.	*Diploneis smithii*	938.041	829.213	624.096	1,166.1	987.504	1,239.18
42.	*Cymbella marina*	130.005	115.3905	85.443	204.51	212.706	215.355
43.	*Nitzschia sigma*	84.932	110.276	66.066	89.465	115.283	146.88
44.	*Nitzschia closterium*	111.886	107.496	95.551	178.89	163.809	197.847
45.	*Ceratium furca*	189.873	150.417	210.99	363.032	324.919	306.243
46.	*Trichodesmium erythraea*	4,548.093	5,749.001	3,372.324	6,302.452	5,808.083	6,345.69
47.	*Chlorella marina*	417.7713	467.64	519.11	477.1067	296.4	499.66

Table 2A.3.7 Inter-relationship between phytoplankton cell carbon and salinity during 2010–2011

Species	Correlation coefficient (r) between phytoplankton cell carbon and salinity			p value		
	Premonsoon	Monsoon	Postmonsoon	Premonsoon	Monsoon	Postmonsoon
Coscinodiscus eccentricus	−0.9673	−0.9088	−0.6294	<0.01	<0.01	<0.01
Coscinodiscus jonesianus	−0.5355	−0.5821	−0.5055	<0.01	<0.01	<0.01
Coscinodiscus lineatus	−0.6674	−0.6504	−0.6250	<0.01	<0.01	<0.01
Coscinodiscus radiatus	−0.6718	−0.5723	−0.6274	<0.01	<0.01	<0.01
Coscinodiscus gigas	0.0654	−0.1021	0.0790	IS	IS	IS
Coscinodiscus oculusiridis	−0.7708	−0.7077	−0.7550	<0.01	<0.01	<0.01
Planktoniella sol	−0.8030	−0.8463	−0.7965	<0.01	<0.01	<0.01
Cyclotella striata	−0.3647	−0.3334	−0.3611	IS	IS	IS
Thalassiosira subtilis	−0.2364	−0.2766	−0.2103	IS	IS	IS
Ceratium tripos	−0.1692	−0.1171	−0.1924	IS	IS	IS
Skeletonema costatum	−0.1591	−0.0072	−0.0927	IS	IS	IS
Paralia sulcata	−0.3471	−0.3547	−0.3450	IS	IS	IS
Rhizosolenia crassipina	−0.0707	−0.4175	−0.3827	IS	IS	IS
Rhizosolenia setigeara	−0.4142	−0.3765	−0.3360	IS	IS	IS
Rhizosolenia alata	−0.3186	0.0466	−0.2770	IS	IS	IS
Ceratium teres	−0.1247	−0.2556	−0.1973	IS	IS	IS
Ceratium trichoceros	−0.3453	−0.1612	−0.2761	IS	IS	IS
Bacteriastrum delicatulum	−0.2859	−0.3076	−0.2709	IS	IS	IS
Bacteriastrum varians	−0.1518	0.0903	0.1061	IS	IS	IS
Bacteriastrum comosum	−0.2296	−0.2015	−0.2596	IS	IS	IS
Chaetoceros didymus	−0.3033	−0.3837	−0.3691	IS	IS	IS
Chaetoceros peruvianus	−0.3026	−0.2977	−0.3872	IS	IS	IS
Chaetoceros compressus	−0.2859	−0.1558	−0.3836	IS	IS	IS
Ditylum sol	−0.0872	−0.0748	−0.0623	IS	IS	IS
Triceratium favus	−0.3937	−0.0605	−0.3720	IS	IS	IS
Triceratum reticulatum	−0.4089	−0.1785	−0.3317	IS	IS	IS
Biddulphia sinensis	−0.3512	−0.0733	−0.3595	IS	IS	IS
Biddulphia mobiliensis	−0.1966	−0.1630	−0.0486	IS	IS	IS
Hemidiscus hardmanninus	−0.3699	0.3786	−0.3962	IS	IS	IS
Climacosphenia elongate	−0.3814	−0.3488	−0.3487	IS	IS	IS
Fragilaria oceanica	0.6087	0.6917	0.6494	<0.01	<0.01	<0.01
Rhaphoneis amphiceros	−0.3125	0.0127	−0.4025	IS	IS	IS
Thalassionema nitzschioides	−0.3351	−0.1317	−0.3896	IS	IS	IS
Thalssiothrix longissima	0.1337	−0.2601	−0.3458	IS	IS	IS
Thalassiothrix fraunfeldii	−0.3426	−0.2797	−0.3882	IS	IS	IS
Asterionella japonica	−0.2804	−0.3653	−0.3392	IS	IS	IS
Ceratium extensum	0.0439	0.3595	−0.3795	IS	IS	IS
Gyrosigma balticum	−0.3085	−0.3851	−0.3732	IS	IS	IS
Pleurosigma normanii	−0.2107	−0.2793	−0.2186	IS	IS	IS
Pleurosigma elongatum	−0.2648	−0.2591	−0.2329	IS	IS	IS
Diploneis smithii	−0.0587	−0.3192	−0.2034	IS	IS	IS
Cymbella marina	0.0415	−0.3165	−0.3705	IS	IS	IS
Nitzschia sigma	−0.3280	−0.3674	−0.3703	IS	IS	IS
Nitzschia closterium	−0.3477	−0.3463	−0.3847	IS	IS	IS
Ceratium furca	−0.2220	−0.3118	−0.3300	IS	IS	IS
Trichodesmium erythraea	0.3508	0.4193	0.3436	IS	IS	IS
Chlorella marina	0.7043	0.7633	0.7201	<0.01	<0.01	<0.01

a

$$y = 0.0629x + 39.312$$
$$R^2 = 0.7416$$

b

$$y = 0.0615x + 45.656$$
$$R^2 = 0.7147$$

c

$$y = 0.19x + 22.707$$
$$R^2 = 0.9972$$

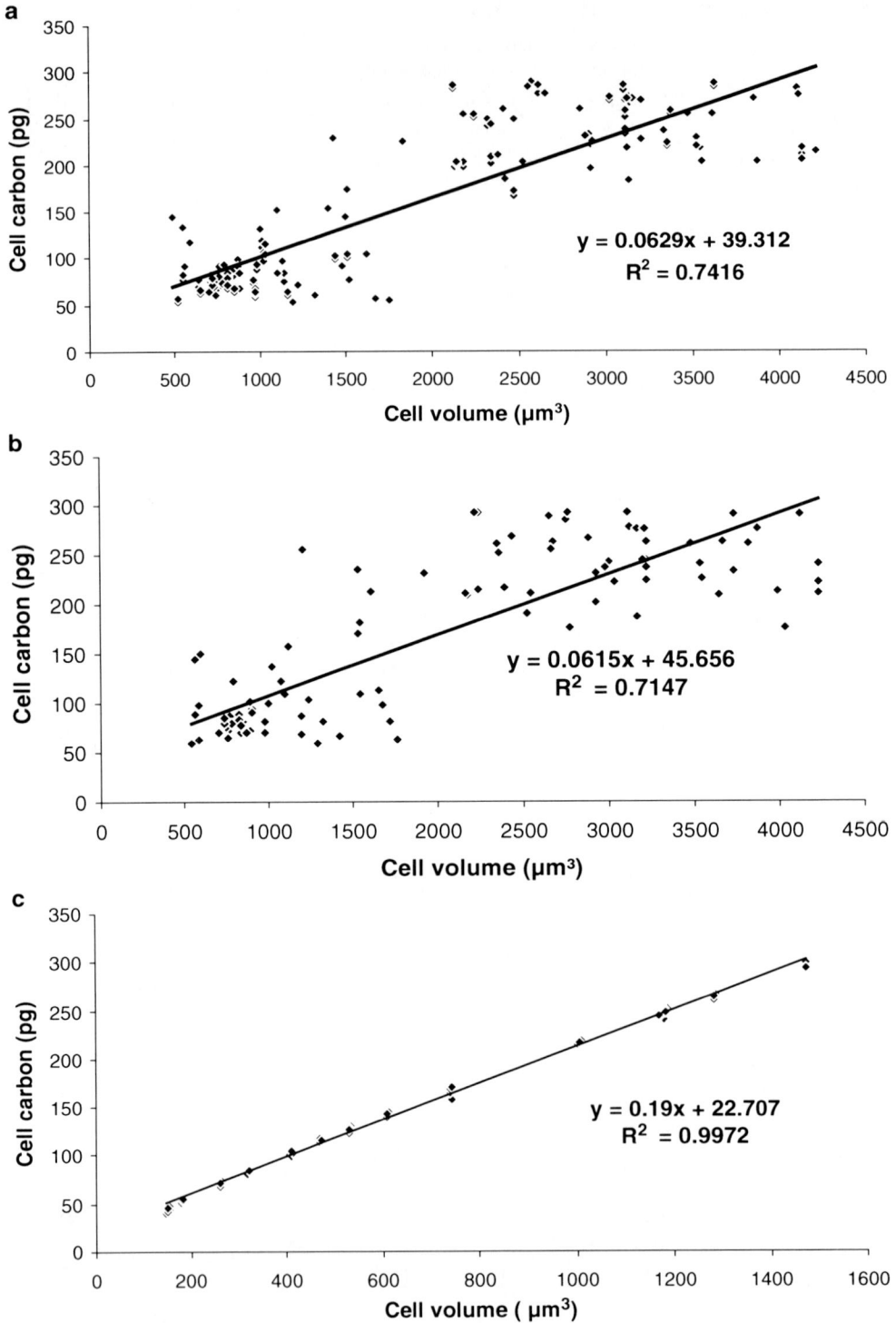

Fig. 2A.3.3 (**a**) Interrelationship between cell carbon content (pg) and cell volume (μm^3) for cyanobacteria and green algae in central sector during 2010–2011. (**b**) Interrelationship between cell carbon content (pg) and cell volume (μm^3) for cyanobacteria and green algae in western sector during 2010–2011. (**c**) Interrelationship between cell carbon content (pg) and cell volume (μm^3) for dinoflagellates in central sector during 2010–2011.

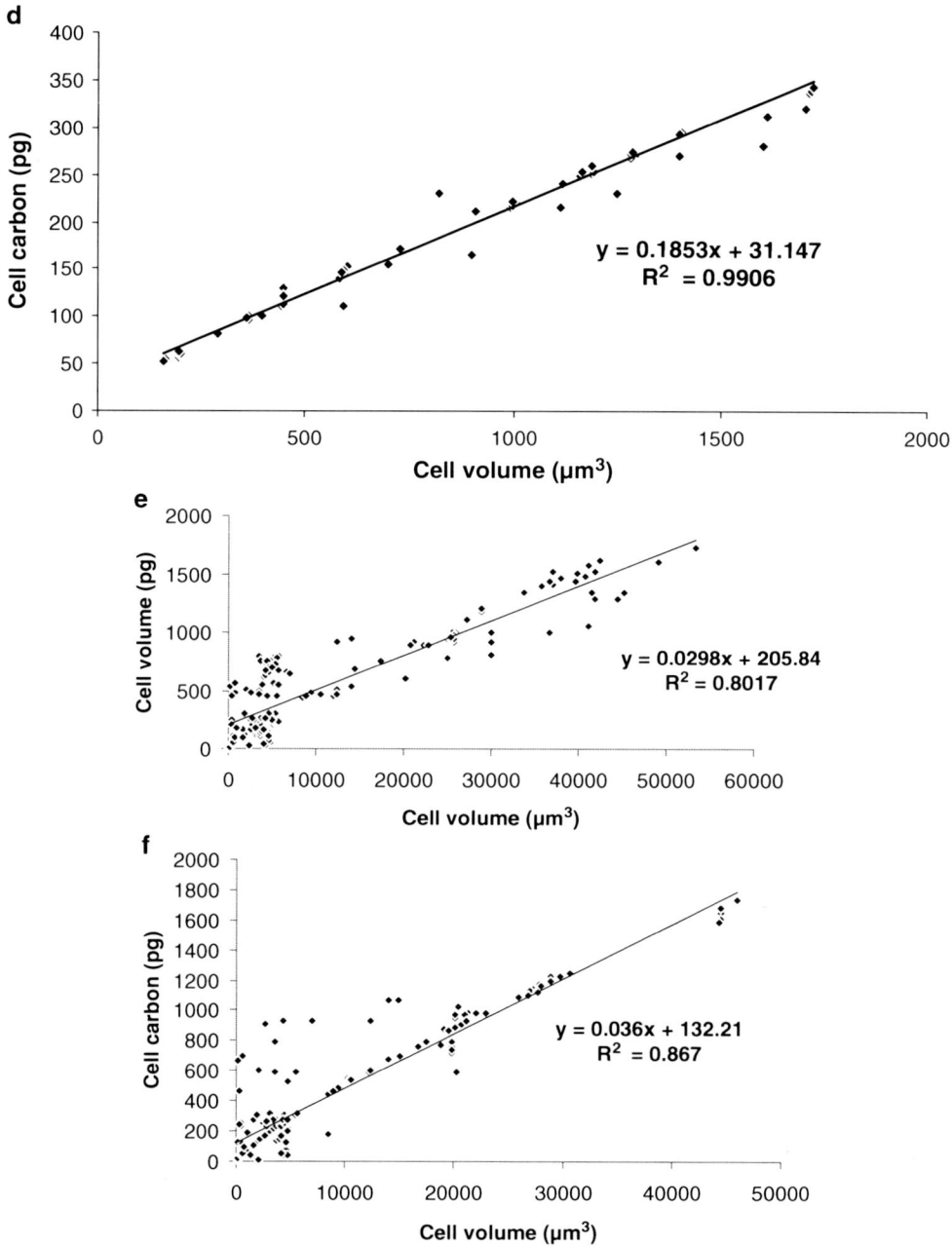

Fig. 2A.3.3 (continued) (**d**) Interrelationship between cell carbon content (pg) and cell volume (μm^3) for dinoflagellates in western sector during 2010–2011. (**e**) Interrelationship between cell carbon content (pg) and cell volume (μm^3) for diatoms in central sector during 2010–2011. (**f**) Interrelationship between cell carbon content (pg) and cell volume (μm^3) for diatoms in western sector during 2010–2011

References of Annexure 2A.3

Banerjee K, Mitra A, Bhattacharyya DP, Choudhury A (2002) Chapter 6: Role of nutrients on phytoplankton diversity in the North east coast of the Bay of Bengal. In: Kumar A (ed) Ecology and ethology of aquatic biota. Daya Publishing House, New Delhi, pp 102–109

Banerjee K, Mitra A, Bhattacharyya DP (2003) Phytopigment level of the aquatic sub-system of Indian Sundarbans at the apex of Bay of Bengal. Sea Explorers 6:39–46

Banerjee K, Senthilkumar B, Purvaja R, Ramesh R (2012) Sedimentation and trace metal distribution in selected locations of Sundarbans mangroves and Hooghly estuary, Northeast coast of India. Environ Geochem Health 34:27–42. doi 10.1007/s 10653-011-93880-0

Botes L (2003) Phytoplankton identification catalogue Saldanha Bay, South Africa. Globallast monograph series no. 7. IMO, London

Boyce DG, Lewis NR, Worm B (2010) Global phytoplankton decline over the past century. Nature 466:591–596

Boyd CM, Johnson CW (1995) Precision of size determination of resistive electronic particle counters. J Plankton Res 17:41–58

Brown LM, Gargantini I, Brown DJ, Atkinson HJ, Govindarajan J, Vanlerberghe GC (1989) Computer-based image analysis for the automated counting and morphological description of microalgae in culture. J Appl Phycol 1:211–225

Chakrabarti PS (1998) Changing courses of Ganga, Ganga–Padma river system, West Bengal, India – RS data usage in user orientation, river behavior and control. J River Res Inst 25:19–40

Chaudhuri AB, Choudhury A (1994) Mangroves of the Sundarbans, India, vol 1. IUCN, Bangkok

Chisholm SW (1992) Phytoplankton size. In: Falkowski PG, Woodhead AD (eds) Primary productivity and biogeochemical cycles in the sea. Plenum Press, New York, pp 213–237

Chowdhury MSN, Hossain MS, Barua P (2011) Environmental functions of the Teknaf peninsula mangroves of Bangladesh to communicate the values of goods and devices. Mesopot J Mar Sci 26:79–97

Cunningham A, Buonnacorsi CA (1992) Narrow-angle forward light scattering from individual algal cells: implications for size and shape discrimination in flow cytometry. J Plankton Res 14:223–234

Davidson K, Roberts EC, Gilpin LC (2002) The relationship between carbon and biovolume in marine microbial mesocosms under different nutrient regimes. Eur J Phys 37:501–507

Estep KW, MacIntyre F, Hjorleifsson E, Sieburth JM (1986) MacImage: a user friendly image-analysis system for the accurate mensuration of marine organisms. Mar Ecol Prog Ser 33:243–253

Gordon R (1974) A tutorial on ART (algebraic reconstruction techniques). IEEE Trans Nucl Sci 2:78–93

Harold F (2002) Force and compliance: rethinking morphogenesis in walled cells. Fungal Genet Biol 37:271–282

Hastings JW, Sweeney BM, Mullin MM (1962) Counting and sizing of unicellular marine organisms. Ann N Y Acad Sci 99:180–289

Hildebrand M, York E, Kelz JI, Davis AK, Frigeri LG, Allison DP, Doktycz MJ (2006) Nanoscale control of silica morphology and three-dimensional structure during diatom cell wall formation. J Mater Res 21:2689–2698

Krambeck C, Krambeck HJ, Overbeck J (1981) Microcomputer-assisted biomass determination of plankton bacteria on scanning electron micrographs. Appl Environ Microbiol 42:142–149

Malone TC (1980) Algal size. In: Morris I (ed) The physiological and ecology of phytoplankton. University of California Press, Berkeley, pp 433–463

Maloney TE, Donovan EJ Jr, Robinson EL (1962) Determination of numbers and sizes of algal cells with an electronic particle counter. Phycologia 2:1–8

McAlice BJ (1971) Phytoplankton sampling with Sedgwick-Rafter Cell-1, Ira C. Darling Center for Research, Teaching and Service, and Departments of Oceanography and Zoology, University of Maine, Walpolc 04573, 16(1):19–28

Menden-Deuer S, Lessard EL, Satterberg J (2001) Effect of preservation on dinoflagellate and diatom cell volume and consequences for carbon biomass predictions. Mar Ecol Prog Ser 222:41–50

Mitra A, Banerjee K, Gangopadhayay A (2004) Introduction to marine phytoplankton. Daya Publishing House, Delhi, 102pp

Mitra A, Gangopadhyay A, Dube A, Schmidt ACK, Banerjee K (2009) Observed changes in water mass properties in the Indian Sundarbans (northwestern Bay of Bengal) during 1980–2007. Curr Sci 97:1445–1452

Mitra A, Chowdhury R, Banerjee K (2011) Concentrations of some heavy metals in commercially important finfish and shellfish of the River Ganga. Environ Monit Asses 184:2219–2230. doi: 10.1007/s10661-011-2111-x

Montanges DJS, Berges JA (1994) Estimating carbon, nitrogen, protein, and chlorophyll a from volume in marine phytoplankton. Limnol Ocean 39:1044–1060

Mullin MM, Sloan PR, Eppley RW (1966) Relationship between carbon content, cell volume, and area in phytoplankton. Limnol Oceanogr 11:307–311

Olson RJ, Vaulot D, Chisholm SW (1985) Marine phytoplankton distributions measured using shipboard flow cytometry. Deep-Sea Res 32:1273–1280

Pickett-Heaps JD, Schmidt AMM, Edgar LA (1990) The cell biology of diatom value formation. Prog Phycol Res 7:1–168

Raha A, Das S, Banerjee K, Mitra A (2012) Climate change impacts on Indian Sundarbans: a time series analysis. Biodivers Conserv 21:1924–2008. doi:10.1007/s10531-012-0260-z

Roemmich D, McGowan J (1995) Climatic warming and the decline of zooplankton in the California. Curr Sci 267:1324–1326

Roubeix V, Lancelot C (2008) Effect of salinity on growth, cell size and silicification of an euryhaline freshwater diatom. Cyclotella meneghiniana Kutz. Transit Waters Bull 1:31–38. ISSN 1825-229X. doi:10.1285/i1825229Xv2n1p31; http://siba2.unile.it/ese/twb

Sabine CL et al (2004) The oceanic sink for anthropogenic CO_2. Science 305:367–371

Sournia A (1981) Morphological base of competition and succession. Can Bull Fish Aquat Sci 210:339–346

Steen HB (1980) Characters of flow cytometers. In: Melamed MR, Lindmo T, Mendelsohn ML (eds) Flow cytometry and sorting, 2nd edn. Wiley-Liss, New York, pp 11–25

Sun J, Liu DY (2003) Geometric models for calculating cell biovolume and surface area for phytoplankton. J Plankton Res 25:1331–1346

Verlencar XN, Desai S (eds) (2004) Phytoplankton identification manual, vol 2. National Institute of Oceanography, Dona Paula

Wood AM, Horan PK, Muirhead K, Phinney DA, Yentsch CM, Waterbury JB (1985) Discrimination between types of pigments in marine Synechococcus spp. by scanning spectroscopy, epifluorescence microscopy, and flow cytometry. Limnol Oceanogr 30:1303–1315

How Mangroves Resist Natural Disaster?

3

We must understand each other or die

William Davenport

A natural disaster (e.g. volcanic eruption, earthquake, landslide and tsunami) is the consequence of transformation of a potential hazard into physical event and its subsequent adverse impact on human beings. Natural disasters are caused by several agents that may be broadly divided into climatic and geological components (Table 3.1).

Human vulnerability, caused by the lack of proper planning and appropriate emergency management, leads to serious financial, structural and human impact. The resulting loss depends on the capacity of the population to cope with or resist the disaster: their *resilience*. Thus, a very common thumb rule in the domain of disaster is 'disasters occur when hazards coincide with vulnerability'. A natural hazard never gets translated into a natural disaster in areas without vulnerability, e.g. strong earthquakes in uninhabited areas. The term *natural* has consequently been disputed and has come under controversy because the events simply do not cause hazards or disasters without human involvement.

To clarify the intersection and confusion of terminologies, it is logical to define the terms appropriately.

Hazard: A hazard involves something which could potentially be harmful to a person's life, health and property or to the environment. There are a number of methods of classifying a hazard, but most systems use some variation on the factors of *likelihood* of the hazard turning into an incident and the *seriousness* of the incident if it were to occur. Hazards are defined as phenomena that pose a threat to people, structures, or economic assets and which may cause a disaster. They could be either man-made or naturally occurring in our environment.

Disaster: Disaster is a sudden calamitous event bringing great damage, loss, and destruction and devastation to life and property. The damage caused by disasters is immeasurable and varies with the geographical location, climate, the type of Earth surface and degree of vulnerability. This influences the mental, socio-economic, political and cultural state of the affected area. The resilience of the affected population is also an important parameter regulating the magnitude of damage caused by disaster.

Catastrophes: The most extreme hazard events create catastrophes, or disasters, which normally arrive without warning. White and Haas (1975) defined catastrophes as any situation in which the damages to people, property or society in general are so severe that recovery and/or rehabilitation after the event is a long and difficult process.

A. Mitra, *Sensitivity of Mangrove Ecosystem to Changing Climate*,
DOI 10.1007/978-81-322-1509-7_3, © Springer India 2013

Table 3.1 Causative agents of natural disaster

Climatic	Geological
Floods	Earthquakes
Storm surges	Tsunamis
Windstorms	Volcanoes
Wildfires	Landslides
Heat waves	Avalanches
Dust storms	
Snowstorms	

Table 3.2 Major types of natural disaster based on causative factors

Geophysical	
Climatic and meteorological	Includes blizzards and snow, droughts, floods, fog, frost, hailstorms, heat waves, hurricanes, lightning and tornadoes
Geological and geomorphological	Includes avalanches, earthquakes, erosion, landslides, shifting sand, tsunamis and volcanic eruptions
Biological	
Flora	Includes fungal diseases (like Dutch elm disease), infestations (like weeds and water hyacinth), hay fever and poison ivy
Fauna	Includes bacterial and viral diseases (such as malaria and rabies), infestations (like rabbits and locusts) and venomous animal bites

Risk: This is a measure of the expected losses due to a hazardous event of a particular magnitude occurring in a given area over a specific time period. Risk is a function of the probability of particular occurrences and the losses each would cause. The level of risk depends on the nature of the hazard, vulnerability of the elements which are affected and the economic values of those elements.

Vulnerability: It is defined as the extent to which a community, structure, service and/or geographic area is likely to be damaged or disrupted by the impact of a particular hazard, on account of their nature, construction and proximity to hazardous terrain or a disaster prone area.

3.1 Natural Disaster: Some Case Studies

Natural disaster encompasses several types of events like thunderstorms, volcanic eruption, cyclones, earthquakes and even rapid spreading of microbes causing diseases in the biotic community. On the basis of causative factors, natural disasters are of two main types (Table 3.2).

In this chapter, we have concentrated on two most common natural disasters, namely, cyclone and tsunami, that have serious negative impact on the coastal ecosystem.

3.1.1 Cyclone

A tropical cyclone is an almost circular storm that has a low surface pressure at the centre of its system, high winds that spiral inwards in a counterclockwise direction and is usually accompanied by heavy rain. A tropical cyclone's primary energy source is the release of the heat of condensation from water vapour condensing at high altitudes, with solar heating being the initial source for evaporation. Therefore, a tropical cyclone can be visualized as a giant vertical heat engine supported by mechanics driven by physical forces such as the rotation and gravity of the Earth. In another way, tropical cyclones could be viewed as a special type of mesoscale convective complex, which continues to develop over a vast source of relative warmth and moisture. Condensation leads to higher wind speeds, as a tiny fraction of the released energy is converted into mechanical energy; the faster winds and lower pressure associated with them in turn cause increased surface evaporation and thus even more condensation. Much of the released energy drives updrafts that increase the height of the storm clouds, speeding up condensation. This positive feedback loop continues for as long as conditions are favourable for tropical cyclone development. Factors such as a continued lack of equilibrium in air mass distribution would also give supporting energy to the cyclone. The rotation of the Earth causes the system to spin, an effect known as the Coriolis effect, giving it a cyclonic characteristic and affecting the trajectory of the storm.

What primarily distinguishes tropical cyclones from other meteorological phenomena is deep convection as a driving force. Because convection is strongest in a tropical climate, it defines

Table 3.3   Details of various types of cyclones that affected the east coast of India during the period 1891–1991

Types of disturbances	Cyclonic disturbances	Depressions/deep depressions	Cyclonic storms	Severe cyclonic storms
Number	1,009	592	268	149
Maximum	158	131	51	38
Minimum	4	1	0	1
Yearly average	10	6	3	1.5
Per cent of total	–	58	27	15
Wind speed (km/h)	31–>118	31–61	61–68	88–>118

Source: Shrestha (1998)

the initial domain of the tropical cyclone. By contrast, midlatitude cyclones draw their energy mostly from pre-existing horizontal temperature gradients in the atmosphere. To continue to drive its heat engine, a tropical cyclone must remain over warm water, which provides the needed atmospheric moisture to maintain the positive feedback loop running. When a tropical cyclone passes over land, it is cut off from its heat source and its strength diminishes rapidly.

In tropical and some subtropical regions of the world, organized cloud clusters form in response to perturbations in the atmosphere. When a cloud cluster forms in an area sufficiently removed from the Equator, then the magnitude of Coriolis accelerations is not negligible and an organized, closed circulation may occur. A tropical system with a developed circulation, but with wind speeds of less than 17.4 m/s (i.e. 63 km/h or 39 mph), is termed as *tropical depression*. Given that conditions are favourable for continued development (basically warm surface waters, little or no wind shear and a high pressure area aloft), this circulation can intensify to the point where sustained wind speeds exceed 17.4 m/s, at which time it is termed a *tropical storm*. If development continues to a level where the maximum sustained wind speed equals or exceeds 33.5 m/s (121 km/h or 75 mph), the storm is termed a *cyclone* (Indian Ocean), *typhoon* (Western Pacific) or *hurricane* (Atlantic and Eastern Pacific). Cyclone is often associated with storm surges, high tidal amplitudes and heavy rainfall. The Bay of Bengal region in the eastern part of India is noted for cyclonic depressions (Table 3.3).

All tropical cyclones are areas of low atmospheric pressure near the Earth's surface. The pressures recorded at the centres of tropical cyclones are among the lowest that occur on Earth's surface at sea level. Tropical cyclones are characterized and driven by the release of large amounts of latent heat of condensation, which occurs when moist air is carried upwards and its water vapour condenses. This heat is distributed vertically around the centre of the storm. Thus, at any given altitude (except close to the surface, where water temperature dictates air temperature), the environment inside the cyclone is warmer than its outer surroundings. A typical tropical cyclone is characterized by the presence of rainbands, eye and eyewalls (Fig. 3.1).

3.1.1.1 Rainbands

Rainbands are bands of showers and thunderstorms that spiral cyclonically towards the storm centre. High wind gusts and heavy downpours often occur in individual rainbands, with relatively calm weather between bands. Tornadoes often form in the rainbands of landfalling tropical cyclones. Intense annular tropical cyclones are distinctive for their lack of rainbands; instead, they possess a thick circular area of disturbed weather around their low-pressure centre. While all surface low-pressure areas require divergence aloft to continue deepening, the divergence over tropical cyclones is in all directions away from the centre. The upper levels of a tropical cyclone feature winds directed away from the centre of the storm with an anticyclonic rotation, due to the Coriolis effect. Winds at the surface are

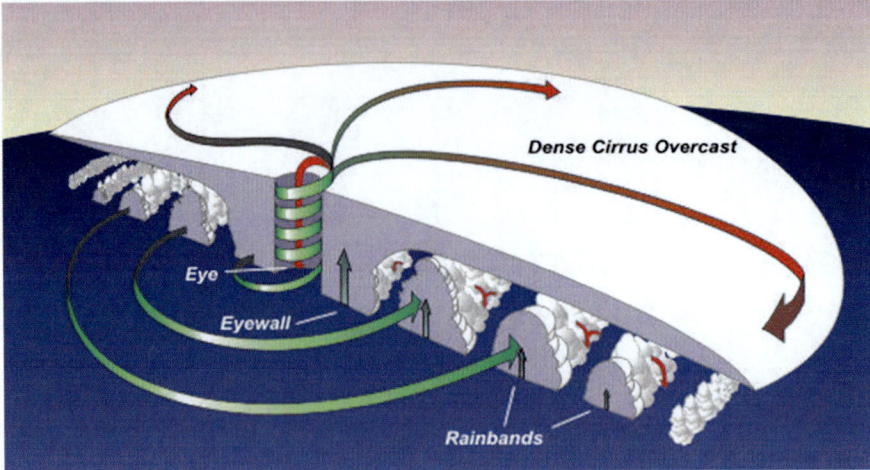

Fig. 3.1 Structure of tropical cyclones

strongly cyclonic, weaken with height and eventually reverse themselves.

3.1.1.2 Eye and Inner Core

A strong tropical cyclone will harbour an area of sinking air at the centre of circulation. If this area is strong enough, it can develop into an eye. Weather in the eye is normally calm and free of clouds, although the sea may be extremely violent. The eye is normally circular in shape and may range in size from 3 km (1.9 miles) to 370 km (230 miles) in diameter. Intense, mature tropical cyclones can sometimes exhibit an inward curving of the eyewall's top, making it resemble a football stadium; this phenomenon is thus sometimes referred to as the *stadium effect*.

There are other features that either surround the eye or cover it. The central dense overcast (CDO) is the concentrated area of strong thunderstorm activity near the centre of a tropical cyclone; in weaker tropical cyclones, the CDO may cover the centre completely. The eyewall is a circle of strong thunderstorms that surrounds the eye; this is the zone where the greatest wind speeds are found, where clouds reach the highest and precipitation is the heaviest. The heaviest wind damage occurs where a tropical cyclone's eyewall passes over land. Eyewall replacement cycles occur naturally in intense tropical cyclones.

When cyclones reach peak intensity, they usually have an eyewall and radius of maximum winds that contract to a very small size, around 10 km (6.2 miles) to 25 km (16 miles). Outer rainbands can organize into an outer ring of thunderstorms that slowly moves inwards and robs the inner eyewall of its needed moisture and angular momentum. When the inner eyewall weakens, the tropical cyclone weakens (in other words, the maximum sustained winds weaken and the central pressure rises). The outer eyewall replaces the inner one completely at the end of the cycle. The storm can be of the same intensity as it was previously or even stronger after the eyewall replacement cycle finishes. The storm may strengthen again as it builds a new outer ring for the next eyewall replacement.

3.1.1.3 Types of Tropical Cyclones

Tropical cyclones are classified into three main groups, based on intensity: tropical depressions, tropical storms and a third group of more intense storms, whose name depends on the region. For example, if a tropical storm in the Northwestern Pacific reaches hurricane-strength winds on the Beaufort scale, it is referred to as a *typhoon*; if a tropical storm passes the same benchmark in the Northeast Pacific Basin, or in the Atlantic, it is called a *hurricane*. Neither 'hurricane' nor 'typhoon' is used in either the Southern Hemisphere or the Indian Ocean. In these basins, storms of tropical nature are referred to as simply 'cyclones'. Comparison between different basins become difficult as each basin uses a separate system of terminology (Table 3.4 and Fig. 3.2). In the Pacific Ocean, hurricanes from the Central North Pacific sometimes cross the International Date Line into the Northwest Pacific, becoming typhoons; on rare occasions, the reverse may occur. It should also be noted that typhoons with sustained winds greater than 67 m per second (130 kn) or 150 miles per hour (240 km/h) are called *super typhoons* by the Joint Typhoon Warning Center.

A *tropical depression* is an organized system of clouds and thunderstorms with a defined, closed surface circulation and maximum sustained winds of less than 17 m per second (33 kn) or 39 miles per hour (63 km/h). It has no eye and does not typically have the organization or the spiral shape of more powerful storms. However, it is already a low-pressure system, hence the name 'depression'. The practice of the Philippines is to name tropical depressions from their own naming convention when the depressions are within the Philippines' area of responsibility.

A *tropical storm* is an organized system of strong thunderstorms with a defined surface circulation and maximum sustained winds between 17 m per second (33 kn) (39 miles per hour (63 km/h)) and 32 m per second (62 kn) (73 miles per hour (117 km/h)). At this point, the distinctive cyclonic shape starts to develop, although an eye is not usually present. Government weather services, other than the Philippines, first assign names to systems that reach this intensity (thus the term *named storm*).

A *hurricane* or *typhoon* (sometimes simply referred to as a tropical cyclone, as opposed to a depression or storm) is a system with sustained winds of at least 33 m per second (64 kn) or 74 miles per hour (119 km/h). A cyclone of this intensity tends to develop an eye, an area of relative calm (and lowest atmospheric pressure) at the centre of circulation. The eye is often visible in satellite images as a small, circular, cloud-free spot. Surrounding the eye is the eyewall, an area about 16 km (9.9 miles) to 80 km (50 miles) wide in which the strongest thunderstorms and winds circulate around the storm's centre. Maximum sustained winds in the strongest tropical cyclones have been estimated at about 85 m per second (165 kn) or 195 miles per hour (314 km/h).

3.1.1.4 Some Incidences of Tropical Cyclones

The 1970 *Bhola cyclone* is the deadliest tropical cyclone on record, killing more than 300,000 people after striking the densely populated Ganges Delta region of Bangladesh on 13 November 1970. Its powerful storm surge was responsible for the high death toll. The *North Indian cyclone basin* has historically been the deadliest basin, with several cyclones since 1900 killing more than 100,000 people, all in Bangladesh. Elsewhere, *Typhoon Nina* killed nearly 100,000 in China due to a devastating flood that caused 62 dams including the Banqiao Dam to fail. The *Great Hurricane* of 1780 is the deadliest Atlantic hurricane on record, killing about 22,000 people in the Lesser Antilles. A tropical cyclone does need not be particularly strong to cause memorable damage, primarily if the deaths are from rainfall or mudslides. *Tropical Storm Thelma* in November 1991 killed thousands in the Philippines, while in 1982, the unnamed tropical depression that eventually became *Hurricane Paul* killed around 1,000 people in Central America.

Hurricane Katrina is estimated as the costliest tropical cyclone worldwide, causing $81.2 billion in property damage with overall damage estimates exceeding $100 billion. Katrina killed at least 1,836 people after striking Louisiana and Mississippi as a major hurricane in August 2005. The *Galveston Hurricane* of 1900 is the deadliest natural disaster in the United States, killing an estimated 6,000–12,000 people in Galveston, Texas. *Hurricane Iniki* in 1992 was the most powerful storm to strike Hawaii in recorded history, hitting *Kauai* as a *Category 4 hurricane*, killing six people and causing US $3 billion in damage. Other destructive Eastern Pacific hurricanes include *Pauline* and *Kenna*,

Table 3.4 Tropical cyclone classifications (all winds are 10-min averages)

Beaufort scale	10-min sustained winds (knots)	N Indian Ocean IMD	SW Indian Ocean MF	Australia BOM	SW Pacific FMS	NW Pacific JMA	NW Pacific JTWC	NE Pacific and N Atlantic NHC, CHC and CPHC
0–6	<28	Depression	Trop. disturbance	Tropical low	Tropical depression	Tropical depression	Tropical depression	Tropical depression
7	28–29	Deep depression	Depression					
	30–33							
8–9	34–47	Cyclonic storm	Moderate tropical storm	Trop. cyclone (1)	Tropical cyclone	Tropical storm	Tropical storm	Tropical storm
10	48–55	Severe cyclonic storm	Severe tropical storm	Tropical cyclone (2)		Severe tropical storm		Hurricane (1)
11	56–63							
12	64–72		Tropical cyclone	Severe tropical cyclone (3)		Typhoon		Hurricane (2)
	73–85							
	86–89	Very severe cyclonic storm	Intense tropical cyclone	Severe tropical cyclone (4)			Typhoon	Major hurricane (3)
	90–99							Major hurricane (4)
	100–106			Severe tropical cyclone (5)				
	107–114							
	115–119		Very intense tropical cyclone				Super typhoon	
	>120	Super cyclonic storm						Major hurricane (5)

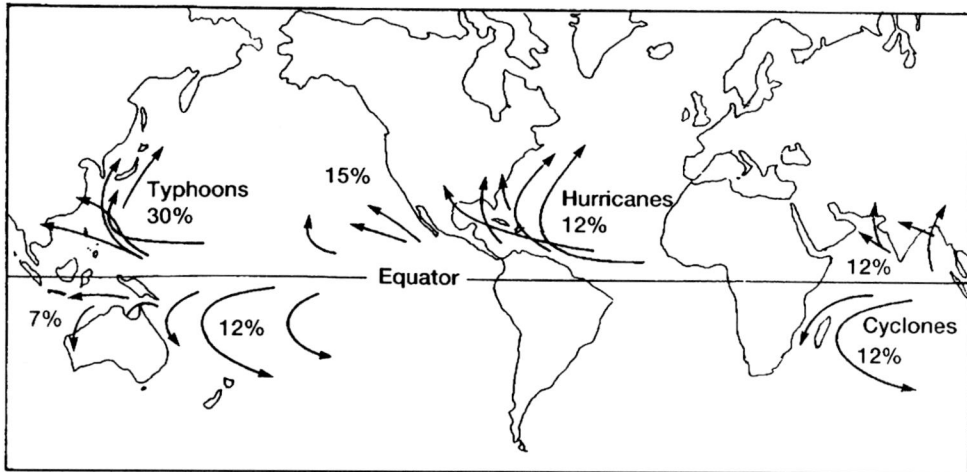

Fig. 3.2 Typical tropical cyclone tracks with global distribution and local names (*Source:* Abbott 2006)

both causing severe damage after striking Mexico as major hurricanes. In March 2004, *Cyclone Gafilo* struck northeastern Madagascar as a powerful cyclone, killing 74, affecting more than 200,000 and becoming the worst cyclone to affect the nation for more than 20 years. The Indian subcontinent faced several tropical cyclones through ages. Chittibabu et al. (2004) compiled a comprehensive list of 128 tropical cyclones that struck the maritime state of Orissa (in the east coast of India) from 1804 to 1999. Included in these strikes was the *Supercyclone* of 29–30 October 1999, which killed approximately 10,000 people and had a 7.5-m storm surge. A location map of Orissa and an inundation map of Orissa districts caused by the supercyclone in 1999 are given in Fig. 3.1. The India Meteorological Department (IMD) classified cyclone on the basis of sustained wind speed (Table 3.5).

The incidence of *AILA* (a severe tropical cyclonic storm) in the Gangetic plain region of India on 25 May 2009 posed severe impacts in terms of embankment damage in areas not protected by mangroves. Although significant changes in water quality were documented due to AILA (Table 3.6), the infrastructural damage was greatly retarded by mangroves that reduced the momentum of waves and tidal surges.

Table 3.5 Tropical cyclone classifications (all winds are 10-min averages)

Storm type	Abbreviation	Wind speed (km/h)
Super cyclone	SC	>221
Very severe cyclonic storm	VSCS	119–221
Severe cyclonic storm	SCS	88–118
Cyclonic storm	CS	63–87
Cyclonic depression	CDP	62 or less
Cyclonic disturbance during monsoon	CD	Not specified

3.1.2 Tsunami: An Incidence That Opened the Eyes of Millions

26 December 2004 was a black day for the planet Earth. The cruel clutches of tsunami severely damaged coastal communities in countries along the Indian Ocean and took the lives of more than 2,00,000 people. The event, however, pointed out the need and importance of coastal vegetation in combating natural disasters like tsunamis, cyclones and devastating tidal surges.

In India, the tsunami waves, triggered by the earthquake in the sea near Sumatra, struck the southern and eastern coastal areas. Walls of water as high as 10 m (33 ft) crashed on the beach and penetrated up to 3 km inland, causing extensive damage in the Andaman and Nicobar Islands and

Table 3.6 Variations of hydrological parameters in different phases before, during and after AILA

Stations	Surface water salinity (Phase A)	Surface water salinity (Phase B)	Surface water salinity (Phase C)	pH (Phase A)	pH (Phase B)	pH (Phase C)	DO (Phase A)	DO (Phase B)	DO (Phase C)
Stn. 1	3.41	3.99	3.45	7.65	7.65	7.65	5.71	4.98	5.23
Stn. 2	4.88	5.93	4.96	7.68	7.69	7.68	5.43	5.00	5.21
Stn. 3	6.10	7.53	6.18	7.70	7.70	7.70	6.63	6.03	6.50
Stn. 4	13.17	16.32	13.98	8.00	8.00	8.00	6.55	5.89	6.11
Stn. 5	12.05	14.95	12.78	8.10	8.11	8.10	4.80	4.74	5.31
Stn. 6	13.84	17.23	14.01	8.00	8.03	8.01	4.91	4.67	5.12
Stn. 7	15.55	19.40	15.98	8.10	8.11	8.11	4.80	4.65	5.32
Stn. 8	14.43	18.04	14.87	8.10	8.12	8.11	4.75	4.43	5.02
Stn. 9	17.22	21.58	17.96	8.10	8.11	8.11	4.60	4.54	5.99
Stn. 10	19.33	24.30	20.05	8.15	8.15	8.16	4.68	4.60	5.06
Stn. 11	20.58	25.93	21.00	8.20	8.21	8.21	5.05	4.99	5.19
Stn. 12	23.08	29.77	24.67	8.20	8.21	8.21	5.02	4.91	5.27

Phase A, pre-AILA period (18.05.2009); Phase B, AILA phase (27.05.2009); Phase C, post-AILA phase (04.06.2009)
Units of surface water salinity and DO are psu and ppm, respectively

the coastal districts of Tamil Nadu, Kerala, Andhra Pradesh and Pondicherry. According to government reports, about 10,880 people lost their lives and nearly 5,800 people could not be traced. Almost 1,54,000 houses were either destroyed or damaged, entailing losses of about Rs. 994 crore or USD 228.5 million. The tsunami destroyed or damaged nearly 75,300 fishing crafts including wooden catamarans and mechanized boats including trawlers worth about Rs. 935 crore (USD 215 million); fishing gears worth of Rs. 65 crore (USD 15 million) were also lost leading to loss of livelihood for thousands and thousands of fishing families. Apart from these, standing crops of paddy, groundnut, coconut, cashew, mango, banana, minor millets and vegetables were totally destroyed in thousands of hectares, and seawater intrusion rendered these productive lands unfit for cultivation. Even in this situation, it has been reported that damage in terms of loss of lives and properties in the villages, which are behind mangrove wetlands and shelter belt plantations such as plantations of Casuarina and of palm trees and other thick coastal vegetation, was limited as the intensity of the tsunami was reduced by these natural bioshields or biobarriers. It has been reported that in the Pichavaram mangrove region of Tamil Nadu,

fishing and farming villages, namely, T.S. Pettai, Vadakku Pichavaram, Killai Fisher Colony, MGR Nagar and Kalaignar Nagar, which are under direct physical coverage of the mangrove wetlands, were protected from the fury of the tsunami. These hamlets are located about 500 m to 2.5 km away from the sea and 50–500 m away from the mangrove forest. Fishers and farmers in these hamlets narrated that mangrove trees along the first few rows bore the brunt of the tsunami waves and the friction created by these trees and the trees of subsequent rows reduced the speed of the water. According to the villagers, seawater flashed into the mangroves by tsunami was distributed into lagoon, tidal creeks and canals associated with mangrove wetland, and therefore, the amount of water reaching a point was very much reduced. This clearly indicates that both the mangrove forests and the associated wetlands together played a crucial role in mitigating the impact of the tsunami. Similar observations have also been made in Indonesia, Sri Lanka and Thailand. In Hambantota District of Southern Sri Lanka, it was observed that undisturbed mangrove stands in areas such as Rekawa, Kahanda and Kalametiya villages had contributed to reduce the damage caused by the tsunami. Even local communities in Rekawa and Kahanda said that their lives and properties were saved by the intact mangrove stands. In areas where the mangrove stands were totally or partially cleared in this district, the damage was high.

Although tsunamis and cyclones are completely different natural disasters in their generation mechanisms or origin, they both cause huge damage to lives and properties.

The hydraulic characteristics of tsunamis are likely to be very different from those of wind waves and tidal waves (Latief and Hadi 2007). The period of a tsunami is usually between 10 min and 1 h as compared with periods of 12–24 h for normal waves (Mazda et al. 2007). A tsunami propagates like a tidal bore and its momentum increases with movement upstream into shallower water. Model simulations using data from hydrological experiments to predict the attenuation of tsunami energy by mangroves were generated by Hiraishi and Harada (2003) based on the 1998 tsunami that destroyed parts of the north coast of Papua New Guinea. The model output suggests a 90 % reduction in maximum tsunami flow pressure for a 100-m wide forest belt at a density of 3,000 trees/ha. Model results obtained by Hamzah et al. (1999), Harada and Imamura (2005), Latief and Hadi (2007) and Tanaka et al. (2007) for various types of coastal vegetation, including mangroves, were very similar. Tanaka et al. (2007) modelled the relationship of species-specific differences in drag coefficient and in vegetation thickness with tsunami height and found that species differed in their drag force in relation to tsunami height, with the palm, *Pandanus odoratissimus*, and *Rhizophora apiculata*, being more effective than other common vegetation, including the mangrove *Avicennia alba*. These data point to the importance of preserving or selecting appropriate species to act as wave barriers to offer sufficient shoreline protection. Tsunamis (like hurricanes and tidal bores) can be categorized as another form of disturbance, albeit massive and infrequent.

Mangrove forests impacted by the 2004 Indian Ocean tsunami were located in sheltered areas (bays, lagoons, estuaries) with very few located on open coast, making it initially difficult to assess whether the areas impacted by the tsunami suffered less because of the intrinsic protective capacity of the forests or because they were sheltered from direct exposure to the open sea (Chatenoux and Peduzzi 2007). However, several reports based on initial post-impact surveys in southeastern India, the Andaman Islands and Sri Lanka (Danielsen et al. 2005; Dahdouh-Guebas et al. 2005; Kathiresan and Rajendran 2005) indicated that mangroves offered a significant defence against the full impact of the tsunami. The conclusions of Kathiresan and Rajendran (2005) and Vermaat and Thampanya (2006) that the presence of mangroves saved lives along the Tamil Nadu coast of southeast India are invalid however as inappropriate statistical tests were used (Vermaat and Thampanya 2007). A more proper test of the same data indicated no significant effect of the presence or absence of mangroves on the human death toll (Kerr et al. 2006) and points to the need for caution to avoid overstating the role of mangroves in tsunami protection. Nevertheless, ground surveys and QuickBird pre-tsunami and IKONOS post-tsunami image analysis (Danielsen et al. 2005) and multivariate analysis of mangrove field data (Dahdouh-Guebas et al. 2005) covering the entire Tamil Nadu coast suggest less destruction of man-made structures located directly behind the most extensive mangroves.

3.2   Resistance of Mangroves Against Cyclone and Tidal Surges

Mangrove forests play a great role in the protection of coastal communities from the fury of cyclones and storms. The example of the supercyclone is very relevant in this context. It hit the Orissa coast on 29 October 1999 with a wind speed of 310 km/h and played havoc largely in the areas devoid of mangroves. On the contrary, practically no damage occurred in regions with luxuriant mangrove growth. Similarly, in the Mahanadi delta, where large-scale deforestation and reclamation of mangrove land for several purposes have been undertaken, maximum losses of life and property have been reported from time to time during stormy weather. These events are periodic confirmatory tests to prove that mangroves can form the best shelter belt against cyclones and storms.

There are several examples on record that can be cited to prove the protective role of mangroves against storms and tidal surges. For instance, the 1970 typhoon and the accompanying tidal waves that claimed about 3 lakhs human lives in Bangladesh probably would not have been so devastative if thousands of hectares of mangrove swamps would not have been cleared and replaced with paddy fields for short-term economic gains. Likewise, in the Kutch areas of Gujarat, where mangrove trees were illegally cut, the effect of the heavy cyclone during 1983 was felt deep inside the human habitation, which took a heavy toll of lives.

Tropical cyclones and storms are more common in the Bay of Bengal. They thus severely affect the south Indian coast as compared to that of the Arabian Sea. According to Koteswaram (1984), there were about 346 cyclones that include 133 severe ones in the Bay of Bengal, whereas the Arabian Sea had only 98 cyclones including 55 severe ones between the years 1891 and 1970. These cyclones with tremendous speed hit the coastline and inundate the shores with strong tidal wave, severely destroying and disturbing coastal life. However, mangroves like *Rhizophora* spp. seem to act as a protective force towards this natural calamity (McCoy et al. 1996).

A series of experiments carried out by the EqTAP project (Development of Earthquake and Tsunami Disaster Mitigation Technologies and Their Integration for the Asia-Pacific Region), funded by the Japanese Government, have shown that mangrove forests and certain other types of coastal vegetation can effectively reduce the impact of tsunamis on coastlines (Hiraishi and Koike 2001; Dinar 2002; Hiraishi 2003). Empirical and field-based evidence is limited, but analytical models show that 30 trees per 100 m^2 in a 100-m wide belt may reduce tsunami flow rate by as much as 90 % (Hiraishi and Harada 2003). EqTAP recommend using a coastal green belt to protect homes, as it is sustainable and much cheaper than artificial barriers (Hiraishi and Koike 2001; Dinar 2002; Hiraishi 2003). Studies in Vietnam also demonstrated the usefulness of mangrove forests in coastal protection (Mazda et al. 1997).

Mangrove ecosystem acts as the natural barrier against cyclonic depressions, surges and erosion activities caused by wave actions. According to Kabir et al. (2006), the mangrove forests are considered a low-cost and natural form of protection for lands subjected to strong currents and surges. The role of mangroves as natural coast guard may be attributed to certain unique properties of mangroves as discussed here.

3.2.1 Wave Attenuation Properties of Mangroves

There are many documented studies on the hydrodynamics within mangrove swamps and their wave attenuation properties. These include the research works undertaken by Mazda et al. (1997, 2005), Liu et al. (2003), Wu et al. (2001), Brinkman et al. (1997) and Massel et al. (1999). Field observations of surface wave attenuation in mangrove forests were undertaken in both Townsville, Australia, and on Iriomote Island, Japan. High-resolution wave gauges were deployed throughout the mangroves along transects in line with the dominant direction of wave propagation. Data were collected to verify a numerical model of wave attenuation. The numerical model was based on the fact that surface waves propagating within a mangrove forest are subject to substantial energy loss due to two main energy dissipation mechanisms: (1) multiple interactions of wave motion with mangrove trunks and roots and (2) bottom friction. The dissipative characteristics of the mangrove forest were estimated from physical parameters such as trunk diameter, spatial density and vegetation structure, which were not necessarily vertically and horizontally uniform. The resulting rate of wave energy attenuation depended strongly on the density of the mangrove forest, the diameter of the mangrove roots and trunks and on the spectral characteristics of the incident waves. The numerical model

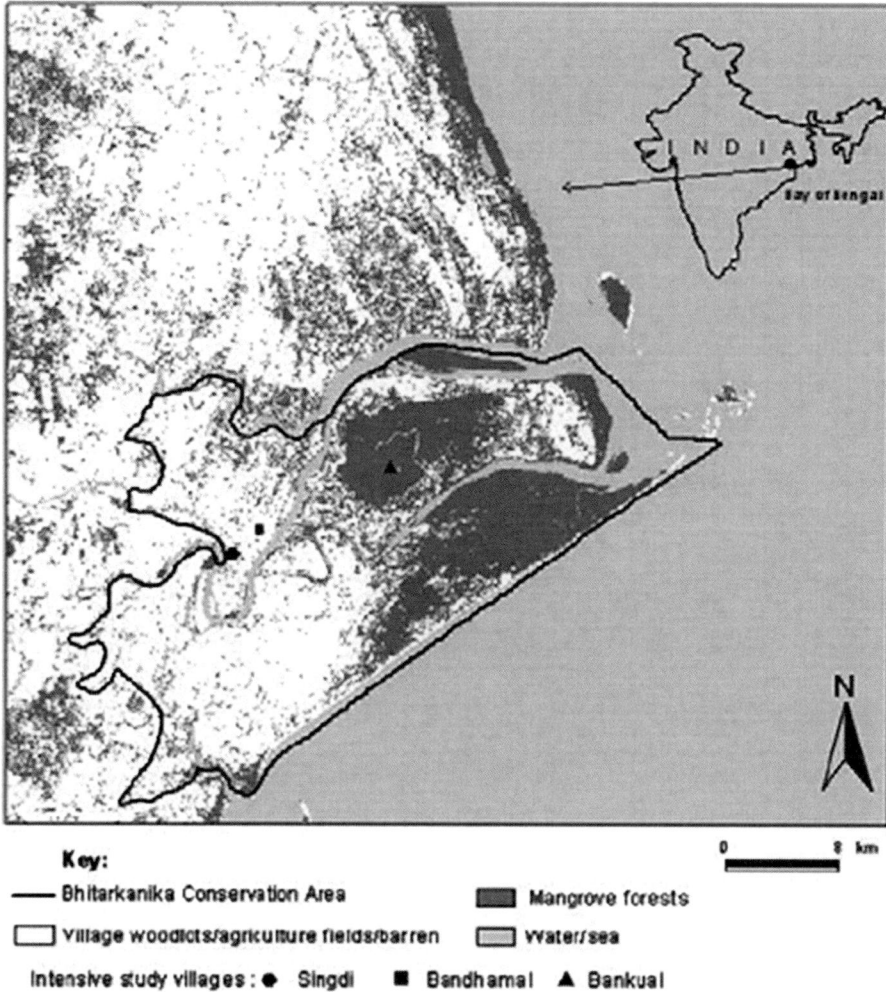

Fig. 3.3 Experimental sites at Orissa to evaluate the defensive role of mangroves against cyclone

results were supported by field observations, which showed substantial attenuation of wave energy within the mangrove forest. Typically, wave energy is attenuated by a factor of 2 within 50 m of the front of the mangrove forest. Hence, the wave heights are typically attenuated by a factor of square root 2 given that the wave energy is related to the square of the wave height. Both field and model results also indicated that longer-period waves, such as swell waves, are subjected to less attenuation, while short-period waves with frequencies typical of locally generated wind waves lose substantial energy due to interactions with the vegetation.

Badola and Hussain (2005a, b) evaluated the protective function of mangroves of Bhitarkanika in the maritime state of Orissa, India. The Bhitarkanika mangrove ecosystem is the second largest mangrove forest of mainland India (Fig. 3.3). Originally around 672 km², it is now limited to an area of 145 km² and is a wildlife sanctuary. This mangrove forest and the associated coast sustain a rich diversity of Indian mangrove flora and fauna. The mangrove forests of Bhitarkanika differ considerably from other mangroves because of the dominant tree species – *Sonneratia apetala*, *Heritiera fomes*, *H. littoralis* and several species of *Avicennia*. In order to critically analyse the defensive role of mangroves, three situations were identified: (1) a village in the shadow of mangroves, (2) a village not in the shadow of mangroves and with no embankment, and (3) a village not in the shadow of mangroves, but with an embankment on the seaward side. Bankual village was in the shadow of a mangrove forest, Singidi village was neither in the shadow of mangroves nor protected by an embankment from storm surge, and Bandhamal village was not in the shadow of mangroves, but had a seaward side embankment. The report indicated that the intensity of the impact of the 1999 cyclone on these villages should have been fairly uniform, as all the three selected villages were equidistant from the seashore and had similar aspects. The two villages outside mangrove cover were located close to each other, but both were far from the mangrove forest in order to eliminate any effect of mangrove forest presence.

It was documented after the study that the loss incurred per household was greatest (US$153.74) in the village that was not sheltered by mangroves but had an embankment, followed by the village that was neither in the shadow of mangroves nor the embankment (US$44.02) and the village that was protected by mangrove forests (US$33.31).

3.2.2 Sediment Trapping Properties of Mangroves

Mangroves trap and stabilize sediment and reduce the risk of shoreline erosion because they dissipate surface wave energy. It is this attribute that makes mangroves a potential natural solution for coastal protection-related problems. The complexity of the mangrove forest habitat, network of root system and presence of vast expanse of pneumatophores increase the residence time of water, which assists in the assimilation of inorganic nutrients and traps suspended particulate matter. Thus, mangroves also function as flood control barrier and binder of sediment particles (http://www.fao.org/gpa/sediments/habitat.htm). This feature of mangroves is very important in raising the land level

relative to the adjacent water level. It has been estimated by Kathiresan (2003a) that mangroves help in trapping the sediment up to 25 % at low tide as compared to high tide. This high efficiency of trapping suspended sediment may be attributed to widespread occurrence of numerous respiratory roots in *Avicennia* and to compactly arching stilt roots of *Rhizophora*. The density of mangrove species and their complexity of root systems thus constitute most important factors, for determining the sedimentation process.

Along with true mangroves, associate floral species also play a major role in sediment accretion and rise in land level. Salt marshes accrete sediments and organic matter and can keep pace with sea-level rise as long as sediment supplies are adequate. Increased flooding related to increased rainfall and runoff can bring more sediment to coastal marshes and potentially result in enhanced carbon sequestration (Reed 1999). However, flooding of marshes alters their chemical conditions (Nayman et al. 1990) and can ultimately result in the loss of marsh area. The loss of 59.5 km^2/year in the Mississippi River deltaic plain due to reduced sediment input is an extreme example of what can happen (Britsch and Kemp 1990). Increased flooding may initially increase C-fixation, however. Metabolic processes altered with increased flooding. Morris et al. (1990) showed a strong positive correlation between mean sea level and net annual above-ground production of *S. alterniflora* and attributed the higher productivities to increased wetland flooding. However, heavy flooding, as is documented on the rapidly subsiding portions of the Mississippi delta, increases benthic respiration and greenhouse gas emissions and decreases net primary productivity (Delaune et al. 1990).

Porteresia coarctata, commonly known as salt-marsh grass, is abundantly available in Indian Sundarbans region (Fig. 3.4). It acts as a pioneer species in the process of ecological succession in an island ecosystem. The species has unique sediment holding capacity and gradually invites other floral species to settle as shown in the present flow chart.

Fig. 3.4 *Porteresia coarctata* is abundant in the mudflats of lower Gangetic delta

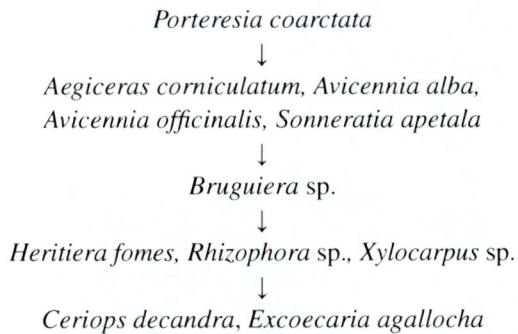

Porteresia coarctata

↓

Aegiceras corniculatum, Avicennia alba, Avicennia officinalis, Sonneratia apetala

↓

Bruguiera sp.

↓

Heritiera fomes, Rhizophora sp., *Xylocarpus* sp.

↓

Ceriops decandra, Excoecaria agallocha

The property of mangroves to impart stability to the shores was successfully used to stabilize the boulders placed to protect the Nayachar Island (Fig. 3.5) in the upper stretch of the Hugli estuary. The island is located about 56 nautical miles south of Kolkata, directly facing the port-cum-industrial complex of Haldia. The island is bounded by latitudes 21°54′41″ N–22°01′28″ N and longitudes 88°03′02″ E–88°08′43″ E. The island has an area of about 47 km^2 and a maximum elevation of 6.5 m above mean sea level (MSL). Length of the island centrally measured from north to south is 15,861 m and from east to west is 4,340 m.

The land topography is generally flat, with elevation ranging from as low as 1.08 m MSL to as high as 3.68 m MSL along the coastal region of the island. Average ground level of the island is 1.366 m MSL. The island is convex upwards with gently sloping eastern and western flanks. The eastern flank is undergoing erosion at present, while the western edge is accreting. According to recent estimation, the present accretion rate is about 1 km^2/year on the Haldia port area, and the erosion rate is about 0.05 km^2/year on the eastern part of Nayachar. The island is exposed to strong scouring action of the running Hugli River, and hence, a programme was undertaken by Calcutta Port Trust in the early 1990s to stabilize the island and increase the draft of the adjacent navigational channel. Eleven true mangrove species were planted inside and along the border of the island on geo-jute matrix, and within a span of 2 years, excellent growth of the species (particularly *Sonneratia apetala*) stabilized the boulders with considerable accretion of silt particles. No damage occurred on the island except tidal inundation and uprooting of few trees during the AILA of 25 May 2009. The biodiversity of the island has also increased gradually over a period of time (Table 3.7).

Fig. 3.5 Location of Nayachar Island in the upper stretch of Hugli River

Table 3.7 Shannon–Wiener species diversity index (H) of Nayachar Island (temporal variation)

Biotic community	H during 1996	H during 2009
Phytoplankton	2.786	2.987
Zooplankton	1.768	1.931
Benthic molluscs	1.097	3.654

3.3 Migration of Mangroves in Response to Sea-Level Rise

Mangroves have adapted special aerial roots, support roots and buttresses to live in muddy, shifting and saline conditions. They possess the ability to adapt to changes in sea level by growing upwards in place or by expanding landward or seaward. Mangroves produce peat from decaying litter fall and root growth and by trapping sediment from the ambient water. The process of building peat helps mangroves keep up with sea-level rise. For example, in western Jamaica, mangrove communities were able to sustain themselves because their rate of sedimentation exceeded the rate of the mid-Holocene sea-level rise (ca. 3.8 mm/year) (Hendry and Digerfeldt 1989). Mangroves can expand their range despite sea-level rise if the rate of sediment accretion is sufficient to keep up with sea-level rise and if migration is not blocked by local conditions, such as infrastructure (e.g. roads, agricultural fields, dikes, urbanization, seawalls, and shipping channels) and topography (e.g. steep slopes). A study in the Adelaide river area of the Northern Territory in Australia showed that the landward zone of mangroves doubled in size in the last 20 years (Jones 2002). Thus, inland wetland migration can offset the losses of coastal wetlands in areas that have low-lying coastal uplands (Houghton et al. 2001). In Moreton Bay of Australia, about 3,807 ha of mangroves were lost due to a combination of natural causes and mangrove clearing over a 24-year period (Manson et al. 2003). However, the majority of these losses were offset by the establishment of 3,590 ha of new mangroves which settled mostly on the landward edge of existing mangroves and by the improved management of coastal wetlands in the Greater Brisbane Area (Dutton 1992). If inland migration or growth cannot occur fast enough to account for the rise in sea level, then mangroves will become progressively smaller with each successive generation and may perish (UNEP 1994). In the lower Gangetic plain, several hectares of land have been reclaimed through continuous afforestation programmes of the Department of Forest, Government of West Bengal, and large number of NGOs have also been involved in such ecorestoration programmes.

3.3.1 Environmental Factors That Affect Mangrove Response to Sea Level

In order to analyse the impact of sea-level rise on mangrove ecosystems, one must take into account the major factors that affect the ecological balance of that ecosystem. These regulatory factors include substrate type of the mangrove ecosystem, coastal processes, local tectonics, availability of freshwater and sediment, and salinity of soil and groundwater (Belperio 1993; Semeniuk 1994; Blasco et al. 1996). Climatic variability (e.g. changes in rainfall and the frequency and intensity of cyclonic storms) can exacerbate the factors affecting mangrove response to sea level because it can alter the freshwater inflow to mangroves, the sediment and nutrient inputs and the salinity regime. In an analysis of the impacts of sea-level rise on estuaries, Kennish (2002) highlighted the importance of local conditions such as the size and shape of the estuary, its orientation to fetch and local currents, the aerial distribution of wetlands, the geology of the neighbouring watersheds and land use in upland areas. These factors often increase the magnitude of tidal surges and intrusion of seawater in the adjacent villages resulting in salinization of agricultural lands and even groundwater. Thus, vulnerability assessments will be important for determining which areas are more likely to survive despite climatic changes.

Tidal range and sediment supply are two critical indicators of mangrove response to sea-level rise. Mangrove communities in macro-tidal, sediment-rich areas (e.g. mangrove communities in northern Australia; Semeniuk 1994; Woodroffe 1995) may be better able to withstand and survive sea-level rise than those in micro-tidal sediment-starved areas (e.g. mangroves in Caribbean islands; Parkinson et al. 1994). The tidal profiles in different Indian estuaries are ideal for the survival of mangroves (Fig. 3.6), but the ideal condition is attained in Ganga–Brahmaputra basin, where the sediment load is extremely rich. The Ganga–Brahmaputra basin comprises an area of about 10.98 lakh km², and the delta area is about 150,000 km². Each year Ganga and Brahmaputra rivers bring around 166.70 crore tonnes of silt, and it is this silt which has created the world's largest delta and the delta-building process is still ongoing (Banerjee 1999).

The migration of mangroves is also a function of geo-physico-chemical nature of the substratum.

It has been observed that carbonate settings are often associated with coral atolls and islands, where landward migration of mangroves to escape the effects of sea-level rise is not possible and sediments are often limited; thus, mangrove communities in carbonate islands are considered extremely vulnerable to sea-level rise (UNEP 1994). Therefore, sea-level rise is expected to

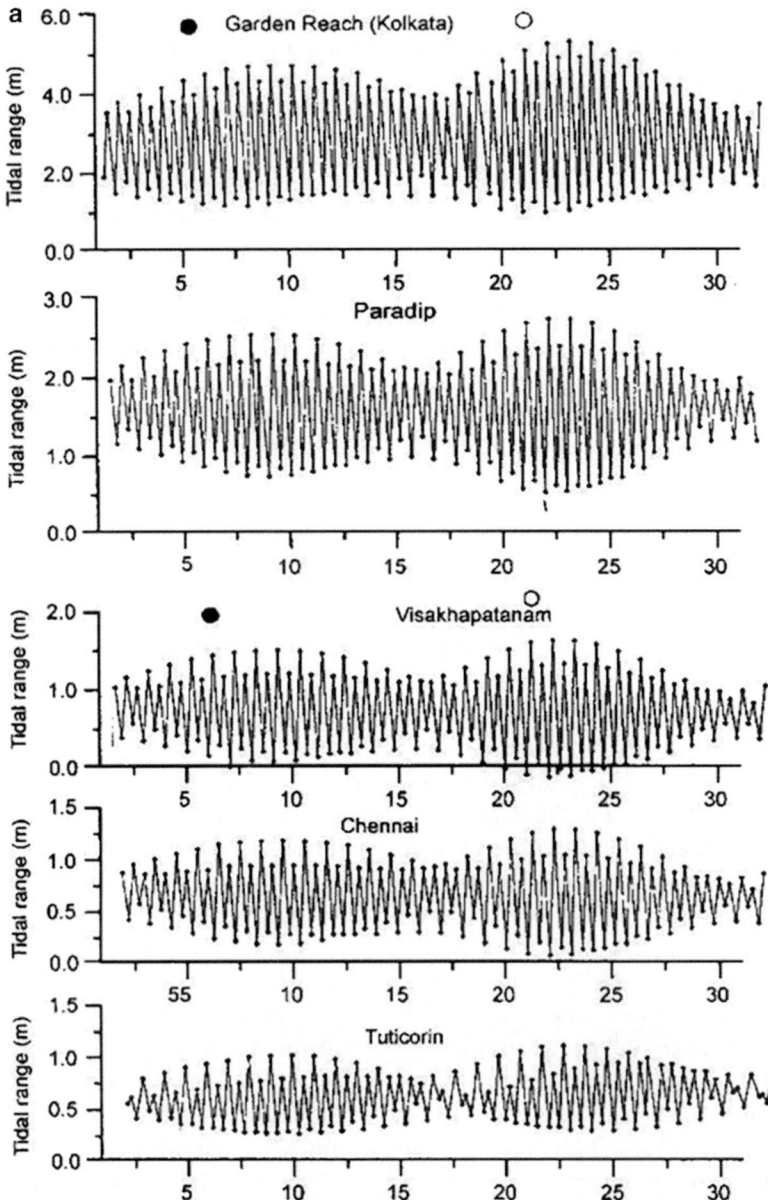

Fig. 3.6 (**a**) Tidal range in the east coast of India. (**b**) Tidal range in the west coast of India (*Source:* Qasim 2004)

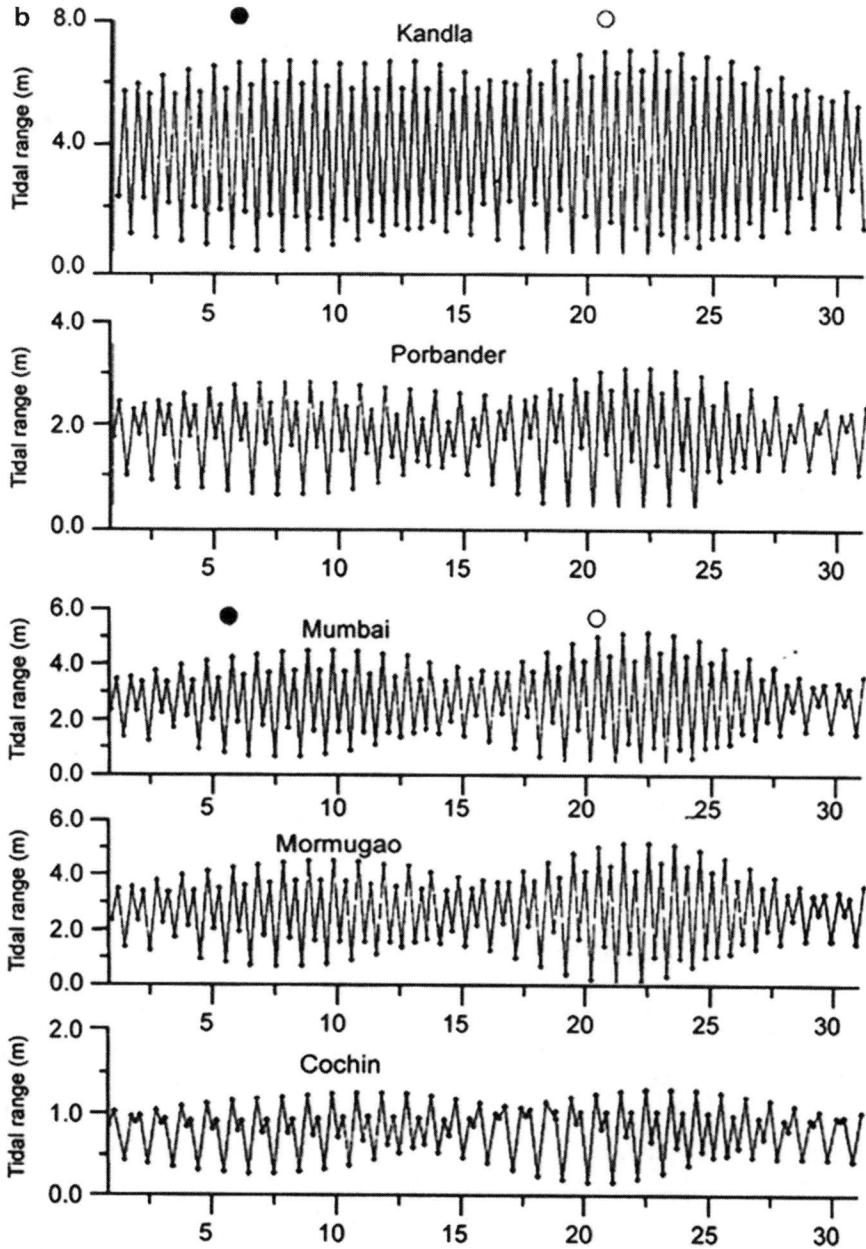

Fig. 3.6   (continued)

decrease the geographic distribution and species diversity of mangroves on small islands with micro-tidal sediment-limited environments (IPCC 1997). Mangroves with access to alloch-thonous sediments, such as riverine mangroves, are more likely to survive sea-level rise than those with low external inputs (Woodroffe 1990; Pernetta 1993). It is important to note that although access to sediment is critical for mangroves to survive sea-level rise, too much sediment (e.g. resulting from poor agricultural practices) can bury their pneumatophores and kill mangroves

(Ellison and Stoddart 1991). In addition to varying sediment input rates, sediment accumulation rates also differ for mangrove ecosystems worldwide. According to Ellison and Stoddart (1991), through accretion, low island mangroves can keep pace with sea-level rise of up to 1.2 mm/year, while high island mangroves can keep pace with rates of 4.5 mm/year, depending on sediment supply. As mentioned above, global projections of sea-level rise are between 1.0 and 8.8 mm/year; thus, mangroves may not survive sea-level rise in some areas. In the low-lying island mangroves in Bermuda, the rate of sediment accretion under mangroves has been 0.8–1.1 mm per 100 years over the last 2,000 years, but the present rate of sea-level rise in Bermuda exceeds 2 mm per year (Ellison 1993), clearly outpacing the sediment accretion rate. Furthermore, the seaward margin of the mangroves has retreated and eroded significantly; Twenty-six per cent of the largest mangrove area at Hungry Bay, Bermuda, has been lost over the last century due to retreat of its seaward edge. On Kosrae Island, Micronesia, most mangrove habitats developed by accumulating mangrove peat with a gradual sea-level rise of 1.0–2.0 mm/year (Fujimoto et al. 1996). A rapid rate of relative sea-level rise of about 10 mm/year occurred between 4,100 and 3,700 BP and caused mangroves to retreat landward and stop accumulating peat (Fujimoto et al. 1996). Therefore, if sea-level rise exceeds 10 mm/year, mangroves on Pacific Islands may move landward and quickly reduce in number. Using Holocene stratigraphic records and sea-level curves for sites around the world, Ellison and Stoddart (1991) found that mangrove ecosystems can keep pace with sea-level rise of 8–9 cm per 100 years, are stressed at 9–12 cm per 100 years and cannot adjust at rates above this level. However, Snedaker (1995) pointed out that over the last 147 years, sea-level rise in Florida has been about 20 cm (double the rate of collapse predicted by Ellison and Stoddart 1991) and mangrove systems in Florida have not collapsed and are even expanding in some areas. Therefore, it is critical to account for site-specific expansion of mangroves in context to sea-level rise.

Individual mangrove species have varying degree of tolerances to the period, frequency and depth of inundation. Mangrove zones are related to shore profile, soils and salinity, and changes in these can lead to alteration in mangrove species composition. Different species may be able to move into new areas at different speeds, making some species capable of accommodating a higher sea-level rise rate than others (Semeniuk 1994). He described how mangroves in Northwest Australia colonize new substrates that become available through erosion, inundation and dilution of hypersaline groundwater of the salt flats. Mangrove zones displace the adjoining zone as sea-level rises. Varying tolerances of inundation and salinity may result in changes in mangrove species composition with changes in inundation and salinity due to sea-level rise. For example, the SELVA-MANGRO model, an integrated landscape model, was used in Florida to predict species regeneration based on probability functions of species and community tolerance to water level and salinity (Ning et al. 2003). In persistently inundated soils, red mangrove seedlings were favoured, whereas in irregularly flooded soils, white and black mangrove seedlings were favoured. This difference in seedling survival may be more due to different tolerances of inundation than to salinity. Although *Rhizophora mangle* (red mangrove), *Avicennia germinans* (black mangrove) and *Laguncularia racemosa* (white mangrove) have different salinity tolerances, their differences are only significant for salinities >50 psu (Menezes et al. 2003). *R. mangle* has the lowest salinity tolerance, about 70 psu of these three mangrove species (Chen and Twilley 1998). Menezes et al. (2003) concluded that pore water salinity had little to no influence on the tree species composition on the forest level. Therefore, the differences in seedling survival may be due to higher tolerance of inundation by *R. mangle*. Some scientists are exploring how different functional root types of mangrove species respond to changes in elevation to determine if certain root structures may be more or less vulnerable to sea-level rise (Vicente 1989; Krauss et al. 2003). In the Caribbean, Vicente (1989) noted that prop roots

Table 3.8 Species-wise peat accumulation rate of true mangroves at Chotomollakhali Island

Species	Peat accumulation rate (mm/year)
Sonneratia apetala	3.9
Avicennia alba	2.8
Avicennia marina	3.5
Avicennia officinalis	3.2
Bruguiera gymnorrhiza	2.1
Excoecaria agallocha	2.9

of *Rhizophora mangle* stand higher above mean sea level than the aerial roots of *Avicennia germinans* which protrude only slightly out of the mud. These authors suggest that rapid sea-level rise may lead to local extinctions of *A. germinans* but have an insignificant effect on *R. mangle*. Ellison and Stoddart's (1991) work in Tonga also suggests that *Rhizophora* communities are better positioned to survive rising sea level due to higher peat accumulation rates beneath *Rhizophora* (5.3 mm/year) than *Bruguiera* and *Excoecaria* (2.6 mm/year). The different rate of peat accumulation by different species of mangroves was also confirmed by the present authors after a rigorous study (during 2004–2009) in the Chotomollakhali Island (88°54′26.71″ longitude and 22°10′40.00″ latitude) of eastern Indian Sundarbans (Table 3.8).

3.3.2   The Oscillating Mangrove Coordinates

The phenomenon of sea-level rise may lead to the phenomenon of shifting the coordinates of mangrove zone. When the force of relative sea-level rise is the predominant component causing change in the position of mangrove belt, landscape-level response of mangroves to relative sea-level rise, over a period of decades and longer, can be predicted based on the reconstruction of paleo-environmental mangrove response to past sea-level fluctuations (Ellison and Stoddart 1991; Woodroffe 1995; Ellison 1993, 2000; Gilman 2004). This prediction can be done on the basis of (a) the sea-level change rate relative to the mangrove

surface, (b) the mangrove's physiographic setting (slope of the land adjacent to the mangrove, slope of the mangrove and presence of obstacles to landward migration) and (c) erosion or progradation rate of the mangrove seaward margin (Ellison and Stoddart 1991; Ellison 1993, 2000; Woodroffe 1995; Alongi 1998; Gilman 2004). There are three general scenarios for mangrove response to relative sea-level rise, given a landscape-level scale and time period of decades or longer (Fig. 3.7).

1. *No Change in Relative Sea Level*
   When sea level is not changing relative to the mangrove surface, the mangrove margins will remain in the same location (Fig. 3.7a), provided other factors (like human intervention) remain constant.

2. *Relative Sea-Level Lowering*
   When sea level is dropping relative to the mangrove surface, the seaward and landward boundaries of mangroves will tend to migrate seaward (Fig. 3.7b), and depending on the topography, the mangrove may also expand laterally.

3. *Relative Sea-Level Rising*
   If sea level is rising relative to the mangrove surface, the mangrove's seaward and landward margins retreat landward (Fig. 3.7c). However, under this situation, the survival of mangroves will depend on getting proper environmental conditions like aquatic salinity, soil quality, frequency and period of tidal inundation. If the optimum conditions are not achieved, the mangroves may get restricted in a narrow patch or fringe or be lost in course of time.

However, over small temporal and spatial scales and where natural and anthropogenic forces other than changing relative sea level exert a larger influence on mangrove margin position, change in mangrove coordinates will be variable (Woodroffe 1995). For example, forces affecting sediment-budget balances, including changes in sediment inputs from rivers, variations in coastal currents and wind directions and strength, and construction of seawalls and other shoreline erosion control structures, can produce erosion or accretion of the mangrove seaward margin irrespective of any change

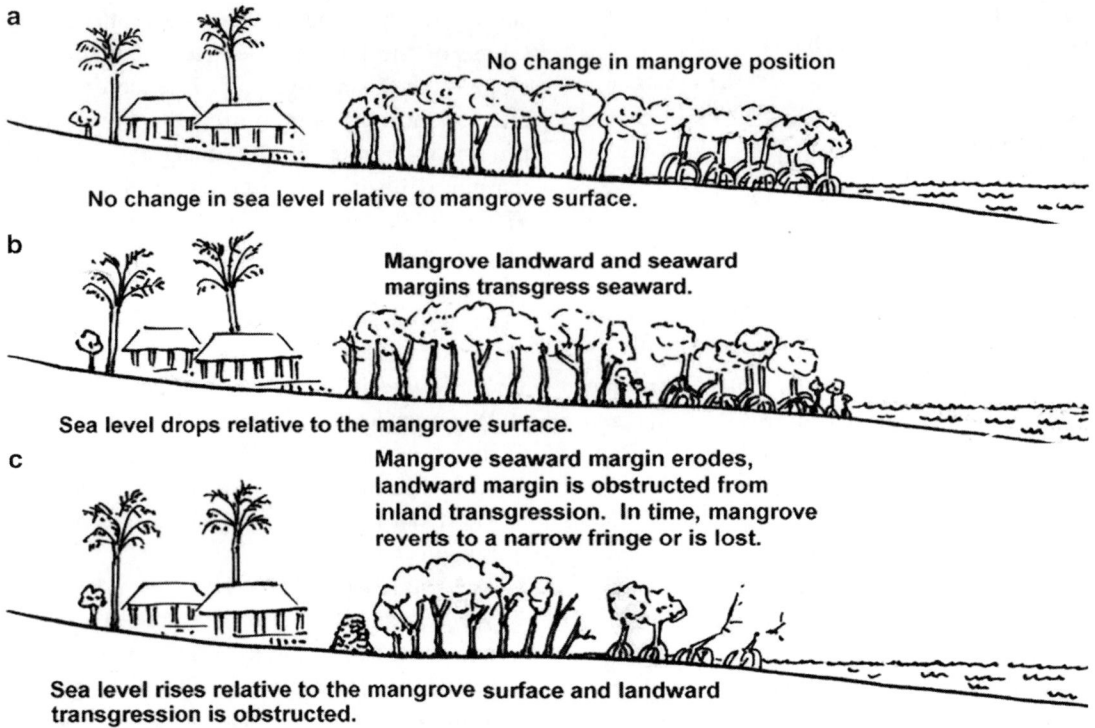

Fig. 3.7 (**a**) Unchanged coordinates of mangrove margin when the sea level is uniform. (**b**) Seaward migration/expansion of mangroves due to dropping of sea level. (**c**) Constriction of mangrove patch due to sea level rise and simultaneous obstruction from landward side

in sea level. A very relevant example in this context can be cited through significant changes in the extent of mangrove patch of Prentice Island (88°20′35.65″ longitude and 21°42′20.26″ latitude) and Lothian Island (88°22′13.99″ longitude and 21°39′01.58″ latitude) in Indian Sundarbans. Survey of Indian maps of different editions (1906–1907, 1920–1921, 1923–1924), Landsat imageries and early literatures of the area (Pal and Bandyopadhyay 1985) provided us to excavate the past pictures of mangrove patch of the two islands (Fig. 3.8).

It is interesting to note how the coordinates of mangrove patch expanded both at Prentice and Lothian Islands over a period of 78 years. The oscillatory nature of mangrove coordinates varies with time, depending on physical, chemical and anthropogenic factors operating in the area. In many cases, the pioneer species (biological agent) provides a congenial environment for the growth and succession of the mangrove species.

Fact Below the Carpet

Tomorrow's weather is never known for certain, but its main features no longer take us by surprise. If storms and tidal surges are forecast, anticipatory responses can be made. We require similar prognostic capabilities for the climate on a decade-to-century time scale, taking account of all natural and anthropogenic factors that are involved and their interactions. There are limitations in the sphere of meteorological forecasting, due to influence of unresolvable small-scale (local) effects within the megascale matrix. Nevertheless, the main features of global climate have shown reasonable stability over the past 5,000 years, with an average surface temperature of $15\pm1°C$ during that period. Considering this, a long-term disaster management plan needs to be formulated giving thrust on bioshield.

PRENTICE & LOTHIAN IN DIFFERENT STAGES OF DEVELOPMENT

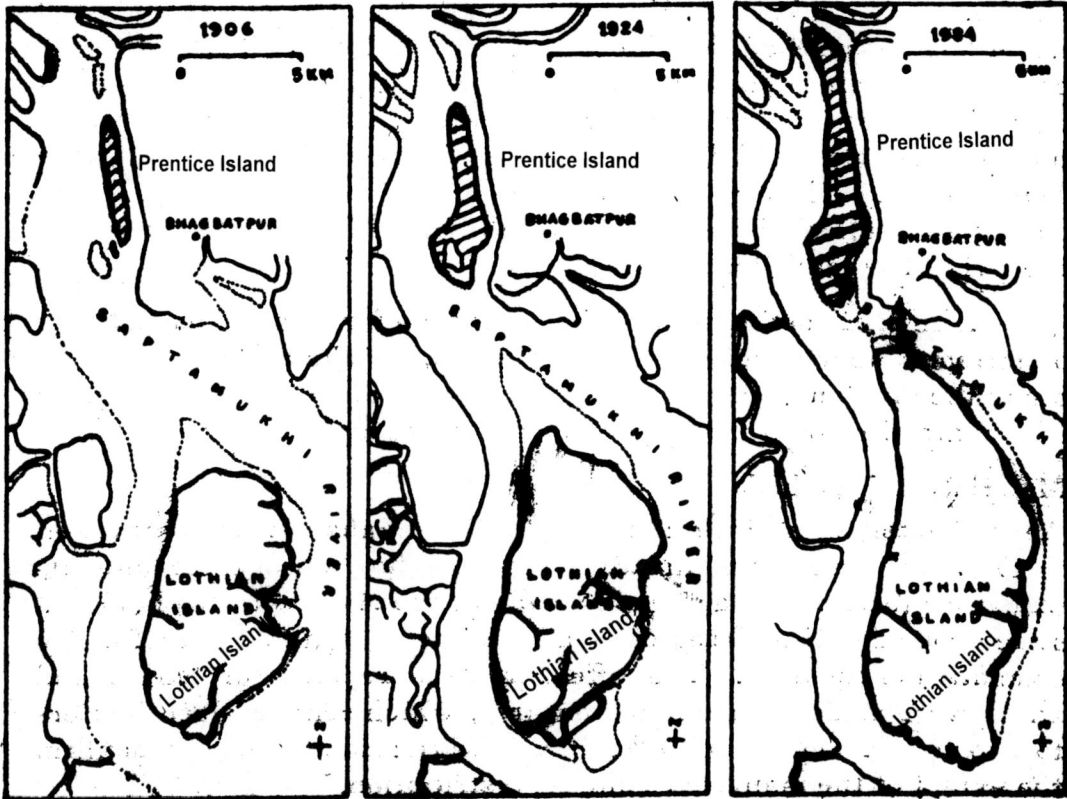

Fig. 3.8 Gradual change of Prentice and Lothian Islands over a period of 78 years

Important References

Abbott PL (2006) Natural disasters. McGraw-Hill Higher Education, Boston

Alongi DM (1998) Coastal ecosystem processes. CRC Press, Boca Raton

Badola R, Hussain SA (2005a) Valuing the storm protection function of Bhitarkanika mangrove ecosystem, India. Environ Conserv 32(1):1–8

Badola R, Hussain SA (2005b) Environmental benefits of mangrove forests: perceptions of local people from the Bhitarkanika Conservation Area, India. Int Forest Rev 7(5):223

Banerjee M (1999) A report on the impact of Farakka barrage on the human fabric. Report submitted to World Commission on Dams on behalf of South Asian Network on Dams, Rivers and People (SANDRP), 29pp

Belperio AP (1993) Land subsidence and sea-level rise in the Port-Adelaide estuary – implications for monitoring the Greenhouse-effect. Aust J Earth Sci 40(4):359–368

Blasco F, Saenger P, Janodet E (1996) Mangroves as indicators of coastal change. Catena 27:167–178

Brinkman RM, Massel SR, Ridd PV, Furukawa K (1997) Surface wave attenuation in mangrove forests. Pac Coasts Ports 97(2):941–946

Britsch LD, Kemp EB (1990) Land loss rates: Mississippi River deltaic plain. US Corps of engineers technical report GL-90-2. U.S. Army Corps of Engineers, New Orleans

Chatenoux B, Peduzzi P (2007) Impacts of the 2004 Indian Ocean tsunami: analysing the potential protecting role of environmental features. Nat Hazards 40:289–304

Chen R, Twilley RR (1998) A gap dynamics model of mangrove forest development along the gradients of soil salinity and nutrient resources. J Ecol 86:37–51

Chittibabu P, Dube SK, MacNabb JB, Murty TS, Rao AD, Mohanty UC, Sinha PC (2004) Mitigation of flooding and cyclone hazard in Orissa, India. Nat Hazards 31:455–485

Dahdouh-Guebas F, Jayatissa LP, Di Nitto D, Bosire JO, Lo Seen D, Koedam N (2005) How effective were mangroves as a defence against the recent tsunami? Curr Biol 15:R443–R447

Danielsen F, Sørensen MK, Olwig MF, Selvam V, Parish F, Burgess ND, Hiraishi T, Karunagaran VM,

Rasmussen MS, Hansen LB, Quarto A, Suryadiputra N (2005) The Asian tsunami: a protective role for coastal vegetation. Science 310:643

Delaune RD, Patrick WH, Lindau CW, Smith CJ (1990) Nitrous oxide and methane emission from Gulf coast wetlands. In: Bouman AF (ed) Soils and the greenhouse effect. Wiley, New York, pp 498–501

Dinar CI (2002) Research on tsunami hazard and its effects on Indonesia coastal region: first year's activities report. In: Proceedings of the fifth multi-lateral workshop on the development of earthquake and tsunami disaster mitigation technologies and their integration for Asia-Pacific Region, Bangkok (5th EqTAP WS)

Dutton IM (1992) Developing a management strategy for coastal wetlands. In: Shafer C, Wang Y (eds) Island environment and coastal development. Nanjing University Press, Nanjing, pp 285–303

Ellison J (1993) Mangrove retreat with rising sea level, Bermuda. Estuar Coast Shelf Sci 37:75–87

Ellison JC (2000) Chapter 15: How South Pacific mangroves may respond to predicted climate change and sea level rise. In: Gillespie A, Burns W (eds) Climate change in the South Pacific: impacts and responses in Australia, New Zealand, and small Islands states. Kluwer Academic Publishers, Dordrecht, pp 289–301

Ellison JC, Stoddart DR (1991) Mangrove ecosystem collapse during predicted sea-level rise: Holocene analogues and implications. J Coast Res 7:151–165

Fujimoto K, Miyagi T, Kikuchi T, Kawana T (1996) Mangrove habitat formation and response to Holocene sea-level changes on Kosrae Island, Micronesia. Mangroves Salt Marshes 1(1):47–57

Gilman E (2004) Assessing and managing coastal ecosystem response to projected relative sea-level rise and climate change. Prepared for the International Research Foundation for development forum on small Island developing states: challenges, prospects and international cooperation for sustainable development. Contribution to the Barbados+10 United Nations international meeting on sustainable development of small Island developing states, Port Louis, 10–14 Jan 2005

Hamzah L, Harada K, Imamura F (1999) Experimental and numerical study on the effect of mangroves to reduce tsunami. Tohoku J Nat Disaster Sci 35:127–132

Harada K, Imamura F (2005) Effects of coastal forest on tsunami hazard mitigation – a preliminary investigation. Adv Nat Technol Hazards Res 23:279–292

Hendry MD, Digerfeldt G (1989) Palaeogeography and palaeoenvironments of a tropical coastal wetland and adjacent shelf during Holocene submergence, Jamaica. Palaeogeography Palaeoclimatol Palaeoecol 73:1–10

Hiraishi T (2003) Tsunami risk and countermeasure in Asia and Pacific Area: applicability of greenbelt Tsunami prevention in the Asia and Pacific Region. Sixth multi-lateral workshop on development of Earthquake and Tsunami Disaster Mitigation Technologies and its Integration for the Asia-Pacific Region (6th EqTAP WS) organized by Earthquake Disaster Mitigation Research Center, NIED, Ise-Kashikojima

Hiraishi T, Harada K (2003) Greenbelt Tsunami prevention in South Pacific region. Report of the Port and Airport Research Institute 42, 1e23. http://eqtap.edm.bosai.go.jp/useful_outputs/report/hiraishi/data/papers/greenbelt.pdf

Hiraishi T, Koike N (2001) Tsunami risk assessment and management: a practical countermeasure to tsunami risk in Asia and Pacific Region. Fourth multi-lateral workshop on development of earthquake and tsunami disaster mitigation technologies and its integration for the Asia-Pacific Region (4th EqTAP WS) Kamakura-City, Kanagawa

Houghton J, Ding Y, Griggs D, Noguer M, van der Linden P, Dai X, Maskell K, Johnson C (eds) (2001) Climate change 2001: the scientific basis. Published for the Intergovernmental Panel on Climate Change. Cambridge University Press, Cambridge/New York, 881pp

Intergovernmental Panel on Climate Change (IPCC) (1997) The regional impacts of climate change: assessment of vulnerability. Cambridge University Press, Cambridge

Jones M (2002) Climate change – follow the mangroves and sea the rise. National Parks J 46(6)

Kabir MM, Ahmed MMZ, Azam MH, Jakobsen F (2006) Effect of afforestation on storm surge propagation: a mathematical model study. Institute of Water Modeling, http://www.iwmbd.org/html/PUBS/publications/P015.PDF. Bangladesh, 10 July 2006

Kathiresan K (2003) How do mangrove forests induce sedimentation? Rev Biol Trop 51(2):355–360

Kathiresan K, Rajendran N (2005) Coastal mangrove forests mitigated tsunami. Estuar Coast Shelf Sci 65:601–606

Kennish MJ (2002) Environmental threats and environmental future of estuaries. Environ Conscrv 29(1):78–107

Kerr AM, Baird AH, Campbell SJ (2006) Comments on "coastal mangrove forests mitigated Tsunami" by K. Kathiresan and Rajendran, N. Estuar Coast Shelf Sci 67:539–541 [Estuarine Coastal Shelf Science, 2005, 65, 601–606]

Koteswaram P (1984) Climate and mangrove forests. Report of the second introductory training course on mangrove ecosystems. Sponsored by UNDP and UNESCO, Goa, pp 29–46

Krauss KW, Allen JA, Cahoon DR (2003) Differential rates of vertical accretion and elevation change among aerial root types in Micronesian mangrove forests. Estuar Coast Shelf Sci 56:251–259

Latief H, Hadi S (2007) The role of forests and trees in protecting coastal areas against Tsunamis. In: Braatz S, Fortuna S, Broadhead J, Leslie R (eds) Coastal protection in the aftermath of the Indian Ocean Tsunami: what role for forests and trees?. Proceedings of the regional technical workshop, Khao Lak, 28–31 Aug 2006, FAO, Bangkok, pp 5–35. http://www.fao.org/forestry/site/coastalprotection/en/

Liu WC, Hsu MH, Wang CF (2003) Modeling of flow resistance in mangrove swamp at mouth of Tidal Keelung River, Taiwan. J Waterw Port Coast Ocean Eng 129(2):86–92

Manson FJ, Loneragan NR, Phinn SR (2003) Spatial and temporal variation in distribution of mangroves in Moreton Bay, subtropical Australia: a comparison of pattern metrics and change detection analyses based on aerial photographs. Estuar Coast Shelf Sci 57:653–666

Massel SR, Furukawa K, Brinkman RM (1999) Surface wave propagation in mangrove forests. Fluid Dyn Res 24:219–249

Mazda Y, Wolanski E, King B, Sase A, Ohtsuka D, Magi M (1997) Drag force due to vegetation in mangrove swamps. Mangroves Salt Marshes 1:193–199

Mazda Y, Kobashi D, Okada S (2005) Tidal-scale hydrodynamics within mangrove swamps. Wetl Ecol Manage 13:647–655

Mazda Y, Wolanski E, Ridd PV (2007) The role of physical processes in mangrove environments: manual for the preservation and utilization of mangrove ecosystems. Terrapub, Tokyo, 598pp

McCoy ED, Mushinsky HR, Johnson D, Meshaka WE (1996) Mangrove damage caused by Hurricane Andrew on the southwestern coast of Florida. Bull Mar Sci 59(1):1–8

Menezes M, Berger U, Worbes M (2003) Annual growth rings and long-term growth patterns of mangrove trees from the Bragança Peninsula, North Brazil. Wetl Ecol Manage 11:233–242

Morris JT, Kjerfve B, Dean JM (1990) Dependence of estuarine productivity on anomalies in mean sea level. Limnol Oceanogr 35:926–930

Nayman JA, Delaune RD, Patrick WH (1990) Wetland soil formation in the rapidly subsiding Mississippi River deltaic plain: mineral and organic matter relationships. Estuar Coast Shelf Sci 31:57–69

Ning ZH, Turner RE, Doyle T, Abdollahi KK (2003) Integrated assessment of the climate change impacts on the Gulf Coast region. Gulf Coast Climate Change Assessment Council (GCRCC) and Louisiana State University (LSU) Graphic Services, Baton Rouge

Pal AK, Bandyopadhyay MK (1985) The role of mangroves in deltaic morphology: a study in Prentice and Lothian Islands, Sunderbans, W.B. In: Bhosale LJ (ed) The mangroves: proceedings on national symposium on biology, utilization and conservation of mangroves. WWF – India, Kolkata, pp 218–221

Parkinson RW, DeLaune RD, White JC (1994) Holocene sea-level rise and the fate of mangrove forests within the wider Caribbean region. J Coast Res 10:1077–1086

Pernetta JC (1993) Mangrove forests, climate change and sea-level rise: hydrological influences on community structure and survival, with examples from the Indo-West Pacific. A marine conservation and development report. IUCN, Gland, vii+46pp

Qasim SZ (2004) Handbook of tropical estuarine biology. Narendra Publishing House, New Delhi, 131pp

Reed DJ (1999) Response of mineral and organic components of coastal marsh accretion to global climate change. Curr Top Wetl Biogeochem 3:90–99

Records of districts of West Bengal, Midnapur and 24 Parganas Districts (1906–1907, 1920–1921, 1923–1924) Survey of India map, No. 79 C/1, C/2, C/6

Semeniuk V (1994) Predicting the effect of sea-level rise on mangroves in Northwestern Australia. J Coast Res 10(4):1050–1076

Shrestha ML (ed) (1998) The impact of tropical cyclones on the coastal regions of SAARC countries and their influence in the region. SAARC Meteorological Research Centre, Agargaon, 329pp

Snedaker SC (1995) Mangroves and climate change in the Florida and Caribbean region: scenarios and hypotheses. Hydrobiologia 295:43–49

Tanaka N, Sasaki Y, Mowjood MIM, Jinadasa KBSN, Homchuen S (2007) Coastal vegetation structures and their functions in tsunami protection: experience of the recent Indian Ocean tsunami. Landsc Ecol Eng 3:33–45

United Nations Environment Programme (UNEP) (1994) Assessment and monitoring of climatic change impacts on mangrove ecosystems. UNEP regional seas reports and studies. Report no. 154. UNEP, Nairobi

Vermaat JE, Thampanya U (2006) Mangroves mitigate tsunami damage: a further response. Estuar Coast Shelf Sci 69:1–3

Vermaat JE, Thampanya U (2007) Erratum to "mangroves mitigate tsunami damage: a further response". Estuar Coast Shelf Sci 75:564 [Estuarine, Coastal and Shelf Science, 69 (1–2) (2006) 1–3]

Vicente VP (1989) Ecological effects of sea-level rise and sea surface temperatures on mangroves, coral reefs, seagrass beds and sandy beaches of Puerto Rico: a preliminary evaluation. Sci Cienc 16:27–39

White GF, Haas JE (1975) Assessment of research on natural hazards. MIT Press, Cambridge, MA

Woodroffe CD (1990) The impact of sea-level rise on mangrove shoreline. Prog Phys Geogr 14:483–502

Woodroffe CD (1995) Response of tide-dominated mangrove shorelines in northern Australia to anticipated sea-level rise. Earth Surf Process Landf 20(1):65–85

Wu Y, Falconer RA, Struve J (2001) Mathematical modeling of tidal currents in mangrove forests. Environ Model Software 16:19–29

Internet Reference

http://www.fao.org/gpa/sediments/habitat.htm

Impact of Climate Change on Mangroves

4

If we love our children, we must love our Earth with tender care and pass it on, diverse and beautiful, so that man, on a warm spring day 10,000 years hence, can feel peace in a sea of grass, can watch a bee visit a flower, can hear a sandpiper call in the sky, and can find joy in being alive.

Hugh H. Iltis

Climate change has several components of varied nature and scale that affect the ecosystems of the planet Earth. For mangroves, however, the most relevant components include changes in sea level, high water events, storminess, precipitation, temperature, atmospheric CO_2 concentration, ocean circulation patterns, health of functionally linked neighbouring ecosystems as well as human responses to climate change. Of all the outcomes from changes in the atmosphere's composition and alterations to land surfaces, relative sea-level rise may be the greatest threat to mangroves (Field 1995). Although, to date, it has likely been a smaller threat than anthropogenic activities such as conversion for aquaculture and filling (IUCN 1989; Primavera 1997; Valiela et al. 2001; Alongi 2002; Duke et al. 2007), relative sea-level rise is a substantial cause of recent and predicted future reductions in the area and health of mangroves and other tidal wetlands (IUCN 1989; Ellison and Stoddart 1991; Ellison 2000; Cahoon and Hensel 2006; McLeod and Salm 2006; Gilman et al. 2006, 2007a, b).

The Intergovernmental Panel on Climate Change's best estimate of global average sea-level change during the twentieth century, based mainly on tide gauge observations, is 1.5 ± 0.5 mm/year (Church et al. 2001), while Church et al. (2004) provide an estimate of 1.8 ± 0.3 mm/year from 1950 to 2000. Global sea-level rise during the twentieth century has been significantly influenced by global warming through thermal expansion of seawater and loss of land ice (Church et al. 2001).

Studies provide an estimate that during the last century, the global sea level rose by 10–14 cm. A rise of 5 cm was attributed to thermal expansion of waters while the remaining 5–10 cm was due to the deglaciation. The sea-level rise is not uniform throughout the entire world. It is extremely region specific. India has been identified as one among 27 countries which are most vulnerable to the impacts of global warming related accelerated sea-level rise (UNEP 1989). The high degree of vulnerability of Indian coasts can be mainly attributed to extensive low-lying coastal area, high population density, frequent occurrence of cyclones and storms, high rate of coastal environmental degradation on account of pollution and non-sustainable development. Based on the sea-level records at Indian cities like

A. Mitra, *Sensitivity of Mangrove Ecosystem to Changing Climate*,
DOI 10.1007/978-81-322-1509-7_4, © Springer India 2013

Bombay, Cochin, Chennai and Visakhapatnam, a rise of 0.67 mm/year is noticed. Similar studies on a global scale suggest a rise of 1.0–1.5 mm/ year. The sea-level rise in the next century is estimated to increase by a metre. Researchers working in the field of climate change forecast that Bangladesh will be the most affected and could lose about 40 % of its territory and create some 28 million ecological refuges. Pakistan, Surinam, Senegal, Thailand, Mozambique and Maldives will also be affected. All these predictions have of course high degree of uncertainties.

High water event is another major associate of climate change that has significant adverse impact on coastal ecosystems. Projected increases in the frequency of high water events (Church et al. 2001, 2004) could affect mangrove health and composition due to changes in salinity, recruitment, inundation and changes in the wetland sediment budget (Gilman et al. 2006). Storm surges can also flood mangroves and, when combined with sea-level rise, lead to mangrove destruction. Flooding, caused by increased precipitation, storms or relative sea-level rise, may result in decreased productivity, photosynthesis and survival (Ellison 2000). Inundation of lenticels in the aerial roots can cause the oxygen concentrations in the mangrove to decrease, resulting in death of the tree (Ellison 2004). Inundation is also projected to decrease the ability of mangrove leaves to conduct water and to photosynthesize (Naidoo 1983).

The sea-level rise has high possibility to affect the floral and faunal communities of coastal and estuarine regions. The mangroves, being the primary coast guard in the estuarine and coastal systems, are the first to bear the effect of sea-level rise. According to Field (1995), sea-level rise is the greatest climate change challenge that mangrove ecosystems will face. Geological records indicate that previous sea-level fluctuations have created both crises and opportunities for mangrove communities, and they have survived or expanded in several refuges (Field 1995).

The rate of change of relative sea level as measured at a tide gauge spot may differ substantially from the relative change of sea-level rate occurring in coastal wetlands due to changing

elevation of the wetland sediment surface. Additional variability might be caused by differences in local tectonic processes, coastal subsidence, sediment budgets and meteorological and oceanographic factors between the section of coastline where the coastal wetland is situated and the position of the tide gauge where it is placed, especially when the tide gauge is distant from the wetland. However, even if the variations exist, the vulnerability of the mangrove ecosystem to sea-level rise cannot be ignored as they are the first line vegetation bordering the land and the sea. Global climate change, specifically changes in temperature, CO_2, precipitation, hurricanes and storms, and sea level, combined with anthropogenic threats, will threaten the resilience of mangroves.

The impacts of climate change on mangroves have been summarized briefly in this chapter, although sea-level rise will be emphasized because it is projected to be the greatest climate change threat to mangroves. It is very important to note in this context that the impacts on mangroves will not occur in isolation; the response of mangroves to climate change will be the net result of all these impacts acting synergistically. Antagonistic effects of different environmental variables acting on mangroves cannot be ignored.

4.1   Effect of Climatic Factors on Mangroves

4.1.1   Effect of Rise in Temperature

The Earth has warmed by 0.6–0.8 °C since 1880 and it is projected to warm 2–6 °C by 2100 mostly due to human activity (Houghton et al. 2001). Mangroves are not expected to be adversely impacted by the projected increase in sea temperature (Field 1995). Most mangroves produce maximal shoot density when mean air temperature rises to 25 °C and stop producing leaves when the mean air temperature drops below 15 °C (Hutchings and Saenger 1987). At temperatures above 25 °C, some species may show a declining leaf formation rate (Saenger and Moverly 1985). Temperatures above 35 °C have led to thermal stress affecting mangrove root structures and establishment of mangrove seedlings (UNESCO 1992). At leaf temperatures of 38–40 °C, almost no photosynthesis occurs (Clough et al. 1982; Andrews et al. 1984). Some scientists have suggested that mangroves will move poleward with increasing air temperatures (UNEP 1994; Field 1995; Ellison 2005). Although it is possible that some species of mangroves will migrate to higher latitudes where such range extension is limited by temperature, Woodroffe and Grindrod (1991) and Snedaker (1995) suggest that extreme cold events are more likely to limit mangrove expansion into higher latitudes.

The oscillation of temperature also affects the mangrove photosynthesis. An optimum temperature range exists for mangroves in which the glucose synthesis exhibits maximum value, but this range is not strictly uniform for all the mangrove species. Andrews and Muller (1985) have shown that the rate of photosynthesis is much reduced at higher leaf temperatures. In few mangrove species examined so far, the rate of photosynthesis appears to be relatively unaffected by leaf temperature over the range 17–25 °C, but falls sharply at temperatures much above 35 °C and is close to zero at 40 °C. The temperature response of photosynthesis in mangroves is thus similar to other C_3 plants and unlike that of C_4 plants, which generally have a higher optimum temperature for photosynthesis. In Florida mangroves, little or no photosynthesis occurred at 40 °C, and the temperature optima for photosynthesis was below 35 °C (Moore et al. 1972). There are some views regarding the influence of leaf temperatures on the process and rate of mangrove photosynthesis. According to Andrews et al. (1984) high leaf temperatures may influence photosynthesis indirectly through its effect on the vapour pressure deficit between the leaf and its environment. Apart from this, high leaf temperature also has an adverse effect on carboxylation reactions, with the result that the CO_2 compensation point rises with increasing leaf temperatures.

Considering the community structure of mangrove flora, the effect of rising temperature is, however, different in some regions. Increases in temperature are predicted to benefit the Pacific Islands, because warming is projected to increase the diversity of marginal mangroves at higher latitudes, currently home to only Avicennia species (Burns 2001). In the Pacific Islands, warming is projected to facilitate mangrove expansion into salt-marsh communities (Burns 2001).

Mangrove species in China have demonstrated varying thermal tolerances. Li and Lee (1997) divided the mangrove species in China into three classes based on thermal tolerance: (1) cold-resistant eurytopic species (e.g. Kandelia candel, Avicennia marina and Aegiceras corniculatum), (2) cold-intolerant (thermophilic) stenotopic species (e.g. Rhizophora mucronata, R. apiculata, Lumnitzera littorea, Nypa fruticans and Pemphis acidula) and (3) thermophilic eurytopic species, (e.g. R. stylosa, Bruguiera sexangula, B. gymnorrhiza, Excoecaria agallocha and Acrostichum aureum) (Zhang and Lin 1984).

Despite the uncertainties of how temperature changes will affect the species composition or the seasonal patterns of reproduction and flowering of mangroves, an increase in sea-surface and air temperatures would likely benefit mangroves living near the poleward limits of current distributions, leading to increased species diversity, greater litter production and larger trees in these mangrove systems (Edwards 1995). Temperature increases may impact mangroves by changing the seasonal patterns of reproduction and the length of time between flowering and the fall of mature propagules (UNEP 1994; Ellison 2000). Soil temperature change is expected to be of the same magnitude and rate of increase as that of sea-surface temperature, although variations in soil temperature are generally much less than those of air temperature based on the large capacity of saturated soils to retain heat (UNEP 1994). Therefore, even if soil temperatures increase, it is not likely to adversely affect mangroves (UNEP 1994). Increased sediment temperature may also cause increased growth rates and multiplication of bacteria resulting in increased rates of nutrient recycling (through microbial degradation) and regeneration.

4.1.2 Effect of Alteration of Precipitation Pattern

Precipitation rates are predicted to increase by about 25 % by 2050 in response to global warming. However, at regional scales, this increase will be unevenly distributed with either increases or decreases projected in different areas (Knutson and Tuleya 1999; Walsh and Ryan 2000; Houghton et al. 2001). Changes in precipitation patterns caused by climate change may have a profound effect on both the growth of mangroves and their aerial extent (Field 1995; Snedaker 1995). Regional climate models predict that precipitation will decrease in certain areas (e.g. Central America during the months of winter, Australia in winter) (Houghton et al. 2001). Decreased precipitation may not only result in less freshwater input to mangroves, but it may also cause less freshwater input into the groundwater which has significant probability to increase salinity of the ambient media. Increase in soil salinity results in the rise of salt content in mangrove tissues. Increased salinity and lack of freshwater is likely to decrease mangrove productivity, growth and seedling survival and may change species composition favouring more salt-tolerant species (Ellison 2000, 2004). The examples of Australian mangroves can be cited in this context. These mangroves are stunted, of narrower margins, and interrupted by salt flats in areas of lower rainfall mainly due to salt stress (Ellison 2000). Decreased rainfall, combined with the increase in evaporation in arid areas, is also likely to result in a decrease in mangrove area, decrease in diversity and projected loss of the landward zone to unvegetated hypersaline flats (Snedaker 1995).

In regions where rainfall is projected to increase due to climate change (e.g. northern mid-latitude regions in winter and in the Pacific Islands north of 17°S; Houghton et al. 2001), mangrove area, diversity and growth rates may increase (Ellison 2000). Maximal growth of mangroves has been linked to low salinities (Burchett et al. 1984; Clough 1984). Thus, if precipitation increases and results in decreased soil salinity, mangrove growth rates may increase in some species (Field 1995). In Australia, mangroves grow taller, more productive, and more diverse in areas of higher rainfall (Ellison 2000). Harty (2004) suggests that increases in rainfall reduce salinity levels within salt marshes which allows mangroves to migrate and outcompete salt-marsh vegetation. This trend of mangrove transgression into salt-marsh habitat has been observed in southeast Australia due to increases in precipitation.

In Indian Sundarbans region, mangrove species like *Heritiera fomes* and *Nypa fruticans* are gradually vanishing from the central region owing to complete cut-off of the freshwater supply due to Bidyadhari siltation. These species are, however, coming up luxuriantly in low saline pockets of Sundarbans particularly in the western part, which is gradually freshening due to more flow of freshwater through Ganga–Bhagirathi–Hugli channel (Mitra et al. 2009). Decrease in salinity in this important World Heritage site, by ways of Bidyadhari dredging and interlinking the Ganga–Hugli–Bhagirathi channel (in the western Indian Sundarbans) with the Rivers of the hypersaline central sectors (like Matla), may increase the mangrove species diversity in and around the Matla River in the central Indian Sundarbans.

4.1.3 Effect of Rise in CO_2 Concentration

Atmospheric CO_2 has increased from 280 ppm by volume (ppmv) in the year 1880 to nearly 370 ppmv in the year 2000 (Houghton et al. 2001), and this trend will continue due to intense industrialization and urbanization throughout the globe. Researchers, however, state that most atmospheric CO_2 resulting from burning of fossil fuels will be absorbed into the ocean affecting ocean chemistry. According to UNEP (1994), the efficiency of mangrove water use will be enhanced, and there will be specific species variation in response to elevated CO_2. Due to the increase in water use efficiency, mangroves in arid regions may benefit because decreased water loss via transpiration will accompany CO_2 uptake (Ball and Munns 1992). Increased salinity may, however, pose hindrance to this benefit. If salinity increases in arid regions, then this advantage may be lost, because increases in CO_2 do not affect mangrove growth when salinity is too high for a species to maintain water uptake (UNEP 1994). Increases in CO_2 are not likely to cause mangrove canopy photosynthesis to increase significantly (UNEP 1994). Several scientists, however, documented the positive influence of rising CO_2 on mangrove vegetation. In an experiment aimed to test the effects of humidity, salinity and increased CO_2 on two Australian mangrove species, *Rhizophora stylosa* and *Rhizophora apiculata*, the rate of photosynthesis showed significant increase with increased levels of CO_2. In this experiment, the mangroves were grown in glasshouses for 14 weeks with different combinations of atmospheric CO_2 (340 and 700 ppm), relative humidity (43 and 86 %) and salinity (25 and 75 % of seawater) to determine the effects of these variables on their development and growth. Although *Rhizophora stylosa* has a slower relative growth rate and greater salt tolerance than *Rhizophora apiculata*, the scientists concluded that elevated CO_2 significantly increased rates of net photosynthesis in both mangrove species, but only when grown at the lower salinity level. In addition, while increased CO_2 levels did not significantly affect the relative growth rate of either species, the average growth rates of both species increased with atmospheric CO_2 enrichment in the lower salt environment. These scientists postulated that increased levels of CO_2 might allow these two mangrove species to expand into areas of greater aridity, thus increasing species diversity in those regions. Snedaker and Araújo (1998) exposed four mangrove species *Rhizophora mangle*, *Avicennia germinans*, *Laguncularia racemosa* and *Conocarpus erectus* to increased CO_2 (361–485 ppm). All four species demonstrated significant decreases in stomatal conductance and transpiration and an increase in instantaneous transpiration efficiency. Only *L. racemosa* demonstrated a significant decrease in net primary productivity when exposed to increased CO_2. Snedaker and Araújo (1998) suggested that increased levels of CO_2 on a global scale may result in a competitive disadvantage of *L. racemosa* in mixed mangrove communities

relative to the other species whose rates of net primary productivity are not significantly affected by increases in CO_2. The results of this study indicate that global increases in CO_2 may result in a competitive advantage of mangroves in arid regions due to their ability to minimize water use during periods of water stress while maintaining relatively high rates of CO_2 uptake (Snedaker and Araújo 1998). Farnsworth et al. (1996) analysed the effects of doubled levels of CO_2 on *Rhizophora mangle* seedlings. The seedlings demonstrated significant increases in biomass, total stem length, branching activity and total leaf area compared to seedlings grown in normal levels of CO_2. In this study, reproduction of *Rhizophora mangle* was achieved after only 1 year of growth in a high CO_2 environment, whereas it typically takes a full 2 years before they are able to reproduce in the field; thus, elevated CO_2 also appeared to accelerate maturation in addition to growth. However, Ellison et al. (1996) predict that whether increased atmospheric CO_2 results in enhanced growth of mangroves, it will likely not be enough to compensate for the negative impacts of sea-level rise. One indirect impact on mangroves of increased temperature and CO_2 is the degradation of coral reefs caused by mass bleaching and impaired growth (Hoegh-Guldberg 1999). Damage to coral reefs may adversely impact mangrove systems (where mangrove-reef couple system exists) that depend on the reefs to provide shelter from wave action.

4.2   Effect of Physico-Chemical Variables on Mangroves

4.2.1   Effect of Sea-Level Rise

In the last century, eustatic sea-level has risen 10–20 cm primarily due to thermal expansion of the oceans and melting of glacial ice caused by global warming (Church et al. 2001). Several climate models project an accelerated rate of sea-level rise over coming decades (Church et al. 2001). Sea-level changes have also been influenced by tectonic and isostatic adjustments (i.e. ocean basin deformation and land subsidence or emergence) (Kennish 2002). Past sea-level change has been measured by tide gauges at different locations around the world. Tide gauges are not evenly distributed around the globe which biases the data and does not provide an accurate picture of the global pattern of sea-level change (Cabanes et al. 2001). However, despite the uncertainties in tide gauge data, scientists estimate that the global average sea level rose at a rate of 1.0–2.0 mm (mm)/year during the twentieth century (Houghton et al. 2001). This increase is an order of magnitude larger than the average rate over the previous several 1,000 years (CSIRO 2001). During the twenty-first century, mean sea-level projections range from 0.09 to 0.88 m (Houghton et al. 2001). In addition to the uncertainties of global sea-level rise, uncertainties also exist for how regions will experience different rates and magnitudes of sea-level rise. Regional sea-level rise is affected by tectonic movements that can cause land subsidence or uplift. Natural and human-induced sediment compaction can also exacerbate the impacts of sea-level rise. Humans contribute to land subsidence through coastal development that causes deficits in the sediment budget, shipping channels that cause bank erosion, groundwater or oil extraction that causes submergence, and dredging and mining that causes losses of land. The combination of global sea-level rise and local impacts that cause land subsidence threatens the existence of mangroves worldwide. Mangrove systems cannot keep pace with the rate of sea-level rise. Hence, they tend to migrate in response to rising sea level. As already been discussed, three situations are observed in mangrove ecosystems in relation to sea-level rise (Table 4.1).

Researchers have observed that migration of mangroves in a fragile environment (towards sea) often causes mortality due to stresses caused by a rising sea level such as erosion resulting in weakened root structures and falling of trees, increased salinity and too high a duration, frequency and depth of inundation (Naidoo 1983; Ellison 1993, 2000, 2006; Lewis 2005). Mangroves migrate landward via seedling recruitment and vegetative reproduction as new habitat becomes available landward through erosion, inundation

Table 4.1 Mangrove shifting in relation to changing sea level

Condition	Effect
Relative to mangrove surface when sea level remains unchanged	Mangrove position remains stable
Relative to mangrove surface when sea level falls	Mangrove margins migrate seaward
Relative to mangrove surface when sea level rises	Mangrove's seaward and landward margins retreat landward

Remark: This migratory pattern holds good when conditions like human intervention, land availability, topography and soil quality of the available land, tidal inundation and water quality are mangrove friendly

and concomitant change in salinity (Semeniuk 1994). Depending on the ability of individual mangrove species to colonize newly available habitat at a rate that keeps pace with the rate of relative sea-level rise (Field 1995; Duke et al. 1998), slope of adjacent land and presence of obstacles to landward migration of the landward mangrove boundary (sea walls, roads, etc.), some mangroves will gradually be reduced in area, may get restricted to a narrow fringe or face mortality. The mangrove-dominated deltaic Sundarbans is noted for intense erosion in and around a number of islands, and hence mangrove afforestation programmes are undertaken on a regular basis by the government departments to stabilize the mudflats encircling these islands. The necessity of space for mangrove expansion is often taken into consideration while planting mangrove seedlings in the mudflats of Sundarbans in fronts of the village dykes. Narrow-width mudflats are generally avoided as the chance of mangrove expansion is less due to erosion activities caused by the tidal waters of the adjacent estuaries.

4.2.2 Effect of Alteration of Aquatic Salinity

The polar ice melting, being a consequence of global warming, results in the intrusion of seawater in the upstream zone of estuaries and connecting bays.

This can create alteration in the physiological conditions of flora and fauna inhabiting the estuarine stretch. It has been observed that increased salinity decreases mangrove net primary productivity, growth and seedling survival and may possibly change competition between mangrove species (Ellison 2000, 2004). The alteration of salinity in the river mouths and estuaries is a function of several factors like sea-level rise, volume of freshwater discharge from barrage (constructed in the upstream zone of rivers or estuaries), evaporation, precipitation and continental run-off. Changes in precipitation patterns are expected to affect mangrove growth and spatial distribution (Field 1995; Ellison 2000) by altering the salinity level of the ambient aquatic phase. Based primarily on links observed between mangrove habitat condition and rainfall trends (Field 1995; Duke et al. 1998), decreased rainfall and increased evaporation will increase salinity, decreasing net primary productivity, growth and seedling survival, altering competition between mangrove species, decreasing the diversity of mangrove zones, causing a notable reduction in mangrove area due to the conversion of upper tidal zones to hypersaline flats. Areas with decreased precipitation will have a smaller water input to groundwater and less freshwater flow to mangroves, increasing salinity. As soil salinity increases, mangrove trees will have increased tissue salt levels and concomitant decreased water availability, which reduces productivity (Field 1995).

Human factors also contribute to the alteration of groundwater salinity, which subsequently affect the overlying soil salinity. In the coastal regions of Tamil Nadu, salinity of groundwater due to the intrusion of seawater into the subsurface aquifer is a major problem (Subramanian 2000). Due to excess withdrawal of groundwater, the water table has fallen too far below thereby allowing seawater to percolate. Similarly, in Gujarat due to uncontrolled withdrawal of groundwater, the groundwater has become highly saline apart from the fact that depth of the water table reaching at places beyond 200 m (Subramanian 2000). It has been predicted that coastal aquifer system will be more contaminated with salinity

bringing greater complicacy to the problem of tapping usable groundwater (Mohanty 1990). In coastal regions of West Bengal and Orissa, the problem of freshwater is fairly acute because of the depth of water table and high cost of lifting the same from the depth of 700–1,000 m. The shallow saltwater table often makes the stored freshwater in ditches and ponds brackish and surface soil saline in nature. This reduces the mangrove species diversity and productivity to a great extent. The impact of climate change on coastal groundwater system and subsequently on mangrove diversity is schematically represented in Fig. 4.1.

High salinity increases the availability of sulphate in seawater and tidal water, which would increase anaerobic decomposition of peat, increasing the mangrove's vulnerability to any rise in relative sea level (Snedaker 1993, 1995). Reduced precipitation can result in mangrove encroachment into salt-marsh and freshwater wetlands (Saintilan and Wilton 2001; Rogers et al. 2005) if salt-marsh system coexists with mangroves. Hypersalinity in mangrove ecosystem is reduced naturally by increased precipitation, which will result in increased growth rates and biodiversity, increased diversity of mangrove zones and an increase in mangrove area, with the colonization of previously unvegetated areas of the landward fringe within the tidal wetland zone (Field 1995; Duke et al. 1998). For instance, mangroves tend to be taller and more diverse on high rainfall shorelines relative to low rainfall shorelines, as observed in most global locations, including Australia (Duke et al. 1998). Areas with higher rainfall have higher mangrove diversity and productivity probably due to higher supply of fluvial sediment and nutrients, as well as reduced exposure to sulphate and reduced salinity (McKee 1993; Field 1995; Ellison 2000). Mangroves will likely increase peat production with increased freshwater inputs and concomitant reduced salinity due to decreased sulphate exposure (Snedaker 1993, 1995).

Reports on the adverse impact of salinity on chlorophyll content of mangrove species are also available (Kotimire and Bhosale 1980; Shinde and Bhosale 1985). An experiment conducted

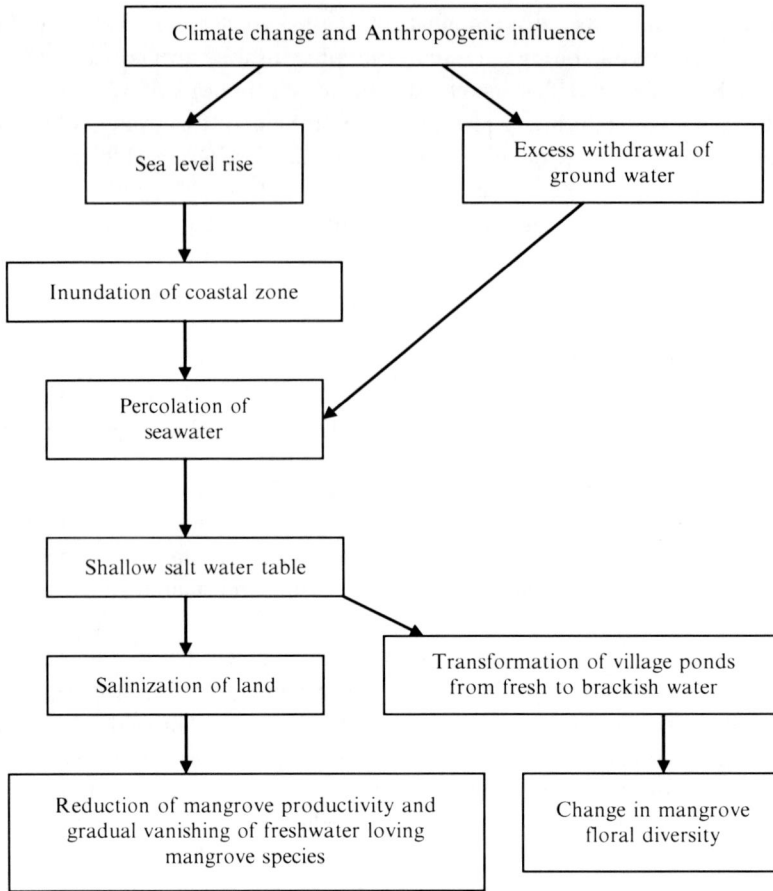

Fig. 4.1 Impact of sea-level rise on mangroves via groundwater system

Table 4.2 Effects of varying concentrations of NaCl on the physical properties and chlorophyll concentration of the leaves of *Sesuvium portulacastrum*

Salt concentration (mM)	Moisture (%)	Dry matter (%)	Mass of leaf (g)	Leaf volume (cm³)	Leaf density (g/cm³)	Chl a (mg/100 g fresh tissue)
Control	68.54	31.46	0.207	0.202	1.024	32.95
10	68.93	31.07	0.225	0.189	1.190	34.60
50	70.02	29.98	0.241	0.230	1.040	39.03
100	78.56	21.44	0.263	0.290	0.904	37.80
150	88.21	11.79	0.269	0.450	0.590	34.87
200	86.98	13.02	0.271	0.460	0.584	30.07
250	87.37	12.63	0.261	0.410	0.636	30.50

Source: Shinde and Bhosale (1985)

by Shinde and Bhosale (1985) on a mangrove associate floral species *Sesuvium portulacastrum* (Table 4.2) showed that succulence, moisture content, mass and volume of the leaf increased with the increasing concentration of NaCl, but the chlorophyll content and leaf density sharply decreased.

The present authors conducted an interesting experiment considering the projected trend of rising salinity due to climate-change-induced sea-level rise.

Table 4.3 Effects of varying salinity level on the physical properties and chlorophyll *a* concentration of the leaves of *Heritiera fomes*

Salinity (psu)	Moisture (%)	Dry matter (%)	Mass of leaf (g)	Leaf volume (cm^3)	Leaf density (g/cm^3)	Chl *a* (mg/100 g fresh tissue)
0	49.98	50.02	0.168	0.183	0.9180	36.89
2.0	56.23	43.77	0.177	0.205	0.8634	34.56
5.0	59.18	40.82	0.180	0.224	0.8035	32.45
8.0	60.05	39.95	0.201	0.256	0.7852	34.09
10.0	60.10	39.90	0.214	0.253	0.8458	30.77
12.0	59.23	40.77	0.153	0.281	0.5445	28.31

The freshwater-loving mangrove species *Heritiera fomes* (locally referred to as Sundari in Sundarbans region) were exposed to 12 different salinities diluting standard seawater (Table 4.3), and physical properties of the leaf along with chlorophyll content were measured after a period of 6 months' treatment. Significant negative correlation values were observed between salinity and chlorophyll content of the leaves and also between leaf density and chlorophyll contents ($r_{salinity \times Chl\,a} = -0.9022$, $p < 0.01$; $r_{salinity \times leaf\ density} = -0.7673$, $p < 0.01$), but with other physical properties of the leaves, the salinity exhibited significant positive relationships ($r_{salinity \times \%\ moisture} = 0.8119$, $p < 0.01$; $r_{salinity \times leaf\ volume} = 0.9847$, $p < 0.01$).

The adverse impact of salinity on leaf chlorophyll of mangrove species significantly affects the rate of photosynthesis. Various studies have shown that a number of mangrove species grow best at salinities between 4 and 15 psu (Connor 1969; Clough 1984, 1985; Downton 1982; Burchett et al. 1984). Till date, there have been few studies on the effect of salinity on photosynthetic gas exchange in mangroves. Clough (1985) stated in his communication that the rate of light-saturated photosynthesis decreases with increasing salinity of ambient media, attributing this to co-limitation of assimilation rate by stomatal conductance and photosynthetic capacity in response to differences in water status induced by the various salinity treatments. Thus, on the evidences available so far, it is most likely that salinity exerts its effect on photosynthesis mainly through changes in leaf water status.

The impact of salinity is not, however, uniform for all mangrove species. In few species, it is observed that high saline condition causes a positive influence on chlorophyll level. A series of experiments conducted by the author on the influence of aquatic salinity on selective mangrove species like *Rhizophora mangle*, *Avicennia* spp., *Nypa fruticans* and *Heritiera fomes* reveal different results from which the adaptability of mangroves to rising salinity can be extrapolated (Tables 4.4, 4.5, 4.6, 4.7, 4.8, and 4.9).

A very interesting picture was generated after statistical analysis of the data sets in Tables 4.4, 4.5, 4.6, 4.7, 4.8, and 4.9. The concentrations of chlorophyll decreased significantly with salinity in *Nypa fruticans*, *Avicennia marina*, *Avicennia officinalis*, *Avicennia alba* and *Heritiera fomes*, but for *Rhizophora mangle* a reverse situation was observed. This species could tolerate salinity up to 30 psu (in the laboratory condition) and could be maintained for more than 30 days. The chlorophyll level also increased significantly with the increase of salinity which confirms the adaptation of the species at biochemical level to high salinity related to rising sea level (Table 4.10). The decrease of chlorophyll pigment in rest four species is a clear indication of the failure of photosynthetic machinery in hypersaline situation, which the central Indian Sundarbans is facing due to absence of head-on freshwater discharge to push back the saline water intrusion from the Bay of Bengal.

4.2.3   Effect of Alteration of Ocean Circulation Pattern

Key oceanic water masses are changing with time. Recent evidence, cited from Indian Sundarbans, indicates that the salinity and density have

Table 4.4 Effects of different salinities on pigment level in *Rhizophora mangle*

Duration of treatment (day)	Salinity (psu)	Chl *a*	Chl *b*	Total chlorophyll	Chl *a:b*	Carotenoid
7	2	0.65 ± 0.03^{ab}	0.20 ± 0.007^{ab}	0.85 ± 0.03^{ab}	3.25	0.19 ± 0.03^{a}
	5	0.62 ± 0.01^{a}	0.18 ± 0.004^{a}	0.80 ± 0.03^{a}	3.44	0.16 ± 0.03^{a}
	10	0.67 ± 0.01^{ab}	0.21 ± 0.006^{ab}	0.88 ± 0.03^{ab}	3.19	0.21 ± 0.03^{a}
	15	0.69 ± 0.02^{ab}	0.22 ± 0.008^{bc}	0.91 ± 0.03^{ab}	3.14	0.18 ± 0.03^{a}
	20	0.71 ± 0.03^{b}	0.23 ± 0.006^{c}	0.94 ± 0.03^{b}	3.08	0.19 ± 0.03^{a}
14	2	0.62 ± 0.02^{ab}	0.18 ± 0.005^{ab}	0.80 ± 0.03^{ab}	3.44	0.20 ± 0.03^{a}
	5	0.65 ± 0.04^{a}	0.19 ± 0.005^{a}	0.84 ± 0.03^{a}	3.42	0.16 ± 0.03^{a}
	10	0.60 ± 0.02^{ab}	0.19 ± 0.004^{ab}	0.79 ± 0.03^{ab}	3.16	0.18 ± 0.03^{a}
	15	0.64 ± 0.03^{ab}	0.21 ± 0.007^{bc}	0.85 ± 0.03^{ab}	3.04	0.17 ± 0.03^{a}
	20	0.68 ± 0.04^{b}	0.22 ± 0.009^{c}	0.90 ± 0.03^{b}	3.09	0.19 ± 0.03^{a}
21	2	0.59 ± 0.01^{ab}	0.18 ± 0.003^{ab}	0.77 ± 0.03^{ab}	3.28	0.16 ± 0.03^{a}
	5	0.57 ± 0.01^{a}	0.17 ± 0.004^{a}	$0.74 \pm 0.03a$	3.35	0.16 ± 0.03^{a}
	10	0.62 ± 0.02^{ab}	0.18 ± 0.003^{ab}	0.80 ± 0.03^{ab}	3.44	0.18 ± 0.03^{a}
	15	0.61 ± 0.02^{ab}	0.19 ± 0.005^{bc}	0.80 ± 0.03^{ab}	3.21	0.15 ± 0.03^{a}
	20	0.66 ± 0.04^{b}	0.21 ± 0.005^{c}	0.87 ± 0.03^{b}	3.14	0.18 ± 0.03^{a}
30	2	0.63 ± 0.04^{ab}	0.18 ± 0.002^{ab}	0.81 ± 0.03^{ab}	3.50	$0.20 \pm 0.03a$
	5	$0.57 \pm 0.01a$	$0.16 \pm 0.002a$	$0.73 \pm 0.03a$	3.56	$0.17 \pm 0.03a$
	10	0.61 ± 0.02^{ab}	0.18 ± 0.004^{ab}	0.79 ± 0.03^{ab}	3.39	$0.19 \pm 0.03a$
	15	0.62 ± 0.04^{ab}	$0.20 \pm 0.007bc$	0.82 ± 0.03^{ab}	3.10	$0.15 \pm 0.03a$
	20	$0.67 \pm 0.04b$	$0.22 \pm 0.009c$	$0.89 \pm 0.03b$	3.04	$0.20 \pm 0.03a$

Units of all pigments are mg/g fresh weigh. Different *letters* besides figures indicate statistically different means as at $p \leq 0.01$

Table 4.5 Effects of different salinity on pigment concentrations in *Avicennia marina*

Duration of treatment (day)	Salinity (psu)	Chl *a*	Chl *b*	Total chl	Chl *a:b*	Carotenoid
7	5	0.61^{a}	0.19^{a}	0.80^{a}	3.21^{a}	0.21^{a}
	20	0.59^{b}	$0.18^{a, b}$	0.77^{b}	3.27^{b}	0.16^{b}
14	5	0.60^{a}	0.19^{a}	0.79^{a}	3.15^{a}	0.20^{a}
	20	0.56^{b}	0.17^{b}	0.73^{b}	3.29^{b}	0.15^{b}
21	5	0.61^{a}	0.18^{a}	0.79^{a}	3.38^{a}	0.21^{a}
	20	0.55^{b}	0.17^{ab}	0.72^{b}	3.23^{b}	0.13^{b}
30	5	0.54^{a}	0.18^{a}	0.72^{a}	3.00^{a}	0.17^{a}
	20	0.52^{b}	0.17^{ab}	0.69^{b}	3.05^{ab}	0.10^{b}

Units of all pigments are mg/g fresh weight. Different *letters* besides figures in *superscript* indicate statistically different means as at $p \leq 0.01$

changed significantly over a period of 27 years (Mitra et al. 2009). The study was conducted by the present author in the western and central sectors of mangrove-dominated Indian Sundarbans in collaboration with University of Massachusetts at Dartmouth (USA) with the premonsoon data set from 1980 to 2007 (Fig. 4.2).

Simple density calculations indicated that the western sector is becoming warmer, fresher and lighter, whereas the central part of the deltaic complex is becoming warmer, saltier and denser. Such horizontal contrast might lead to changes in (1) circulation, (2) ecosystem and (3) intensity, extent and paths of monsoonal storms due to

Table 4.6 Effects of different salinity on pigment concentrations in *Avicennia officinalis*

Duration of treatment (day)	Salinity (psu)	Chl *a*	Chl *b*	Total chl	Chl *a:b*	Carotenoid
7	5	0.63[a]	0.15[a]	0.78[a]	4.20[a]	0.27[a]
	20	0.59[b]	0.14[ab]	0.73[b]	4.21[ab]	0.19[b]
14	5	0.65[a]	0.16[a]	0.81[a]	4.06[a]	0.25[a]
	20	0.58[b]	0.14[b]	0.72[b]	4.14[b]	0.18[b]
21	5	0.60[a]	0.14[a]	0.74[a]	4.28[a]	0.22[a]
	20	0.54[b]	0.13[ab]	0.67[b]	4.15[b]	0.20[b]
30	5	0.58[a]	0.14[a]	0.72[a]	4.14[a]	0.21[a]
	20	0.57[b]	0.13[ab]	0.70[b]	4.38[b]	0.15[b]

Units of all pigments are mg/g fresh weight. Different *letters* besides figures in *superscript* indicate statistically different means as at $p \leq 0.01$

Table 4.7 Effects of different salinity on pigment concentrations in *Avicennia alba*

Duration of treatment (day)	Salinity (psu)	Chl *a*	Chl *b*	Total chl	Chl *a:b*	Carotenoid
7	5	0.53[a]	0.18[a]	0.71[a]	2.94[a]	0.19[a]
	20	0.49[b]	0.17[ab]	0.66[b]	2.88[b]	0.15[b]
14	5	0.55[a]	0.19[a]	0.74[a]	2.89[a]	0.16[a]
	20	0.50[b]	0.17[b]	0.67[b]	2.94[b]	0.14[b]
21	5	0.58[a]	0.20[a]	0.78[a]	2.90[a]	0.18[a]
	20	0.52[b]	0.18[b]	0.70[b]	2.89[b]	0.13[b]
30	5	0.56[a]	0.19[a]	0.75[a]	2.95[a]	0.17[a]
	20	0.53[b]	0.18[ab]	0.71[b]	2.94[ab]	0.11[b]

Units of all pigments are mg/g fresh weight. Different *letters* besides figures in *superscript* indicate statistically different means as at $p \leq 0.01$

differential in available energy distribution from the ocean surface. Studies reveal that the frequency of destructive cyclones has increased from the seventeenth century to the twenty-first century (Mitra 2000). Also, the severity of cyclones has increased in the Bay area from 1980 to 1999 (Fritz and Blount 2007).

Seawater density in the central sector was lower ($\sigma_\theta = 4.7$) than that of the west ($\sigma_\theta = 6.5$) in 1980. In 2007, after the effect of warming and resulting differential effects of salinity decrease/increase due to dilution/silting in the western/central sectors, the central waters have become denser ($\sigma_\theta = 9.1$) than the western water ($\sigma_\theta = -1.2$). Thus, the density contrast has not only increased between the central and the west by almost 12 sigma units, but it has changed signs (Table 4.11). Such a change might have serious impact on

the pressure-gradient-driven currents and might induce reverse flow during premonsoon times. Further detailed studies involving high-resolution numerical modelling for this region are needed to understand and quantify such circulation variability which might affect the path and amplitude of cyclones passing through this region (Mitra et al. 2009).

The Intergovernmental Panel on Climate Change, however, reports that at present, there is no clear evidence for ocean circulation change (Bindoff et al. 2007). However, there have been observations of long-term trends in changes in global- and basin-scale ocean heat content and salinity, which are linked to changes in ocean circulation (Gregory et al. 2005; Bindoff et al. 2007). Changes to ocean surface circulation patterns may affect mangrove propagule dispersal

Table 4.8 Effects of different salinity on pigment concentrations in *Nypa fruticans*

Duration of treatment (day)	Salinity (psu)	Chl *a*	Chl *b*	Total chl	Chl *a:b*	Carotenoid
7	2	0.83[a]	0.29[a]	1.12[a]	2.86[a]	0.21[a]
	5	0.76[b]	0.30[a]	1.06[b]	2.53[b]	0.18[b]
	10	0.65[c]	0.25[b]	0.90[c]	2.60[c]	0.16[b]
	15	0.60[d]	0.24[b]	0.84[d]	2.50[b]	0.15[b]
	20	0.41[e]	0.16[c]	0.57[e]	2.56[b]	0.09[c]
14	2	0.79[a]	0.31[a]	1.10[a]	2.55[a]	0.19[a]
	5	0.75[b]	0.29[a]	1.04[b]	2.59[a]	0.18[a]
	10	0.67[c]	0.26[b]	0.93[c]	2.58[a]	0.13[b]
	15	0.53[d]	0.21[c]	0.74[d]	2.52[b]	0.10[c]
	20	–	–	–	–	–
21	2	0.73[a]	0.30[a]	1.03[a]	2.43[a]	0.16[a]
	5	0.68[b]	0.27[b]	0.95[b]	2.52[b]	0.14[b]
	10	0.51[c]	0.20[c]	0.71[c]	2.55[c]	0.12[c]
	15	0.50[c]	0.19[c]	0.69[d]	2.63[d]	0.11[c]
	20	–	–	–	–	–
30	2	0.66[a]	0.26[a]	0.92[a]	2.54[a]	0.18[a]
	5	0.52[b]	0.20[b]	0.72[b]	2.60[b]	0.13[b]
	10	0.48[c]	0.19[a]	0.67[c]	2.53[c]	0.12[c]
	15	0.47[c]	0.18[c]	0.65[d]	2.61[b]	0.06[d]
	20	–	–	–	–	–

Units of all pigments are mg/g fresh weight. Different *letters* besides figures indicate statistically different means as at $p \leq 0.01$

Table 4.9 Effects of different salinity on pigment concentrations in *Heritiera fomes*

Duration of treatment (day)	Salinity (psu)	Chl *a*	Chl *b*	Total Chl	Chl *a:b*	Carotenoid
7	2	0.71[a]	0.22[a]	0.93[a]	3.23[a]	0.19[a]
	5	0.63[b]	0.19[b]	0.82[b]	3.32[b]	0.17[b]
	10	0.50[c]	0.15[c]	0.65[c]	3.33[b]	0.17[b]
	15	0.43[d]	0.13[d]	0.56[d]	3.31[b]	0.14[c]
	20	0.26[e]	0.08[e]	0.34[e]	3.25[a]	0.12[d]
14	2	0.58[a]	0.17[a]	0.75[a]	3.41a	0.21[a]
	5	0.49[b]	0.15[b]	0.64[b]	3.27b	0.19[b]
	10	0.41[c]	0.13[c]	0.54[c]	3.15[c]	0.18[c]
	15	0.32[d]	0.10[d]	0.42[d]	3.20[b]	0.17[d]
	20	0.15[e]	0.05[e]	0.20[e]	3.00[a]	0.14[e]
21	2	0.64[a]	0.19[a]	0.83[a]	3.37[a]	0.18[a]
	5	0.53[b]	0.16[b]	0.69[b]	3.31[b]	0.16[b]
	10	0.47[c]	0.14[c]	0.61[c]	3.36[a]	0.14[c]
	15	0.23[d]	0.07[d]	0.30[d]	3.29[b]	0.12[d]
	20	–	–	–	–	–
30	2	0.55[a]	0.17[a]	0.72[a]	3.24[a]	0.18[a]
	5	0.51[b]	0.15[b]	0.66[b]	3.40[b]	0.15[b]
	10	0.45[c]	0.15[b]	0.60[c]	3.00[c]	0.15[b]
	15	0.20[d]	0.06[c]	0.26[d]	3.33[b]	0.13[c]
	20	–	–	–	–	–

Units of all pigments are mg/g fresh weight; Different *letters* besides figures indicate statistically different means as at $p \leq 0.01$

Table 4.10 Interrelationships between salinity and chlorophyll in selected mangrove species

Species	Combination	'r' value	'p' value
Nypa fruticans	Salinity × Chl. a	−0.7986	<0.01
	Salinity × Chl. b	−0.7746	<0.01
	Salinity × Total Chl.	−0.7947	<0.01
Avicennia marina	Salinity × Chl. a	−0.5416	<0.05
	Salinity × Chl. b	−0.8006	<0.01
	Salinity × Total Chl.	−0.6161	<0.01
Avicennia officinalis	Salinity × Chl. a	−0.6964	<0.01
	Salinity × Chl. *b*	−0.6742	<0.01
	Salinity × Total Chl.	−0.6976	<0.01
Avicennia alba	Salinity × Chl. a	−0.6777	<0.01
	Salinity × Chl. *b*	−0.6342	<0.01
	Salinity × Total Chl.	−0.7120	<0.01
Rhizophora mangle	Salinity × Chl. *a*	0.6045	<0.01
	Salinity × Chl. *b*	0.7803	<0.01
	Salinity × Total Chl.	0.6764	<0.01
Heritiera fomes	Salinity × Chl. *a*	−0.9111	<0.01
	Salinity × Chl. *b*	−0.9018	<0.01
	Salinity × Total Chl.	−0.9096	<0.01

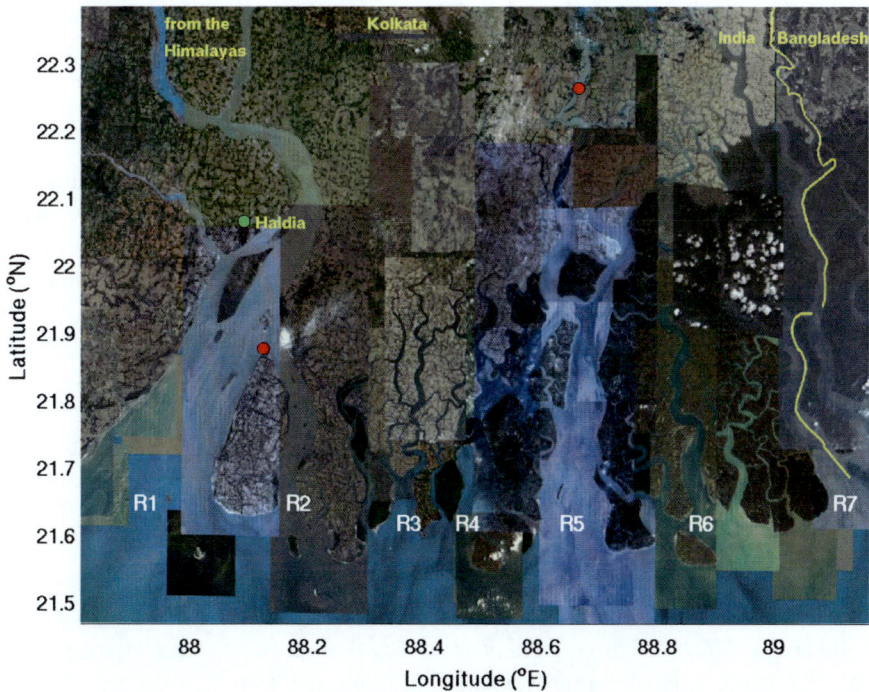

Fig. 4.2 Map of the study region with two stations marked in *red*. The western station is located at 21°52′20.78″ N, 88°7′29.73″ E, at the northern tip of Sagar Island. The seven rivers marked by R1 through R7 from west to east are Hugli, Muriganga, Saptamukhi, Thakuran, Matla, Gosaba and Harinbhanga

Table 4.11 Summary observations during premonsoon over the last three decades

	West			Central			
	1980	2007	Trend (/decade)	1980	1995	2007	Trend[a] (/decade)
Temp (°C)	31	32.6	+0.5	31.25	31.65	32.7	+0.27/+0.67
Salinity (psu)	15	10	−1.67	13	13.5	19	0/+3.67
pH	8.325	8.28	−0.015	8.305	8.3	8.29	0/0
DO (ppm)	5.1	6.0	+0.3	5.1	5.1	4.5	0/−0.4
Transparency (cm)	27	20	−2.3	30	27	23	−2.3/−2.3
Water quality index	75	91	+16	75	76	64	+0.7/−8
Density ($\sigma_\theta = \rho - 1,000$)	+6.5	+3.0	−1.2	+4.7	+5.0	+9.1	+0.2/+2.9

[a]The trend for the central sector is given in two bands: the first for the period 1980–1995 and the second for the period 1995–2007, separated by '/'. Note that 'Zero trend' indicates that there was no significant trend obtained

and the genetic structure of mangrove populations, with concomitant effects on mangrove community structure (Duke et al. 1998; Benzie 1999). This has considerable positive impact on the environment as increasing gene flow between currently separated populations and increasing mangrove species diversity could increase mangrove resistance and resilience.

4.3   Mangrove Resistance to Climate Change: Some Case Studies

Any student of biology or oceanography, who has spent few weeks in mangroves, can easily realize that these forests are some of the toughest ecosystems on the Earth. Exposed to rapid daily, monthly and annual variation in their physical environment, the mangrove vegetations have developed unique ability to cope with extraordinary levels and types of stress. The innate resilience of mangroves to cope with change is a requirement of their niche. Unfortunately, we have not yet developed a comprehensive conservation scheme considering all the upstream and downstream threats including the projected sea-level rise to protect this very valuable reservoir of the planet Earth. The damage caused by the tragic 2004 Asian tsunami was exacerbated by over-clearing of mangroves and other coastal 'bioshields', inappropriate coastal development and inadequate information and preparedness. We have not yet designed any

policy to increase the resistance and resilience of mangroves.

The term *resistance* is used here to refer the ability of mangroves to keep pace with the rising sea level without alteration to its functions, processes and structure (Odum 1989; Bennett et al. 2005). *Resilience*, on the other hand, refers to the capacity of a mangrove to naturally migrate landward in response to rising sea level, such that the mangrove ecosystem absorbs and reorganizes from the effects of the stress to maintain its functions, processes and structure (Carpenter et al. 2001; Nystrom and Folke 2001).

Mangrove systems cannot keep pace with changing sea level when the rate of change in elevation of the mangrove sediment surface is exceeded by the rate of change in relative sea level. There are several interconnected surface and subsurface processes that influence the elevation of mangroves' sediment surface. Mangroves of low-relief islands in carbonate settings that lack rivers were thought to be the most sensitive to sea-level rise, owing to their sediment-deficit environments (Thom 1984; Ellison and Stoddart 1991; Woodroffe 1987, 1995, 2002). However, recent studies have shown that subsurface controls on mangrove sediment elevation can offset high or low sedimentation rates (Cahoon et al. 2006; Cahoon and Hensel 2006), such that sedimentation rates alone provide a poor indicator of vulnerability to rising sea level.

The different processes regulating the elevation/depression of mangrove sediment surfaces are discussed here in brief.

Fig. 4.3 Mangrove litter trap silt particles during the period of inundation

4.3.1   Sediment Accretion and Erosion

Sediment accretion and erosion are determined by a mangrove's geomorphic setting, which affects the sources of sediment, sediment composition and method of delivery (Furukawa and Wolanski 1996; Furukawa et al. 1997; Woodroffe 1990, 2002). Fine sediment particles are carried in suspension into mangrove systems from coastal waters during tidal inundation, form large flocs (cohesive clay and fine silt) and finally settle in the forest during slack high tide as the friction caused by the high mangrove vegetation density slows tidal currents. Wrack or mangrove plant litter on the soil surface (Fig. 4.3) can also trap mineral sediment and contribute to vertical accretion (Cahoon et al. 2006).

Water currents during ebb tides are too low to re-entrain the sediment. Thus, the mangrove structure and the intricate root system cause sediment accumulation (Furukawa and Wolanski 1996). Storms and extreme high water events can alter the mangrove sediment elevation through soil erosion and deposition (Cahoon et al. 2003, 2006). Sedimentation varies with the mangrove species and their root types (Furukawa and Wolanski 1996; Krauss et al. 2003).

Mangrove researchers are investigating seriously to know the mechanism of response of different functional root types to changes in elevation in order to identify if certain root structures may be more or less vulnerable to sea-level rise (Vicente 1989; Krauss et al. 2003). In the Caribbean region, Vicente (1989) documented that prop roots of *Rhizophora mangle* stand are higher above mean sea level than the aerial roots of *Avicennia germinans* which protrude only slightly out of the mud. These authors suggest that rapid sea-level rise may lead to local extinctions of *A. germinans* but have an insignificant effect on *R. mangle*. Ellison and Stoddart's (1991) work in Tonga also suggests that *Rhizophora* communities are better positioned to survive rising sea level due to higher peat accumulation rates beneath *Rhizophora* (5.3 mm/year) than *Bruguiera* and *Excoecaria* (2.6 mm/year). Krauss et al. (2003) compared the vertical accretion rates and elevation change in mangrove forests in Micronesia with three different functional root types, namely, prop roots in *Rhizophora* spp., root knees in *Bruguiera gymnorrhiza* and pneumatophores in *Sonneratia alba*. Prop roots trapped more sediment than either pneumatophores or bare soil control sites.

However, when erosion and shallow subsidence were factored into the elevation, bare soil control sites or pneumatophores showed the highest relative rates of positive elevation change. Therefore, these authors concluded that above-ground analysis of root area or below-ground fine root density cannot be used to determine rates of vertical accretion or elevation change.

4.3.2   Biotic Contributions

Biotic contributions to soil elevation vary from low (allochthonous mineral soils) to very high (autochthonous peat soils), where surface processes include the accumulation of decaying organic matter such as leaf litter and the formation of living benthic microbial, algal or root mats (Woodroffe 1992, 2002; Cahoon et al. 2006). Fine-grained sediment particles are often trapped within the network of algal mat during tidal inundations, and even water currents cannot force these particles to get rid from these network-like structures (Fig. 4.4). The thread-like structures of *Enteromorpha* spp. and *Ulva* sp. are very effective in such natural capturing process.

The litter contributed by the above-ground biomass of mangroves (stem, leaves, twigs, branches, flowers, fruits, etc.) also plays a crucial role in soil elevation. The accumulation of leaf litter is controlled by above-ground production, consumption by detritivores, microbial decomposition and tidal flushing (Middleton and McKee 2001; Cahoon et al. 2006). A research programme was undertaken in five different stations of Indian Sundarbans (Table 4.12 and Fig. 4.5) covering all the three seasons during 2008, and it was observed that the magnitude of litter fall (Table 4.13) is a direct function of species diversity (Table 4.14) and above-ground biomass of mangroves (Table 4.15).

The present case study clearly indicates that mangrove plantation should be emphasized and ambient ecological conditions should be maintained (preferably with respect to salinity/dilution factor, tidal flushing, etc.) for accelerating the above-ground biomass of mangrove trees to trap sediments through litters.

4.3.3   Below-Ground Primary Production

When below-ground root growth exceeds root decomposition, soil organic matter accumulates, causing a net increase in soil volume and contributes to a rise in sediment elevation. Root growth, or

Fig. 4.4 *Ulva lactuca* on mangrove mudflats can effectively trap silt particles

Table 4.12 Sampling stations with coordinates and salient features (Vide Fig. 4.7)

Station	Coordinates	Salient features
Henry's Island (Stn.1)	88°15′24″ E 21°45′24″ N	Faces River Muriganga, which is a branch of Hugli River; located in the western sector of Indian Sundarbans
Sagar South (Stn.2)	88°01′47″ E 21°39′04″ N	Situated at the confluence of River Hugli and Bay of Bengal on the western sector of Indian Sundarbans
Chemaguri (Stn.3)	88°09′11″ E 21°39′49″ N	Located in the western part of Indian Sundarbans and faces River Muriganga on the eastern side
Gosaba (Stn. 4)	88°39′46″ E 22°15′45″ N	Located in the Matla Riverine stretch in the central sector of Indian Sundarbans
Canning (Stn.5)	88°40′36″ E 22°18′37″ N	Located in the upstream of River Matla in the central sector of Indian Sundarbans

the lack thereof, has been observed to be a substantial controlling factor on mangrove soil elevation at some sites (Cahoon et al. 2003, 2006; Cahoon and Hensel 2006; McKee et al. 2007). In particular, mangroves in carbonate settings, such as on low oceanic islands remote from continental sources of sediment, have autochthonous soil, composed primarily of mangrove roots, where below-ground primary productivity and organic matter accumulation are primary controls on sediment elevation (Cahoon et al. 2006; McKee et al. 2007).

4.3.4 Autocompaction

Autocompaction, the lowering of the sediment surface and reduction in sediment volume, is caused by the oxidation (decomposition) and compression of organic material, and inorganic processes, including rearrangement of the mineral architecture, silica solution, clay dehydration and other diagenetic processes (Pizzuto and Schwendt 1997; Cahoon et al. 1999; Allen 2000; Woodroffe 2002; Cahoon and Hensel 2006). Autocompaction is understood

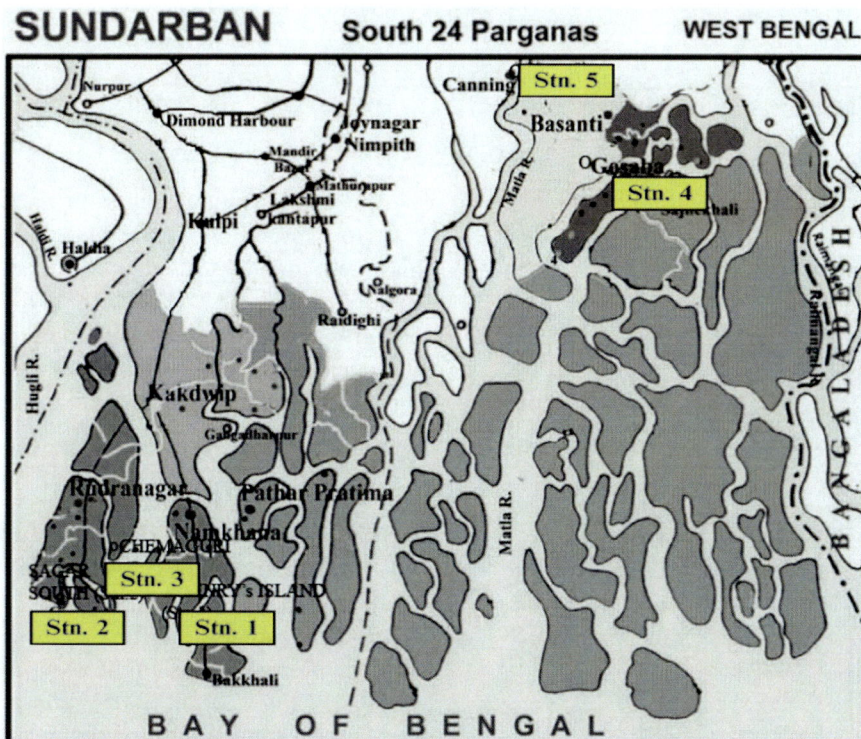

Fig. 4.5 Sampling stations selected for litter fall study in mangrove-dominated Indian Sundarbans

Table 4.13 Monthly variation of physico-chemical variables at the sampling stations

Month	Season	Surface water temperature (°C)	Surface water salinity (‰)	Surface water pH	DO (ppm)	Total litter (g/m²/day)
Henry's Island	Premonsoon	34.0	26.08	8.30	4.56	1.544
	Monsoon	32.9	10.02	8.05	5.98	1.186
	Postmonsoon	29.9	20.10	8.10	5.13	1.447
Sagar South	Premonsoon	34.0	24.15	8.28	4.55	1.605
	Monsoon	32.8	9.66	8.00	4.02	1.055
	Postmonsoon	29.8	18.50	8.10	4.85	1.368
Chemaguri	Premonsoon	33.7	21.80	8.28	5.13	1.417
	Monsoon	32.5	6.59	7.99	5.81	0.996
	Postmonsoon	28.0	12.48	8.04	4.93	1.370
Gosaba	Premonsoon	33.6	22.40	8.10	4.68	1.500
	Monsoon	32.6	12.68	7.95	4.93	0.905
	Postmonsoon	29.6	19.59	8.00	5.01	1.050
Canning	Premonsoon	33.8	22.01	8.10	4.85	1.200
	Monsoon	32.6	9.23	7.80	5.76	0.719
	Postmonsoon	28.9	18.45	7.90	5.32	0.908

Table 4.14 Composition of mangrove flora in the selected stations

Species	Henry's Island	Sagar South	Chemaguri	Gosaba	Canning
Sonneratia apetala	++	++	++	+	+
Avicennia alba	++	++	++	++	++
Avicennia marina	++	++	++	++	++
Avicennia officinalis	++	++	++	++	++
Excoecaria agallocha	+	+	++	++	++
Bruguiera gymnorrhiza	+	+	+	–	–
Aegiceras rotundifolia	+	+	+	–	–
Ceriops tagal	+	+	+	–	–

++ Highly abundant, + available, – absent

Table 4.15 Total litter and AGB of mangroves in five sampling stations in Indian Sundarbans

Station	Season	Total litter (g/m²/day)	AGB (t/ha)
Henry's Island	Premonsoon	1.544	35.89
	Monsoon	1.186	32.09
	Postmonsoon	1.447	33.66
Sagar South	Premonsoon	1.605	41.67
	Monsoon	1.055	29.87
	Postmonsoon	1.368	32.44
Chemaguri	Premonsoon	1.417	32.98
	Monsoon	0.996	20.53
	Postmonsoon	1.370	32.97
Gosaba	Premonsoon	1.500	34.78
	Monsoon	0.905	19.56
	Postmonsoon	1.050	30.11
Canning	Premonsoon	1.200	31.23
	Monsoon	0.719	15.37
	Postmonsoon	0.908	21.06

to decrease asymptotically with the age of the mangrove (Woodroffe 2002). Mangroves suffering mass tree mortality, caused by storms or other acute sources of stress, at sites with substrate composed primarily of peat or organic mud, are susceptible to substantial lowering in elevation of their sediment surface through peat collapse and soil compression (Cahoon et al. 2003).

4.3.5 Fluctuations in Water Table Levels and Pore Water Storage

Hydrology directly affects wetland elevation through processes of compression and dilation storage (Cahoon et al. 2006). The more water that is incorporated into the sediment below the water table, referred to as 'dilation storage' or 'shrink–swell', the more the sediment dilates, increasing sediment volume, which subsequently accelerates the elevation of the wetland sediment surface (Cahoon et al. 2006). The amount of dilation storage and degree of change in elevation of the sediment surface varies with soil type. Changes in groundwater inputs, such as from long-term changes in precipitation levels resulting from climate change, would result in a long-term change in mangrove elevation. Short-term cyclical influences include variability in precipitation and tidal range. Research conducted to date has demonstrated the short-term effects of groundwater recharge on mangrove elevation (Rogers et al. 2005; Whelan et al. 2005). Research is lacking to demonstrate effects of long-term trends in changes in groundwater inputs.

4.3.6   Approaches to Increase the Mangrove Resistance and Resilience

Mangroves have strong power of resistance against all odds. These halophytes were able to persist through the quaternary despite substantial disruptions from large sea-level fluctuations, demonstrating that mangroves are highly resilient to change over historic time scales (Woodroffe 1987, 1992). However, over coming decades, mangrove vulnerability and responses to climate change will be highly influenced by anthropogenic disturbances, including direct sources of degradation such as clearing and filling, and human responses to climate change that adversely affect mangroves. Many of these threats are extremely localized (e.g. the tiger prawn seed catch in Sundarbans that destroys millions of fish juveniles of other species) and can only be tackled through local level awareness. The dragging of nets along the shore for trapping tiger prawn seeds uproots the mangrove seedlings on the mudflats.

To reduce the risk of adverse outcomes from predicted mangrove responses to projected climate change, adaptation activities can be introduced as an attempt to increase the resistance and resilience of ecosystems to climate change stressors (Scheffer et al. 2001; Turner et al. 2003;

Tompkins and Adger 2004; Julius and West 2007). Some of these approaches are summarized here.

4.3.6.1   Management of Activities in Catchments Regulating Mangrove Sediment

In order to accelerate the resistance of mangroves to sea-level rise relative to the mangrove sediment surface, activities within the mangrove catchments can be managed to minimize long-term reductions in mangrove sediment elevation or enhance sediment elevation. For instance, limiting development of impervious surfaces within the mangrove catchments and managing rates and locations of groundwater extraction can reduce alteration to natural groundwater recharge to the mangrove systems, which might be an important control on mangrove elevation. Also, avoiding and limiting human activities that reduce mangrove soil organic matter accumulation, such as the diversion of sediment inputs to mangrove systems, nutrient and pollutant inputs into mangroves and mangrove timber harvesting, can contribute to maintaining relatively natural controls on trends in sediment elevation. Depending on the tree species and nutrient added, nutrient enrichment can affect mangrove productivity by changing root production and organic material inputs thereby changing the rate of change in sediment elevation (Feller et al. 2003; McKee et al. 2002, 2007). Enhancement of mangrove sediment accretion rates, such as through the beneficial use of dredge spoils, could augment mangrove sediment elevation (Lewis 1990), but would need to avoid excessive or sudden sediment deposition.

Management activities are not similar everywhere, rather they are site specific depending on the pattern of threats and needs of the local inhabitants. In Indian Sundarbans region, dragging of nets along the shore to trap tiger prawn seeds disturbs the underlying sediment bed and often uproots mangrove seedlings (Fig. 4.6). A considerable fraction of the island dwellers are involved in this job as there is great demand of tiger prawn seeds for the local level culture. Such activities need to be inhibited through implementation of laws and providing alternative livelihood to the prawn seed collectors.

Fig. 4.6 Tiger prawn seed collectors in Indian Sundarbans dragging nets along the shore

Collection of sandy sediments from beach for construction-related activities is another major threat often observed in the lower stretch of Gangetic delta (Fig. 4.7) that not only reduces the elevation of the sediment bed but also changes the pattern of canalization of water in the bed from the adjacent estuary and bay. Such destructive activities should be stopped through the implementation of strict laws involving port authorities, coast guard and local political parties in the loop.

4.3.6.2 Management of Coastal Activities Through Policy Implementation

Site planning for some sections of shoreline containing mangroves, such as areas that are not highly developed, may facilitate long-term retreat with relative sea-level rise (Dixon and Sherman 1990; Mullane and Suzuki 1997; Gilman 2002). 'Managed retreat' concept involves implementing land-use planning mechanisms before the effects of rising sea level become apparent, which can be planned carefully with sufficient lead time to enable economically viable, socially acceptable and environmentally sound management measures. Coastal development could remain in use until the eroding coastline becomes a safety

hazard or begins to prevent landward migration of mangroves, at which time the development can be abandoned or moved inland. Adoption of legal tools, such as rolling easements, can help make eventual abandonment more acceptable (Titus 1991). Zoning rules for building setbacks and permissible types of new development can be used to reserve zones behind current mangroves for future mangrove habitat. Managers can determine adequate setbacks by assessing site-specific rates for landward migration of the mangrove landward margin. Construction codes can plan for mangrove landward migration based on a desired lifetime for coastal development (Mullane and Suzuki 1997). Any new construction of minor coastal development structures, such as sidewalks and boardwalks, could be required to be expendable with a lifetime based on the assessed sites' erosion rate and selected setback. Rules could prohibit construction of coastal engineering structures, which obstruct natural inland migration of mangroves. This managed coastal retreat will allow mangroves to migrate and retain their natural functional processes. Coastal zones and waterfronts are extremely delicate systems in terms of vulnerability to sea-level rise. These zones also need a congenial environment for the mangroves to grow, thrive and act as effective bioshield.

Fig. 4.7   Sand excavation from the beach for construction jobs

Keeping in view the degradation of the coastal environment and rampant construction activities along the coastal areas of Indian subcontinent, the Ministry of Environment and Forests, Government of India (MoEF) issued a draft Coastal Regulation Zone (CRZ) notification twice, inviting suggestions and objections from the public on 27 June 1990 and 18 December 1990. Based on the suggestions and objections received, the ministry issued the CRZ notification declaring coastal stretches as CRZ and regulating activities in the CRZ. As per this, the CRZ area is defined as coastal stretches of seas, bays, estuaries, creeks, rivers and backwaters which are influenced by tidal action (in the landward side). As per the notification, 500 m on the landward side from the High Tide Line (HTL) and the land area between the Low Tide Line (LTL) and HTL including 500 m along the tidal-influenced water bodies subject to a minimum of 100 m on the width of the water body, whichever is less, is declared as CRZ area. Based on the ecological sensitivity, geomorphological feature and demographic distribution, the CRZ area has been classified into four categories, namely, CRZ-I (sensitive and intertidal), CRZ-II (urban or developed), CRZ-III

(rural or undeveloped), CRZ-IV (Andaman and Nicobar and Lakshadweep Island).

The notification regulates developmental activities in the CRZ area by prohibiting certain activities and permitting the essential activities. The prohibited activities include setting up of new industries and expansion of existing industries, manufacture or handling or storage and handling of hazardous substances (except specified petroleum products in port areas), fish processing units, disposal of wastes and effluents, mining of sands, rocks and other rare minerals and mechanized withdrawal of groundwater. The permissible activities include those activities that require waterfront and foreshore facilities such as construction activities related to defence requirements for which foreshore facilities are essential (e.g. slipways and jetties), operational construction for ports and harbours and construction of hotels and resorts in specified areas.

After the issue of the CRZ notification with the aim to protect and preserve the coastal ecosystems, many of the developmental activities were hindered in several of the maritime states. The Ministry of Environment and Forests has been receiving series of proposals from coastal States/

Central Ministries, industry associations, local communities and NGOs requesting for amendment to CRZ notification on certain specific issues. The ministry, after examining the proposals, had constituted committees to examine the specific issues. Based on the recommendations of the committee/request made by the various agencies, the ministry had amended the CRZ Notification, 1991, as per the provisions laid down in the Environment (Protection) Act, 1986. Some of the amendments constituted to look into specific issues are:

- SO 595(E), dated 18 August 1994 – Relaxed Coastal Regulation Zone area to 50 m along the tidal-influenced water bodies. This was based on the BB Vohra Committee's report. However, the Supreme Court of India in the Writ Petition 664 of 1993 quashed the above amendment.
- SO 73(E), dated 31 January 1997 – Permitted mining of sand and withdrawal of groundwater in the Coastal Regulation Zone area in Andaman and Nicobar.
- SO (E), dated 9 July 1997 – Permitted reclamation within port limits, construction for the operation expansion and modernization of ports. Development of public utilities within Sundarbans areas and storage of 13 POL products within port limits.
- SO 73(E), dated 4 August 2000 – Permitted storage of LNG in the intertidal area and exploration and extraction of oil and gas in Coastal Regulation Zone areas.
- SO 329(E), dated 12 April 2001 – Permitting setting up of projects of Department of Atomic Energy, pipelines and conveying systems in Coastal Regulation Zone areas.
- SO 550(E), dated 21 May 2002 – Permitted non-polluting industries in the Coastal Regulation Zone area of special economic zones. Housing schemes of State Urban Development Authorities initiated prior to 19.2.1991 was also permitted.
- SO 110(E), dated 19 October 2002 – Permitted nonconventional energy facilities, desalination plants, air strips in Coastal Regulation Zone of Andaman and Nicobar and also of

Lakshadweep islands. Storage of nonhazardous cargo such as edible oil, fertilizer and food grain was also permitted.
- SO 460(E), dated 22 April 2003 – Project costing more than Rs. 5 crores requires clearance from Ministry of Environment and Forests.
- SO 636(E), dated 30 May 2003 – Permitted construction of embarkation facilities for Lakshadweep in Coastal Regulation Zone-I areas.
- SO 725(E), dated 24 June 2003 – Permitted construction of trans-harbour sea links passing through Coastal Regulation Zone-I areas.
- SO 838(E), dated 24 July 2003 – Relaxed *No Development Zone* to 50 m from 200 m from HTL in Andaman and Nicobar and Lakshadweep for promoting tourism based on Integrated Coastal Zone Management study.

4.3.7 Fortification

While mangroves provide natural coastal protection that is expensive to replace with artificial structures (Mimura and Nunn 1998), for some sections of highly developed coastline adjacent to mangroves, site planning may justify use of hard engineering technology (e.g. groynes, sea walls, revetments, bulkheads) and other shoreline erosion control measures (e.g. surge breakers, dune fencing, detached breakwaters) to minimize or prevent erosion. Examples of few hard engineering structures used to minimize coastal erosions and increase land elevation through accretion are highlighted in Table 4.16.

All these engineering structures will gradually reduce mangrove ecosystem services no doubt, but may act effectively to prevent the mangroves' natural landward migration and may also increase the depth of the adjacent aquatic subsystem (Tait and Griggs 1990; Fletcher et al. 1997; Mullane and Suzuki 1997; Mimura and Nunn 1998). The construction of a 2.8-km-long guidewall in the mid-1990s adjacent to Nayachar Island (88°15′24″ E and 21°45′24″ N) in the upper stretch of the Hugli estuary not only amplified

Table 4.16 Common coastal engineering structures to prevent erosion

Structure	Description
Groynes	Groynes are wooden, concrete and/or rock barriers or walls at right angles to the sea. Beach materials build up on the updrift side, where littoral drift is predominantly in one direction, creating a wider and a more plentiful beach, thereby enhancing the protection for the coast. However, there is a corresponding loss of beach material on the downdrift side, requiring that another groyne be built there. Moreover, groynes do not protect the beach against storm-driven waves and if placed too close together will create currents, which will carry sand materials offshore
Sea walls	Walls usually of masonry, concrete or rock, built at the base of a cliff or at the back of a beach or used to protect a settlement against erosion or flooding. Older style vertical sea walls used to reflect all the energy of the waves back out to sea, and for this purpose recurved crest walls were often given, which also increase the local turbulence, and thus increasing entrainment of sand and sediment, during storms. Modern sea walls aim to destroy most of the incident energy, resulting in low reflected waves and much reduced turbulence and thus take the form of sloping revetments
Revetments	These are wooden slanted or upright blockades, built parallel to the sea on the coast, usually towards the back of the beach to protect the cliff. The most basic revetments consist of timber slants with a possible rock infill. Waves break against the revetments, which dissipate and absorb the energy. The cliff base is protected by the beach materials held behind the barriers, as the revetments trap some of the materials. They may be watertight, covering the slope completely, or porous, to allow water to filter through after the wave energy has been dissipated
Rip rap	Large rocks are piled or placed at the foot of cliffs, which are placed with native stones of the beach. Rip raps are generally used in areas prone to erosion and absorb the wave energy and hold beach materials
	Construction of rip raps cannot hinder long shore drift. Rip raps have a limited lifespan and are not effective in storm condition. They also reduce the recreational value of a beach
Gabions	Boulders and rocks are wired into mesh cages and usually placed in front of areas vulnerable from heavy to moderate erosion. Gabions are placed sometimes at cliffs edges or jag out at a right angle to the beach like a large groyne. Wire cages filled with crushed stones are used to reduce erosion
Offshore breakwater	These are enormous concrete blocks and natural boulders that are sunk offshore to alter wave direction and to filter the energy of waves and tides. The waves break further offshore and therefore reduce their erosive power. This leads to wider beaches, which absorb the reduced wave energy, protecting cliff and settlements behind

the accretion process but also increased the depth of the adjacent navigational channel.

4.3.8 Establishment of Protected Areas (PA)

Protected areas can be established and managed to implement mangrove representation, replication and refugia. Ensuring representation of all mangrove community types when establishing a network of protected areas and replication of identical communities to spread risk can increase chances for mangrove ecosystems surviving climate change and other stresses (Julius and West 2007). Ensuring that a portfolio of each different community type is represented is a strategy for optimizing climate change resilience as this representation increases the change that at least one of these communities with disparate physical and biological parameters will survive climate change stressors and provide a source for recolonizing. Replication, through the protection of multiple areas of each mangrove community type, by protecting multiple examples of each vegetation zone and geomorphic setting can help avoid the loss of a single community type (Roberts et al. 2003; Salm et al. 2006; Wells et al. 2006). Protected area selection can include mangrove areas that act as climate change refugia, communities that are likely to be more resistant to climate change stresses (Palumbi et al. 1997; Bellwood and Hughes 2001; Salm et al. 2006). For instance, mature mangrove communities will be more resistant and resilient to stresses, including those from climate change, than recently established forests. Protecting refugia areas that resist and/or recover quickly from disturbance in general or that are predicted to be able to keep pace with projected relative sea-level rise can serve as a source of recruits to recolonize areas that are lost or damaged. Protected area site selection should account for predicted ecosystem responses to climate change (Barber et al. 2004). For instance, planners need to account for the likely movements of habitat boundaries and species ranges over time under different sea-level and climate change scenarios, as well as consider an areas' resistance and resilience to projected sea level and climate changes and contributions to adaptation strategies. Site-specific analysis of resistance and resilience to climate change when selecting areas to include in new protected areas should include how discrete coastal habitats might be blocked from natural landward migration and how severe are threats not related to climate change in affecting the site's health.

A system of networks of protected areas can be designed to protect connectivity between coastal ecosystems, including mangroves (Crowder et al. 2000; Stewart et al. 2003; Roberts et al. 2003). Protecting a series of mature, healthy mangrove sites along a coastline could increase the probability of waterborne seedlings to recolonize sites that are degraded. Thus, protected areas are very similar to insurance policies to tide over unfavourable environmental conditions and assure safety to future mangroves. Protected area designs should include all coastal ecosystems to maintain functional links (Mumby et al. 2004).

4.3.9 Mangrove Rehabilitation

In recent years, the pressures of increasing population, food production, industrial and urban development and wood chipping have caused a significant proportion of the world's mangrove resource to be destroyed. There is a necessity to rehabilitate mangrove lands to support human communities. This may involve a landscape mosaic composed of mature forests, logged forests, aquaculture and agriculture. The use of such a mosaic can be sustainable. Another impetus behind the rehabilitation of mangrove ecosystems is the spectacular rise of environmental consciousness over the past 30 years. The rehabilitation of mangrove lands for conservation remains a matter of choice. Such a choice should be determined by the socio-economic priorities of the communities involved, but it is often influenced by pressure from conservation organizations advocating preservation of nature. Four main reasons for rehabilitating mangroves have been identified. These are conservation, landscaping, sustainable production and coastal protection. Out of these the last reason is particularly applicable to climate-change-induced sea-level rise and subsequent erosion.

Mangrove enhancement (by removing stresses that caused their decline) can augment resistance and resilience to climate change, while mangrove restoration (ecological restoration, restoring areas where mangrove habitat previously existed) (Kusler and Kentula 1990; Lewis 2005; Lewis et al. 2006) can offset anticipated losses from climate change.

4.3.10 Regional Monitoring Network

Ecosystems exhibit changes with time which are accelerated by human interventions or compositional changes of the atmosphere. Given uncertainties about future climate change and responses of mangroves and other coastal ecosystems to these changes, there is a need to monitor and study the changes systematically. Establishing mangrove baselines and monitoring gradual changes through regional networks using standardized techniques will enable the separation of site-based influences from global changes to provide a better understanding of mangrove responses to sea level and global climate change, and alternatives for mitigating adverse effects (CARICOMP 1998; Ellison 2000). For instance, coordinated observations of regional phenomena such as a mass mortality event of mangrove trees, or trend in reduced recruitment levels of mangrove seedlings, might be linked to observations of changes in regional climate such as reduced precipitation. The monitoring system, while designed to distinguish climate change effects on mangroves, would also therefore show local effects, providing coastal managers with information to abate these sources of degradation.

It is a hard task to isolate the influence of local level factors (noise) from the matrix of climate change. In this context the example of the impact of climate change on the water quality of the western sector of Indian Sundarbans is extremely relevant as it is difficult to segregate the amalgamated freshwater input in the Hugli channel from the sources like glacier melting, precipitation or Farakka barrage discharge. To avoid such noises in the data set, it is extremely important to establish an archive where all related data must be stored and may be accessed by the users whenever needed.

4.3.11 Awareness Programmes

Recent environment conservation approaches suggest that local people are effective stewards of forest resources. Local restoration and management of mangrove forests, in particular, are now widely advocated as a solution to achieve both economic and environmental conservation goals. Today, governments are no longer viewed as the sole or even primary stewards of forest resources. Increasingly, policies and programmes are crafted with the intent of enlisting local people as partners in forest land management. To make this partnership more strong, outreach and educational programmes are very effective.

Outreach and education activities can augment community support for adaptation actions. The value of wetlands conservation is often underestimated, especially in less developed countries with high population growth and substantial development pressure, where short-term economic gains that result from activities that adversely affect wetlands are often preferred over the less-tangible long-term benefits that are obtained through sustainable use of wetlands.

Fact Below the Carpet
Climate of the planet Earth is changing, and there is no doubt about it, but there is considerable doubt in the pace of change. Mangroves are the first line of defence to protect a nation from the adverse impact of climate-change-related natural disasters. The local communities in the coastal zone are the greatest sufferers of climate change, and hence they should be given priority in the awareness programme. Education and outreach programmes are investments to bring about changes in behaviour and attitudes by having a better informed community of the value of mangroves and other ecosystems. This awareness on the importance of mangroves provides the local community the real knowledge base on the basis of which they can take decisions about the rational use of their mangrove resources. Such awareness also results in grassroots support, formation of educated vote banks and increased political will for measures to conserve and sustainably manage mangroves.

Important References

Commonwealth Scientific and Industrial Research Organisation (CSIRO) (2001) Climate change impacts for Australia. CSIRO impacts and adaptation working group. CSIRO Sustainable Ecosystems, Aitkenvale

United Nations Environment Programme (UNEP) (1994) Assessment and monitoring of climatic change impacts on mangrove ecosystems. UNEP regional seas reports and studies. Report no. 154. UNEP, Nairobi

Allen JR (2000) Morphodynamics of Holocene salt marshes: a review sketch from the Atlantic and Southern North Sea coasts of Europe. Quat Sci Rev 19:1155–1231

Alongi DM (2002) Present state and future of the world's mangrove forests. Environ Conserv 29:331–349

Andrews TJ, Muller GJ (1985) Photosynthesis gas exchange of the mangrove. *Rhizopora stylosa* Griff in its natural environment. Oecologia (Berl) 65:449–455

Andrews TJ, Clough BF, Muller GJ (1984) Photosynthetic gas exchange and carbon isotope ratios of some mangroves in North Queensland. In: Teas HJ (ed) Physiology and management of mangroves, Tasks for vegetation science. Dr. W. Junk Publishers, The Hague/Boston, pp 15–23

Ball MC, Munns R (1992) Plant responses to salinity under elevated atmospheric concentrations of CO_2. Aust J Bot 40:515–525

Barber CV, Miller K, Boness M (eds) (2004) Securing protected areas in the face of global change: issues and strategies. IUCN, Gland

Bellwood DR, Hughes T (2001) Regional-scale assembly rules and biodiversity of coral reefs. Science 292:1532–1534

Bennett EM, Cumming GS, Peterson GD (2005) A systems model approach to determining resilience surrogates for case studies. Ecosystems 8:945–957

Benzie JAH (1999) Genetic structure of coral reef organisms, ghosts of dispersal past. Am Zoo 39:131–145

Bindoff NL, Willebrand J, Artale V, Cazenave A, Gregory J, Gulev S, Hanawa K, Le Quéré C, Levitus S, Nojiri Y, Shum C, Talley L, Unnikrishnan A (2007) Observations: oceanic climate change and sea level. In: Solomon S, Qin D, Manning M, Chen Z, Marquis M, Averyt K, Tignor M, Miller H (eds) Climate change 2007: the physical science basis. Contribution of working group I to the fourth assessment report of the intergovernmental panel on climate change. Cambridge University Press, Cambridge/New York

Burchett MD, Field CD, Pulkownik A (1984) Salinity, growth and root respiration in the grey mangrove Avicennia marina. Physiol Plant 60:113–118

Burns WCG (2001) The possible impacts of climate change on Pacific Island State ecosystems. Int J Glob Environ Issue 1(1):56–72

Cabanes C, Cazenave A, Le Provost C (2001) Sea-level rise during the past 40 years determined from satellite and in situ observations. Science 294:840–842

Cahoon DR, Hensel P (2006) High-resolution global assessment of mangrove responses to sea-level rise: a review. In: Gilman E (ed) Proceedings of the symposium on mangrove responses to relative sea level rise and other climate change effects, 13 July 2006, Catchments to coast, Society of Wetland Scientists 27th international conference, 9–14 July 2006, Cairns Convention Centre, Cairns. Western Pacific Regional Fishery Management Council, Honolulu. ISBN 1-934061-03-4, pp 9–17

Cahoon DR, Day JW, Reed DJ (1999) The influence of surface and shallow subsurface soil processes on wetland elevation, a synthesis. Curr Top Wetl Biogeochem 3:72–88

Cahoon DR, Hensel P, Rybczyk J, McKee K, Proffitt CE, Perez B (2003) Mass tree mortality leads to mangrove peat collapse at Bay Islands, Honduras after Hurricane Mitch. J Ecol 91:1093–1105

Cahoon DR, Hensel PF, Spencer T, Reed DJ, McKee KL, Saintilan N (2006) Coastal wetland vulnerability to relative sea-level rise: wetland elevation trends and process controls. In: Verhoeven JTA, Beltman B, Bobbink R, Whigham DF (eds) Wetlands and natural resource management. Springer, Berlin, pp 271–272

CARICOMP (1998) Caribbean Coastal Marine Productivity (CARICOMP): a Cooperative Research and Monitoring Network of Marine Laboratories, Parks, and Reserves. CARICOMP methods manual level 1. Manual of methods for mapping and monitoring of physical and biological parameters in the Coastal Zone of the Caribbean, CARICOMP Data Management Center, Centre for Marine Sciences, University of the West Indies, Mona, Kingston

Carpenter S, Walker B, Anderies JM, Abel N (2001) From metaphor to measurement: resilience of what to what? Ecosystems 4:765–781

Church J, Gregory J, Huybrechts P, Kuhn M, Lambeck K, Nhuan M, Qin D, Woodworth P (2001) Chapter 11. Changes in sea level. In: Houghton J, Ding Y, Griggs D, Noguer M, van der Linden P, Dai X, Maskell K, Johnson C (eds) Climate change 2001: the scientific basis. Published for the Intergovernmental Panel on Climate Change. Cambridge University Press, Cambridge/New York, pp 639–693, 881pp

Church J, Hunter J, McInnes K, White N (2004) Sea level rise and the frequency of extreme events around the Australian coastline. In: Coast to coast '04 – conference proceedings, Australia's national coastal conference, Hobart, 19–23 Apr 2004, 8pp

Clough BF (1984) Growth and salt balance of the mangroves Avicennia marina (Forsk.) Vierh, and Rhizophora stylosa griff. in relation to salinity. Aust J Plant Physiol 11:419–430

Clough BF (1985) Photosynthesis in mangroves. In: The mangroves: proceedings of national symposium on biology, utilization and conservation of mangroves, Kohlapur, Maharashtra, India, pp 80–88

Clough BF, Andrews TJ, Cowan IR (1982) Primary productivity of mangroves. In: Clough BF (ed) Mangrove ecosystems in Australia – structure function and management. AIMS with ANU Press, Canberra

Connor DJ (1969) Growth of grey mangrove (Avicennia marina) in nutrient culture. Biotropica 1:36–40

Crowder LB, Lyman S, Figueira W, Priddy J (2000) Source-sink population dynamics and the problem of siting marine reserves. Bull Mar Sci 66:799–820

Dixon JA, Sherman PB (1990) Economics of protected areas. A new look at benefits and costs. Island Press, Washington, DC

Downton WJS (1982) Growth and osmotic relations of the mangrove Avicennia marina, as influenced by salinity. Aust J Plant Physiol 9:519–528 (different names in text)

Duke NC, Ball MC, Ellison JC (1998) Factors influencing biodiversity and distributional gradients in mangroves. Glob Ecol Biogeogr 7:27–47

Duke NC, Meynecke JO, Dittmann S, Ellison AM, Anger K, Berger U, Cannicci S, Diele K, Ewel KC, Field CD, Koedam N, Lee SY, Marchand C, Nordhaus I, Dahdouh-Guebas F (2007) A world without mangroves? Science 317:41–42

Edwards A (1995) Impact of climate change on coral reefs, mangroves, and tropical seagrass ecosystems.

In: Eisma D (ed) Climate change impact on coastal habitation. Lewis Publishers, Boca Raton

Ellison JC (1993) Mangrove retreat with rising sea level, Bermuda. Estuar Coast Shelf Sci 37:75–87

Ellison JC (2000) Chapter 15: How South Pacific mangroves may respond to predicted climate change and sea level rise. In: Gillespie A, Burns W (eds) Climate change in the South Pacific: impacts and responses in Australia, New Zealand, and Small Islands States. Kluwer Academic Publishers, Dordrecht, pp 289–301

Ellison JC (2004) Vulnerability of Fiji's mangroves and associated coral reefs to climate change. Review for the world wildlife fund. University of Tasmania, Launceston

Ellison JC (2005) Impacts on mangrove ecosystems. In: The great greenhouse gamble: a conference on the impacts of climate change on biodiversity and natural resource management: conference proceedings, Sydney

Ellison J (2006) Mangrove paleoenvironmental response to climate change. In: Gilman E (ed) Proceedings of the symposium on mangrove responses to relative sea-level rise and other climate change effects, society of wetland scientists 2006 conference, 9–14 July 2006, Cairns. Western Pacific Regional Fishery Management Council and United Nations Environment Programme Regional Seas Programme, Honolulu and Nairobi. ISBN 1-934061-03-4, pp 1–8

Ellison JC, Stoddart DR (1991) Mangrove ecosystem collapse during predicted sea-level rise: Holocene analogues and implications. J Coast Res 7:151–165

Ellison AM, Farnsworth EJ, Twilley RR (1996) Facultative mutualism between red mangroves and root-fouling sponges in Belizean mangal. Ecology 77:2431–2444

Farnsworth EJ, Ellison AM, Gong WK (1996) Elevated CO_2 alters anatomy, physiology, growth and reproduction of red mangrove (*Rhizophora mangle* L.). Oecologia 108:599–609

Feller IC, McKee KL, Whigham DF, O'Neill JP (2003) Nitrogen vs. phosphorus limitation across an ecotonal gradient in a mangrove forest. Biogeochemistry 62:145–175

Field CD (1995) Impacts of expected climate change on mangroves. Hydrobiologia 295(1–3):75–81

Fletcher CH, Mullane RA, Richmond B (1997) Beach loss along armored shorelines of Oahu, Hawaiian Islands. J Coast Res 13:209–215

Fritz HM, Blount C (2007) Thematic paper: role of forests and trees in protecting coastal areas against cyclones; chapter 2: protection from cyclones (English). In: Coastal protection in the aftermath of the Indian Ocean Tsunami: what role for forests and trees? Proceedings of the regional technical workshop, Khao Lak, 28–31 Aug 2006, RAP Publication (FAO), No. 2007/07

Furukawa K, Wolanski E (1996) Sedimentation in mangrove forests. Mangroves Salt Marshes 1:3–10

Furukawa K, Wolanski E, Mueller H (1997) Currents and sediment transport in mangrove forests. Estuar Coast Shelf Sci 44:301–310

Gilman EL (2002) Guidelines for coastal and marine site-planning and examples of planning and management intervention tools. Ocean Coast Manage 45:377–404

Gilman E, Van Lavieren H, Ellison J, Jungblut V, Adler E, Wilson L, Areki F, Brighouse G, Bungitak J, Dus E, Henry M, Kilman M, Matthews E, Sauni I, Teariki-Ruatu N, Tukia S, Yuknavage K (2006) Living with Pacific Island mangrove responses to a changing climate and rising sea level. United Nations Environment Programme. UNEP regional seas reports and studies

Gilman E, Ellison J, Coleman R (2007a) Assessment of mangrove response to projected relative sea-level rise and recent historical reconstruction of shoreline position. Environ Monit Assess 124:112–134

Gilman E, Ellison J, Sauni JI, Tuaumu S (2007b) Trends in surface elevations of American Samoa mangroves. Wetl Ecol Manage 15:391–404

Gregory JM, Dixon KW, Stouffer RJ, Weaver AJ, Driesschaert E, Eby M (2005) A model intercomparison of changes in the Atlantic thermohaline circulation in response to increasing atmospheric CO_2 concentration. Geophys Res Lett 32:L12703

Harty C (2004) Planning strategies for mangrove and salt-marsh changes in Southeast Australia. Coast Manage 32:405–415

Hoegh-Guldberg O (1999) Climate change, coral bleaching and the future of the world's coral reefs. Mar Freshw Res 50:839–866

Houghton J, Ding Y, Griggs D, Noguer M, van der Linden P, Dai X, Maskell K, Johnson C (eds) (2001) Climate change 2001: the scientific basis. Published for the Intergovernmental Panel on Climate Change. Cambridge University Press, Cambridge/New York, 881pp

Hutchings P, Saenger P (1987) Ecology of mangroves. University of Queensland Press, St. Lucia

IUCN (1989) The impact of climatic change and sea level rise on ecosystems. Report for the Commonwealth Secretariat, London

Julius SH, West JM (eds) (2007) Draft. Preliminary review of adaptation options for climate-sensitive ecosystems and resources. Synthesis and assessment product 4.4. U.S. climate change science program. U.S. Environmental Protection Agency, Washington, DC

Kennish MJ (2002) Environmental threats and environmental future of estuaries. Environ Conserv 29(1): 78–107

Knutson TR, Tuleya RE (1999) Increased hurricane intensities with CO_2-induced warming as simulated using the GFDL hurricane prediction system. Clim Dyn 15:503–519

Kotimire SY, Bhosale LJ (1980) Ind J Mar Sci 9:299–301

Krauss KW, Allen JA, Cahoon DR (2003) Differential rates of vertical accretion and elevation change among aerial root types in Micronesian mangrove forests. Estuar Coast Shelf Sci 56:251–259

Kusler JA, Kentula ME (eds) (1990) Wetland creation and restoration: the status of the science. Island Press, Washington, DC

Lewis RR III (1990) Creation and restoration of coastal plain wetlands in Florida. In: Kusler JA, Kentula ME (eds) Wetland creation and restoration: the status of the science. Island Press, Washington, DC, pp 73–101

Lewis RR III (2005) Ecological engineering for successful management and restoration of mangrove forests. Ecol Eng 24:403–418

Lewis RR III, Erftemeijer P, Hodgson A (2006) A novel approach to growing mangroves on the coastal mud flats of Eritrea with the potential for relieving regional poverty and hunger: comment. Wetlands 26:637–638

Li MS, Lee SY (1997) Mangroves of China: a brief review. For Ecol Manage 96:241–259

McKee K (1993) Soil physiochemical patterns and mangrove species distribution-reciprocal effects? J Ecol 81:477–487

McKee KL, Feller IC, Popp M, Wanek W (2002) Mangrove isotopic (delta 15N and delta 13C) fractionation across a nitrogen vs. phosphorus limitation gradient. Ecology 83:1065–1075

McKee KL, Cahoon DR, Feller I (2007) Caribbean mangroves adjust to rising sea level through biotic controls on change in soil elevation. Glob Ecol Biogeogr 16:545–556

McLeod E, Salm R (2006) Managing mangroves for resilience to climate change. IUCN, Gland

Middleton BA, McKee KL (2001) Degradation of mangrove tissues and implications for peat formation in Belizean island forests. J Ecol 89:818–828

Mimura N, Nunn P (1998) Trends of beach erosion and shoreline protection in rural Fiji. J Coast Res 14:37–46

Mitra A (2000) The north-west coast of the Bay of Bengal and deltaic Sundarbans. In: Seas at the millennium: an environmental evaluation, vol II. Pergamon, Amsterdam, pp 145–160

Mitra A, Gangopadhyay A, Dube A, Schmidt ACK, Banerjee K (2009) Observed changes in water mass properties in the Indian Sundarbans (Northwestern Bay of Bengal) during 1980–2007. Curr Sci 97(10):1445–1452

Mohanty M (1990) Sea level rise: background, global concern and implications for Orissa coast, India. In: Victor Rajamanickam G (ed) Sea level variation and its impact on coastal environment. Tamil University Press, Thanjavur

Moore RT, Miller PC, Albright D, Tieszen LL (1972) Comparative gas exchange characteristics of three mangrove species during the winter. Photosynthetica 6:387–393

Mullane R, Suzuki D (1997) Beach management plan for Maui. University of Hawaii Sea Grant Extension Service and County of Maui Planning Department, Maui

Mumby P, Edwards A, Arlas-Gonzalez J, Lindeman K, Blackwell P, Gall A, Gorczynska M, Harbone A, Pescod C, Renken H, Wabnitz C, Llewellyn G (2004) Mangroves enhance the biomass of coral reef fish communities in the Caribbean. Nature 427:533–536

Naidoo G (1983) Effects of flooding on leaf water potential and stomatal resistance in *Bruguiera gymnorrhiza* (L.) Lam. New Phytol 93:369–376

Nystrom M, Folke C (2001) Spatial resilience of coral reefs. Ecosystems 4:406–417

Odum EP (1989) Ecology and our endangered life-support systems. Sinauer Associates Inc., Sunderland

Palumbi SR, Grabowsky G, Duda T, Geyer L, Tachino N (1997) Speciation and population genetic structure in tropical Pacific sea urchins. Evolution 51:1506–1517

Pizzuto JE, Schwendt AE (1997) Mathematical modeling of autocompaction of a Holocene transgressive valley-fill deposit. Wolfe Glade, Delaware

Primavera J (1997) Socio-economic impacts of shrimp culture. Aquacult Res 28:815–827

Roberts CM, Branch G, Bustamante RH, Castilla JC, Dugan J, Halpern BS, Lafferty KD, Leslie H, Lubchenco J, McArdle D, Ruckelshaus M, Warner RR (2003) Application of ecological criteria in selecting marine reserves and developing reserve networks. Ecol Appl 13:S215–S228

Rogers K, Saintilan N, Heijnis H (2005) Mangrove encroachment of salt marsh in Western Port Bay, Victoria: the role of sedimentation, subsidence, and sea level rise. Estuaries 28:551–559

Saenger P, Moverly J (1985) Vegetative phenology of mangroves along the Queensland coastline. Proc Ecol Soc Aust 13:257–265

Saintilan N, Wilton K (2001) Changes in the distribution of mangroves and saltmarshes in Jervis Bay, Australia. Wetl Ecol Manage 9:409–420

Salm RV, Done T, McLeod E (2006) Marine protected area planning in a changing climate. In: Phinney JT, Hoegh-Guldberg O, Kleypas J, Skirving W, Strong A (eds) Coral reefs and climate change: science and management. American Geophysical Union, Washington, DC, pp 207–221

Scheffer M, Carpenter S, Foley J, Folke C, Walker B (2001) Catastrophic shifts in ecosystems. Nature 413:591–596

Semeniuk V (1994) Predicting the effect of sea-level rise on mangroves in northwestern Australia. J Coast Res 10:1050–1076

Shinde LS, Bhosale LJ (1985) Studies on salt tolerance in *Aegiceros corniculatum* (L.) Blanco and *Sesuvium portulacastrum* (L.). In: The mangroves: proceedings of national symposium on biology, utilization and conservation of mangroves, Shivaji University, Kolhapur, pp 300–304

Snedaker SC (1993) Impact on mangroves. In: Maul GA (ed) Climate change in the intra-American seas: implications of future climate change on the ecosystems and socio-economic structure of the marine and coastal regimes of the Caribbean Sea, Gulf of Mexico, Bahamas and N. E. Coast of South America. Edward Arnold, London, pp 282–305

Snedaker SC (1995) Mangroves and climate change in the Florida and Caribbean region: scenarios and hypotheses. Hydrobiologia 295:43–49

Snedaker SC, Araujo RJ (1998) Stomatal conductance and gas exchange in four species of Caribbean mangroves exposed to ambient and increased CO_2. Mar Freshw Res 49:325–327

Stewart RR, Noyce T, Possingham HP (2003) Opportunity cost of ad hoc marine reserve 5 design decisions: an example from South Australia. Mar Ecol Prog Ser 253:25–38

Subramanian V (2000) Water: quantity-quality perspective in South Asia. Kingston International Publishers, Surrey, p 49

Tait JF, Griggs G (1990) Beach response to the presence of a seawall. Shore Beach 58:11–28

Thom BG (1984) Coastal landforms and geomorphic processes. In: Snedaker SC, Snedaker JG (eds) The mangrove ecosystem: research methods. UNESCO, Paris, pp 3–15

Titus JG (1991) Greenhouse effect and coastal wetland policy: how Americans could abandon an area the size of Massachusetts at minimum cost. Environ Manage 15:39–58

Tompkins EL, Adger NW (2004) Does adaptive management of natural resources enhance resilience to climate change? Ecol Soc 19:10

Turner BL, Kasperson R, Matsone P, McCarthy J, Corell R, Christensene L, Eckley N, Kasperson J, Luerse A, Martello M, Polsky C, Pulsipher A, Schiller A (2003) A framework for vulnerability analysis in sustainability science. PNAS Early Edn 100:8074–8079

UNEP (1989) Criteria for assessing vulnerability to sea level rise: a global inventory to high risk area. Delft Hydraulics, Delft, 51pp

UNESCO (1992) Coastal systems studies and sustainable development. In: Proceedings of the COMAR interregional scientific conference, UNESCO, Paris, 21–25 May 1991, 276pp

Valiela I, Bowen J, York J (2001) Mangrove forests: one of the world's threatened major tropical environments. Bioscience 51:807–815

Vicente VP (1989) Ecological effects of sea-level rise and sea surface temperatures on mangroves, coral reefs, seagrass beds and sandy beaches of Puerto Rico: a preliminary evaluation. Sci Cienc 16:27–39

Walsh KJE, Ryan BF (2000) Tropical cyclone intensity increase near Australia as a result of climate change. J Climate 13:3029–3036

Wells S, Ravilous C, Corcoran E (2006) In the front line: shoreline protection and other ecosystem services from mangroves and coral reefs. United Nations Environment Programme World Conservation Monitoring Centre, Cambridge

Whelan KRT, Smith TJ III, Cahoon DR, Lynch JC, Anderson GH (2005) Groundwater control of mangrove surface elevation: shrink–swell of mangrove soils varies with depth. Estuaries 28:833–843

Woodroffe CD (1987) Pacific island mangroves: distributions and environmental settings. Pac Sci 41:166–185

Woodroffe CD (1990) The impact of sea-level rise on mangrove shorelines. Prog Phys Geogr 14:483–520

Woodroffe CD (1992) Mangrove sediments and geomorphology. In: Alongi D, Robertson A (eds) Tropical mangrove ecosystems. Coastal and estuarine studies. American Geophysical Union, Washington, DC, pp 7–41

Woodroffe CD (1995) Response of tide-dominated mangrove shorelines in northern Australia to anticipated sea-level rise. Earth Surf Process Landf 20(1):65–85

Woodroffe C (2002) Coasts: form, process and evolution. Cambridge University Press, Cambridge

Woodroffe CD, Grindrod J (1991) Mangrove biogeography: the role of quaternary environmental and sea-level change. J Biogeogr 18:479–492

Zhang RT, Lin P (1984) Studies on the flora of mangrove plants from the coast of China. J Xiamen Univ (Nat Sci) 23:232–239, in Chinese, with English abstract

Climate Change and Plankton Spectrum of Mangrove Ecosystem

5

Plankton sustain the world fishery. Don't let someone to crush the plankton community and destroy the natural protein bank.

The Author

5.1 Plankton Community of Tropical Mangroves

5.1.1 Phytoplankton

The pelagic environment of the ocean supports two basic types of marine organisms. One type comprises the *plankton*, or those organisms whose powers of locomotion are such that they are incapable of making their way against the current and thus are passively transported by currents in the aquatic system, and the other type includes the *nekton* (free swimmers), which are free-floating animals that, in contrast to plankton, are strong enough to swim against currents and are therefore independent of water movements. The category of nekton includes fish, squid and marine mammals.

The word plankton has come from the Greek word *planktos*, meaning that which is passively drifting or wandering. Depending upon whether a planktonic organism is a plant or animal, a distinction is made between *phytoplankton* and *zooplankton*. Although many planktonic species are of microscopic dimensions, the term is not synonymous with small size as some of the zooplankton include jellyfish of several metres in diameter. It is not necessary that all plankton are completely passive, most of them are capable of swimming too.

Phytoplankton are free-floating tiny floral components that are widely distributed in the marine and estuarine environments. Like land plants, these tiny producers require sunlight, nutrients or fertilizers, carbon dioxide gas and water for growth. The cells of these organisms contain the pigment *chlorophyll* that traps the solar energy for use in *photosynthesis*. The photosynthetic process uses the solar radiation to convert carbon dioxide and water into sugars or high energy organic compounds from which the cell forms new materials. The synthesis of organic material by photosynthesis is termed *primary production*. Since phytoplankton are the dominant producers in the ocean, their role in the marine food chain is of paramount importance. Approximately, 4,000 species of marine phytoplankton have been described and new species are continually being added to this total (Lalli and Parsons 1997). Phytoplankton exhibit remarkable adaptations to remain in floating condition in the seawater. In fact, all marine phytoplankton tend to stay in the photic zone to utilize the solar radiation for performing the process of photosynthesis. In order to retard the process of sinking, this group of organisms adopts various

A. Mitra, *Sensitivity of Mangrove Ecosystem to Changing Climate*,
DOI 10.1007/978-81-322-1509-7_5, © Springer India 2013

mechanisms. These include their small size and general morphology, as the ratio of cell surface area to volume determines frictional drag in the water. Colony or chain formation also increases surface area and slows sinking. Most species carry out ionic regulation, in which the internal concentration of ions is reduced relative to their concentration in seawater. Diatoms also produce and store oil, and this metabolic by-product further reduces cell density. In experimental conditions, living cells tend to sink at rates ranging from 0 to 30 m/day, but dead cells may sink more than twice as fast. In nature, turbulence of surface waters is also an important factor in maintaining phytoplankton near the surface where they receive abundant sunlight.

Phytoplankton present in the marine and estuarine environments use carbon dioxide for photosynthesis and hence play an important role in maintaining the carbon dioxide budget of the atmosphere. The larger the world's phytoplankton population, the more carbon dioxide gets pulled from the atmosphere. This lowers the average temperature of the atmosphere due to lower volumes of this greenhouse gas. Scientists have found that a given population of phytoplankton can double its numbers in the order of once per day. In other words, phytoplankton respond very rapidly to changes in their environment.

Phytoplankton sometimes may cause adverse impact on the marine and estuarine environment. During excessive bloom of phytoplankton, the light energy is intercepted, which could otherwise reach fixed plants like eel grass (*Zostera* spp.) and kelp. Furthermore, when the phytoplankton eventually die back and break down, an excessive amount of oxygen is required to fuel this process, and hence areas may become deprived of oxygen. Excessive nutrients, and/or changes in their relative concentrations, may be one factor in a chain of events leading to changes in the species composition of the phytoplankton communities. Increased occurrence of toxic algal blooms may accelerate toxin production. Toxic phytoplankton, when consumed by shellfish or other species, can affect the marine food chain, including poisoning of seabirds, mammals and

even humans (WWF, Marine Update 50 2001). It has been established that phytoplankton naturally contains DMS (dimethyl sulphide), which is released from dead phytoplankton into the atmosphere. This compound can transform into sulphuric acid, which eventually may contribute to acid rain (http://oceanlink.island.net/ask/pollution.html).

Nearly all marine plants, whether unicellular or multicellular, even those attached to substrata (sessile) or free floating, pass some part of their life cycle in floating condition as phytoplankton. However, those organisms which always remain planktonic throughout the life cycle are (1) diatoms, (2) dinoflagellates, (3) coccolithophores, (4) selective species of blue-green algae and (5) some species of green algae.

5.1.1.1 Diatoms

These floating plants are all microscopic in size and are characterized by the presence of shell or frustule. The shell or frustule is composed of translucent silica. The cell wall of diatom has two parts resembling a pillbox bottom and lid. The lid is called the *epitheca* and the bottom is known as *hypotheca*. These shells have great importance from the geological point of view and constitute the diatomaceous crust. The diatoms exhibit remarkable varieties and forms, and many species possess beautifully sculptured shells.

Depending on the nature of valves and pattern of ornamentation in the valve surface, the diatoms are grouped into *centric* and *pennate* diatoms. The major differences between these two groups are given in Table 5.1.

5.1.1.2 Dinoflagellates

These are important producers of the marine environment and rank second in the importance in the economy of the sea. Typically, these are unicellular, some are naked while others are armoured with plates of cellulose. The dinoflagellates possess two flagella for locomotion. Several of them are luminescent and produce light. *Skeletonema costatum* and *Coscinodiscus eccentricus* are common dinoflagellates in marine and estuarine system.

Table 5.1 Differences between centric and pennate diatoms

Point	Centric diatom	Pennate diatom
Cell shape	Discoid, solenoid or cylindrical	Elongated and fusiform, oval, sigmoid or roughly circular
Ornamentation	Radial in nature, i.e. the arrangement of markings is radiating from the centre	Bilateral in nature, i.e. the arrangement of the markings is on either side of the apical (main) axis

5.1.1.3 Coccolithophores

These are among the smallest category of phytoplankton having a size range between 5 and 20 μm. Some coccolithophores have flagella while others are devoid of them. Their soft bodies are shielded by tiny, calcified circular plates or shields of various designs. These are normally found in the open sea, but their profuse occurrence has been recorded in coastal waters. They form important diet components of filter-feeding animals. Examples are *Isochrysis galbana* and *Coccolithus* sp.

5.1.1.4 Blue-Green Algae

These include both unicellular and multicellular organisms. The blue colour in them is due to the presence of a pigment known as phycocyanin. Of the various organisms belonging to this category, the most important is *Trichodesmium erythraeum* because in certain seasons of the year, its biomass increases greatly resulting in the formation of clumps.

5.1.1.5 Green Algae

Microscopic green algae present in the planktonic community largely occur in coastal waters. The green colour in them is due to the presence of chloroplasts. They are widely distributed in the warmer (tropical) seas, and only few species are found in the Arctic and Antarctic oceans. *Chlorella marina* and *Chlorella salina* are common green algae in estuarine waters.

5.1.1.6 Classification of Phytoplankton

The phytoplankton community consists of a variety of organisms, namely, diatoms, dinoflagellates, blue-green algae, silicoflagellates and coccolithophores, which ranges in terms of size from 0.001 to 0.2 mm.

Table 5.2 Classification of phytoplankton on the basis of size

Plankton category	Maximum dimension (μm)
Ultraplankton	<2
Nanoplankton	2–20
Microplankton	20–200
Macroplankton	200–2,000
Megaplankton	>2,000

Phytoplanktons may be classified variously from different angles as stated here.

1. On the basis of size, the phytoplankton may be grouped under five categories (Table 5.2).
2. Phytoplankton may also be classified on the basis of the cell characteristics (Table 5.3).

5.1.2 Zooplankton

The zooplankton occupy the tier next to phytoplankton in the marine food web. They either graze on phytoplankton or feed on other members of zooplankton. Foraminifera and radiolarians are members of zooplankton community who possess only one cell. Copepods and euphausiids are widely distributed zooplankton in the marine and estuarine environments. Zooplankton exhibit the phenomenon of *vertical migration* in which some species migrate towards the sea surface during the night-time and return back to their original depth during the day. The amount of daily migration varies between 10 and 500 m, and this change of zone is keenly related to their nutritional requirement or maintenance of their light level. The zooplankton are mainly filter feeding in nature and are the prey of many shore-dwelling species. The zooplankton also exhibit various adapta-

Table 5.3 Classification of phytoplankton on the basis of cell characteristics

Class	Common name	Area(s) of predominance	Common genera
Cyanophyceae (cyanobacteria)	Blue-green algae	Tropical	*Oscillatoria, Synechococcus*
Rhodophyceae	Red algae	Cold temperate	*Rhodella*
Cryptophyceae	Cyptomonads	Coastal	*Cryptomonas*
Chrysophyceae	Chrysomonads	Coastal	*Aureococcus*
	Silicoflagellates	Cold waters	*Dictyocha*
Bacillariophyceae (Diatomophyceae)	Diatoms	All waters, especially coastal waters	*Coscinodiscus, Chaetoceros, Rhizosolenia*
Raphidophyceae	Chloromonads	Brackish	*Heterosigma*
Xanthophyceae	Yellow-green algae	Brackish	Very rare
Eustigmatophyceae	Yellow-green algae	Estuarine	Very rare
Prymnesiophyceae	Coccolithophorids	Oceanic	*Emiliania*
	Prymnesiomonads	Coastal	*Isochrysis*
			Prymnesium
Euglenophyceae	Euglenoids	Coastal	*Eutreptiella*
Prasinophyceae	Prasinomonads	All waters	*Tetraselmis*
			Micromonas
Chlorophyceae	Green algae	Coastal	Rare
Pyrrophyceae (Dinophyceae)	Dinoflagellates	All waters, esp. warm	*Ceratium*
			Gonyaulax
			Protoperidinium

tions to keep them in floating condition. In many protozoa, the cilia act as blade of the rotor of a helicopter which enable the ciliates to move up and down the water. The scyphomedusae floats and moves in the direction of current by the rhythmic contraction and relaxation of their umbrella. Although these are large and heavy, but 99 % of their mass is due to the presence of a jelly-like substance, due to which their density is reduced. *Velella* sp. has chambers filled with air that keeps it afloat. The zoeae of many arthropods have well-developed uropods, and these help them to remain in floating condition or buoy up by swimming. The molluscan larvae are important component of zooplankton community. The *Lanthina* sp. remains afloat by creating froth of bubbles, which contract the surface film of water and the animal hangs on froth. The zooplankton usually produce 3–5 generations in warm water where there is plenty of food supply and optimum temperature. At higher latitudes, where the phytoplankton growth is restricted

within a brief period of time, the zooplankton produce only a single generation in a year.

The most abundant members of marine zooplankton are crustaceans called copepods. In coastal waters, the population density of copepods can be as high as 1,00,000 individuals per cubic metre of water. The main reasons for such large numbers are the tremendous reproductive capacity of these animals and the rich food supply. Copepods are the primary consumers of diatoms. A single copepod can consume as many as 1,20,000 diatoms per day.

In the northwestern Bay of Bengal, adjacent to deltaic Sundarbans, the zooplankton community comprises a heterogeneous assemblage of animals covering many taxonomic groups including copepods, mysids, lucifers, gammarid, cladocerae, ostracods, cumacea, hydromedusae, ctenophores and chaetognaths among holoplankters and polychaete larvae, molluscan and echinoderm larvae, crustacean larvae and fish eggs and larvae among meroplankters. Among the zooplankton, calanoid copepods constitute the major bulk, followed by cyclopoid copepods and harpecticoid (Chaudhuri and Choudhury 1994). In general, the higher abundance of zooplankton is encountered during the premonsoon periods. The main cause of the high species diversity index of zooplankton during the premonsoon months may be attributed to the coincidence of the breeding period of the coastal and estuarine fishes in the months of late February and March in the tropical mangrove ecosystem (Mitra 2000). A 10-year survey conducted on the ichthyoplankton (a major component of zooplankton) diversity in and around mangrove-dominated Indian Sundarbans revealed the standing stock and community diversity in the order premonsoon > postmonsoon > monsoon (Table 5.4).

Some common zooplankton (other than fish juveniles/ichthyoplankton) found in the coastal and estuarine waters of Bay of Bengal are listed in Table 5.5.

5.1.2.1 Classification of Marine and Estuarine Zooplankton

Zooplankton are found in almost all the layers of the photic zone of the ocean. They are potentially limited by two factors in the coastal and estuarine zones: firstly, by turbidity which can limit phytoplankton production and thus restrict the ration supply for the zooplankton community and, secondly, by currents which, particularly in small estuaries, are dominated by high river flow that usually carries the zooplankton out to the sea. The zooplankton biomass can increase the fishery productivity because they chiefly consume the primary producers (phytoplankton) and form the major food source for members of higher trophic levels in which several species of Osteichthyes and Chondrichthyes exist. Zooplankton are classified according to their habitat, depth distribution, size and duration of planktonic life (Tables 5.6, 5.7, 5.8, and 5.9).

5.2 Importance of Plankton

Phytoplankton form the foundation stone of world fishery. It has been estimated that the total annual primary production of the world seas is around 20×10^9 tonnes of carbon, which has the capacity to yield 240 million tonnes of fish although our present harvest is only about 60 million tonnes.

The distribution and abundance of the commercially important finfish and shellfish and their larvae are dependent on some species of the phytoplankton besides serving as their main food source. Among the species of diatoms, *Fragilaria oceanica* and *Hemidiscus hardmannianus* have been recorded to indicate the abundance of the clupeid fish, Hilsa in the Hugli estuarine system (adjacent to Bay of Bengal) and oil sardine *Sardinella longiceps* in the west coast of India. The 'white water phenomenon' due to abundance of the coccolithophores has been a good omen for the herring fisheries in the British waters. Similarly the colonial diatom, *Fragilaria antarctica* is known to indicate the abundance of the Antarctic krill, *Euphausia superba*.

Although elaborate studies have been made on marine and estuarine phytoplankton in many countries of the world, knowledge of this group of marine flora is very scanty in India. Hence, large-scale investigations on the phytoplankton especially on their dynamics in Indian seas are imperative to locate new fishing grounds and to understand their role in sustaining the blue revolution.

Phytoplankton are the sources of fossil fuels. It has been reported that some species of phytoplankton store their extra food as oil rather than starch to get an advantage in terms of buoyancy. When oil-storing forms of phytoplankton die and descend to the bottom of the sea and undergo microbial degradation, they are often covered by huge sediment load and subjected to enormous pressure. Such conditions are supposed to change the deposited phytoplankton into fossil fuel after hundreds and millions of years. Thus, human world is benefited from these tiny floral components even after their death.

The diatoms are commercially important as 'diatomaceous Earth' (diatomite or kieselguhr), a congregation of dead silicon-rich diatom frustules in the seabeds like protozoan oozes. This material is employed in the filtration of fruit juice, syrup and varnish; in the boilers, electric

Table 5.4 Seasonal variations of standing stock and Shannon Wiener species diversity index of finfish juvenile in and around Indian Sundarbans during a span of 10 years (1994–2003)

Year	Monthly average standing stock (N) of finfish juvenile species									Monthly average of Shannon–Wiener species diversity index of finfish juvenile species (H)								
	stn 1			stn2			stn3			stn1			stn2			stn3		
	Prm	Mon	Pom	Prm	Mon	Pom	Prm	Mon	Pom	Prm	Mon	Pom	Prm	Mon	Pom	Prm	Mon	Pom
1994	392	161	245	1,598	134	522	1,593	260	767	2.9151	2.2994	2.7801	3.5291	1.7486	3.0118	3.4839	2.2867	3.2634
1995	321	133	291	1,315	122	422	1,303	213	625	2.9178	2.3014	2.7791	3.5297	1.5557	3.0087	3.4825	2.2792	3.2665
1996	200	82	183	1,273	99	416	1,290	213	631	2.8946	2.2715	2.7675	3.5291	1.6186	3.0187	3.4842	2.2894	3.2781
1997	195	79	175	1,279	99	413	1,229	201	596	2.8816	2.2440	2.7362	3.5097	1.5998	3.0374	3.4795	2.2942	3.2786
1998	215	89	197	1,090	84	361	1,055	172	651	2.9286	2.3276	2.8140	3.5277	1.4978	3.0275	3.4813	2.2973	3.2809
1999	217	89	198	716	56	233	679	123	333	2.9295	2.3268	2.8054	3.5234	1.6254	3.0432	3.4616	2.2846	3.2878
2000	224	93	204	639	51	209	601	103	291	2.9197	2.3078	2.7919	3.5353	1.6378	3.0421	3.4647	2.2933	3.2508
2001	221	92	201	545	42	180	521	89	255	2.9253	2.3098	2.7977	3.5415	1.6021	3.0330	3.4847	2.3029	3.2921
2002	225	93	204	910	178	442	934	153	440	2.9310	2.3078	2.7919	3.4058	2.5097	3.1565	3.4986	2.2077	3.3045
2003	389	38	67	936	73	304	1,020	79	367	2.8682	1.4809	1.9693	3.5312	1.6087	3.0308	3.4273	1.8401	3.0912

stn refers to station, *Prm* refers to premonsoon, *Mon* refers to monsoon, *Pom* refers to postmonsoon

Table 5.5 Common zooplankton around northwestern Bay of Bengal coast

Serial no.	Name	Systematic position	Salient features
1.	*Globigerina triloculinoides*	Phylum – Protozoa	Composed of subglobular chambers arranged in two whorls
		Class – Sarcodina	Wall is calcareous and coarsely perforated
		Subclass – Rhizopoda	Sutures are distinct, depressed and umbilical with a prominent lip
		Order – Foraminifera	
		Family – Orbulinidae	
		Subfamily – Globigerinina	
		Genus – *Globigerina*	
		Species – *triloculinoides*	
2.	*Globigerina parva*	Phylum – Protozoa	Test is small and highly lobate
		Class – Sarcodina	Wall is calcareous, finely perforated
		Subclass – Rhizopoda	Presence of 12 chambers which are spherical and arranged in 2½ whorls
		Order – Foraminifera	
		Family – Orbulinidae	
		Subfamily – Globigerinina	
		Genus – *Globigerina*	
		Species – *parva*	
3.	*Globigerina opima*	Phylum – Protozoa	Equatorial periphery is slightly lobulate and periphery rounded
		Class – Sarcodina	Spherical chambers are arranged in 2½ whorls of which last consists of 5½ chambers
		Subclass – Rhizopoda	Sutures are radially depressed and aperture is interomarginal and extra umbilical
		Order – Foraminifera	
		Family – Orbulinidae	
		Subfamily – Globigerinina	
		Genus – *Globigerina*	
		Species – *opima*	
4.	*Acanthometron* sp.	Phylum – Protozoa	Protoplasmic strands project all directions as thin, long, straight filaments
		Class – Sarcodina	Possesses beautiful and siliceous skeleton
		Subclass – Radiolaria	Central capsule is perforated and horny
		Genus – *Acanthometron* sp.	
5.	*Tintinnopsis beroidea*	Phylum – Protozoa	Usually bullet shaped with slightly pointed end
		Class – Polyhymenophora	Encrustations heavy
		Subclass – Spirotrichia	Aboral part of lorica remains spherical, while oral portion sometimes starts narrowing
		Order – Oligochaeta	
		Suborder – Tintinnida	
		Family – Codonellidae	
		Genus – *Tintinnopsis*	
		Species – *beroidea*	
6.	*Tintinnopsis tombulosa*	Phylum – Protozoa	Lorica possesses two regions – bowl and column
		Class – Polyhymenophora	Oral flare absent
		Subclass – Spirotrichia	Bowl and column uniformly agglomerated
		Order – Oligochaeta	
		Suborder – Tintinnida	
		Family – Codonellidae	
		Genus – *Tintinnopsis*	
		Species – *tombulosa*	

(continued)

Table 5.5 (continued)

Serial no.	Name	Systematic position	Salient features
7.	*Brachionus angularis*	Phylum – Rotifera	Lorica circular in shape
		Class – Monogononta	Presence of four anterior spines, from which two marginal and two medium spines come out
		Order – Ploima	Presence of deep u-shaped sinus franked by dorsal median spines
		Family – Brachionidae	
		Subfamily – Brachioninae	
		Genus – *Brachionus*	
		Species – *angularis*	
8.	*Brachinus calyciflorus*	Phylum –Rotifera	Length and breadth of lorica, more or less the same
		Class – Monogononta	Four occipital spines are present
		Order – Ploima	V-shaped sinus with median spines longer than lateral spines
		Family – Brachionidae	
		Subfamily – Brachioninae	
		Genus – *Brachionus*	
		Species – *calyciflorus*	
9.	*Nannocalanus minor*	Phylum – Rotifera	First antennae reach up to caudal rami by about half of the body length
		Class – Monogononta	Left leg distinctly longer than the right leg
		Order – Ploima	External marginal spines are greatly enlarged on left exopodite
		Suborder – Calanoida	
		Family – Calanidae	
		Genus – *Nannocalanus*	
		Species – *minor*	
10.	*Acrocalanus gracilis*	Phylum – Rotifera	Body is parallel sided, almost 3 times as long as broad
		Class – Monogononta	Cephalosome is evenly rounded
		Order – Ploima	Antennae longer than the body and having two long hairs at their ends
		Suborder – Calanoida	
		Family – Paracalanoidae	
		Genus – *Acrocalanus*	
		Species – *gracilis*	
11.	*Paracalanus parvus*	Phylum – Rotifera	Urosome 4 segmented in male and asymmetrical in female
		Class – Monogononta	5th leg symmetrical in male and asymmetrical in female
		Order – Ploima	Left foot much longer, bulb-like eminence on cephalosome
		Suborder – Calanoida	
		Family – Paracalanoidae	
		Genus – *Paracalanus*	
		Species – *parvus*	
12.	*Acartia spinicauda*	Phylum – Rotifera	Spines are present at the corners of the metasome
		Class – Monogononta	Terminal claw of the 5th leg straight and scarcely widened at base
		Order – Ploima	Male have spines at 3rd urosome segment which overreach the 4th segment
		Suborder – Calanoida	
		Family – Acartiidae	
		Genus – *Acartia*	
		Species – *spinicauda*	

(continued)

Table 5.5   (continued)

Serial no.	Name	Systematic position	Salient features
13.	*Acartia erythraea*	Phylum – Rotifera	Terminal claw of 5th leg is slightly curved and smooth
		Class – Monogononta	5th leg has a thickened terminal claw
		Order – Ploima	Presence of two pairs of prominent spines in 2nd urosome
		Suborder – Calanoida	
		Family – Acartiidae	
		Genus – *Acartia*	
		Species – *erythraea*	
14.	*Acartia danae*	Phylum – Rotifera	5th leg of female is straight
		Class – Monogononta	Presence of crowd of small teeth near the tip
		Order – Ploima	Length of the species is 1.0–1.2 mm
		Suborder – Calanoida	
		Family – Acartiidae	
		Genus – *Acartia*	
		Species – *danae*	
15.	*Oithona similis*	Phylum – Rotifera	1st antenna reaches up to end of the last metasome
		Class – Monogononta	Body highly pellucid
		Order – Ploima	First segment bears a small semicircular process
		Suborder – Cyclopoida	
		Family – Oithonidae	
		Genus – *Oithona*	
		Species – *similis*	
16.	*Leuckartiara* sp.	Phylum – Cnidaria	Medusae may reach a height of mm and posses highly folded margins
		Class – Hydrozoa	Presence of about 20 marginal tentacles
		Order – Anthomedusae	Gonads are horseshoe shaped with folds directed outwards
		Genus – *Leuckartiara* sp.	
17.	*Aequorea* sp.	Phylum – Cnidaria	Presence of saucer-shaped medusae of 20 cm diameter
		Class – Hydrozoa	Presence of thick jelly in the bell
		Order – Anthomedusae	Numerous radial canals are presently attached to the gonads
		Genus – *Aequorea* sp.	
18.	*Cosmetira* sp.	Phylum – Cnidaria	Presence of numerous marginal tentacles (about 100) with swollen bases
		Class – Hydrozoa	Presence of light marginal vesicles without ocelli
		Order – Anthomedusae	Gonads present on the radial canals are long
		Genus – *Cosmetira* sp.	
19.	*Obelia medusa*	Phylum – Cnidaria	Medusae are saucer shaped with a very flat bell
		Class – Hydrozoa	There are 4 radial canals and one ring canal from which 24 solid marginal tentacles originate
		Order – Anthomedusae	Body is transparent, medusae luminescent with 8 aboral marginal vesicles
		Genus – *Obelia*	
		Species – *medusa*	

(continued)

Table 5.5 (continued)

Serial no.	Name	Systematic position	Salient features
20.	*Phialidium* sp.	Phylum – Cnidaria	Diameter of the umbrella may reach 20 mm but is normally between 5 and 10 mm
		Class – Hydrozoa	Presence of 32 hollow marginal tentacles and up to 3 marginal vesicles found between each pair of tentacles
		Order – Anthomedusae	Presence of four radial canals
		Genus – *Phialidium* sp.	
21.	*Physalia physalis*	Phylum – Cnidaria	Presence of enormous coloured float formed simply as a hollow pocket by infolding of outer layer
		Class – Hydrozoa	At regular intervals, the whole body twists over to wet itself first on the side and then on the other
		Order – Siphonophora	Presence of hydrophyllia or bracts which protect the gonophores and gastrozooids
		Suborder – Physophorida	
		Family – Physalia	
		Genus – *Physalia*	
		Species – *physalis*	
22.	*Porpita porpita*	Phylum – Cnidaria	It is a free-floating-modified hydroid polyp up to 8 cm in diameter
		Class – Hydrozoa	Numerous clavate tentacles bearing powerful stinging cells hang from the rim of horny central disc-shaped float which support the animal on the water surface
		Order – Athecata	Body tissues contain symbiotic algae
		Suborder – Capitata	
		Family – Veliidae	
		Genus – *Velella*	
		Species – *velella*	
23.	*Aurelia aurita*	Phylum – Cnidaria	It is a complex colony made up of a large number of polyps crowding the undersurface of an oval-shaped float
		Class – Scyphozoa	Presence of central mouth with large number of smaller polyps
		Order – Semaeostomeae	Deep blue in colour with a translucent sail protruding above the water surface
		Family – Ulmaridae	
		Genus – *Aurelia*	
		Species – *aurita*	
24.	*Rhizostoma* sp.	Phylum – Cnidaria	Presence of dome-shaped bell up to 1 m in diameter
		Class – Scyphozoa	Deep purplish blue in colour with margins lobed and devoid of tentacles
		Order – Rhizostomeae	Presence of multi-ciliated mouthy that acts as suckers for nourishment
		Genus – *Rhizostoma* sp.	
25.	*Beroe* sp.	Phylum – Cnidaria	Presence of eight rows of beating plates down the two sides
		Class – Scyphozoa	It is about 15 cm in length and shaped like vegetable marrow or thimble
		Order – Rhizostomeae	Mouth opens into a large space that fills the entire mass of the animal body
		Family – Beroidae	
		Genus – *Beroe* sp.	

(continued)

Table 5.5   (continued)

Serial no.	Name	Systematic position	Salient features
26.	*Tomopteris* sp.	Phylum – Annelida	Body is transparent and parapodia are flat and translucent
		Class – Polychaeta	Parapodia are paddle-shaped, biramous and lack setae
		Order – Phyllodocida Family – Tomopteridae Genus – *Tomopteris* sp.	Presence of two long tentacles, one on each side of the head
27.	*Sagitta maxima*	Phylum – Chaetognatha	Head is wide and oval with seminal vesicles distinct
		Class – Sagittoidea	Posterior fin merges with the anterior one
		Genus – *Sagitta* Species – *maxima*	Characterized by the presence of an anterior pair of lateral fins which are small and rounded
28.	*Sagitta enflata*	Phylum – Chaetognatha	Commonly known as 'arrow worm' because of its long, straight, slender body
		Class – Sagittoidea	Body is divisible into short head, long trunk and a short tail piece
		Genus – *Sagitta* Species – *enflata*	Mouth is with stout curved bristles and body with two
29.	*Sagitta lyra*	Phylum – Chaetognatha	Body is translucent and somewhat flabby with greater width at the middle
		Class – Sagittoidea	Head is large and wider than length with eyes oval and pigmented; neck is well defined
		Genus – *Sagitta* Species – *lyra*	Anterolateral fins are longer than the posterior fins and tail fin is lobate
30.	*Sagitta hispida*	Phylum – Chaetognatha	Body is stout, rigid and opaque
		Class – Sagittoidea	Head is wider than the body and there is no distinct neck
		Genus – *Sagitta* Species – *hispida*	Anterior fines are shorter and narrower than the posterior fins
31.	*Sagitta robusta*	Phylum – Chaetognatha	Head not broad but opaque due to the presence of strong longitudinal muscles
		Class – Sagittoidea	Anterior and posterior lateral fins are about equal in length and fully rayed
		Genus – *Sagitta* Species – *robusta*	Eyes are very large and eye pigment is concentrated in an eclipse.
32.	*Pleurobranchia pileus*	Phylum – Ctenophora	Popularly known as 'comb jelly' or 'sea gooseberry'
		Class – Tentaculata	Beautiful, delicate, bio-luminating planktonic organism
		Order – Cydippida Family – Pleurobrachiidae Genus – *Pleurobrachia* Species – *pileus*	Presence of eight rows of comb plates and a balancing organ (statocyst)
33.	*Podon* sp.	Phylum – Arthropoda	Head demarcated from the body by a deep trance groove
		Class – Crustacea	Bivalved carapace forms a distinct dorsal chamber which acts as brood pouch
		Subclass – Branchiopoda Order – Cladocera Family – Polyphenidae Genus – *Podon* sp.	Presence of biramous setae antennae behind the prominent compound eye

(continued)

Table 5.5 (continued)

Serial no.	Name	Systematic position	Salient features
34.	*Gigacuma halei*	Phylum – Arthropoda	Presence of large swollen cephalothoraxes and long slender abdomen ending in style-like caudal rami
		Class – Crustacea	Possesses carapace which covers three or four thoracic segments and anteriorly first thoracic limb
		Subclass – Branchiopoda Order – Cumacea Genus – *Gigacuma* Species – *halei*	First three pairs of thoracic limbs are modified to maxillipeds and uropods are slender
35.	*Euphausia diomedae*	Phylum – Arthropoda	Possesses external gills attached to endopods of the biramous thoracic limbs
		Class – Crustacea	Presence of prominent pair of spines on each laterofrontal border and smaller one on each lateral margin of the carapace
		Subclass – Branchiopoda Order – Euphausiacea Genus – *Euphausia* Species – *diomedae*	They are of about 40 mm in length
36.	*Lucifer hanseni*	Phylum – Arthropoda	Body is laterally compressed
		Class – Crustacea	It is about 1 cm in length
		Subclass – Branchiopoda Order – Decapoda Suborder – Dendrobranchiopoda Genus – *Lucifer* Species – *hanseni*	The 5th periopod is absent or vestigial and luminescent cells are present in the telson
37.	*Macrosetella gracilis*	Phylum – Arthropoda	Body is fusiform with length about 1.2–1.3 mm (males) and 1.4–1.5 mm (females)
		Class – Crustacea	Caudal setae are very long
		Order – Harpacticoida Genus – *Macrosetella* Species – *gracilis*	Caudal rami are slender, cylindrical and four times as long as broad
Larval forms			
38.	Trochophore larvae of polychaete	Phylum – Annelida	Almost spherical body with little tuft of long cilia and sensory cells at the upper pole
		Class – Polychaeta	Mouth is located equatorially
		Specimen – Trochophore larvae of polychaete	A main ciliated girdle runs around the sphere just above the equator
39.	Spirorbis larvae	Phylum – Annelida	Larvae exhibit bilateral symmetry and can be divided into head, thorax and abdomen
		Class – Polychaeta	The prototroch consists of a complete circle of cilia
		Specimen – Spirorbis larvae	Branchial rudiments are present under the dorsoventral sides of the head

(continued)

Table 5.5   (continued)

Serial no.	Name	Systematic position	Salient features
40.	Sabellarid larvae	Phylum – Annelida	Characterized by the presence of four red eyespots and a pair of posteriorly projecting tentacles
		Class – Polychaeta	Presence of two bundles of barbed provisional setae
		Specimen – Sabellarid larvae	Presence of a dorsal hump posterior to the eyespot and two dorsal longitudinal rows of highly clustered cilia
41.	Copepod nauplii	Phylum – Arthropoda	Body kite shaped with pointed end and single distinct eye on the front side
		Class – Crustacea	Presence of three pairs of limbs
		Order – Copepoda	First pair of limb is uniramous and others biramous
		Specimen – Copepod nauplii	
42.	Nauplii of *Penaeus indicus*	Phylum – Arthropoda	Ocellus present at the anterior median region of the body
		Class – Crustacea	Presence of a pair of dorsally curved caudal setae at the posterior end of the body
		Order – Decapoda	First pair of appendage is uniramous; 2nd and 3rd pairs are biramous, with the third pair shorter than the other appendages
		Suborder – Macrura	
		Specimen – Nauplii of *Penaeus indicus*	
43.	Nauplii of Balanus	Phylum – Arthropoda	Body is triangular
		Class – Crustacea	Presence of a pair of posterior spines
		Order – Decapoda	Tip of the rostrum is truncated
		Suborder – Macrura	
		Specimen – Nauplii of Balanus	
44.	Cypris of Balanus	Phylum – Arthropoda	Presence of a bivalve shell
		Class – Crustacea	Presence of a compound eye and 6 pairs of thoracic appendages
		Order – Decapoda	Consists of a short abdomen
		Suborder – Macrura	
		Specimen – Cypris of Balanus	
45.	Larva of *Squilla hieroglyphica*	Phylum – Arthropoda	Carapace large and broad
		Class – Crustacea	Presence of three spinules on lateral margin of the carapace-1st near the base of anterolateral spines, 2nd at the junction between 5th and 6th thoracic somites and 3rd at the posterolateral corner
		Order – Stomatopoda	Eyestalk is ¾ as long as the cornea
		Specimen – Larva of *Squilla hieroglyphica*	
46.	*Penaeus indicus*	Phylum – Arthropoda	Presence of well-developed curved rostrum
		Class – Crustacea	Presence of bifurcated supraorbital spines and stalked compound eyes
		Order – Decapoda	Length is 1.5 mm and there are no frontal organs
		Suborder – Macrura	
		Specimen – Protozoea of *Penaeus indicus*	

(continued)

Table 5.5 (continued)

Serial no.	Name	Systematic position	Salient features
47.	Mysis of *Penaeus indicus*	Phylum – Arthropoda	Presence of spine on the scaphocerite and an unsegmented pleopod bud
		Class – Crustacea	Cleft of telson extends to the origin of the penultimate pair of lateral telsonic setae
		Order – Decapoda	Length of the organism is 3.5 mm
		Suborder – Macrura	
		Specimen – Mysis of *Penaeus indicus*	
48.	Zoea of mud crab, *Scylla serrata*	Phylum – Arthropoda	Presence of laterally compressed carapace
		Class – Crustacea	Characterized by one dorsal, one rostral and two lateral spines, all of which are elongated
		Order – Cirripedia	Eyes sessile, abdomen 5 segmented and telson is furcated
		Suborder – Brachyura	
		Specimen – Zoea of mud crab, *Scylla serrata*	
49.	Megalopa of mud crab, *Scylla serrata*	Phylum – Arthropoda	All appendages are well developed
		Class – Crustacea	Abdomen with five pairs of swimming pleopods
		Order – Cirripedia	Carapace depressed and a pair of chelipeds also developed
		Suborder – Brachyura	
		Specimen – Megalopa of mud crab, *Scylla serrata*	
50.	Veligers of *Saccostrea cucullata*	Phylum – Mollusca	It is a D-shell larval stage with semi-transparent velum and is about 70 μm size
		Class – Bivalvia	Oval shape at early 'umbo stage' when measures 100 μm
		Specimen – Veligers of *Saccostrea cucullata*	Presence of irregular eyespot at 'eye stage' when it measures 300 μm
51.	Veligers of *Littorina littorea*	Phylum – Mollusca	Larvae id 1½ whorled shape
		Class – Bivalvia	Characterized by two dark pigmented particles
		Specimen – Veligers of *Littorina littorea*	It is pale yellowish and has no sculpturing
52.	Bipinnaria larva of starfish	Phylum – Echinodermata	Presence of a wavy band of cilia at the equatorial girdle
		Class – Asteroidea	This band runs down on each side and loops round the front above and below where the mouth begins to form
		Specimen – Bipinnaria larva of starfish	Before the mouth opens, gut is developed with a tiny stomach and intestinal opening at the lower end of larva
53.	Brachiolaria larva	Phylum – Echinodermata	Composed of 3 special appendages in the preoral region called branchiolarian arms
		Class – Asteroidea	Presence of adhesive papillae on the arms except the medio-dorsal one
		Specimen – Brachiolaria larva	The arms are unicilliated except the medio-dorsal one that bears circum oral ciliary band
54.	Auricularia larva of sea cucumber	Phylum – Echinodermata	The folding of the ciliated band is very long
		Class – Holothuroidea	The circumoral ciliary band is drawn out into paired preoral lobes anteriorly and paired anal lobes posteriorly
		Specimen – Auricularia larva of sea cucumber	Circumoral ciliary field is continuous

(continued)

Table 5.5 (continued)

Serial no.	Name	Systematic position	Salient features
55.	Brachyuran zoea	Phylum – Rotifera	Single dorsal and rostral spines elongated
		Class – Monogononta	A pair of lateral spines present, one pair pointed downwards and other pointed upwards and forward
		Order – Ploima	Ends of spines are flattened like spear head
		Suborder – Brachyura	
		Family – Portunidae	
		Specimen – Brachyuran zoea	

Table 5.6 Classification of zooplankton on the basis of habitat

Type	Description
Oceanic plankton	These are marine zooplankton that inhabit beyond the continental shelf
Neritic plankton	These zooplankton inhabit waters overlying continental shelves. These waters are often very productive as they receive the run-off from the adjacent landmasses that triggers the phytoplankton growth in these regions
Brackish water plankton	These zooplankton inhabit estuarine regions, where there is a continuous mixing of freshwater and seawater. The zooplankton species of this category have wide range of tolerance to different dilution factors. Such zooplankton are very common in the shrimp culture farms and form important diet of the prawns

Table 5.7 Classification of zooplankton on the basis of depth distribution

Type	Description
Neuston	The zooplankton of this category are restricted at the top few millimetres (usually 10 mm) of the surface micro layer
Pleuston	These are widely distributed at the surface of the sea (with parts of the body sometime projecting above the water)
Epipelagic	These are distributed between 0 and 300-m water column, *e.g.* siphonophores and arrow worms
Mesopelagic	The zooplankton of this category are restricted within the depth 300–1,000 m, *e.g.* euphausiids and chaetognath
Bathypelagic	These are restricted within the depth 1,000 and 3,000 m, *e.g.* foraminifera and euphausiids
Abyssopelagic	The waters overlying the vast abyssal plains of the ocean are inhabited by a variety of zooplankton species, which are often referred to as abyssopelagic zooplankton. These zooplankton are thus restricted between 3,000 and 4,000 m

Source: Santhanam and Srinivasan (1998)

ovens and refrigerators as insulators and as polztion of paraffin wax from petroleum; and in the manufacture of concrete. Because of the presence of the hydrocarbon compound 'diatomin', the diatoms serve as the indicators of rich petroleum grounds. Present-day scientists are of the opinion that these tiny free-floating producers are the chief precursors of the petroleum-rich fields of Venezuela at Los Angeles.

The dinoflagellates also at times play an important role in the fishery sector as by virtue of their power of luminescence, they can emit 'cold living light' which, though lasts for about one-tenth of a second, helps in identifying certain fish shoals during night.

Phytoplankton can also be used to reflect the water quality in terms of certain pollutants other

Table 5.8 Classification of zooplankton on the basis of size

Type	Size range
Nanozooplankton	<20 μm
Microzooplankton	20–200 μm
Mesozooplankton	200 μm–2 mm
Macrozooplankton	2–20 mm
Megazooplankton	>20 mm

Source: Santhanam and Srinivasan (1998)

Table 5.9 Classification of zooplankton on the basis of duration of planktonic life

Type	Description
Holoplankton	This group includes organisms which are planktonic throughout their life cycle, *e.g.* tintinnids, cladocerans, copepods and chaetognaths
Meroplankton	This group encompasses those organisms which remain planktonic only for a portion of their life cycle, *e.g.* larvae of benthic invertebrates and fish larvae (ichthyoplankton)

Source: Santhanam and Srinivasan (1998)

than sewage (that contribute considerable amount of nitrate and phosphate in the ambient water). It has been reported that an increase in cell volume occurs in certain species (like *Dunaliella tertiolecta* and *Phaeodactylum tricornutum*) due to copper pollution (Stauber and Florence 1987).

Phytoplankton play an important role in maintaining the global carbon cycle. The elemental composition of phytoplankton is $C:N:P = 106:16:1$, which is commonly referred to as the 'Redfield Ratio' (Redfield 1934, 1958). This ratio clearly explains that about 100 units of carbon are delivered to the deep sea for every 16 units of nitrogen and 1 unit of phosphorus. As such, the biological pump delivers carbon from the atmosphere to the deep sea, where it is concentrated and sequestered for centuries. The larger the world's phytoplankton population, the more carbon dioxide gets pulled from the atmosphere. This lowers the average temperature of the atmosphere due to lower volumes of this green house gas. Scientists have found that a given population of phytoplankton can double its numbers in the order of once per day. In other words, phytoplankton respond very rapidly to changes in their environment. Large populations of this organism, sustained over long periods of time, would significantly lower atmospheric carbon dioxide levels and, in turn, lower average temperatures. According to Burke Hales, an assistant professor in the College of Oceanic and Atmospheric Sciences at Oregon State University, 'in order for the phytoplankton to be a long term sink for carbon, they somehow have

to get deposited in the deep ocean, but this does not always happen. If the phytoplankton are just eaten at the surface, or do not sink to any great depth, then the carbon is eventually released back in the atmosphere'. New research has revealed that phytoplankton may be one of the main historic controls on global warming, and that fertilizing the oceans with iron results in increased phytoplankton productivity, a hypothetical way to offset the effects of global warming. Through photosynthesis, these tiny, free-floating aquatic plants can convert carbon dioxide to organic carbon, and there appears to be a prehistoric relationship between iron in the ocean and atmospheric levels of carbon dioxide. The carbon cycle is a complicated system of causes and effects that are not completely understood, but researchers have long suspected that the oceans are the main regulators of the Earth's atmosphere. During the ice ages, more of the water of this planet was locked up in glaciers that created arid, windy condition and a lot of dust. This iron-rich dust was blown out to the sea, stimulating productivity of phytoplankton throughout the world's oceans, which ultimately resulted in the reduction of carbon dioxide levels.

Phytoplankton also influence cloud formation and hence play a regulatory role on the climate of the Earth. It was given expression through the CLAW hypothesis (published two decades ago by R. Charlson, J. Lovelock, M. Andreae and S. Warren), which supposes that the gas dimethyl sulphide produced by marine plankton influences cloud formation and hence albedo and climate. However, direct evidence for a link between plankton and clouds has been slow to emerge (Charlson et al. 1987). A recent paper (Meskhidze and Nenes 2006) states a remarkably close seasonal and spatial associations between seasurface chlorophyll (an indicator of biological activity) and atmospheric properties for a 6-year period over a substantial area of the Southern Ocean. Over high-chlorophyll areas, the number of cloud droplets doubled whereas the droplets size decreased by 30 % compared with other regions, leading to an atmospheric cooling comparable to that over highly polluted regions.

Diatoms are important components contributing to the stability of intertidal mudflats. They excrete polysaccharides, which trap sediment grains and stabilize the sediment (http://www.nioo.know.hl/cemo/ecoflat/work.htm).

Mass culture of phytoplankton may be undertaken in closed systems of the space capsules. The nitrogenous wastes of the travellers could be diluted and circulated as algal nutrients and the respired air as a source of carbon dioxide. In exchange, the travellers could utilize the oxygen released by the photosynthesis of algae and benefit from the food value of cultured phytoplankton.

The recent inclination of the scientists towards the phytoplankton community is mainly due to their paramount importance in various spheres of life, as pointed below (http://www.bigelow.org/sci_overview):

1. Oceans cover more than 70 % of the Earth's surface, containing all but 3 % of the world's water, and this saline water mass supports huge quantum of phytoplankton.
2. Dissolved gases in seawater regulate the composition of our atmosphere through phytoplankton compartment.
3. The world's phytoplankton generate half of the oxygen we breathe and absorb half of the carbon dioxide that may be contributing to global warming.
4. Phytoplankton are the sources of fossil fuel.
5. Phytoplankton constitute the base of the marine and estuarine food web and are the dominant producers of the aquatic ecosystem.
6. Phytoplankton act as the sink of anthropogenic wastes (specially nitrate and phosphate) that are contributed by sewage and agricultural operations.

There are major contributions of zooplankton to the marine and estuarine environments:

1. Among the various levels of production in the sea, the secondary production contributed by zooplankton is an important linkage between the primary and tertiary productions. The zooplankton mainly consumes the primary producers and forms the major food source for tertiary producers. An important food for the Antarctic baleen whales is the krill, which are important zooplankton in the Antarctic region.
2. In certain parts of the world, the zooplankton are directly consumed as food, *e.g.* certain species of *mysids* are consumed as food in West Indies. A deep-water copepod (*Euchaeta norvegica*) is a delicious food item in West Indies.
3. In many corners of the globe, zooplankton are treated as source of bioactive substances.
4. Certain species of zooplankton are used as bio-indicators of water quality.
5. The zooplankton contribute substantial biogenic material to ooze formation, which has wide application in instrument-related industry where the ooze is used as thermal insulators, chromatographic column filters, etc.

5.3   Impact of Climate Change on Plankton Community

Water mass of estuaries, bays and oceans is mostly heated from the top, by sunlight, which means the formation of a layer of warm water on the top of a huge depth of cold water. Phytoplankton can only grow near the surface (photic zone), because they need sunlight and they quickly use up the nutrients there. For continuous supply of nutrients, mixing or churning of water is very important. As a result, regions of upwelling water, such as off the coasts of Peru and Antarctica, have particularly rich sea life. The ocean currents that drive them are not expected to stop any time soon, but warmer temperatures at the surface could be reducing smaller-scale mixing. Water temperatures strongly influence the growth rates of phytoplankton. Phytoplankton in warmer equatorial waters grow much faster than their cold-water counterparts. With worldwide temperatures predicted to increase over the next century, it is important to gauge the reactions of phytoplankton species. A study done by the Indian scientists (with the Department of Science and Technology, Govt. of India funding) reveals the increase in the population density of the phytoplankton over a period of 30 years in the same location (21°52′41.3″ N; 88°11′50.1″ E) in Indian Sundarban mangrove ecosystem. The gradual

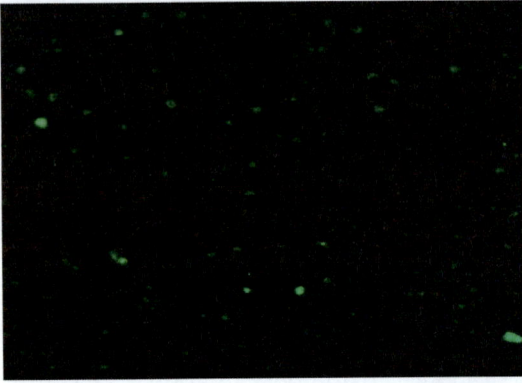

Fig. 5.1 Year: 1980 (average surface water temperature =~31.2 °C)

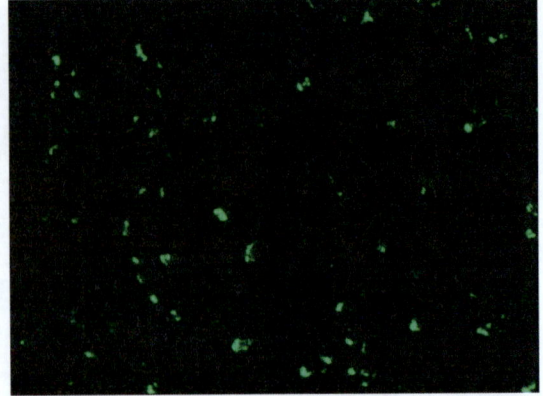

Fig. 5.2 Year: 1990 (average surface water temperature =~31.7 °C)

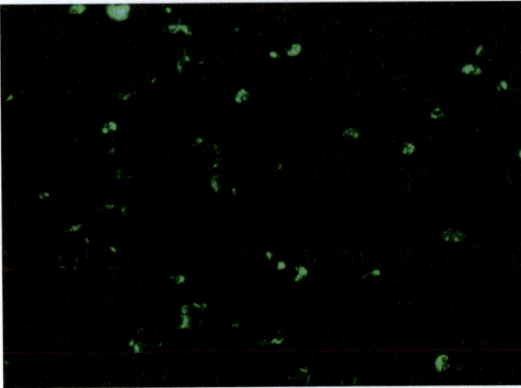

Fig. 5.3 Year: 2000 (average surface water temperature =~31.9 °C)

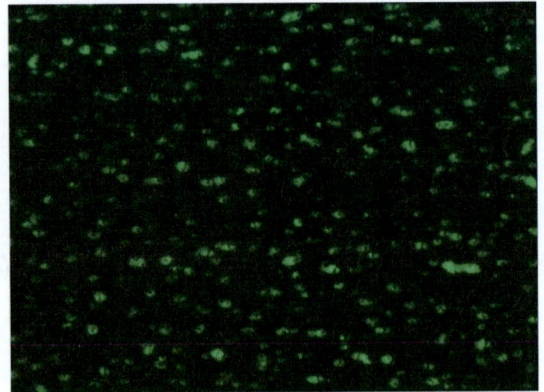

Fig. 5.4 Year: 2010 (average surface water temperature =~32.3 °C)

increase of phytoplankton with temperature is observed through fluorescence microscopic study (Figs. 5.1, 5.2, 5.3, and 5.4). The nutrients in the study area originate from fish landing stations, sewage and several aquacultural farms existing in the area. The rich mangrove vegetation (survival ground of some 34 true mangrove species) also contributes nutrients through detritus and litters in the deltaic Sundarbans region.

In Australia several researches were initiated on the impact of climate change on plankton. Between the early 1970s and 2000–2009 off eastern Tasmania, abundances of key cold-water zooplankton species have declined, and warm-water species have increased. Another study suggested

thinning and increased porosity of shells of two pteropod snails in NW and NE Australia over the past 40 years as ocean pH declined. These changes are likely to be the first of many in the future that could have profound effects on marine food webs. It has been reported by a group of Australian scientists with high confidence that the distribution of smaller and less abundant subtropical and tropical zooplankton species will expand poleward, and that larger and more abundant temperate and polar species will retract poleward. The scientists also reported that the phenology of temperate and polar species will move earlier as temperatures warm. With medium confidence, the researchers reported that calcifying plankton will be

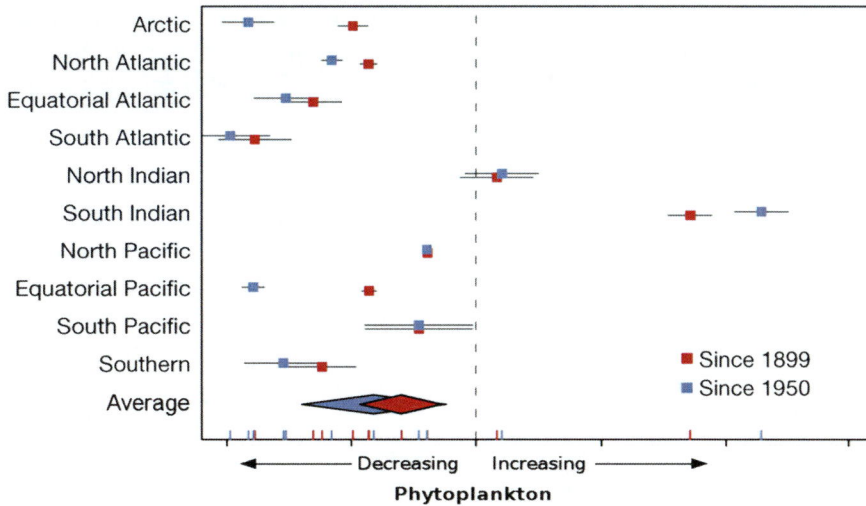

Fig. 5.5   Global phytoplankton decline over the past century

detrimentally affected by ocean acidification as pH declines. Moreover the researchers state that the abundance of zooplankton will change as temperature warms, with a mosaic of increases and decreases in zooplankton abundance in response to climate change. Collectively, these changes are likely to cause spatial reorganization of food webs in temperate and polar regions, trophic mismatch in temperate regions, changes in nutrient enrichment and direct effects on zooplankton calcifiers. These impacts of climate change on zooplankton will cause widespread and significant changes (mostly adverse, but sometimes positive) to higher trophic levels (takluyverwordpress.com/2010/08/06-climate-change-affecting-phytoplankton/).

Report published in online journal *Science Express* (http://www.ijpsi.org) shows that by the end of the twenty-first century, warmer oceans will cause populations of these marine microorganisms to thrive near the poles and shrink in equatorial waters. In the tropical oceans, it is predicted that there will be a 40 % drop in potential diversity preferably in the number of strains of phytoplankton (http//www.nsf.gov). Thomas et al. (2012), an Australian scientist, says 'If the oceans continue to warm as predicted, there will be a sharp decline in the diversity of phytoplankton in tropical waters and a poleward shift in species'

thermal niches – if they don't adapt'. 'The research is an important contribution to predicting plankton productivity and community structure in the oceans of the future', says David Garrison (http//www.nsf.gov), programme director in the National Science Foundation's (NSF) Division of Ocean Sciences.

Based on projections of ocean temperatures in the future, however, many phytoplankton may not adapt quickly enough. It is a fact that phytoplankton cannot regulate their cell temperatures or migrate, and therefore, their extinction is not very uncommon in the matrix of extreme temperature. It is said by Kremer (2012) that 'We've shown that a critical group of the world's organisms has evolved to do well under the temperatures to which they're accustomed'. But warming oceans may significantly limit their growth and diversity, with far-reaching implications for the global carbon cycle. 'Future models that incorporate genetic variability within species will allow us to determine whether particular species can adapt', says Klausmeier (2012), 'or whether they will face extinction'. Researches conducted by scientific experts (Boyce et al. 2010) exhibit a clear decline of plankton in the Atlantic ocean and the polar oceans, a smaller decline in the Pacific, while plankton actually increased in the Indian ocean (Fig. 5.5). The scientists separated the year-by-year variation in plankton levels from the overall trends and compared that variation to various ocean 'oscillations'. These are roughly regular patterns in temperature and pressure, the best known of which is the El Niño/ Southern Oscillation in the Pacific. In most areas with oscillations, there was less plankton in warmer years (the pattern did not fit for the North Indian Ocean, perhaps due to the effects of the monsoon rains).

The present author conducted a study in three sectors of Indian Sundarbans (western, central and eastern) jointly with Dr. Atanu Kumar Raha, Former PCCF, Department of Forest, Govt. of West Bengal (India), in 2012 and arrived at few core findings related to impact of climate change on the plankton community in the mangrove-dominated Indian Sundarbans. These are:

- Surface water temperature in the mangrove-dominated deltaic complex of Indian Sundarbans exhibited a gradual increase in all the three sectors of Indian Sundarbans (Mitra et al. 2009).
- Salinity of the territorial waters of Indian Sundarbans has exhibited considerable variation over a span of 27 years (Mitra et al. 2009, 2011). Whether this is due to heavy siltation, decreased freshwater supply through Hugli estuary, tectonic tilt of the Bengal basin towards eastward direction or ingression of seawater from Bay of Bengal (due to sea-level rise) is still under investigation (Raha et al. 2012).
- Surface water pH has shown considerable variation, which may be attributed to climate change. The decreasing trend may be correlated to more absorption of atmospheric CO_2 and increased rate of sewage discharge.
- Dissolved oxygen did not exhibit any marked variations over time.
- Transparency of the aquatic phase has reduced due to increased erosion of the adjacent landmasses, which may be a direct consequence of sea-level rise and subsequent tidal amplitude.
- The nutrient load has increased with the passage of time (except silicate), which may be due to hike up in the sewage discharge from the adjacent cities, towns and villages or unplanned mushrooming of shrimp farms in

and around the Sundarbans Biosphere Reserve (that generate nitrate, phosphate, etc.). The erosion of the adjacent landmasses and mangrove forests is also a major contributor of nutrients.

- Phytoplankton species diversity exhibited significant variation since 1990 (Tables 5.10, 5.11, and 5.12), with more numbers of stenohaline species in the upstream waters of central Indian Sundarbans, where the salinity has increased maximum compared to western and eastern sectors (Fig. 5.6).
- Since 1990, the overall diversity of phytoplankton has exhibited an increasing trend in all the three sectors reflecting the abundance of nutrients. Interestingly in the central sector, the standing stock and diversity exhibits the lowest value which may be attributed to hypersaline condition in this zone in and around the Matla River.

The standing stock of phytoplankton is best reflected through ocean colour remote sensing. It is a widely recognized tool for monitoring the optically active biogeochemical parameters like phytoplankton pigments (*e.g.* chlorophyll), suspended particulate matter and yellow substance. Measurements of ocean colour and the fate of light in the ocean are extremely useful for describing biological dynamics in surface waters. The ocean colour monitor (OCM) of the Indian Remote Sensing Satellite IRS-P4 is optimally designed for the estimation of chlorophyll in coastal and oceanic waters, detection and monitoring of phytoplankton blooms, studying the suspended sediment dynamics and the characterization of the atmospheric aerosols. In the present study, the phytoplankton standing stock monitored in the chlorophyll images derived from the IRS-P4 OCM data for aquatic system in and around mangrove-dominated Indian Sundarbans clearly shows more pigment level in the western sector of Indian Sundarbans, compared to the central sector (Fig. 5.7). The coincidence of high pigment level with low saline region and vice versa is a distinct signal of the adverse impact of seawater intrusion (may be due to sea-level rise) on the tiny free-floating producer community of the marine and estuarine waters.

Table 5.10 Population density (No. × 10⁵/l) and Shannon–Wiener species diversity index (H) of phytoplankton during 1990 in three sectors of Indian Sundarbans

S. No.	Phytoplankton species	Western	Central	Eastern
1.	*Coscinodiscus eccentricus*	19.71	14.54	17.81
2.	*Coscinodiscus jonesianus*	13.91	11.24	12.81
3.	*Coscinodiscus lineatus*	10.91	7.74	11.01
4.	*Coscinodiscus radiatus*	18.11	14.34	18.11
5.	*Coscinodiscus gigas*	8.91	1.14	8.91
6.	*Coscinodiscus oculusiridis*	4.61	0.84	6.11
7.	*Planktoniella sol*	5.61	3.14	7.71
8.	*Cyclotella striata*	8.71	4.01	8.41
9.	*Thalassiosira subtilis*	11.21	4.93	9.91
10.	*Ceratium tripos*	2.81	0.04	0.61
11.	*Skeletonema costatum*	4.44	0.74	1.11
12.	*Paralia sulcata*	4.54	3.04	1.41
13.	*Rhizosolenia crassispina*	2.84	0.84	1.01
14.	*Rhizosolenia setigera*	1.34	1.04	0.94
15.	*Rhizosolenia alata*	4.24	2.04	2.54
16.	*Ceratium teres*	1.14	0.01	0.54
17.	*Ceratium trichoceros*	2.74	1.14	2.14
18.	*Bacteriastrum delicatulum*	4.04	2.74	4.44
19.	*Bacteriastrum varians*	7.04	2.24	6.64
20.	*Bacteriastrum comosum*	2.54	0.14	1.84
21.	*Chaetoceros didymus*	2.48	0.08	1.68
22.	*Chaetoceros peruvianus*	1.28	0.07	1.12
23.	*Chaetoceros compressus*	0.58	0.45	0.18
24.	*Ditylum sol*	5.08	1.68	4.48
25.	*Protoperidinium crassipes*	0.38	0.24	0.88
26.	*Triceratium favus*	9.58	6.38	8.78
27.	*Triceratium reticulatum*	8.18	2.58	1.98
28.	*Biddulphia sinensis*	3.98	1.28	3.28
29.	*Biddulphia mobiliensis*	5.78	2.98	1.88
30.	*Hemidiscus hardmannianus*	4.48	0.48	0.08
31.	*Climacosphenia elongate*	0.01	0.01	0.01
32.	*Fragilaria oceanica*	5.08	2.18	1.18
33.	*Rhaphoneis amphiceros*	0.12	0.01	0.07
34.	*Thalassionema nitzschioides*	2.28	0.22	0.53
35.	*Thalassiothrix longissima*	5.88	1.98	4.18
36.	*Thalassiothrix fraunfeldii*	4.78	2.58	3.98
37.	*Asterionella japonica*	0.28	0.24	1.1
38.	*Ceratium extensum*	2.28	0.76	1.18
39.	*Gyrosigma balticum*	3.28	0.78	1.88
40.	*Pleurosigma normanii*	10.08	4.48	8.48
41.	*Pleurosigma elongatum*	3.38	0.66	1.88
42.	*Diploneis smithii*	6.38	3.08	5.28
43.	*Cymbella marina*	3.98	1.58	2.58
44.	*Nitzschia sigma*	1.88	0.29	1.28
45.	*Nitzschia closterium*	2.98	0.18	1.58
46.	*Ceratium furca*	1.18	0.08	0.12
47.	*Trichodesmium erythraea*	19.18	12.88	17.18
48.	*Dicrateria gilva*	8.08	4.58	4.88
49.	*Chlorella marina*	6.28	1.98	4.68
N		268.59	130.7	210.4
H		3.3993	2.9446	3.1922

Table 5.11 Population density (No. × 10^5/l) and Shannon–Wiener species diversity index (H) of phytoplankton during 2000 in three sectors of Indian Sundarbans

S. No.	Phytoplankton species	Western	Central	Eastern
1.	*Coscinodiscus eccentricus*	24.6	19.2	21.8
2.	*Coscinodiscus jonesianus*	18.8	15.9	16.8
3.	*Coscinodiscus lineatus*	15.8	12.4	15.0
4.	*Coscinodiscus radiatus*	23.0	19.0	22.1
5.	*Coscinodiscus gigas*	13.8	5.8	12.9
6.	*Coscinodiscus oculusiridis*	9.5	5.5	10.1
7.	*Planktoniella sol*	10.5	7.8	11.7
8.	*Cyclotella striata*	13.6	9.0	12.4
9.	*Thalassiosira subtilis*	16.1	9.5	13.9
10.	*Ceratium tripos*	5.7	3.5	4.6
11.	*Skeletonema costatum*	6.4	2.7	5.1
12.	*Paralia sulcata*	6.5	5.0	5.4
13.	*Rhizosolenia crassispina*	4.8	2.8	5.0
14.	*Rhizosolenia setigera*	3.3	3.0	2.9
15.	*Rhizosolenia alata*	6.2	4.0	5.1
16.	*Ceratium teres*	3.1	1.2	2.5
17.	*Ceratium trichoceros*	4.7	3.1	4.1
18.	*Bacteriastrum delicatulum*	6.0	4.7	6.4
19.	*Bacteriastrum varians*	9.0	4.2	8.6
20.	*Bacteriastrum comosum*	4.5	2.1	3.8
21.	*Chaetoceros didymus*	4.8	2.4	4.0
22.	*Chaetoceros peruvianus*	3.6	1.1	2.2
23.	*Chaetoceros compressus*	2.9	1.9	2.5
24.	*Ditylum sol*	7.4	4.0	6.8
25.	*Protoperidinium crassipes*	2.7	1.1	1.5
26.	*Triceratium favus*	11.9	8.7	11.1
27.	*Triceratium reticulatum*	10.5	4.9	4.3
28.	*Biddulphia sinensis*	6.3	3.6	5.6
29.	*Biddulphia mobiliensis*	8.1	5.3	4.2
30.	*Hemidiscus hardmannianus*	6.8	2.8	2.4
31.	*Climacosphenia elongate*	0.7	0.15	0.3
32.	*Fragilaria oceanica*	7.4	4.5	3.5
33.	*Rhaphoneis amphiceros*	0.43	0.08	0.15
34.	*Thalassionema nitzschioides*	4.6	1.9	1.8
35.	*Thalassiothrix longissima*	8.2	4.3	6.5
36.	*Thalassiothrix fraunfeldii*	7.1	4.9	6.3
37.	*Asterionella japonica*	2.6	0.6	2.1
38.	*Ceratium extensum*	4.6	1.6	3.5
39.	*Gyrosigma balticum*	5.6	3.1	4.2
40.	*Pleurosigma normanii*	12.4	6.8	10.8
41.	*Pleurosigma elongatum*	5.7	1.5	4.2
42.	*Diploneis smithii*	8.7	5.4	7.6
43.	*Cymbella marina*	6.3	3.9	4.9
44.	*Nitzschia sigma*	4.2	1.9	3.6
45.	*Nitzschia closterium*	5.3	2.5	3.9
46.	*Ceratium furca*	3.5	1.8	2.1
47.	*Trichodesmium erythraea*	21.5	15.2	19.5
48.	*Dicrateria gilva*	10.4	6.9	7.2
49.	*Chlorella marina*	8.6	4.3	7.0
N		436.90	23.04	333.90
H		3.6894	3.5890	3.6232

Table 5.12 Population density (No. × 10⁵/l) and Shannon–Wiener species diversity index (H) of phytoplankton during 2010 in three sectors of Indian Sundarbans

S. No.	Phytoplankton species	Western	Central	Eastern
1.	*Coscinodiscus eccentricus*	27.85	22.45	25.05
2.	*Coscinodiscus jonesianus*	22.05	19.15	20.05
3.	*Coscinodiscus lineatus*	19.05	15.65	18.25
4.	*Coscinodiscus radiatus*	26.25	22.25	25.35
5.	*Coscinodiscus gigas*	17.05	9.05	16.15
6.	*Coscinodiscus oculusiridis*	14.27	10.27	14.87
7.	*Planktoniella sol*	15.27	12.57	16.47
8.	*Cyclotella striata*	18.37	13.77	17.17
9.	*Thalassiosira subtilis*	20.87	14.27	18.67
10.	*Ceratium tripos*	10.47	8.27	9.37
11.	*Skeletonema costatum*	11.17	7.47	9.87
12.	*Paralia sulcata*	11.27	9.77	10.17
13.	*Rhizosolenia crassispina*	9.57	7.57	9.77
14.	*Rhizosolenia setigera*	7.53	7.23	7.13
15.	*Rhizosolenia alata*	10.43	8.23	9.33
16.	*Ceratium teres*	7.33	5.43	6.73
17.	*Ceratium trichoceros*	8.93	9.35	8.33
18.	*Bacteriastrum delicatulum*	10.23	8.93	10.63
19.	*Bacteriastrum varians*	13.23	8.43	12.83
20.	*Bacteriastrum comosum*	8.73	6.33	8.03
21.	*Chaetoceros didymus*	9.03	6.63	8.23
22.	*Chaetoceros peruvianus*	5.57	3.07	4.17
23.	*Chaetoceros compressus*	4.87	3.87	4.47
24.	*Ditylum sol*	9.37	5.97	8.77
25.	*Protoperidinium crassipes*	4.67	4.89	3.47
26.	*Triceratium favus*	13.87	10.67	13.07
27.	*Triceratium reticulatum*	12.47	6.87	6.27
28.	*Biddulphia sinensis*	8.67	5.97	7.97
29.	*Biddulphia mobiliensis*	10.47	7.67	6.57
30.	*Hemidiscus hardmannianus*	9.17	5.17	4.77
31.	*Climacosphenia elongate*	3.07	4.44	2.67
32.	*Fragilaria oceanica*	9.77	6.87	5.87
33.	*Rhaphoneis amphiceros*	2.8	2.45	2.52
34.	*Thalassionema nitzschioides*	6.97	4.27	4.17
35.	*Thalassiothrix longissima*	10.57	6.67	8.87
36.	*Thalassiothrix fraunfeldii*	9.47	7.27	8.67
37.	*Asterionella japonica*	4.97	2.97	4.47
38.	*Ceratium extensum*	6.97	3.97	5.87
39.	*Gyrosigma balticum*	9.89	8.97	8.49
40.	*Pleurosigma normanii*	16.69	11.09	15.09
41.	*Pleurosigma elongatum*	9.99	6.79	8.49
42.	*Diploneis smithii*	12.99	9.69	11.89
43.	*Cymbella marina*	10.59	8.19	9.19
44.	*Nitzschia sigma*	8.49	6.19	7.89
45.	*Nitzschia closterium*	9.59	8	8.19
46.	*Ceratium furca*	7.79	6.09	6.39
47.	*Trichodesmium erythraea*	25.79	19.49	23.79
48.	*Dicrateria gilva*	14.69	14	11.49
49.	*Chlorella marina*	12.89	8.59	11.29
N		572.06	433.22	507.28
H		3.7822	3.8999	3.7563

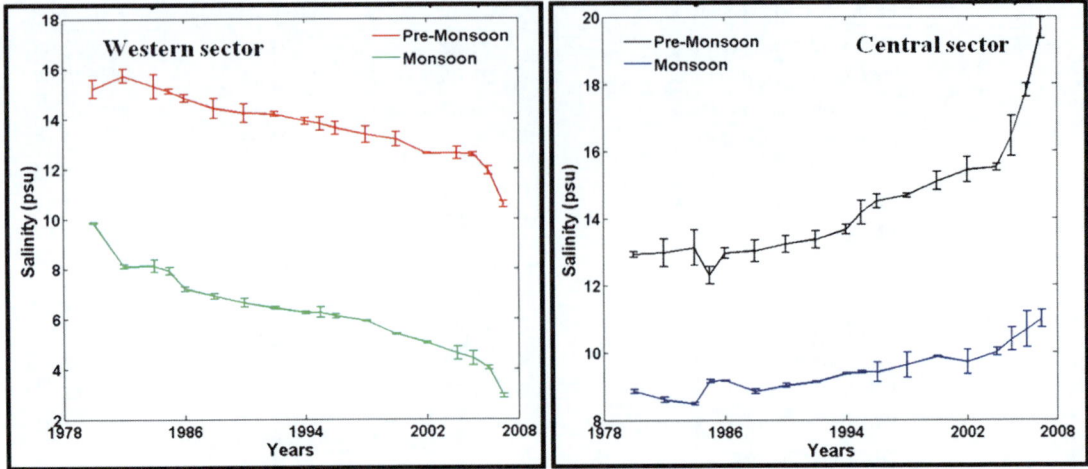

Fig. 5.6 Salinity decreased during last 30 years on the western (*left*) sector, while it increased on the central (*right*) sector of the Indian Sundarbans. Trends are similar in both premonsoon and monsoonal periods. The *left panel* shows the increasing freshening in the western station due to meltwater, while the *right panel* shows the increasing salinity to the central. Note the 5 psu bias between the monsoon and premonsoon salt content in the water in both stations

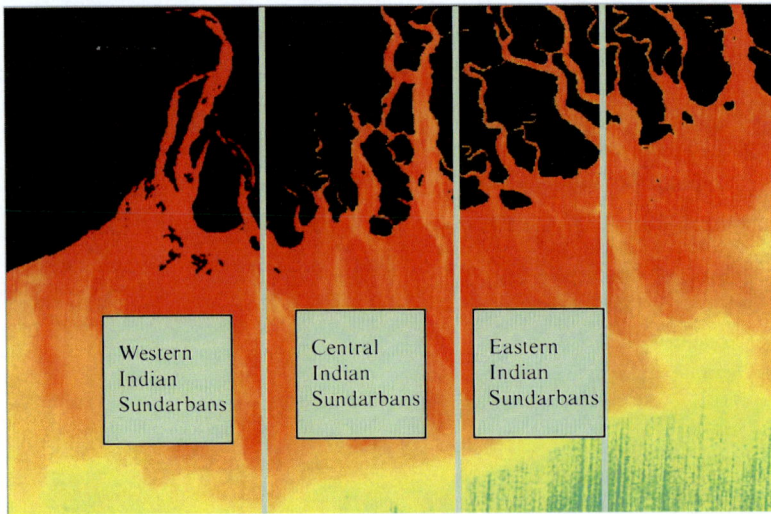

Fig. 5.7 Chlorophyll *a* level in western, central and eastern Indian Sundarbans

Zooplankton are beacons of climate change because of several major reasons. Firstly, zooplankton are poikilothermic, so their physiological processes, such as ingestion, respiration and reproductive development, are highly sensitive to temperature, with rates doubling or tripling with a 10 °C temperature rise (Mauchline 1998). Secondly, most zooplankton species are short lived (<1 year), and hence there can be tight coupling of climate and population dynamics (Hays et al. 2005). In fact, some evidence suggests that plankton are more sensitive indicators of change than even environmental variables themselves, because the non-linear responses of plankton communities can amplify subtle environmental signals (Taylor et al. 2004). Thirdly, unlike other marine groups, such as fish and many intertidal organisms, zooplankton are generally not commercially exploited (exceptions include krill and some jellyfish species), so studies of long-term trends in response to environmental change are generally not confounded with trends in exploitation, and therefore *noise factor* is minimum. Fourthly, the distribution of zooplankton can accurately reflect temperature and ocean currents as plankton are free floating and mostly remain so for their entire life. In contrast, terrestrial organisms are either rooted to their substratum or need to spend considerable energy to migrate or move. Further, reproductive products of zooplankton are distributed by currents and not by vectors, making changes in zooplankton distribution with climate change easier to understand than the more complex responses of terrestrial species. Lastly, because ocean currents provide an ideal mechanism for dispersal over large distances, almost all marine animals have a planktonic stage in their life cycles; therefore, alterations in the distribution of many marine groups are at least partially determined while floating in the zooplankton. Recent evidence suggests that many of the meroplanktonic life stages are even more sensitive to climate change than their holozooplanktonic neighbours living permanently in the plankton. All of these attributes combine to make zooplankton sensitive beacons of climate change. Scientists also claim that the zooplankton community is also affected by the episode of climate change. Several macroscale researches have shown that the increase in seawater temperature has triggered a major reorganization in calanoid copepod species composition and biodiversity. During the last 40 years, there has been a northerly movement of warmer-water plankton by 10° latitude in the northeast Atlantic and a similar retreat of colder-water plankton to the north. This geographical movement is much more pronounced than any documented terrestrial study, presumably due to advective processes. Over the last decade, there has been a progressive increase in the presence of warm-water/subtropical species into the more temperate areas of the northeast Atlantic.

In terms of marine phonological changes and climate, the plankton of the North Sea has been extensibly studied using Continuous

Plankton Recorded Data. Using 66 taxa, it was found that the plankton community was responding to changes in SST by adjusting their seasonality (in some cases the shift in seasonal cycles of over 6 weeks was detected), but more importantly the response to climate warming varied between different functional groups and trophic levels, leading to mismatch. It is thought that temperate marine environments are particularly vulnerable to phenological changes caused by climate warming because the recruitment success of higher trophic levels is highly dependent on synchronization with pulsed planktonic production. The rapid changes in plankton communities observed over the last few decades in the North Atlantic, related to regional changes, have enormous consequences for other trophic levels and biogeochemical processes.

Copepods form the most important component of zooplankton in almost all of Indian estuaries (Qasim 2003). The occurrence of other organisms is largely dependent on the site of observation in the estuary showing an assortment of marine, brackish water and freshwater forms. For example, in the Hugli estuary in the upper regions, which largely remain freshwater-dominated rotifers, protozoans and cladocerans were found in large numbers. In the middle zone with brackish water copepods, mysids, ostracods and cladocerans were abundant, and in the marine zone, many other species of crustaceans including lucifers, gammarids, cumaceans, amphipods, isopods and other forms such as hydromedusae and chaetognaths were present (Shetty et al. 1963). Copepods form the dominant groups of zooplankton throughout the year, constituting 73.0–96.4 % of the total biomass. During the high saline period in the estuary (premonsoon), which generally lasts from March to early June, copepods show considerable diversity and abundance. During the monsoon months, their biomass decline to a minimum, and many species of copepods disappear from estuary. With the rise in salinity from November onwards, repopulation and proliferation in the copepod fauna occur (Sarkar et al. 1986a, b).

A long-term study (1990–2010) conducted on the zooplankton community in three sectors of

Table 5.13 Counts (No./m³) and percentage composition (in parentheses) of major zooplankton in India Sundarban mangrove ecosystem during 1990–2012

Zooplankton	Western			Central			Eastern		
	1990	2000	2012	1990	2000	2012	1996	2000	2012
Hydromedusa	2.5	3.1	3.8	1.2	1.1	0.6	2.3	2.7	3.4
	(0.15)	(0.14)	(0.24)	(0.08)	(0.04)	(0.02)	(0.3)	(0.13)	(0.17)
Siphonophora	0.01	0.1	0.4	0.5	0.06	0.01	0.0	0.05	0.2
	(0.00)	(0.0)	(0.03)	(0.03)	(0.00)	(0.00)	(0.0)	(0.00)	(0.01)
Ctenophora	0.1	0.4	0.6	1.4	0.1	0.04	0.1	0.3	0.4
	(0.01)	(0.02)	(0.04)	(0.09)	(0.00)	(0.00)	(0.01)	(0.01)	(0.02)
Chaetognath	10.6	12.8	23.7	21.1	3.8	1.9	28.3	10.7	12.9
	(0.67)	(0.58)	(1.51)	(1.43)	(0.15)	(0.06)	(3.31)	(0.51)	(0.64)
Cladocera	4.1	6.8	11.2	10.4	3.3	2.1	0.3	7.7	14.9
	(0.26)	(0.31)	(0.72)	(0.73)	(0.13)	(0.06)	(0.03)	(0.37)	(0.73)
Ostracoda	0.0	0.3	1.1	0.2	0.0	0.0	0.01	0.2	0.8
	(0.0)	(0.01)	(0.07)	(0.01)	(0.00)	(0.00)	(0.00)	(0.01)	(0.04)
Copepoda	1,354	1,879	1,134	1,242	2,356	3,281	696	1,777	1,609
	(86.11)	(85.20)	(72.42)	(84.35)	(92.98)	(95.48)	(81.39)	(85.17)	(79.36)
Amphipoda	0.2	0.9	2.4	0.1	0.2	0.1	0.2	0.5	2.3
	(0.01)	(0.04)	(0.15)	(0.01)	(0.01)	(0.00)	(0.02)	(0.02)	(0.11)
Lucifers	33.3	45.1	44.2	43.4	41.1	34.9	13.5	42.0	62.0
	(2.12)	(2.04)	(2.82)	(2.95)	(1.62)	(1.02)	(1.57)	(2.01)	(3.06)
Mysidacea	1.1	2.1	3.9	2.5	0.4	0.1	0.7	5.4	4.3
	(0.06)	(0.10)	(0.25)	(0.16)	(0.02)	(0.00)	(0.08)	(0.26)	(0.21)
Cumacea	0.2	0.6	1.8	0.5	0.3	0.0	0.3	0.5	1.4
	(0.01)	(0.03)	(0.11)	(0.03)	(0.01)	(0.00)	(0.03)	(0.02)	(0.07)
Invertebrate eggs	0.5	0.9	1.8	0.9	0.3	0.2	0.1	0.8	1.6
	(0.03)	(0.04)	(0.11)	(0.06)	(0.01)	(0.01)	(0.01)	(0.04)	(0.08)
Invertebrates larvae	154.3	231.0	302.8	131.6	120.0	111.3	109.8	221.1	287.5
	(9.81)	(10.47)	(19.34)	(8.93)	(4.74)	(3.24)	(12.83)	(10.60)	(14.18)
Pteropoda	0.0	1.9	6.5	0.1	0.0	0.0	0.0	0.5	4.2
	(0.0)	(0.09)	(0.42)	(0.0)	(0.00)	(0.00)	(0.0)	(0.02)	(0.21)
Copelata	4.5	3.8	5.9	9.7	2.2	0.9	0.8	2.9	4.8
	(0.29)	(0.17)	(0.38)	(0.65)	(0.09)	(0.03)	(0.09)	(0.14)	(0.24)
Fish eggs	4.0	11.0	13.4	4.7	3.2	1.8	0.04	10.4	12.1
	(0.25)	(0.50)	(0.86)	(0.32)	(0.13)	(0.05)	(0.0)	(0.50)	(0.60)
Fish larvae	2.8	4.9	6.6	2.0	1.6	1.3	2.7	3.3	5.6
	(0.19)	(0.22)	(0.42)	(0.13)	(0.06)	(0.04)	(0.31)	(0.16)	(0.28)
Miscellaneous	0.1	0.7	1.8	0.1	0.3	0.02	0.1	0.3	0.6
	(0.01)	(0.03)	(0.11)	(0.01)	(0.01)	(0.00)	(0.01)	(0.01)	(0.03)

Indian mangrove ecosystem (Table 5.13) exhibits interesting results that may attributed to salinity fluctuation.

Few important findings (that can be interpreted from Table 5.13) in relation to climate-change-induced sea-level rise and subsequent salinity fluctuations are listed here:

1. Significant spatial variation is observed in zooplankton community ($p < 0.01$), which may be due difference in salinity.

2. The lowest zooplankton standing stock in the central Indian Sundarbans is due to the hyper-saline nature of the aquatic phase owing to freshwater cut-off (from the upstream region) due to siltation of the Bidyadhari channel since fifteenth century (Chaudhuri and Choudhury 1994; Mitra et al. 2009, 2011; Raha et al. 2012).

3. The copepod community projects as an exception to the normal rule of adverse impact of

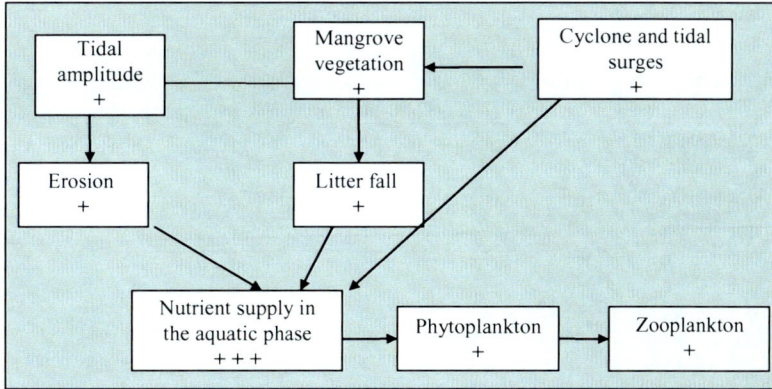

Fig. 5.8 '+' denotes increased magnitude of an event. Sundarbans delta experiences a sea-level rise of 3.14 mm/year (measurement taken at Sagar Island, the largest island in Indian Sundarbans with coordinates 88°03′ 06.17″ longitude and 21°38′ 54.37″ latitude) compared to the global average of ~2.50 mm/year. This has resulted in increased tidal amplitude. The frequency of cyclonic depression has also increased. All these natural processes have enhanced erosion and litter fall leading to increased level of nutrients in the aquatic phase. The end result is an overall increase in plankton community (*Sources*: Hazra et al. 2002; Mitra et al. 2009; Raha et al. 2012)

climate change on mangrove plankton community. The population of copepod was highest in the central sector and also increased with time in this sector which exhibits significant positive correlation of copepod population with salinity.

4. The overall increase of zooplankton since 1990 coincides with the increase of phytoplankton diversity in the same location, which clearly reflects the abundance of nutrients with the passage of time since 1990.

The present finding thus reflects a positive impact of climate change-induced salinity rise on the plankton community of the tropical mangrove ecosystem (Fig. 5.8). However, if the salinity exceeds 35 psu (mean salinity of Bay of Bengal = ~35 psu) in the present system, a negative impact cannot be ignored.

The phenomenon of reduced growth rate of the dinoflagellate species *Oxyrrhis marina* with increasing salinity was reported by Kain and Fogg (1958) and Buskey et al. (1998). Several studies have demonstrated higher mortalities for copepods (Moreira 1975) and crab zoeae (Vernberg and Vernberg 1975) under hypersaline conditions. Several species of phytoplankton have extremely narrow range of salinity tolerance that allows maximum growth and exhibits reduced growth and mortality under hypersaline conditions (Kain and Fogg 1958,

1960). In the tropical mangrove ecosystem like Sundarbans, the lowest cell volume (Table 5.14) and standing stock of *Coscinodiscus* spp. during premonsoon (the period characterized by highest salinity due to excessive evaporation and minimum rainfall) is an indication of adverse impact of salinity on tropical mangrove phytoplankton species at salinity around 35 psu (Mitra et al. 2012).

Fact Below the Carpet

The tiny microscopic communities of the world ocean are the plankton. They can convert solar energy into several utilizable forms of energy and thus sustain the faunal community of the oceans. These plankton depend on nutrients that are mostly contributed by coastal vegetation. To serve the ocean biotic community, it is necessary to save the coastal vegetation. The industry houses must feel this basic principle and consider the clause of coastal vegetation conservation under CSR. Common mass must raise voice against clearing the coastal vegetation patches for short-term benefits from aquaculture or tourism.

Table 5.14 Cell volume and cell carbon of six major species of *Coscinodiscus* from Indian Sundarbans mangrove water

No.	Species	Salinity	Average cell volume 'V' (in μm^3)	Average cell carbon (in picogram)
1	*Coscinodiscus eccentricus*	Mean salinity = 32 psu Mean salinity = 14 psu Mean salinity = 20 psu	10,637.39 44,968.71 12,477.54	531.38 1,709.60 604.44
2	*Coscinodiscus jonesianus*	Mean salinity = 32 psu Mean salinity = 14 psu Mean salinity = 20 psu	12,132.22 24,996.13 12,336.21	590.84 1,061.90 598.88
3	*Coscinodiscus lineatus*	Mean salinity = 32 psu Mean salinity = 14 psu Mean salinity = 20 psu	4,196.70 10,778.1 4,317.48	249.79 536.80 255.60

(continued)

Table 5.14   (continued)

No.	Species	Salinity	Average cell volume 'V' (in μm³)	Average cell carbon (in picogram)
4	*Coscinodiscus radiatus*	Mean salinity = 32 psu Mean salinity = 14 psu Mean salinity = 20 psu	44,451.19 70,918.19 45,240.20	1,693.66 2,473.80 1,717.99
5	*Coscinodiscus gigas*	Mean salinity = 32 psu Mean salinity = 14 psu Mean salinity = 20 psu	4,266.90 8,042.07 4,387.07	253.17 423.30 258.94
6	*Coscinodiscus oculusiridis*	Mean salinity = 32 psu Mean salinity = 14 psu Mean salinity = 20 psu	3,839.30 8,551.3 3,978.08	232.39 444.90 239.18

Important References

Boyce D, Lewis M, Worm B (2010) Global phytoplankton decline over the past century. Nature 466(7306): 591–596

Buskey EJ, Wysor B, Hyatt CJ (1998) The role of hypersalinity in the persistence of the Texas "brown tide" in the Laguna Madre. J Plankton Res 20:1553–1565

Charlson RJ, Lovelock JE, Andreae MO, Warren SG (1987) Oceanic phytoplankton, atmospheric sulfur, cloud albedo and climate. Nature 326:655. doi:10.1038/326655A0

Chaudhuri AB, Choudhury A (1994) Mangroves of the Sundarbans, vol I, India. IUCN – The World Conservation Union, Bangkok

Garrison D (2012) Long term study reveals potential results of climate change. American Institute for Biological Sciences. www.Internet.edu. Accessed on 5 Oct 2010

Hays GC, Richardson AJ, Robinson C (2005) Climate change and marine plankton. Trends Ecol Evol 20:337–344

Hazra S, Ghosh T, Dasgupta R, Gautam S (2002) Sea level and associated changes in the Sundarbans. Sci Cult 68(9–12):309–321

Kain JM, Fogg GE (1958) Studies on the growth of marine phytoplankton I. Asterionella japonica Gran. J Mar Biol Assoc UK 37:397–413

Kain JM, Fogg GE (1960) Studies on the growth of marine phytoplankton III. Prorocentrum micans Ehrenburg. J Mar Biol Assoc UK 39:33–50

Klausmeier CA (2012) Successional dynamics in the seasonally forced diamond food web. Am Nat 180:1–16

Kremer A (2012) How well can existing forests withstand climate change? In: EUFORGEN climate change and forest genetic diversity: implications for sustainable forest management in Europe. pp 3–17

Lalli CM, Parsons TR (1997) Energy flow and nutrient cycling. In: Biological oceanography: an introduction, 2nd edn. Open University/Elsevier. University of British Columbia, Vancouver, Canada, pp 112–146

Mauchline J (1998) The biology of calanoid copepods. Adv Mar Biol 33:710

Meskhidze N, Nenes A (2006) Phytoplankton and cloudiness in the Southern Ocean. Science 314(5804): 1419–1423. doi:10.1126/science.1131779

Mitra A (2000) The Northeast coast of the Bay of Bengal and deltaic Sundarbans. In: Sheppard C (ed) Seas at the millennium – an environmental evaluation. Pergamon, Amsterdam, pp 143–157

Mitra A, Gangopadhyay A, Dube A, Schmidt ACK, Banerjee K (2009) Observed changes in water mass properties in the Indian Sundarbans (Northwestern Bay of Bengal) during 1980–2007. Curr Sci 97: 1445–1452

Mitra A, Sengupta K, Banerjee K (2011) Standing biomass and carbon storage of above-ground structures in dominant mangrove trees in the Sundarbans. For Ecol Manag 261(7):1325–1335

Mitra A, Zaman S, Kanti Ray S, Sinha S, Kakoli Banerjee K (2012) Inter-relationship between phytoplankton cell volume and aquatic salinity in Indian Sundarbans.

Nat Acad Sci Lett 35:485–491. Springer doi:10.1007/s40009-012-0083-1

Moreira GS (1975) Studies on the salinity resistance of the copepod Euterpina acutifrons (Dana). In: Vernberg FJ (ed) Physiological ecology of Estuarine organisms. University of South Carolina Press, Columbia, pp 73–80

Qasim SZ (2003) Indian estuaries. Allied Publisher Pvt. Limited, New Delhi, 420pp

Raha A, Das S, Banerjee K, Mitra A (2012) Climate change impacts on Indian Sundarbans: a time series analysis (1924–2008). Biodivers Conserv 21:1289–1307. Springer doi:10.1007/s10531-012-0260-z

Redfield AC (1934) In: Danial RJ (ed) James Johnstone memorial volume. On the proportions of organic derivations in sea water and their relation to the composition of plankton, University of Liverpool Press, Liverpool, 176pp

Redfield AC (1958) The biological control of chemical factors in the environment. Am Sci 46:205–207

Santhanam R, Srinivasan A (1998) A manual of marine zooplankton. Oxford and IBH Publishing company Pvt. Ltd, New Delhi, pp 1–4

Sarkar SK, Singh BN, Choudhury A (1986a) Composition and variation in abundance of zooplankton in the Hooghly Estuary, West Bengal, India. Proc Indian Acad Sci 95:125–134

Sarkar SK, Singh BN, Choudhury A (1986b) The ecology of copepods from Hooghly Estuary, West Bengal, India. Mahasagar – Bull Nat Inst Oceanogr 19:103–112

Shetty HPC, Saha SB, Ghosh BB (1963) Observations on the distribution and fluctuations of plankton in the Hoogly-Matla estuarine system, with notes on their relation to commercial fish landings. Indian J Fish 8:326–363

Stauber JL, Florence TM (1987) Mechanism of toxicity of ionic copper and copper complexes to algae. Mar Biol 94:511–519

Taylor KC, White J, Severinghaus J, Brook E, Mayewski P, Alley R, Steig E, Spencer M (2004) Abrupt climate change around 22 ka on the Siple Coast of Antarctica. Quat Sci Rev 23(1–2):7–15

Thomas MK, Kremer CT, Klausmeier CA, Litchman E (2012) A global pattern of thermal adaptation in marine phytoplankton. Science 338:1085–1088. doi:10.1126/science.1224836

Vernberg FJ, Vernberg WB (1975) Adaption to extreme environments. In: Vernberg FJ (ed) Physiological ecology of estuarine organisms. University of South Carolina Press, Columbia, pp 165–180

WWF News Letter (2001) Impacts of climate change on life in Africa http://wwf.panda.org/?3866/Newsletter-March-2001. Accessed on 3 July 2013

Internet References

http://www.bigelow.org/sci_overview
http://www.ijpsi.org
http://www.nioo.know.hl/cemo/ecoflat/work.htm
http//www.nsf.gov
http://oceanlink.island.net/ask/pollution.html
www.nature.com/news/2010/100728/full/news.2010.379.html

Climate Change and Its Impact on Brackish Water Fish and Fishery

6

We should not tackle vast problem with half vast concepts.

Preston E. Cloud Jr.

6.1 Major Categories of Fishes in the Marine and Estuarine Waters

The neritic zone of the ocean, coastal and estuarine waters is extremely important from the production point of view as these areas receive major nutrients (ammonia, nitrate, phosphate and silicate – the raw materials for primary production) from adjacent landmasses and sustain the foundation community of marine and estuarine biodiversity – the *phytoplankton*. This community comprises of diverse species of tiny free-floating floral components like *Coscinodiscus* sp., *Chaetoceros* sp., *Fragilaria* sp. and *Biddulphia* sp. The upwelling areas of the marine environment also support large population of several types of phytoplankton due to the presence of nutrient-rich water that are transported from the bottom of the ocean to the surface layer. The estuaries flowing through mangrove forests and salt-marsh grass ecosystems are also saturated with nutrients sourced from these coastal vegetations through microbial degradation of litter. The nutrients are basically the building blocks of phytoplankton biomass. Phytoplankton provide food to the *zooplankton* (the major groups include copepods, chaetognaths and harpacticoids) of the pelagic zone, which are finally consumed by fishes (like herring, cod, flounder, Bombay duck and Hilsa) that comprise the nekton community of the marine and estuarine ecosystems (Fig. 6.1). Any change in the lower tiers of food web due to sea-level rise, saline water intrusion into the bays and estuaries, ocean acidification (as a result of lowering of pH) or temperature rise is likely to be transmitted to the members of higher trophic level (nekton).

By definition, pelagic animals that swim actively are known as nekton. Most of the animals under this category are vertebrates (animals with backbones such as fishes, reptiles, marine birds and marine mammals), but a few representatives are invertebrates (like squids and nautiluses and some species of arthropods). Marine and estuarine fishes are the dominant vertebrate nekton that live in salt water, brackish water or even migrate in the freshwater system for breeding purpose. Fishes may also live near the surface or at great depths both in extreme warm and cold conditions. Like other ectothermic (cold-blooded) organisms, fishes are incapable of generating and maintaining a steady internal temperature from metabolic heat; so the internal body temperature of a fish is usually the same as that of the surrounding environment. This is one of the reasons why fishes are highly vulnerable to oscillation of environmental parameters due

A. Mitra, *Sensitivity of Mangrove Ecosystem to Changing Climate*,
DOI 10.1007/978-81-322-1509-7_6, © Springer India 2013

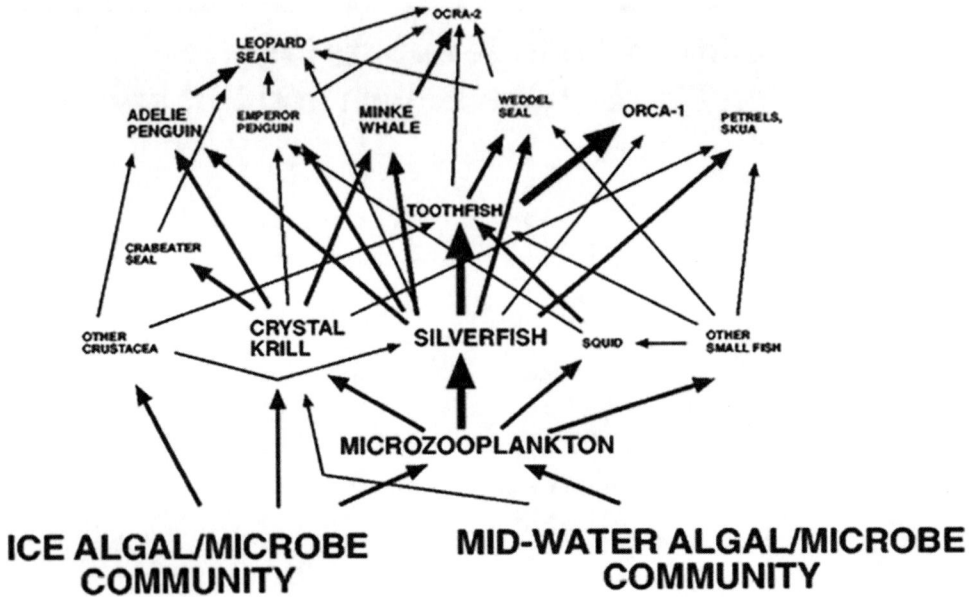

Fig. 6.1 Food web of marine ecosystem

to climate change. The degree of vulnerability is, however, functions of *exposure* of the fish species to a particular parameter (variable), the *sensitivity* of the species to the parameter and the *adaptive capacity* of the species to cope with the stress. Mathematically the concept may be expressed as

$$\text{Vulnerability} = f\left(\begin{array}{c}\text{exposure, sensitivity,} \\ \text{adaptive capacity}\end{array}\right)$$

Fishes are divided into two major groups based on the nature of their skeletons: the cartilaginous fishes (*Chondrichthyes*) and the bony fishes (*Osteichthyes*). Both bony (*e.g. Tenualosa ilisha* and *Scatophagus argus*) and cartilaginous fishes (like whale shark, dog shark, sting rays and guitar fishes) constitute the major sectors of the nekton community in the Bay of Bengal region and adjacent estuaries, which is the case study area in the present chapter.

6.1.1   Osteichthyes

Bony fishes are found at all depths and in all the oceans, but their distribution is determined directly or indirectly by the abundance and bio-

mass of primary producers. This is the basis of evaluating potential fishing zone (PFZ), which is detected from the satellites. The PFZs are not static. They exhibit significant temporal and spatial variations as function of environmental variables (Figs. 6.2, 6.3, and 6.4). It is very important to note in this context that satellites do not observe fish stocks directly, but measurements, such as sea-surface temperature (SST), sea-surface height (SSH), ocean colour, ocean winds and sea ice, characterize critical habitat that influences marine resources including fish stocks. Most of the spatial features that are important to marine ecosystems like ocean fronts, eddies, convergence zones, river plumes and coastal processes cannot be adequately resolved without satellite data. Chlorophyll, present in the phytoplankton, is the only biological component of the marine ecosystem accessible to remote sensing (via ocean colour) and as such provides a key metric for evaluating the health and productivity of marine ecosystems on a global scale. Long-term ocean colour satellite monitoring provides an important tool for better understanding of the marine processes, ecology, fish stock and the coastal environmental changes (Tang and Kawamura 2001). Modern oceanographic vessels are, therefore, linked to satellites via computers

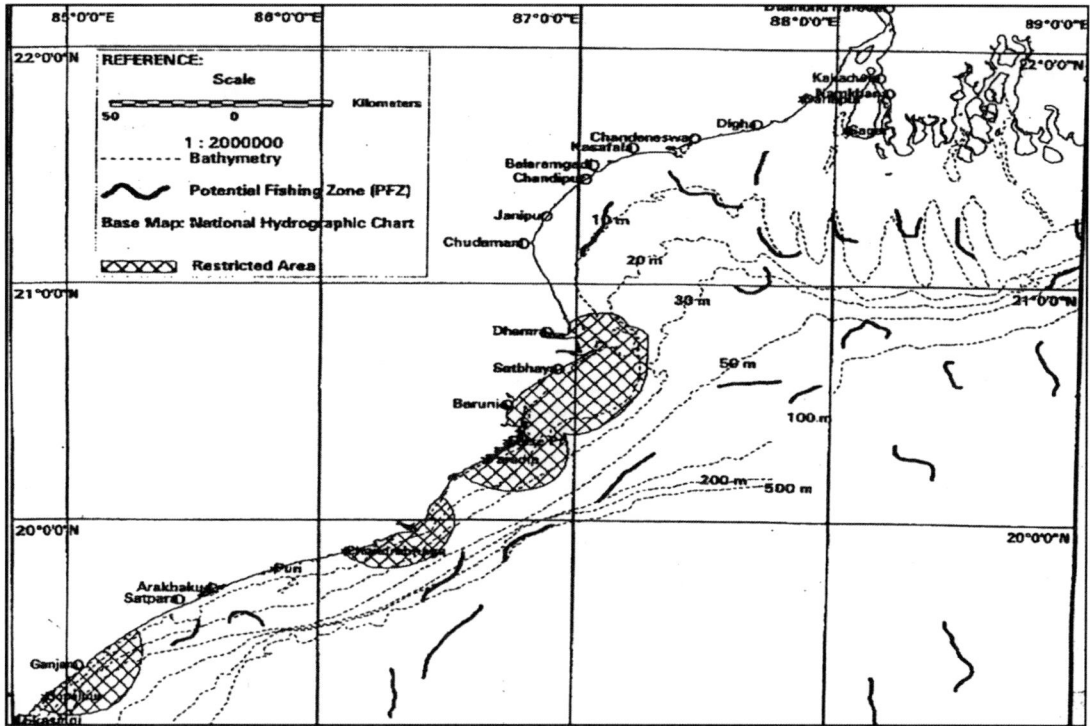

Fig. 6.2 Map showing potential fishing zone off Orissa and West Bengal coast; map based on SST/chlorophyll composite of 08–09 November 2003

allowing scientists to use immediate data to plan their sampling programmes while at sea.

Fishes are mostly concentrated in upwelling areas, shallow coastal zones and estuaries. The surface waters support much greater populations of fish per unit volume of water than the deeper zones, where food resources are very less in terms of quality (diversity) and quantity. The presence of mangroves and other associate floral species also regulate the distribution of fishes in the marine and estuarine compartments. The rich bony fish (osteichthyes) diversity in the waters of mangrove-dominated Indian Sundarbans is a relevant example in this context. The aquatic subsystem of deltaic Sundarbans is the dwelling spot, nursery and breeding ground of a wide variety of finfish and shellfish. Hamilton-Buchanan (1922) carried out an extensive survey in this Gangetic delta complex with respect to fish fauna. The works of Hora (1933, 1934, 1936, 1943), Pearse (1932), Prashad et al. (1940),

Talwar (1991), and Talwar et al. (1992) reveal valuable information on the fish resources of Indian Sundarbans. The catch composition of finfish in and around this tropical mangrove ecosystem usually exhibits the pattern: *Bombay duck > Cat fishes > Clupeids > Prawns > Croakers > Pomfret > Ribbon fish > Sardines > Elasmobranchs > Eels.* The annual landing data of finfish from Indian Sundarbans region indicates an increasing trend (Table 6.1) in spite of the existence of several threats in this deltaic lobe.

Several workers have depicted the taxonomic diversity of fish species in and around the aquatic subsystem of mangrove-dominated Sundarbans deltaic complex. Jhingran (1982) documented a total of 172 species and stated that the species diversity is comparatively more in the high saline zone (lower estuarine stretch) of Indian Sundarbans. His estimate revealed 73 species of freshwater origin and 99 species of marine/higher salinity origin. Mandal and Nandi (1989) documented 141 species

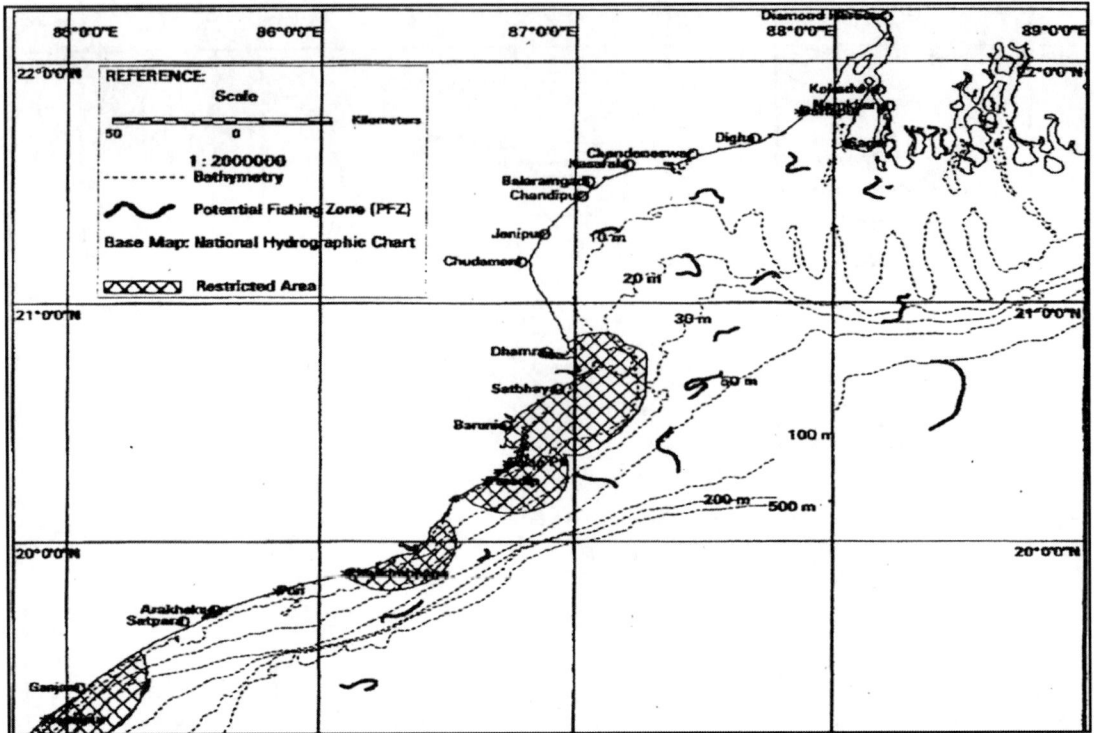

Fig. 6.3 Map showing potential fishing zone off Orissa and West Bengal coast; map based on SST/chlorophyll composite of 23–24 November 2003

of finfish under 100 genera, while Chaudhuri and Choudhury (1994) recorded 250 species under 96 genera in the waters of Indian Sundarbans. Khan (2003) recorded 107 species from Sundarbans Biosphere Reserve region, but this figure does not include the species restricted in the low saline upper zone of the Hugli–Matla estuarine complex.

With respect to ecological tolerance, a large fraction of the finfish species in the mangrove-dominated estuarine complex is euryhaline in nature and moves freely from the upper stretch of minimum salinity to lower stretch of maximum salinity. Such species will be affected at a smaller magnitude by climate change as they have a wide span of adaptation to salinity. Species that are restricted only to lower stretch of the Hugli–Matla estuarine complex and adjacent to Bay of Bengal region or in the upper freshwater-dominated zone will be most affected on account of climate change due to their narrow range of ecological tolerance.

The fish fauna of the estuarine waters in and around Indian Sundarbans has been classified into

residents and *transients* (migrants). The species whose individuals of different sizes are present during all the months of the year in any zone of the estuary are referred to as resident species. The important resident finfish species are *Mugil parsia, Mugil tade, Polynemus paradiseus, Polydactylus indicus, Otolithoides biauritus, Lates calcarifer, Hilsa toli, Arius jella, Harpodon nehereus, Setipinna taty, Ilisha elongata, Setipinna phasa, Coilia ramcarati, Pama pama* and *Sillaginopsis panijus*. The transient or migratory fishes enter and stay in the Bay of Bengal associated estuaries for a short period. Depending on their migratory pattern and direction, the migrants are divided into three categories (Jhingran 1982):

1. Marine forms that migrate upstream and spawn in freshwater areas of the estuary like *Tenualosa ilisha, Polynemus paradiseus, Sillaginopsis panijus* and *Pama pama*
2. Freshwater species, which spawn in saline area of the estuary like *Pangasius pangasius*

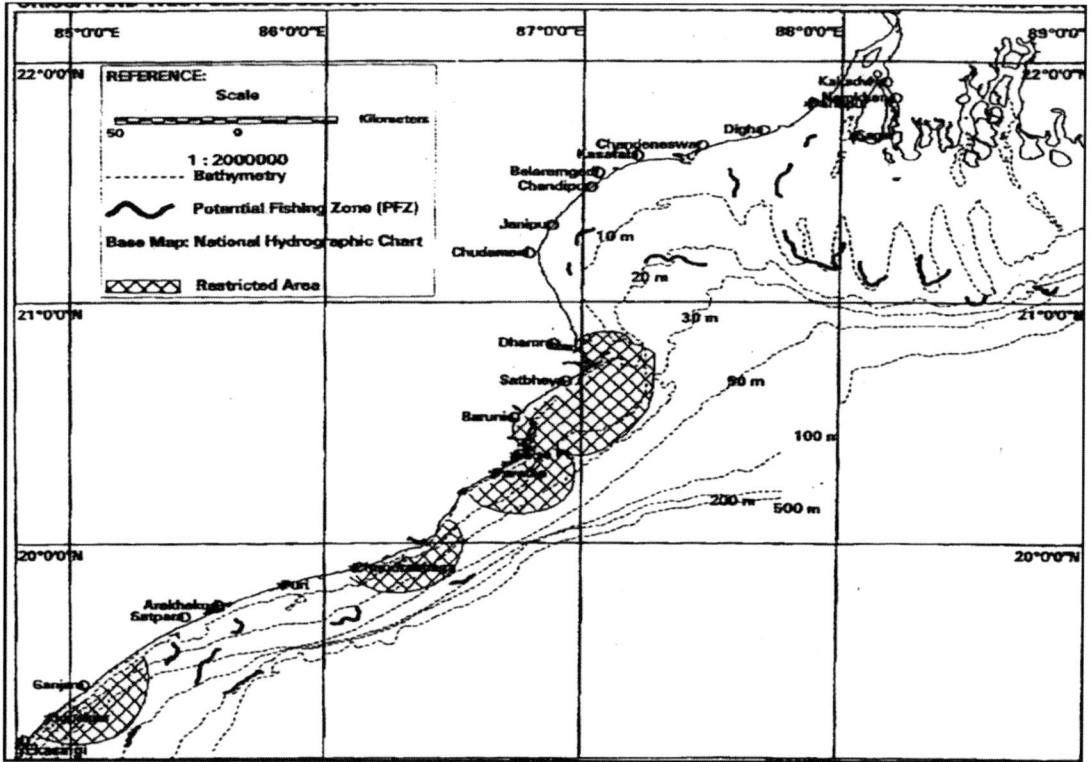

Fig. 6.4 Map showing potential fishing zone off Orissa and West Bengal coast; map based on SST/chlorophyll composite of 07–08 December 2003

Table 6.1 Mean annual landings of important fish species (in metric tonnes) from Indian Sundarbans estuarine system during 1964–1965 to 1975–1976 and 1987–1988 to 1990–1991

Fish species	1964–1965 to 1972–1973	1973–1974 to 1975–1976	1987–1988 to 1990–1991
Tenualosa ilisha	–	–	2,679.45
Harpodon nehereus	1,242.2	2,579.8	4,842.52
Pama pama	73	320.1	4,547.47
Setipinna spp.	422.1	780.6	6,221.4
Arius jella	5.3	23.3	915.9
Pampus spp.	5.7	9.8	902.77
Polynemus paradiseus	3.6	21.8	191.4
Coilia spp.	1.3	2.3	987.72
Ilisha elongata	26.9	102.2	674.3
Otolithoides biauritus	103.8	156.8	322.07
Polydactylus indicus	0.6	3	154.15
Polydactylus sp.	–	–	115.15
Trichiurus sp.	246.5	806.4	3,026.97
O. militaris	5.3	44.7	–
Others	551.6	921.8	8,699.42
Total	2,622.2	5,772.6	34,280.88

Source: Jhingran (1982)

3. Marine species that spawn in less saline water of the estuary like *Arius jella, Osteogeneious militaris* and *Polydactylus indicus*

6.1.2   Chondrichthyes

Elasmobranchs (chondrichthyes) constitute a vital segment of marine and estuarine nekton and are of great commercial importance all over the globe, apart from being a major component in marine food web. About 350 species of sharks and 320 species of rays are known to exist. Nearly all are marine, although a few species inhabit estuaries and a very few are permanent inhabitants of freshwater. It has been observed that sharks usually prefer swimming in open waters, whereas rays tend to be found on or near the bottom. The annual catch of elasmobranchs in India is around 70,000 tonnes, which is over 4 % of total marine fish landings (http://www.mpeda. com/FisheryResources/Elasmobranchs/ Elasmobranchs.htm).

According to MPEDA (1999), the annual average landings of elasmobranchs during 1995–1998 was 75,037 tonnes although the estimated potential of this group is 168,000 tonnes. The major elasmobranchs contributing to Indian fisheries are *Rhizoprionodon oligolinx, Isurus oxyrinchus, Sphyrna blochii, Sphyrna mokarran, Rhynchobatus djiddensis, Rhinobatos granulatus, Rhina ancylostoma, Dasyatis sephen, Dasyatis uarnak, Dasyatis imbricatus, Dasyatis marginatus, Himantura alcockii, Aetobatus narinari, Aetomylaeus mehotii, Aetomylaeus maculatus, Rhinoptera javanica, Gymnura poecilura* and *Mobula diabolus.* Sharks account for 60–70 % of this total figure. Maritime states like Tamil Nadu, Gujarat, Maharashtra, Kerala, Karnataka and Andhra Pradesh contribute around 85 % of shark landings in India. About 65 species of sharks that contribute to fishery stock are sighted in Indian waters, and over 20 of them belong to Carcharhinidae and Sphyrnidae families (http://www.mpeda.com/FisheryResources/ Elasmobranchs/Elasmobranchs.htm). Sharks are thus dominant species of elasmobranchs and play a major role in both ecology and economics. The

various direct and indirect products obtained from sharks are today used in food, tourism and pharmaceutical industries. Because of such multiple uses, the community is presently under threat due to overexploitation. Deterioration of water quality due to anthropogenic activities and industrial discharge has increased the magnitude of threat. Few species of sharks (particularly *Glyphis gangeticus*) are so sensitive in nature that they cannot withstand the alteration of water quality caused by rapid industrialization and urbanization. Such species are extremely vulnerable to oscillation of hydrological parameters often induced by climate change. The drastic reduction of *Glyphis gangeticus* population in recent times in the Ganga-Bhagirathi-Hugli riverine stretch is an indication of change in water quality, but how far it is related to climate-change-induced alteration needs to be critically investigated.

6.2   Impact of Climate Change on Fish and Fishery

Climate change has both direct and indirect impacts on fish stocks which are exploited commercially. Direct effects act on physiology and behaviour of fishes and alter their growth, reproduction, mortality and distribution. Indirect effects encompass events like alteration of aquatic productivity, biotic community structure and composition of the marine and estuarine ecosystems on which fishes depend for food and survival. Changes in primary and secondary production will obviously have a major effect on fisheries production, but it is not possible in the current state of knowledge to make accurate quantitative predictions of changes in global marine primary production solely due to climate. The comparative study (Sarmiento et al. 2005), in which six different Atmosphere-Ocean General Circulation Models (AOGCMs) were used, indicates that production may increase by not more than 10 % over the period to 2050, but the level of confidence in this estimate is low and the baseline for the comparison is the 'pre-industrial' state. In contrast, observations from satellite and large-scale plankton sampling show declines in phytoplank-ton and chlorophyll all over the past 20–50 years, which are consistent with the expected conse-quences of reduced nutrient supply due to strengthening of vertical density gradients.

Although global aggregated marine primary production is not expected to change substan-tially over the next four or five decades, there is a stronger basis for predicting changes in produc-tion at regional level and also good observational evidence, particularly for the North Pacific and North Atlantic. In both cases, changes in produc-tion are driven mainly by regime-scale and event-scale (*e.g. El Nino*) changes. These are important components of the climate system whose predict-ability and impacts are gradually getting revealed. Further improvements in modelling and monitor-ing at these time scales, with better resolution of regional impacts, are likely to yield greater benefits to fisheries forecasting and management.

Qualitative changes in production may have major impacts on food chains leading to altera-tion of fish diversity regardless of changes in the absolute level of primary production. Examples of this include the observed switch over from krill to salps as the major nektonic species in parts of the Antarctic (Atkinson et al. 2004) and the ascendance of gelatinous species to a domi-nant position in areas such as the Black Sea. In the former case, climate change was probably a major factor, but in the latter it was not.

Global, regional and smaller-scale impacts of climate change on biological production are ultimately the sum of processes which act on individual organisms. There are very few detailed, experimental studies on responses to climate change at individual level. Increasing tempera-ture interacts with other environmental changes, including declining pH and increasing nitrogen and ammonia to increase metabolic costs of organisms. The consequences of these interac-tions are speculative and complex. An experi-mental study on rainbow trout (*Oncorhynchus mykiss*) showed positive effects on appetite, growth, protein synthesis and oxygen consump-tion due to a 2 °C increase in winter, but negative effects of the same temperature increase in summer. Thus, rising temperature may not only

cause seasonal increases in growth but also increases risks to fish populations living towards the upper end of their thermal tolerance zone (Morgan et al. 2001). The outbreak of microbial diseases particularly in shellfish (shrimps, crabs, oysters, etc.) is also related to rising temperature of the surrounding environment. Climate change has been implicated in mass mortalities of many aquatic species, including plants, fish, corals and mammals, but lack of standard epidemiological data and information on pathogens generally makes it difficult to attribute causes (Harvell et al. 1999).

Global climate change has the potential to alter fish and fisheries dramatically. Kawasaki (2001) provided specific evidence of the effects of global warming on world fisheries production, particularly on tuna and sardines. The major consequences of climate change on marine and estuarine fish spectrum of the planet Earth are now well understood at the population level, physiological level and even at the biochemical level as discussed here in details. Few conse-quences may not be backed up with sound case studies/examples, but definite scientific knowl-edge has been forwarded to tunc the concepts.

6.2.1   Effect on Biogeographic Distribution

Temperature of the ambient environment is a critical parameter and has long been a focus of biogeographic studies because of its overwhelm-ing influence on the physiology of exothermic organisms. Fishes are exothermic organisms, and their survival, growth, egg development and even competitive ability are all dependent on ambient temperature. Biogeographic distributions often provide insight into thermal limits for ectotherms, such as fish whose physiology and reproductive success are strongly influenced by temperature. These thermal limits can be used to protect distri-butional changes following climate change by assuming fish migration along isotherms to remain within a suitable (optimum) thermal envelope or temperature regime. The distributional shifts can include abandonment of areas currently occupied

if future temperature exceeds physiological tolerances as well as colonization of new areas if
previously unsuitable temperature conditions
are ameliorated. This approach has been termed
'forecasting from historical analogy'. Some
important case studies from different corners of
the globe may confirm the biogeographic distribution of fishes in response to oscillation of temperature. Shifts in the relative abundance of
finfish in Long Island Sound bear the signature of
ocean warming. Like the lobster, winter flounder
is also at the southern end of its distribution and
it is also exhibiting extremely severe declines.
Commercial landings in New York are only 15 %
of what they were 50 years ago. According to
annual resource assessment surveys conducted
since 1984 by the Connecticut Department of
Environmental Protection (CTDEP), winter
flounder abundance in Long Island Sound is now
less than 10 % of what it was in 1990. More
researches are, however, needed to determine if
winter flounder are declining due to warming
temperatures. When the finfish community of
Long Island Sound is visualized as a whole, evidence of warming as the causative factor becomes
much stronger (CTDEP et al. 2006). Most of the
cold-water species of Long Island Sound have
been declining over the past 15 years (*e.g.* lobster, winter flounder, Atlantic herring, cunner,
longhorn sculpin, sea raven, ocean pout, winter
skate, little skate), while the populations of the
warm-water fishes (like striped bass, weakfish,
summer flounder, menhaden, scup, striped sea
robin, butterfish, Atlantic moonfish, hickory
shad) have been increasing. Thus, for predicting
response of finfish species to climate change,
thermal limits based on biogeographic distribution are important approaches.

Rising sea level is a direct arm of climate
change, which may inundate wetlands and other
low-lying lands, erode beaches, intensify flooding and increase the salinity of rivers, bays and
estuaries. Salinity alteration due to seawater
intrusion has been reported from various parts of
the world. The rising trend of salinity over a
period of 27 years has also been documented in
the Matla River in the central part of Indian
Sundarbans (Mitra et al. 2009). The biogeo-

graphic distribution of fish species is influenced
by such alteration of aquatic salinity. An interesting example in this context is the extension of the
spawning ground of *Tenualosa ilisha* in more
upstream zone after the construction of the
Farakka barrage on the Gangetic stretch of India.
The studies conducted during 1982–1992 clearly
reveal that the ecology of Hugli estuary (downstream zone of Ganga–Bhagirathi system) has
undergone a major change due to construction of
the Farakka barrage in 1975. The increased freshwater discharge in the western part of Indian
Sundarbans has resulted in considerable decrease
in salinity throughout the estuary. The freshwater
zone has extended towards the mouth of the estuary, and the marine and estuarine zone has been
pushed almost towards the end of the lower
stretch of the estuary (Sinha et al. 1996). This has
resulted in the expansion of the spawning ground
of *Tenualosa ilisha* in the post-Farakka barrage
period (De and Saigal 1989). Prior to 1975
(before the installation of the Farakka barrage),
the spawning ground was restricted from Calcutta
to Medgachi (Fig. 6.5a), but survey conducted
during 1987–1989 showed that the spawning
ground has expanded more in the upstream zone
almost towards Nabadwip (Fig. 6.5b). This case
study clearly confirms that biogeographic distribution of fish (preferably migratory fish) and
even their spawning ground coordinates are regulated by aquatic salinity.

Alteration of aquatic salinity due to climate
change has profound influence on fish distribution and their migratory range, but the direction
of salinity shift has a major role to play in determining extent and pattern of migration
(Table 6.2).

6.2.2 Effect on Fish Community Structure

Global, regional and local effects of climate
change on biotic community are ultimately the
sum of processes which act on individual organisms.
Perceptible changes are evident in the past few
decades in the climate of the planet Earth as mani-
fested by increase in air and water temperature.

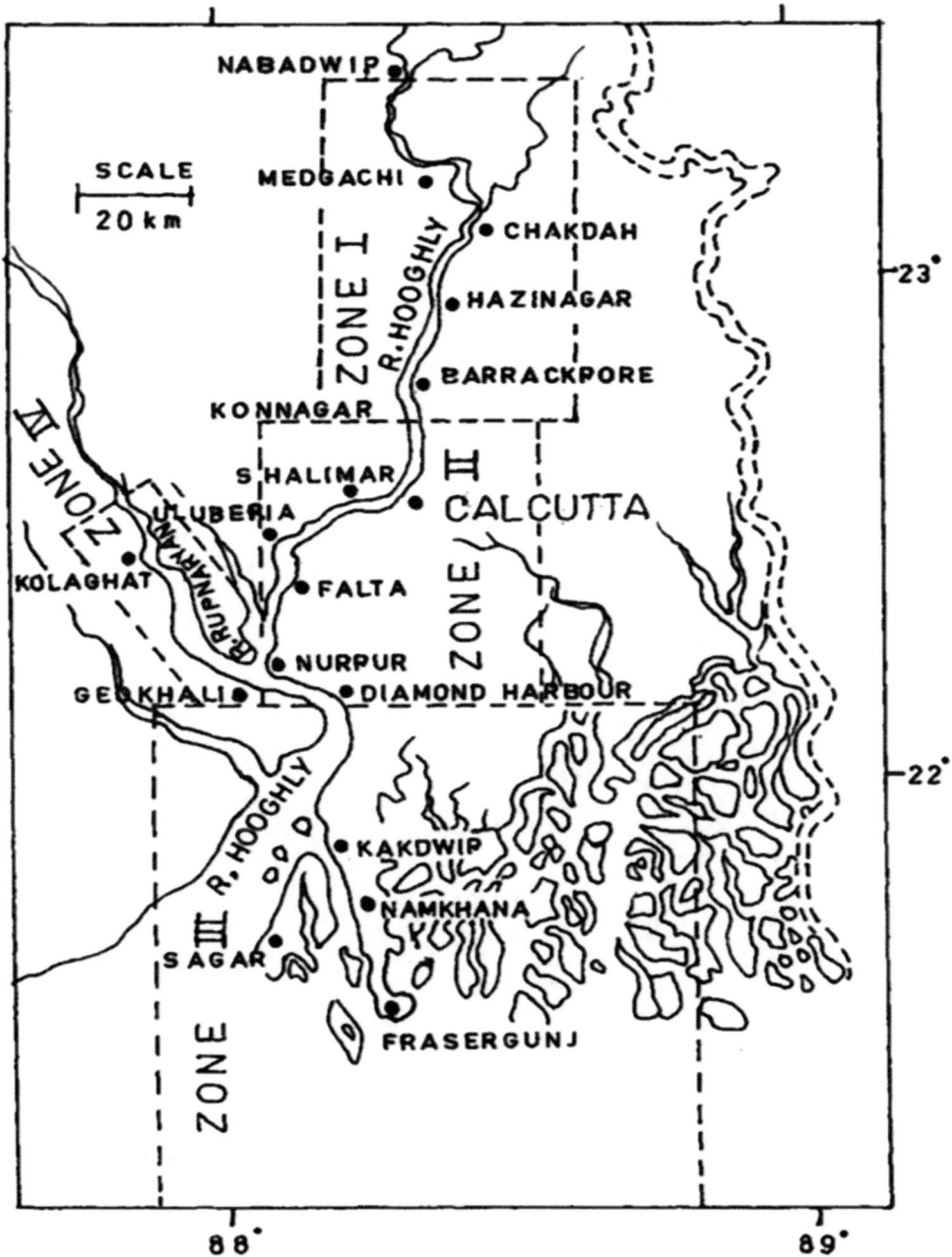

Fig. 6.5 (a) Horizontal zonation of Hugli estuary before the construction of Farakka barrage; the marine and estuarine zone (ZONE III) was from Diamond Harbour to the mouth of the estuary and the spawning ground of *Tenualosa ilisha* was up to Medgachi (ZONE I). (b) Horizontal zonation of Hugli estuary after the construction of Farakka barrage; the marine and estuarine zone (ZONE III) has been pushed downstream from Kakdwip to the mouth of the estuary and the spawning ground of *Tenualosa ilisha* has extended up to Nabadwip (ZONE I)

This has triggered the alteration of salinity profile in marine and estuarine compartment and more specifically at the river mouth and estuarine systems that are connected to glaciers. The fish community thriving in these dynamic systems shifts or orients (adapt) themselves in response to

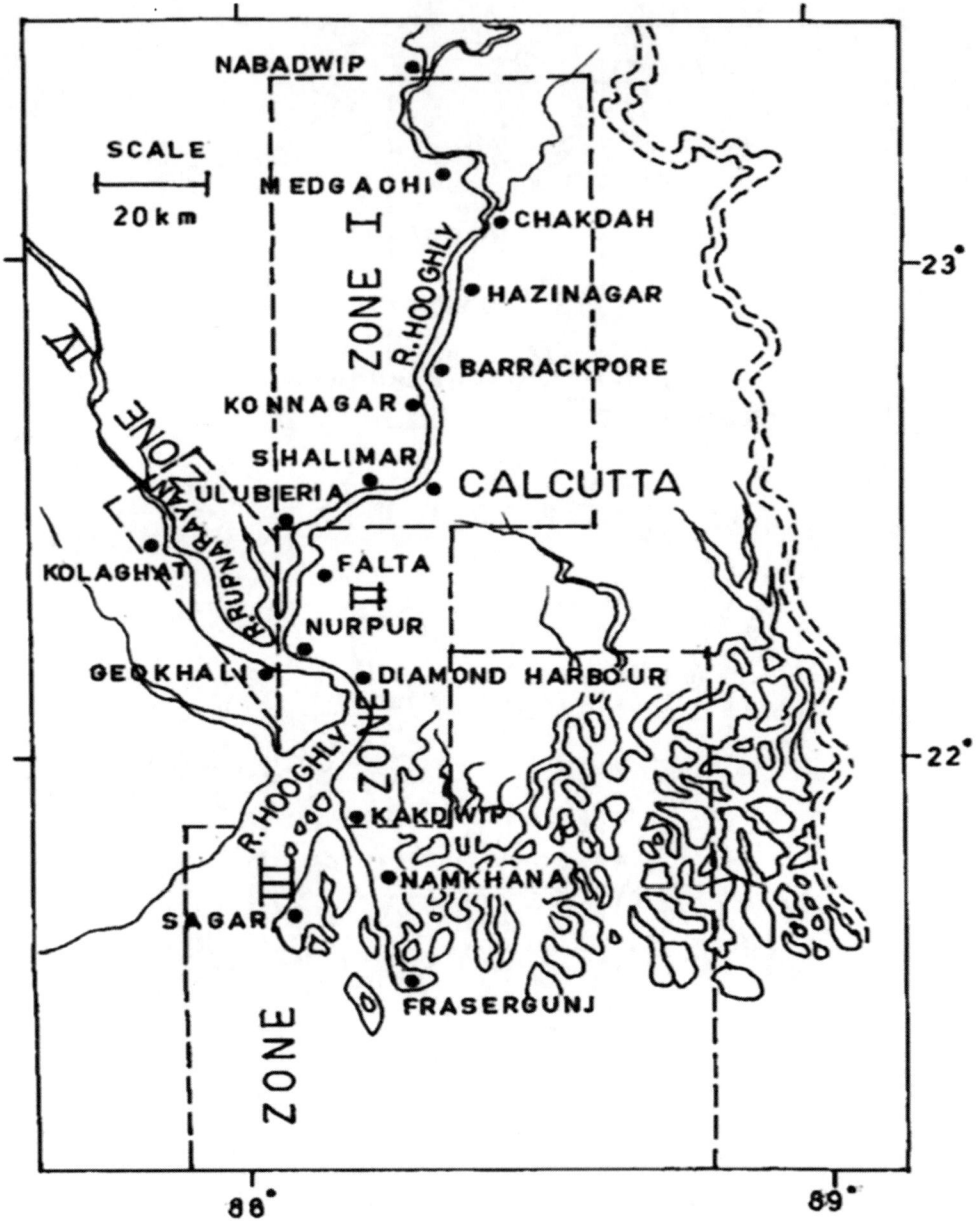

Fig. 6.5 (continued)

ecological changes. The mangrove ecosystem of Indian Sundarbans is an ideal zone for such study as the western and central sectors of the deltaic lobe are drastically different from each other with respect to salinity. The rivers in the western zone (Hugli and Muriganga) receive the freshwater from Himalayan glaciers, after being regulated by Farakka barrage. The construction of the Farakka barrage on the Ganga River was done in April 1975 to augment water supply to the Calcutta port. The project has brought about a significant increase in freshwater discharge in its distributary, the Hugli estuary (Sinha et al. 1996). The rivers in the central sector (Saptamukhi, Thakuran, Matla and Gosaba), on the other hand, have lost their connections with Ganga–Bhagirathi system in course of time and are now tide fed in nature. This variation probably caused a compositional variation in fish community. The western sector showed the presence of more economically important fish species in comparison to trash fishes that may be attributed to decreasing trend of salinity in this zone (Table 6.3) in recent times.

In the central sector, the ingression of seawater and resultant salinity increase has completely reversed the picture with more quanta of trash fishes in comparison to economically important species (Table 6.4). We computed the Shannon–Wiener Species diversity index (H) with the collected samples (collection time was 18.02.2009 to 17.02.2010) for both commercial and trash varieties with a sample size of 100 and 50 kg for commercially important and trash fishes, respectively, and documented more diversity of trash fishes in the central sector, which may be attributed to intrusion of saline water from the adjacent Bay of Bengal region in the south.

The rate of climate change may thus be a major determinant of the abundance and distribution of new fish populations. Rapid change from physical forcing usually will favour production

Table 6.2 Pattern and extent of fish migration as function of aquatic salinity

Fish type	Salinity (high to low: an effect of deglaciation, barrage discharge, high precipitation, etc.)	Salinity (low to high: an effect of seawater intrusion, sea-level rise, high evaporation rate, minimum precipitation, etc.)
Resident fish	Migration towards optimum salinity, preferably in the downstream zone	Migration towards optimum salinity, preferably in the upstream region
Anadromous fish	Congenial condition for spawning resulting in the expansion of the spawning zone	Difficult situation resulting in the shrinkage of spawning area
Catadromous fish	Difficult situation for spawning resulting in the shrinkage of spawning area; more migration towards the downstream zone	Congenial condition for spawning resulting in the expansion of the spawning zone

Table 6.3 Salinity (in ‰) variation in the western and central Indian Sundarbans after the commissioning of the Farakka barrage

Place	Pre-Farakka (1960–1961)	Post-Farakka (1985)	Post-Farakka (1995)	Post-Farakka (2005)[a]
Kakdwip (in the western sector)	32.80	15.10	13.93	8.56
Canning (in the central sector)	28.00	28.90	26.70	26.84

Source: Mukherjee and Kashem (2007)
[a]Survey conducted by the authors

Table 6.4 Mean value of Shannon–Wiener species diversity index (H) computed from 1-year survey (18.02.2009 to 17.02.2010) on fish catch by local fisherman in Indian Sundarbans

Station	Commercial variety	H (sample size = 100 kg)	Trash variety	H (sample size = 50 kg)
Namkhana (in western Indian Sundarbans)	*Pama pama, Polynemus paradiseus, Arius jella, Tenualosa ilisha, Sillaginopsis panijus, Osteogeneious militaris* and *Polydactylus indicus*	3.187	*Thryssa* sp., *Stolephorus* sp., *Harpodon nehereus, Cynoglossus* sp.	1.895
Bali island (in central Indian Sundarbans)	*Pama pama, Polynemus paradiseus, Arius jella, Sillaginopsis panijus*	2.014	*Thryssa* sp., *Stolephorus* sp., *Harpodon nehereus, Cynoglossus* sp.	3.961

of smaller, low-priced, opportunistic species that discharge large number of eggs over long periods. Reports of decline of fish species numbers due to increase of salinity have been published by several workers (Carpelan 1967; Copeland 1967; Hammer 1986). The main causes behind the alteration of fish community structure due to increase in salinity (a consequence of seawater ingression because of warming effect) are:

- Reproductive failure
- Interaction of other environmental parameters with salinity to cause excessive mortality (synergistic effect)
- Loss of primary food supply (mainly plankton) due to adverse impact of salinity tolerance for that organism (plankton)
- Direct mortality due to extreme saline condition

In the western sector of Indian Sundarbans, lowering of salinity due to supply of freshwater through Hugli (after the commissioning of the Farakka barrage) has made the environment congenial particularly for fish like Hilsa (*Tenualosa ilisha*). The landing volume of fish has also increased during the post-Farakka period (Table 6.5). A long-term study of few years is, however, needed to discriminate the seasonal effect (feature) of fish diversity from the impact of salinity fluctuation on fish catch in the present geographical locale.

6.2.3   Effect on Fish Stock Dynamics and Larval Recruitment

Warming of water often alters the organisms of lower trophic level (like plankton), which are the preferred diet for fish juveniles and adults. Under this condition, the growth rate of the fish species is reduced and the recruitment also faces an adverse environmental situation. The case study of Baltic Sea is very important in this context. Several species of zooplankton have faced elimination in recent times. *Calanus* spp. are now virtually absent from the Baltic. The copepod *Pseudocalanus elongatus* is a key species and a major food organism for fish larvae and adult pelagic planktivorous fish, which is also exhibit-

ing a decreasing trend. Large interannual and interdecadal changes in the hydrographic environment of the Baltic, in particular the decreasing salinity of the deep basins, are thought to be responsible for fluctuations in the standing stock of *P. elongatus* (Mollmann et al. 2000). This has an effect on the diet and condition of herring, resulting in considerable fluctuations in the growth rate of herring between 1977 and 1998 (Mollmann et al. 2003a). Oscillation of hydrographic features has also affected the stock dynamics of cod in the Baltic (Mollmann et al. 2003b). Variability in the seasonal timing and spatial distribution of zooplankton and cod larval production has exhibited significant consequences in the survival rate of cod larvae and their subsequent recruitment (Hinrichsen et al. 2002; Kuster et al. 2003). The decadal changes in temperature and salinity in the Baltic, which cause these changes in the fish stocks, can be linked to the regional climate indicators such as the North Atlantic Oscillation (NAO) and in turn to global climate. The related regional changes which have taken place in the Baltic and in the North Sea are sufficiently abrupt and persistent to be described as regime shifts (Beaugrand 2004).

6.2.4   Effect on Fish Diet (Prey) Spectrum

Incidence of change in diet composition of fishes is now reported due to change in aquatic salinity. Biologically, the sea is a relatively simple system. Food webs are short and lack diversity (Walker 1961). In such a system, changes in the lower trophic levels can dramatically affect abundance of top predators as seen in popular game fish, orange mouth corvina (*Cynoscion xanthulus*) occurring in the Salton Sea, California. The major source of food for corvina has been *Bairdiella*. *Bairdiella* depend almost entirely on the benthic worm, *Nereis* (Neanthes). Adult corvina may tolerate salinities as high as 75‰ and adult *Bairdiella* as high as 58‰. If reproductive success of *Nereis* is threatened at 50‰, both these species will disappear at salinities over 50‰ (in fact reproductive failure of *Bairdiella*

Table 6.5 Annual average fish landings by species in the Hugli estuary (in metric tonnes)

Species	Pre-Farakka period (1966–1967 to 1974–1975)	Post-Farakka period (1984–1985 to 1989–1990)
Tenualosa ilisha	1,077.1	1,017.7
(Family Clupeidae, SF Alosinae)	(14.4)	(3.6)
Setipinna spp.	651.1	3,243.5
(Family Engraulidae)	(8.7)	(11.6)
Harpodon nehereus	1,929.1	4,932.3
(Family Harpodontidae)	(25.8)	(17.7)
Trichiurus spp.	452.0	2,996.5
(Family Trichiuridae)	(6.0)	(10.7)
Pama pama	172.7	3,700.4
(Family Sciaenidae)	(2.3)	(13.3)
Silago panijus	17.4	15.3
(Family Sillaginidae)	(0.2)	(*)
Tachysurus jella	176.0	597.7
(Family Ariidae)	(2.3)	(2.1)
Polynemus paradiseus	18.6	91.6
(Family Polynemidae)	(0.2)	(0.3)
Coilia spp.	70.9	800.3
(Family Engraulidae)	(0.9)	(2.9)
Tenualosa toli	16.9	49.2
(Family Clupeidae, SF Alosinae)	(0.2)	(0.2)
Ilisha elongata	164.2	447.3
(Family Pristigasteridae)	(2.2)	(1.6)
Eleutheronema tetradactylum	24.8	14.6
(Family Polynemidae)	(0.3)	(*)
Sciaena biauritus	188.8	303.3
(Family Sciaenidae)	(2.5)	(1.1)
Pangasius pangasius	76.1	3.7
(Family Pangasiidae)	(1.0)	(*)
Liza parsia	42.9	18.0
(Family Mugilidae)	(0.6)	(0.1)
Lates calcarifer	24.1	6.2
(Family Centropomidae)	(0.3)	(*)
Pampus argenteus	–	452.5
(Family Stromateidae)	–	(1.6)
Prawns	751.1	1,856.6
	(10.1)	(6.6)
Freshwater fishes	–	–
Others	1,609.3	7,351.3
	(21.6)	(26.3)
Total	7,463.1	27,898.0
	(78.7)	(93.9)

Source: Sinha et al. (1996)
Figures in parentheses indicate percentage of the catch
(*) Less than 0.1 % of catch

will likely occur first). In such a scenario, corvina may be able to switch to an alternate prey source such as *Tilapia*. Such a switch over is apparently happening in the sea now. Because of behavioural or distributional characteristics of predator and prey, switching may not always be possible (Hagar 1984; Kitchell and Crowder 1986; Jude et al. 1988). In simple food chain, the fate of the top predator is often governed by the weakest link in the chain.

Antarctic krill (*Euphausia superba*) is among the most abundant animal species on Earth, providing the main food supply for fish, birds and even whales. They have declined since 1976 in the high-latitude SW Atlantic sector, probably due to reduction in winter sea ice extent around the Western Atlantic Peninsula (Atkinson et al. 2004). Krill is dependent on the highly productive summer phytoplankton blooms in the area east of the Antarctic Peninsula and south of the polar front. Salps, by contrast, occupy the extensive lower productivity regions of the Southern Ocean and can tolerate warmer water than krill. Their populations have increased abruptly in abundance. These changes exerted profound effects within the Southern Ocean food web. Penguins, albatrosses, seals and whales have wide foraging ranges but are still prone to krill shortage.

6.2.5   Effect on Reproductive System

All the stages of reproduction in fish, namely, gametogenesis and gamete maturation, ovulation/spermiation, spawning and early development stages, are affected by aquatic temperature. Imbalance or rapid change in temperature are stressful to fish and may also be linked with other stress factors. The primary effect of stress is the activation of sympathetico-chromaffin tissue and hypothalamic–pituitary pathways, resulting in the release of respective catecholamines and corticosteroid hormones in the bloodstreams. These will increase the metabolic processes to reduce the stress response in fish. If stress is maintained, then the effects start manifesting by the inhibition of reproductive function, cessation of ovulation and depression of reproductive hormones in blood and ovarian failure. Temperature change modulates the hormone action at all levels of reproductive endocrine cascade. A very relevant example is provided by prediction from the response and community change of fish that might be expected with global warming in Great Lakes Basin (Casselman 2002). In an increasing temperature regime, cold-water and even cool-water fishes would be expected at least in the short run to lose condition. This would directly affect reproductive potential and reduce recruitment. On the other hand, warm-water habitat would increase sustainability, enhancing growth, condition, production and reproductive potential of warm-water fishes. It has also been forecasted that global warming, which is affecting temperature increase in Great Lake Basin, will substantially decrease recruitment of cold-water fishes species and increase relative recruitment of warm-water species, twofold with 1 °C increase and threefold with 2 °C increase in water temperature.

Ocean acidification due to increased CO_2 level is a well-accepted concept. Most of our knowledge on the direct effects of ocean acidification on marine organisms focuses on species known as 'marine calcifiers' (*e.g.* corals and molluscs) that build skeletons or shells made of calcium carbonate. Many of these species will suffer impaired ability to build skeletons as pH decreases. We know less about the direct impacts of acidification on harvested species like fishes and squids. In these species, the response to acidification is likely to involve physiological diseases including acidosis of tissue and body fluids, leading to impaired metabolic function. Egg and larval stages are likely to be much more susceptible than adults, suggesting that reduced reproductive success will be among the first symptoms to appear. The indirect effects of acidification on fisheries will include loss of reef habitat and breeding ground constructed by marine calcifiers. Many fishes depend on the physical structure provided by coral skeletons or shell-building organisms such as oyster reefs as essential habitat for one or more life stages. Adverse impact on these physical structures due to lowering of aquatic pH will compel the fish species to lose their preferred niches in different stages of their life cycle.

Increase of aquatic salinity particularly at the river mouth and estuary is a direct impact of sea-level rise due to global warming. An interesting case study from Salton Sea depicts the adverse effect of rising salinity on reproduction-related phenomena of fish fauna. Experiments with Salton Sea fish eggs and larvae led researchers to conclude that salinities in excess of 40‰ adversely influence the development of embryos and larvae of *Bairdiella* and *Sargo*.

6.2.6 Effect on Osmoregulation of Migratory Fishes

Global-warming-induced sea-level rise and subsequent increase of salinity in river mouth, estuaries, bays and coastal waters not only pose an adverse impact on the physiology of resident fish species but also affect the migratory route and span of anadromous and catadromous species. Fishes in saline waters must maintain proper concentration of salts in their body fluid and prevent excessive loss of water. This requires various adaptive mechanisms and the expenditure of energy since the osmotic concentration in fishes is different than that of sea water. Osmoregulatory mechanisms include drinking water, excretion and secretion of accumulating salts. These mechanisms are aided by limited skin permeability of fishes adapted to saline waters of marine and estuarine environment.

The degree of adaptation is not similar in all the fish species. It varies significantly from species to species. The highest salinity at which fishes have been reported to survive is the occurrence of *Cyprinodon variegates* at 142.4‰. Some species can cope with wide range of salinities (euryhaline) by tolerating a certain degree of change in the body fluids or well-developed osmoregulatory mechanisms, while many others are stenohaline in nature.

The anadromous species such as the Indian shad (*Tenualosa ilisha*) have a characteristic of early development in freshwater followed by seaward movement, and again they (matured form) return to freshwater system for spawning. This migratory pattern is, thus, a direct function of ambient aquatic salinity, and therefore adult

Table 6.6 Seasonal variations of physico-chemical variables in the surface waters of Hugli estuarine system

Season	Dissolved oxygen (mg/l)	Surface water temp. (°C)	Dilution factor	Salinity (‰)
Upper stretch				
A	6.0	34.0	0.96	1.5
B	5.9	33.5	1.00	0.0
C	5.8	22.1	0.97	1.0
Middle stretch				
A	6.3	33.5	0.70	10.5
B	6.2	33.0	1.00	0.0
C	5.9	25.1	0.90	3.3
Lower stretch				
A	6.9	33.9	0.23	26.1
B	6.8	33.2	0.67	8.5
C	6.3	26.0	0.40	20.0

Source: Mitra (2000)
A premonsoon, *B* monsoon, *C* postmonsoon

(gravid) fishes are seen in schools migrating upstream of Hugli estuary (at the apex of Bay of Bengal) during monsoon. A unique characteristic feature of the Hugli estuarine system is the attainment of zero salinity in the upstream during monsoon (Table 6.6) – a condition that is extremely preferred by adult *T. ilisha* for spawning. However, after the commissioning of Farakka barrage in the upstream of Hugli estuarine system, the freshwater flow has increased in the system and now the so-called upstream box has expanded in range. This has caused the shifting of the spawning regime more upstream from Medgachi to Nabadwip (Fig. 6.2A, B). A time series analysis of *T. ilisha* landing reveals an increasing trend (Fig. 6.6) – an indication of positive influence of dilution of the Hugli estuarine waters by Farakka barrage discharge.

In American shad, high-salinity tolerance develops at the time of larval–juvenile metamorphosis (July), several months before the peak of downstream migration (October). At the end of the migratory period, ion losses occur in laboratory reared and wild fish, coincident with increased gill Na^+, K^+- ATPase activity. Ion losses are delayed in fish maintained at elevated temperature (summer), indicating that higher temperature compels a longer period of freshwater residence for shad. Less is known about the

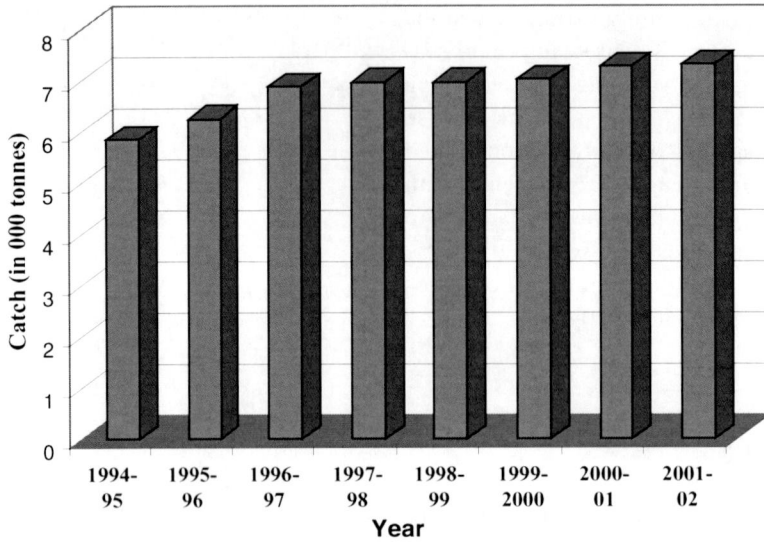

Fig. 6.6 Trend of *T. ilisha* catch at Kakdwip fish landing stations in the Hugli estuarine stretch

impact of global warming on the osmoregulatory function of anadromous fishes. More research is needed on the salinity tolerance and physiological changes that occur during migration.

6.2.7 Effect on Heavy Metal Toxicity

It is difficult to relate the magnitude of pollution with the climate change-related phenomena except the fact that sea-level rise will accelerate the erosion of the adjacent landmasses, resulting in more input of terrestrial matter into the aquatic system. The inputs from the adjacent landmasses include diverse types of organic and inorganic substances mainly from agricultural and industrial activities. The relative toxicity of a typical pollutant, such as the heavy metal copper, is strongly found to be temperature dependent. Sublethal exposure of copper at all concentrations together with all temperature combinations cause less severe damage, but the changes in some species like winter trout exposed to 0.47 μmol/l copper are quite substantial.

The heavy metals in the brackish water system generally deposit on the sediment bed or remain in dissolved state in the water column, depending on the nature of chemical species, which are influenced by factors like aquatic salinity and pH. Since climate change has direct effect on salinity and pH, the indirect influence of climate change on heavy metal level cannot be ruled out completely. In countries like India, development of housing complex, tourist centres, recreational units and shrimp farms has posed a negative impact on the positive health of the ecosystem particularly estuaries and coastal zones. The aquatic phase in and around Indian Sundarbans is no exception to this rule. The area is presently stressed with unplanned mushrooming of shrimp culture units, industries and hotels, which in most cases release their respective wastes without any adequate treatment. These wastes containing appreciable concentrations of heavy metals find their way into the coastal water and adjacent estuaries (particularly Hugli–Matla estuarine complex). Use of antifouling paints for conditioning fishing vessels and trawlers is another major source of Cu in the present geographical locale. Official records state that the plying of mechanized boats and trawlers increased by 37.7 %, since 1998–1999 (Table 6.7).

Table 6.7 Number of mechanized boats and trawlers

Year	Number
1998–1999	3,362
1999–2000	3,585
2000–2001	3,622
2001–2002	3,763
2002–2003	4,175
2003–2004	4,481
2004–2005	4,521
2005–2006	4,575
2006–2007	4,630

Source: Mukherjee and Kashem (2007)

The Hugli estuarine complex is considered possibly the most polluted estuary in the world with almost a large number of main factories located close to the mouth discharging almost half a billion litres a day of untreated waste including the effluent from pulp and paper mills, pesticides manufacturing plants, distilleries, thermal power plants, yeast, rayon, cotton, vegetable oils, soap, fertilizers, leather manufacturing units and antibiotic plants (Table 6.8). According to UNEP report, 1,125 million litres of wastewater is discharged per day in Hugli estuary (Mukherjee and Kashem 2007). A vital ingredient of the released wastes is Cu, and it is sourced from paint industries, galvanizing units, antifouling paints, algicide used in shrimp culture ponds, etc.

Our databank since 1980 reflects an increasing trend of dissolved Cu in western Indian Sundarbans (Table 6.9), which is an indication of water quality variation. It is very difficult to relate the oscillation of dissolved Cu with climate change due to noise created by recent industrialization and urbanization in the Haldia industrial complex in the upstream zone. We also lack data on the effect of temperature increase on Cu toxicity. It has been documented that lowering of pH (which is a consequence of CO_2 increase) favours transference of Cu from the sediment (bed material) to aquatic phase (Mitra et al. 1994; Mitra 1998), and thus an indirect, but feeble string of Cu toxicity and elevation of atmospheric CO_2 can be sketched.

Statistical analyses of our data set on aquatic pH, dissolved Cu and biologically available Cu in sediment of western Indian Sundarbans (Table 6.9) confirm the process of transference of Cu from the sediment bed to the upper aquatic phase ($r_{\text{dissolved Cu} \times \text{sediment Cu}} = -0.903$, $p < 0.01$; $r_{\text{dissolved Cu} \times \text{pH}} = -0.887$, $p < 0.01$ and $r_{\text{sediment Cu} \times \text{pH}} = 0.747$, $p < 0.01$) in a condition of low pH value. Considering dissolved Cu as variable 1 (VAR_1), biologically available Cu as variable 2 (VAR_2) and pH as variable 3 (VAR_3), the scatter plot generated through SYSTAT reflects that higher pH favours the precipitation of Cu from water to sediment, whereas lower pH increases the concentrations of dissolved Cu (Fig. 6.7). The vulnerability of the increasing trend of dissolved heavy metals cannot be ignored as the process is related to bioaccumulation and biomagnification leading to human health hazard.

Researches have confirmed an increasing trend of heavy metals in the fish tissue of Gangetic delta since 1998 (Tables 6.10, 6.11, and 6.12). A long-term research needs to be undertaken to investigate the interrelationship between the heavy metal level in fish tissue and indicators of climate change (like water temperature or pH). Simulation experiment may be an effective approach in this context.

6.2.8   Effect on Fish Disease

Environmental factors play a key role in the initiation and spread of fish diseases. Kawasaki (2001) provided specific evidence of the effects of global warming on world fisheries production, particularly on tuna and sardines. The triggering of the fish disease due to warming of atmosphere was seen in American lobster. Massive, catastrophic summer–fall mortalities of lobsters in Long Island Sound began in August 1999, and the incidence continued to occur to a greater or lesser degree in subsequent summers. An extensive federally sponsored research programme has identified summer warming of Long Island Sound bottom waters, coupled with hypoxia, as the most likely causes of the outbreak of disease. One of these diseases called 'excretory calcinosis', discovered by scientists at Stony Brook University, is a gill tissue blood disorder resulting directly from warm-water temperatures

Table 6.8 Major industries in the upstream zone of Hugli estuary

| Name of the industries | Size and category | Status | Water use (m³/day) | Waste generation | | Status of disposal of waste effluent | | Waste |
				Effluent (m³/day)	Solid (MT/day)	Status of effluent (m³/day)	Disposal of effluent discharged in the sea Direct/indirect	Solid waste
Shaw Wallace & Co.	Large pesticides	Private	265	50	0.007	Treated	Indirect	NA
Hindustan Lever Ltd.	Large chemicals	Private	3,750	1,065	NA	Treated	Indirect	NA
Consolidated Fibres & Synthetic Chemicals Ltd.	Large fibre	Private	3,185	2,303	0.002	Treated	Indirect	NA
Haldia Dock Complex	Large dock	Private	2,880	640	25	Untreated	Indirect	NA
IOC Ltd.	Large oil and refinery	Private	14,650	13,800	NA	Treated	Indirect	NA
Chloride Industries	Large lead battery	Private	446	402	0.008	Treated	Indirect	NA
HFC Ltd. (main plant fertilizer not operating since 1978)	Large fertilizer	Private	3,400	3,400	NA	Partly treated	Indirect	NA

Table 6.9 Concentrations of dissolved Cu, biologically available Cu in sediment and surface water pH in western Indian Sundarbans

Year	Dissolved Cu (ppb)	Biologically available Cu (ppm)	pH
1980	36.88	22.81	8.33
1982	39.99	18.15	8.33
1984	49.08	13.96	8.32
1985	47.80	16.39	8.32
1986	48.09	18.81	8.32
1988	48.56	15.11	8.31
1990	53.22	14.80	8.32
1992	54.02	12.85	8.31
1994	53.00	13.34	8.31
1995	61.11	11.12	8.31
1996	64.17	11.56	8.3
1998	64.10	12.44	8.3
2000	59.27	13.87	8.3
2002	67.88	10.13	8.29
2004	65.08	11.98	8.29
2005	55.55	14.55	8.29
2006	65.11	12.84	8.28
2007	68.09	10.02	8.28

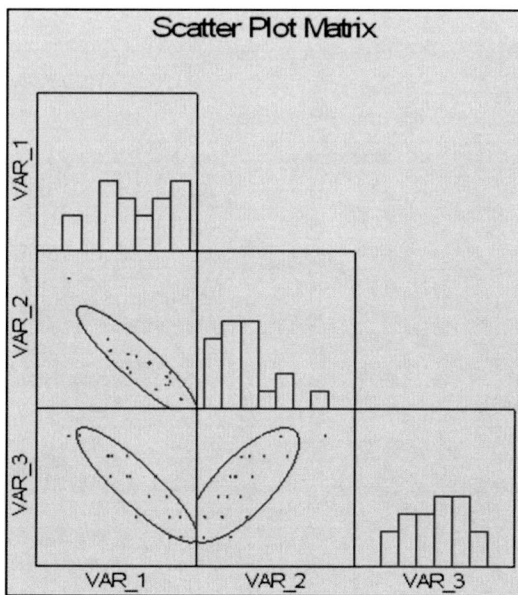

Fig. 6.7 Scatter plot matrix showing the interrelationship between dissolved Cu (VAR_1), biologically available Cu in sediment (VAR_2) and surface water pH (VAR_3)

(Dove et al. 2004). Other lobster diseases also appear to result from the stress of high temperature and hypoxia. The result of these multiple stresses has been a 75 % reduction in total landings and 85 % reduction in the overall abundance of the population. These diseases now appear to be moving northward.

In Indian subcontinent, particularly in the Gangetic delta region, the microbial diseases of tiger prawn (*Penaeus monodon*) are witnessed during the summer months in the shrimp farms. A significant positive correlation ($r=0.8297$; $p<0.01$) has been observed between water temperature and percentage of affected shrimps at Satjelia island ($22°04'35.17''$ N latitude and $88°44'55.70''$ E longitude) in central Indian Sundarbans (Table 6.13).

Warming of atmosphere and hydrosphere, thus, may have a disastrous impact on the livelihood of the coastal population in countries like India, where a considerable percentage of the people are engaged in shrimp farming by improved traditional method.

Another example of climate-induced effects on fisheries involves the northward expansion of a disease known as 'dermo' that affects the oyster. It is caused by *Perkinsus marinus*, a parasite that kills about 50 % of oysters yearly as recorded in the Gulf of Mexico. Prior to the late 1980s, the parasite was known to occur only south of lower Chesapeake Bay. In the early 1990s, however, dermo underwent a 500-km northward range expansion extending all the way into the Gulf of Maine. Researchers at Rutgers University have demonstrated that the range expansion occurred during years when winters were unusually warm (Ford and Smolowitz 2007). The prevalence of dermo is now high from Delaware Bay to Cape Cod, with no signs of abating.

6.2.9 Effect on Protein Synthesis

The acute and evolutionary effects of small changes in ambient temperature on protein structure and function of fishes have been studied by many researchers around the globe. Analyses have revealed that at upper limits of thermal

Table 6.10 Zn concentrations (in ppm dry wt.) in finfish muscles collected from Hugli estuary

Species	1998	2000	2004	2006	2008
Coilia sp.	12.56	15.67	21.02	17.87	23.79
Thryssa hamiltonii	9.05	13.44	17.39	11.87	29.18
Rhinomugil corsula	10.67	16.90	19.55	26.89	34.78
Mugil cephalus	16.65	15.55	20.41	30.10	41.56
Sillago sihama	4.29	7.08	9.29	12.48	17.80
Pisodonophis boro	20.14	26.98	31.43	45.95	51.67
Tenualosa ilisha	13.34	16.67	18.90	21.76	30.76
Liza parsia	20.78	23.17	28.56	31.09	35.88
Liza tade	24.55	33.88	39.65	41.05	45.33
Stolephorus commersonii	5.67	8.02	3.09	6.40	7.52

Table 6.11 Cu concentrations (in ppm dry wt.) in finfish muscles collected from Hugli estuary

Species	1998	2000	2004	2006	2008
Coilia sp.	2.15	3.08	5.13	6.15	8.99
Thryssa hamiltonii	5.64	7.01	8.66	5.50	10.34
Rhinomugil corsula	4.78	3.99	5.26	7.00	8.45
Mugil cephalus	9.45	11.78	17.90	14.43	19.60
Sillago sihama	2.04	2.82	3.40	4.77	4.05
Pisodonophis boro	12.56	14.83	15.92	17.01	20.22
Tenualosa ilisha	3.80	4.18	6.28	5.99	8.39
Liza parsia	4.38	6.00	7.19	6.35	9.01
Liza tade	5.45	7.56	9.25	11.78	8.11
Stolephorus commersonii	2.79	BDL	3.90	2.83	4.01

Table 6.12 Pb concentrations (in ppm dry wt.) in finfish muscles collected from Hugli estuary

Species	1998	2000	2004	2006	2008
Coilia sp.	BDL	BDL	2.12	2.68	BDL
Thryssa hamiltonii	2.00	3.03	2.96	2.17	4.36
Rhinomugil corsula	3.17	4.31	4.89	2.01	BDL
Mugil cephalus	6.23	7.21	4.99	5.25	7.15
Sillago sihama	BDL	BDL	BDL	2.02	BDL
Pisodonophis boro	3.48	3.00	5.88	6.56	4.59
Tenualosa ilisha	3.12	BDL	2.28	4.13	BDL
Liza parsia	BDL	2.35	1.99	BDL	1.76
Liza tade	1.78	BDL	3.24	2.15	1.75
Stolephorus commersonii	BDL	BDL	BDL	BDL	BDL

tolerance ranges, protein function and structure might be impaired significantly in fishes. Like minor and reversible changes in protein struc-ture, there may be change in critical functional traits like inhibition of effective substrate bind-ing sites on protein molecule if environmental

Table 6.13 Record of microbial attack in *Penaeus monodon* (at 45 DOC) as function of ambient water temperature in Satjelia island, central Indian Sundarbans

Water temperature (°C)	% of affected shrimp at 45 DOC
31.5 (March 2007)	41.6
33.2 (April 2007)	45.0
36.8 (May 2007)	47.8
41.0 (June 2007)	63.2
32.1 (March 2008)	40.4
33.6 (April 2008)	49.1
37.0 (May 2008)	48.9
41.1 (June 2008)	78.9
36.3 (July 2008)	40.7

change leads to conformational change in protein. Larger yet still reversible changes in conformation may require the activities of heat shock proteins (HSPs) to prevent inappropriate aggregation of partially unfolded proteins and to restore native conformation.

In fish it is an accepted fact that increase in temperature, within the thermal range for a particular species, will lead to increased growth rate provided that food is not a limiting factor. Protein growth occurs when protein synthesis exceeds protein degradation. Considering the structure of protein as a function of ambient temperature, the effect of climate-change-related warming on the fish community cannot be ignored. As a result of global warming and increase in water temperature, if food supply is not limiting, it will result in higher food consumption rates and higher protein synthesis rates. This is the reason behind the increased rates of protein synthesis and growth in the northern salmon stocks as the annual growing season increases, resulting in a decrease in the average age at smolting.

6.3   Managing Fish and Fisheries

The majority of the world's 36 million fisher folk and 200 million fishery-dependent people live in areas vulnerable to climate change or depend for a major part of their livelihood on resources whose distribution and productivity are known to be influenced by climate variation. Coastal West Bengal, which is our case study area, is situated in the northeast coast of the Indian subcontinent. The state has 2,212,019 fisher folk populations, of which 269,565 depend on marine and estuarine fishes. The presence of Bay of Bengal in the southern part of the state with its tentacle stretching towards the funnel-shaped Hugli estuary in the north has made the geographic locale highly fragile and dynamic from the geo-physico-chemical point of view (Fig. 6.8).

In this framework, very few researches have addressed the issue of climate change with an insight of trend analysis in the fishery sector. Our study concludes that the changes in the marine and estuarine fish community spectrum in response to oscillation of climate may be both direct and indirect in nature. The magnitude of these changes will vary depending on the type of ecosystem and fishery. Few important impacts of climate change are listed here:

1. Shifting the range of fish species as functions of temperature and salinity.
2. Oscillation of upwelling zone fisheries due to change in ocean currents.
3. Qualitative change in fish diversity due to alteration of salinity and members of lower trophic level (preferably plankton). This has been observed in case of Indian Sundarbans, where the quantum and diversity of trash fishes have increased in the central sector around Matla River due to increase of salinity (Fig. 6.9).
4. Significant adverse impact on reef fisheries as consequences of temperature rise and ocean acidification.
5. Disruption to fish reproductive system.
6. Alteration of migratory route (in case of migratory fish species like shad) and range of spawning ground as a function of salinity (see Fig. 6.5a, b).
7. Alteration of residence time of migratory fishes in fresh/brackish water system.
8. Alteration of fish community structure with dominancy of trash fishes.
9. Increased mortality (particularly in the sector of shrimp culture) due to microbial disease.

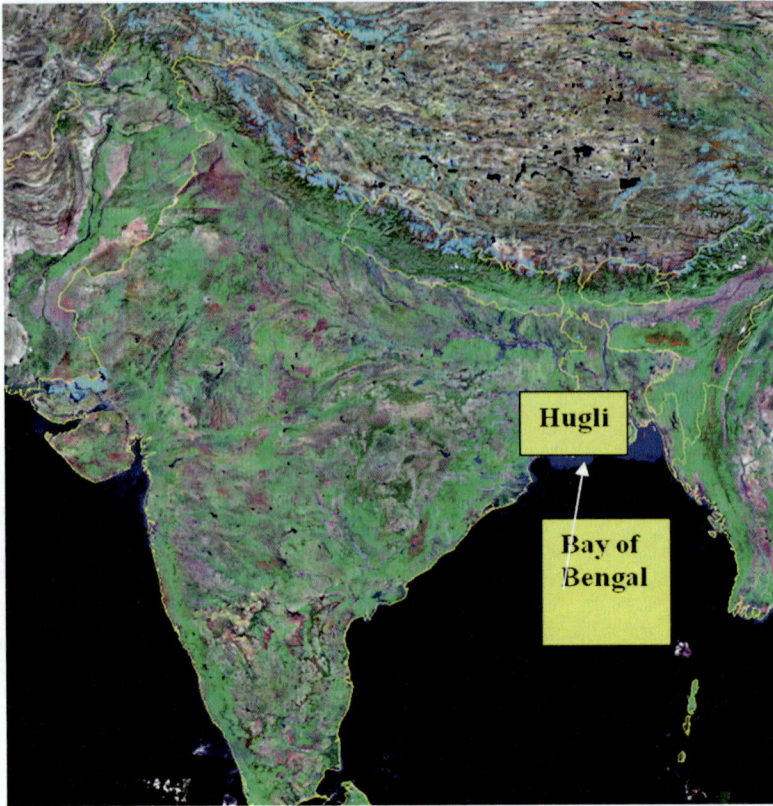

Fig. 6.8 Bay of Bengal with its arm towards Hugli estuary in the state of West Bengal

Fig. 6.9 Dominancy of trash fishes in the catch from central Indian Sundarbans

It is extremely difficult to judge at a global level who will be the main losers and winners from changes in fisheries as a result of climate change, aside from the obvious advantages of being well informed, well capitalized and able to shift to alternative areas or kinds of fishing activity (or other non-fishery activities) as circumstances change. Some of the most affected systems may be in the mega-deltas of rivers in Asia, such as the Mekong, where 60 million people are linked in various tiers to fishery activities. These are mainly seasonal floodplain fisheries, which, in addition to overfishing, are increasingly threatened by changes in the hydrological cycle and in land use, damming, irrigation and channel alteration. The Gangetic delta of Indian subcontinent is another threatened ecosystem experiencing salinity fluctuation and sea-level rise at an alarming rate. The deltaic complex sustaining luxuriant mangrove vegetation is the nursery and breeding ground of a variety of finfish and shellfish. The habitat is, however, deteriorating rapidly due to intense industrialization in the upstream zone (Fig. 6.10), which is the breeding site of *Tenualosa ilisha*.

Continuous long-term monitoring on fish landing with reference to hydrological parameters (preferably of PFZ) can provide an in-depth picture of the impact of climate change on fishery. Monitoring not only provides the information required to measure the current state (*e.g.* of a fish population, ecosystem and pollutant), but it also provides the measurements for validating and checking the sensitivity of models, which are being used to predict future states. Monitoring programmes need to be designed with these objectives in mind.

Research is another important wing to know the unknown, hear the unheard and see the unseen. The magnitude of uncertainty related to climate change can also be revealed through research. A great deal of research is already underway to examine the performance of fisheries management system as a whole and to bring about improvements. To date, the issue of climate-change-induced changes has often been mentioned, but not addressed in such research. Assessment frameworks which include the consequences of climate change and the uncertainty resulting from them in a complete and transparent way need to be developed.

Development of AOGCMs should include the specific kinds of information and output needed to evaluate climate change impacts on marine systems. The relationship between expected long-term changes and decadal (and shorter) variability is extremely important in considering climate impacts on fisheries. Downscaling and regional modelling of ocean climate change is also critical in making realistic regional forecasts of impacts.

To reduce the level of carbon dioxide in the atmosphere, ocean fertilization has become a popular technology. The resulting pulses of phytoplankton growth sequester carbon from the atmosphere and may help reduce the build-up of atmospheric CO_2. Although this possibility deserves serious scrutiny, the ecosystem impacts of fertilization in most aquatic ecosystems almost always contain undesirable consequences for water quality, food webs and fisheries. Hypoxia in Long Island Sound, for example, results largely from over-fertilization by nitrogen, which is the limiting nutrient in many coastal waters. Sometimes the blooms produced by enrichment turn out to be harmful algal species like 'red tide' or 'brown tide'. The ecological consequences of ocean fertilization on a scale sufficient to minimize the build-up of greenhouse gases need further research to evaluate the potential risks of unintended negative impacts.

Considering the impact of climate change as just one of a number of pressures on brackish water fish and fishery, a sound management needs to be undertaken. At short time scales (~1–5 years), the effects of long, slow changes in climatic means are relatively unimportant. However, interannual variability, climate events such as *El Nino* and regime shifts can have large effects which need to be critically analyzed through continuous research and monitoring. At short-term scales, the main anthropogenic impacts are due to excessive level of fishing, fisheries induced damage to the marine ecosystem, degradation or loss of

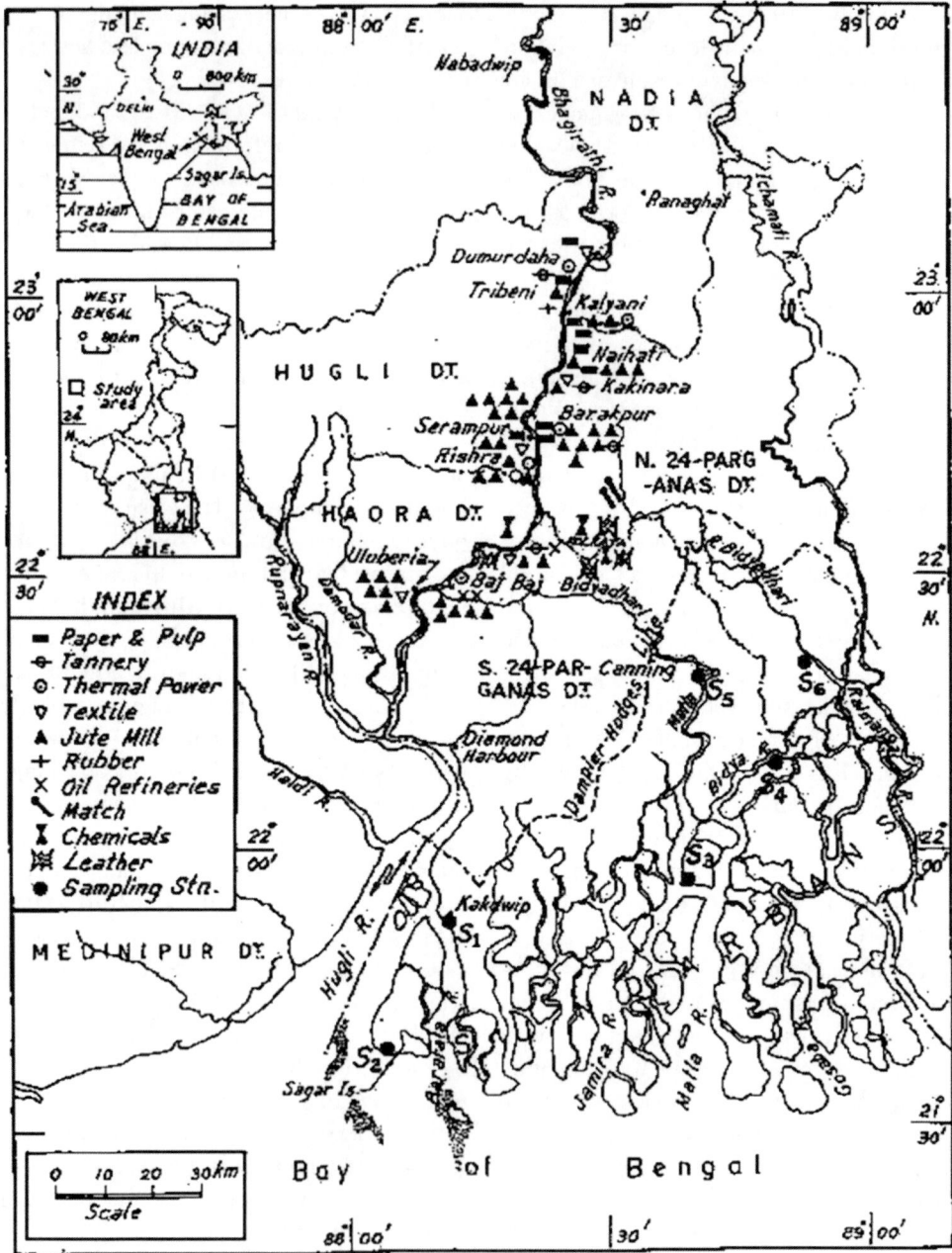

Fig. 6.10 Map showing the industrial units in the upstream of Hugli River

coastal habitat, pollution, introduction of exotic species and undesirable side effects of aquaculture. The collection of tiger prawn seeds in the coastal areas of West Bengal, particularly in the Indian Sundarbans region, is another major local level threat to fish stock of the delta region. A survey done on this aspect revealed that for every tiger prawn seed

Table 6.14 Number of finfish juveniles wasted per seed of tiger prawn/net/day

Year	Stn. 1			Stn. 2			Stn. 3		
	Prm	Mon	Pom	Prm	Mon	Pom	Prm	Mon	Pom
1994	392	161	245	1,598	134	522	1,593	260	767
1995	321	133	291	1,315	122	422	1,303	213	625
1996	200	82	183	1,273	99	416	1,290	213	631
1997	195	79	175	1,279	99	413	1,229	201	596
1998	215	89	197	1,090	84	361	1,055	172	651
1999	217	89	198	716	56	233	679	123	333
2000	224	93	204	639	51	209	601	103	291
2001	221	92	201	545	42	180	521	89	255
2002	225	93	204	910	178	442	934	153	440
2003	389	38	67	936	73	304	1,020	79	367

Source: Banerjee et al. (2005)
Stations 1, 2, and 3 are the three sampling sites about 1 km apart in the western Indian Sundarbans
Prm premonsoon, *Mon* monsoon, *Pom* Postmonsoon

collection, about 38–1,598 finfish juveniles of other commercially important species are wasted per net per day (Table 6.14) by the seed collectors as they are not remunerative unlike tiger prawn seeds whose price per 1,000 varies from $6 to $36 depending on the seasons. Considering all these adverse impacts, the management plan must include:

1. Continuous monitoring of hydrological parameters and condition index of fish species to understand the degree of stress operating on the species
2. Awareness programmes on regular basis (preferably in local language) to minimize the threats that stems out of ignorance
3. Establishment of seed bank/fish sanctuary like Sundarban Tiger Reserve, India etc. to ensure safety to the larvae and brooders/gravid females
4. Coastal vegetation conservation as it is the direct source of nutrients to ichthyoplankton
5. Strict implementation of laws to control the point-source pollution
6. Thrust on alternative livelihood schemes with government subsidy
7. Development of strategies for immediate damage inspection after natural hazards like cyclones and tidal surges
8. Establishment of disaster control centres well equipped with warning/forecasting/prediction facilities

Fact Below the Carpet

Paul and Anne Ehrlich (1970) have remarked: 'Judging from the fishing industry's bahaviour towards the sea, one might conclude that if it was to go into the chicken farming business it would plan to eat up all the feed, all the eggs, all the chicks, and all the chickens simultaneously, while burning down the henhouses to keep itself warm'. Basically, in the sphere of fishery, there is always a tendency to cut the goose with the knife of technology to get all the golden eggs at a time. This overexploitation along with the adverse impact of climate change and other anthropogenic factors may put the ultimate nail to the coffin of the global fishery if not properly managed in the right way and right time. Management is not just setting few rules or policy or introduction of a philosophical concept, but it becomes action oriented only when like-minded persons play together in a common tune of harmony and cooperation.

Important References

Atkinson AA, Siegel V, Pakhomov E, Rothery P (2004) Long-term decline in krill stock and increase in salps within the Southern Ocean. Nature 432:100–103

Banerjee K, Bhattacharyya DP, Mitra A (2005) Ichthyoplankton community spectrum in coastal West Bengal: threats and conservation. Indian Sci Cruiser 19(5):34–40

Beaugrand G (2004) The North Sea regime shift: evidence, causes, mechanisms and consequences. Prog Oceanogr 60:245–262

Carpelan LH (1967) Invertebrates in relation to hypersaline habitats. Invertebrates in supersaline waters. Univ Tex Contrib Mar Sci 12(21):219–229

Casselman JM (2002) Effect of temperature, global extremes, and climate change on year class production of warm water, cool water, and cold water fish in the Great Lakes Basin. Am Fish Soc Symp 32:29–60, American Fisheries Society

Chaudhuri AB, Choudhury A (1994) Mangroves of the Sundarbans, India. IUCN, Bangkok

Copeland BJ (1967) Environmental characteristics of hypersaline lagoons. Univ Tex Contrib Mar Sci 12:207–218

CTDEP, Bureau of Natural Resources, Marine Fisheries Division (2006) A study of marine recreational fisheries in Connecticut. Federal aid in sport fish restoration, F-54-R-25, Annual performance report

De TK, Saigal BN (1989) Spawning of Hilsa, *Tenualosa ilisha* (Hamilton) in the Hooghly estuary. J Inland Fish Soc India 21:46–48

Dove ADM, LoBue C, Bowser P, Powell M (2004) Excretory calcinosis: a new fatal disease of wild American lobsters *Homarus americanus*. Dis Aquat Organ 58(2–3):215–221

Ehrlich Paul R, Ehrlich AH (1970) Population, resources, environment – issues in perfect human ecology, 2nd edn. W.H. Freeman, San Francisco

Ford SE, Smolowitz R (2007) Infection dynamics of an oyster parasite in its newly expanded range. Mar Biol 151:119–133

Hagar JM (1984) Diets of Lake Michigan salmonids: an analysis of predator-prey interaction. M.S. thesis, University of Wisconsin, Madison, 74 p

Hamilton-Buchanan (1922) An account of the fishes found in river Ganges and its branches. Edinbrough/London, 405pp

Hammer UT (1986) Saline lake ecosystems of the world. Dr. W Junk Publishers, Dordrecht

Harvell CD, Kim K, Burkholder JM, Colwell RR, Epstein PR, Grimes DJ, Hofmann EE, Lipp EK, Osterhaus ADME, Overstreet RM, Porter JW, Smith GW, Vasta GR (1999) Emerging marine disease – climate links and anthropogenic factors. Science 285:1505–1510

Hinrichsen HH, Mollmann C, Voss R, Koster FW, Kornilovs G (2002) Biophysical modeling of larval Baltic cod (*Gadus morhua*) survival and growth. Can J Fish Aquat Sci 59:1858–1873

Hora SL (1933) Animals in Brackish water at Uttarbhag, Lower Bengal. Curr Sci 1:12–38

Hora SL (1934) Brackish water animals of Gangetic Delta. Curr Sci 2:426–427

Hora SL (1936) Ecology and bionomics of Gobioid fishes of Gangetic Delta. C R Congr Int Zool 12:841–864

Hora SL (1943) Evidence of distribution of fishes regarding rise in salinity of the river Hugli. Curr Sci 12:69–90

Jhingran VG (1982) Fish and fisheries of India. Hindustan Publishing Corporation, Delhi

Jude DJ, Tesar FJ, Deboe SF, Miller TJ (1988) Diet and selection of major prey species by Lake Michigan salmonines, 1973–1982. Trans Am Fish Soc 116(5):677–691

Kawasaki T (2001) Climate change, regime shift and -stock management. Presented to international commemorative symposium of the 70th anniversary of the Japanese Society of Fisheries Science, Yokohama, 1–5 Oct 2001, 6pp

Khan RA (2003) Fish faunal resources of Sunderban estuarine system with special reference to the biology of some commercially important species. Zoological Survey of India, Occ. Paper No. 209, Kolkata

Kitchell JF, Crowder LB (1986) Predator-prey Interaction in Lake Michigan: model predictions and recent dynamics. Environ Biol Fish 16(1):205–211

Koster FW, Neuenfeldt S, Mollmann C, Vinther M, St. John MA, Tomkiewicz J, Voss R, Kraus G, Schnack D (2003) Fish stock development in the Central Baltic sea (1976–2000) in relation to variability in the physical environment. ICES Mar Sci Symp 219:294–306

Mandal AK, Nandi NC (1989) Fauna of Sundarban mangrove ecosystem, West Bengal, India, Fauna of conservation areas. Zoological Survey of India, Calcutta

Mitra A (1998) Status of coastal pollution in West Bengal with special reference to heavy metals. J Indian Ocean Stud 5(2):135–138

Mitra A (2000) Chapter 62: The Northeast coast of the Bay of Bengal and deltaic Sundarbans. In: Sheppard C (ed) Seas at the millennium – an environmental evaluation. Elsevier Science, Oxford, pp 143–157

Mitra A, Trivedi S, Choudhury A (1994) Inter-relationship between trace metal pollution and physico-chemical variables in the frame work of Hooghly estuary. Indian Ports 10:27–35

Mitra A, Gangopadhyay A, Dube A, Andre CKS, Banerjee K (2009) Observed changes in water mass properties in the Indian Sundarbans (Northwestern Bay of Bengal) during 1980–2007. Curr Sci 97:1445–1452

Mollmann C, Kornilovs G, Sidrevics L (2000) Long-term dynamics of main mesozooplankton species in the Central Baltic Sea. J Plankton Res 22:2015–2038

Mollmann C, Kornilovs G, Fetter M, Koster FW, Hinrichsen HH (2003a) The marine copepod, *Pseudocalanus elongatus*, as a mediator between climate variability and fisheries in the Central Baltic Sea. Fish Oceanogr 12:360–368

Mollmann C, Koster FW, Kornilovs G, Sidrevics L (2003b) Interannual variability in population dynamics of calanoid copepods in the Central Baltic Sea. ICES Mar Sci Symp 219:294–306

Morgan I, McDonald DG, Wood CM (2001) The cost of living for freshwater fish in a warmer, more polluted world. Glob Chang Biol 7:345–355

MPEDA (1999) Guidelines for green certification of freshwater ornamental fishes, Panampilly Avenue, Cochin, India 122pp

Mukherjee M, Kashem A (2007) Chapter IX: Pollution threats to Sundarban wetlands. In: Mukherjee M (ed) Sundarbans wetlands. Department of Fisheries, Aquaculture, Aquatic Resources, and Fishing Harbours, Govt. of West Bengal, Kolkata, pp 150–165

Pearse AF (1932) Observation on the ecology of certain fishes and crustaceans along the bank of Matlah river at Port Canning. Rec Ind Mus 34:289–298

Prashad B, Hora SL, Nair KK (1940) Observation on seaward migration of so called Indian Shad, *Hilsa ilisha* (Hamilton). Rec Indian Mus 42:529–552

Sarmiento JL, Slater R, Barber R, Bopp L, Doney SC, Hirst AC, Kleypas J, Matear R, Mikolajewicz U, Monfray P, Orr J, Soldatov V, Spall SA, Stouffer R (2005) Response of ocean ecosystems to climate warming. Glob Biogeochem Cycl 18:123–148

Sinha M, Mukhopadhyay MK, Mitra PM, Bagchi MM, Karmakar HC (1996) Impact of Farakka Barrage on the hydrology and fishery of Hooghly Estuary. Estuaries 19(3):710–722

Talwar PK (1991) Pisces. In: Arun Gopal Jhingran (ed) Faunal resources of Ganga, Part 1. Zoological Survey of India, Calcutta, pp 59–145

Talwar PK, Mukherjee P, Saha D, Pal SN, Kar S (1992) Marine and estuarine fishes. In: Arun Gopal Jhingran (ed) State fauna series three: fauna of West Bengal, Part 2. Zoological Survey of India, Calcutta, pp 243–342

Tang DL, Kawamura H (2001) Long-term series satellite ocean colour products on the Asian waters. In: Proceedings of the 11th PAMS/JECSS workshop. Hanrimwon Publishing (CD-ROM: 0112-PO3), Seoul, pp 49–52

Walker BW (1961) The ecology of the Salton Sea, California, in relation to the sport fishery. Fish Bulletin No. 113. State of California, Department of Fish and Game, San Diego

Internet Reference

http://www.mpeda.com/FisheryResources/Elasmobranchs/Elasmobranchs.htm

Climate Change and Livelihood: Are We Approaching Towards an Inevitable Change?

We can't ignore the security threats from climate change. The decisions we make in coming months will determine whether we meet this challenge head-on and prevail or if we are to suffer the worst consequences of a warming planet. This time we have to connect the dots before we face catastrophe.

John Kerry

According to the World Resources Institute, 2.2 billion people, or 39 % of the world's population, live on or within 100 km (60 miles) of seashore. Have we ever imagined the fate of this considerable chunk of population due to sea-level rise or natural disasters (like tsunami and cyclone)? These obvious problems cannot be escaped – either the problems will ruin away the population or the population will have to overcome the problems by mitigation and adaptation. Adaptation to climate disruptions involves developing ways of protection from climate impacts, such as building sea walls to protect communities from rising sea levels or relocating coastal communities. Mitigation involves implementing ways to reduce the rise of sea level, such as reducing greenhouse gas concentrations in the atmosphere that lead to thermal expansion of the ocean.

An important asymmetry, however, exists between adaptation and mitigation. Unlike mitigation, adaptation in most cases provides local benefits and is the only response available for the impacts that will occur over the next several decades before mitigation measures can have an effect. Adaptation – such as building sea walls, relocating residents, altering the variety of crops

planted, increasing water reservoir capacity or transforming from freshwater pisciculture to brackish water pisciculture system – benefits those locally who pay for it. The benefits of mitigation effort are much more diffuse and global. Mitigation efforts such as lowering emissions in the United States will reduce atmospheric greenhouse gas concentrations in India as well, but will not be effective unless widely adopted.

Further, the asymmetry extends to the time scale differences between adaptation and mitigation impacts. Mitigation measures take much longer to have an impact than adaptation measure. A mitigation response such as reforestation will take much longer to realize its full benefit of sequestering carbon. An adaptation to rising water such as sea wall will protect a coastline immediately.

Additionally, the asymmetry includes the different capacities between the developed and the developing nations of the world to undertake adaptations. In places where adequate money is available, adaptation may occur without relying on government intervention. Conversely, in less affluent places, little or no adaptation may be undertaken. This is because financial assistances are needed to initiate the adaptation-related programmes in several tiers. The example of initiating oyster culture in Sundarbans is very pertinent in this context. The island dwellers do not have any idea on the edible value of these bivalves. Therefore, a series of trainings and workshops are needed to develop the mind set-up of the local people. In the second phase, additional fund is required for technology transfer, and finally to create a market linkage, institutional set-up should be well equipped to promote the new livelihood products.

Every sector of human activity from growing food to transportation offers opportunities for adapting to climate disruption. Worldwatch Institute summarizes some of the adaptation strategies in seven different sectors in Table 7.1.

Table 7.1 Examples of planned livelihood in different sectors

Sector	Adaptation strategy
Water	Water storage and conservation techniques
	Desalination
	Increased irrigation efficiency
Agriculture	Adjustment of planting dates and crop variety
	Crop relocation
	Improved land management (such as erosion control and soil protection through tree planting)
Infrastructure and settlement	Relocation
	Improved sea walls and storm surge barriers
	Creation of wetlands as buffer against sea-level rise and flooding
Human health	Improved climate-sensitive disease surveillance and control
	Improved water supply and sanitation services
Tourism	Diversification of tourism attractions and revenues
Transport	Realignment and relocation of transportation routes
	Improved standard and planning for infrastructure to cope with warming and damage
Energy	Strengthening of infrastructure
	Improved energy efficiency
	Increased use of renewable resources

Source: Engelman, State of the World (2009) Worldwatch

The livelihood sector is greatly influenced by the events of climate change. If the population is adapted to new livelihood, the cost of mitigation is reduced. It has been documented that the cost of adaptation is inversely proportional to the costs of mitigation and to that of the cost of unmitigated climate disruption impacts. The livelihood is thus a vital sector in the sphere of climate change-associated economics.

7.1 Influence of Climate Change on Livelihood

Livelihood is a broad term encompassing ecological, social and economic assets of a nation being driven by human resource (capital and capabilities). In other words, it may be stated that the livelihood comprises of human efficiencies, assets (including both material and social resources) and activities required for means of living.

A well-standard livelihood can be achieved if the utilization of available resources is maximum with minimum wastage. This can be expressed as

$$Y = \frac{X_{ru}}{X_{ra}}$$

where Y = livelihood standard, X_{ra} = resources available to the society/community/individual and X_{ru} = resources utilized. In most cases, the available resources remain unutilized due to lack of perfect technology, management, policy, etc. and sometimes social and political interventions and hence the value of Y hardly becomes 1. It is important to note in this context that well-standard livelihood does not mean exploitation of resources. Hence, the author is very much conscious about the term *available resources* (X_{ra}) and not the *total resources*. The total resources need to be conserved through proper policy, appropriate technology and harnessing renewable energy sources.

A livelihood may be regarded as *sustainable* when it can cope with and recover from stresses and shocks and maintain or enhance its capabilities and assets both for present and future generations, without undermining the natural resource base. Livelihood assets may be of different categories like human, social, physical, financial and natural resources. These assets are usable by the community in maintaining a considerable (satisfactory, which again is a relative term) living standard. Lack or insufficient quantity of any one or more of these assets would give rise to a situation where, if left uncompensated, it would pose a substantial constraint to attain livelihood goals.

Climate change, which is an inevitable truth of the present world, has already initiated constraints on livelihoods by affecting the assets directly and indirectly as stated here:

7.1.1   Direct Impact of Climate Change on Livelihood Assets

- Temperature rise affecting crop production, human health (directly by heat waves), etc.
- Sea-level rise causing salinization of coastal land, groundwater aquifers, etc.
- Sea-level rise posing adverse impact on agriculture, freshwater fishery, etc.
- Increased tidal surges causing crop loss, mangrove and coastal vegetation loss, etc.
- Reduced availability of economically important fishes from coastal waters

7.1.2   Indirect Impact of Climate Change on Livelihood Assets

- Alteration of salinity influencing plankton community (the food for fish) thereby affecting fish population.
- pH level depression causing coral and oyster reef deterioration – thus threatening the shelter ground of fishes.
- Sea-level rise causing shrinkage of mangrove areas and landward migration of mangroves, thereby threatening the nursery of finfish and shellfish juveniles.
- Temperature rise enhancing microbial diseases in aquaculture sector by accelerating plankton bloom. It has been documented that *Vibrio cholerae* is known to live in sea-borne plankton that blooms as the sea-surface warms.

- Temperature rise increasing the probability of outbreak of human infectious diseases. During a recent examination of satellite data, National Aeronautics and Space Administration (NASA) scientists have discovered an association between the height and temperature of sea surface and outbreak of cholera in Bangladesh in 1992 and 1995. A research programme conducted by Rawlings under the supervision of Rita Colwell at the University of Maryland during 2005 forwarded the view that cholera cases in Bangladesh surge in spring and autumn when coastal waters are at their warmest.

The concept of livelihood has been gaining weightage in recent *era* and is now seen as fundamental to poverty reduction approaches around the world. The emergence of livelihood approaches has led to new understandings on how poverty, and the ability to move out of poverty, reflects the (lack of) capabilities and assets available to the poor. This includes material assets such as access to land, other natural resources, financial capital and credit, tools and inputs into productive activities and others. It also reflects human capabilities (the knowledge and skills of the family), social and political factors such as contact networks and the openness of government institutions and, critically, the capability to withstand the effects of shocks such as natural disasters. For most households, and especially for poor people, these assets are deployed in a series of livelihood activities: the means through which a household gains an income and meets its basic needs. This includes paid employment, but for poor people in particular, it includes the ability to farm and to exploit common property resources for livestock, fishing, gathering fuel wood and many other things. Reliable and secure access to these resources, to land, water and biotic resources, is fundamental to the livelihoods of the poor. Climate change has potential to pose an adverse impact on resource flow, which can seriously affect the economic profile of marginal farmers, fisherman and poor landless people, who directly depend on the natural resource base. The negative role of climate change on livelihoods is also a function of geographical features of the area. The increased melting of ice, which is consistent with the accelerating rise in temperature that has occurred since 1980, is of great concern in low-lying regions of coastal countries and low-lying island countries. Perhaps the most easily measured effect (indicator) of rising sea level is the inundation of coastal areas. Donald F. Boesch (2000), with the University of Maryland's Center for Environmental Sciences, estimated that for each 1-m rise in sea level, the shoreline will retreat by an average of 1,500 m or nearly a mile (Boesch 2000).

In 2000, the World Bank published a map showing that a 1-m rise in sea level would inundate half of Bangladesh's riceland. With a rise in sea level of up to 1 m forecast for this century, tens of millions people from Bangladesh would be forced to migrate in the neighbouring country of India, which has a total population of 1,129,866,154 (15 % of world's total population) (as per July 2007) (http://www.the_world_factbook.html). In a country with such a dense population, this would be a traumatic experience. Rice-growing river floodplains in other Asian countries would also be affected, including India, Thailand, Vietnam, Indonesia and China. With a 1-m rise in sea level, more than a third of Shanghai, a city of 13 million people, would be under water (World Development Report 1999/2000). The developed nations of the world would also be affected by such increase in sea levels. It has been estimated that such a rise would cost the United States 36,000 km^2 (14,000 mile^2) of land, most of it in the middle Atlantic and Mississippi Gulf states. With a 50-year storm surge, large portions of Lower Manhattan and the National Mall in the centre of Washington, D.C., would be flooded with seawater (James et al. 2000). While public attention focuses on the effect of ice melting on sea-level rise, the thermal expansion of the oceans as a result of rising temperature is also raising sea level. At present, scientists estimate the relative contributions of ice melting and thermal expansion to sea-level rise to be about the same. Together, the two are raising sea level at a measurable rate. It has become an indicator to watch – a trend that could force a

human migration of unimaginable dimensions. It also raises questions about responsibility to future generations that humanity has never before faced (IPCC 2007). Worst situation will be faced by people living below poverty line in deltaic and estuarine regions, as the basic livelihood options of these people (like paddy cultivation and freshwater fish culture in homestead ponds) will shrink due to seawater intrusion. The mangrove ecosystem and coastal vegetation, being important natural resource bank (ecological assets), would also lose their potential to meet the needs of the island dwellers and coastal population on account of its landward shifting due to sea-level rise.

7.2 Indian Subcontinent: A Unique Test Bed

India has been identified as one among 27 countries which are most vulnerable to the impact of global warming-related accelerated sea-level rise (UNEP 1989). The high degree of vulnerability of Indian coasts can be mainly attributed to extensive low-lying coastal area, high population density, frequent occurrence of cyclones and storms, high rate of coastal environmental degradation on account of pollution (Table 7.2) and non-sustainable development.

The Indian subcontinent has a total population of 1,129,866,154, of which 43.3 %

Table 7.2 Type and quantum of pollutants entering annually into the coastal waters of India

Input/pollutant	Quantum-annual
Sediments	1,600 million tonnes
Industrial effluents	50×10^6 m^3
Sewage – largely untreated	0.41×10^9 m^3
Garbage and other solids	34×10^6 tonnes
Fertilizer residue	5×10^6 tonnes
Synthetic detergent residue	1,30,000 tonnes
Pesticide residue	65,000 tonnes
Petroleum hydrocarbons (Tar balls residue)	3,500 tonnes
Mining rejects, dredged spoils and sand extractions	0.2×10^6 tonnes

Source: Agarwal and Lal (2001); www.survus.mdx.ac.uk/pdfs/3dikshas.pdf

(approximately 489,422,974) is concentrated in the 12 maritime states of India (Table 7.3). Within 100 km of the coastline, 26 % of the total maritime's state population (approximately 127,249,973) lives (http://country_profiles.pdf), who are directly or indirectly dependent on the coastal and estuarine resources for their livelihood.

It has been recorded that about 5,958,744 persons sustain their life through fishery or aquaculture (http://www.coastalpopulation.html), and the figure of such population in the east coast of India alone is 950,000. Any global warming-induced climatic change such as increase in sea-surface temperature, change in frequency, intensity of tracks of cyclones, sea-level rise may aggravate the potential risks to coastal zones. The rise in sea level could result in the loss of

cultivable land due to inundation, saltwater intrusion into coastal ecosystems and into groundwater systems and loss of terrestrial and marine biodiversity.

Past observations on the mean sea level along the Indian coast indicate a long-term rising trend of about 1.0 mm/year on an annual mean basis. However, the recent data suggests a rising trend of 2.5 mm/year in sea-level rise along Indian coastline. Model simulation studies based on an ensemble of four A-O GCM outputs indicate that the oceanic region adjoining the Indian subcontinent is likely to warm up at its surface by about 1.5–2.0 °C by the middle of this century and by about 2.5–3.5 °C by the end of the century. The corresponding thermal expansion-related sea-level rise is expected to be between 15 and 38 cm by the middle of century and between 46 and 59 cm by the end of the century (Lal and Aggarwal 2000). This simulated rise in sea level by 46–59 cm along Indian coastline is comparable with the projected global mean sea-level rise of 50 cm by the end of this century and may have significant impact on coastal zones of India (Table 7.4).

The climate-change-induced sea-level rise may cause inundation of considerable percentage of land in coastal zone, and millions of *climatic refugees* will be the ultimate outcome of this invisible terrorism (Table 7.5), who will have no food security, no proper shelter and being a constant prey to all the diseases.

The possible adverse effects on the livelihood sectors of Indian subcontinent in relation to climate change are discussed here in points.

Table 7.3 Population of maritime state of India

Sl. no	Name of state	Population
1.	Gujrat	5,05,96,992
2.	Maharashtra	9,67,52,247
3.	Goa	13,43,998
4.	Karnataka	5,27,33,958
5.	Kerala	3,18,38,619
6.	Tamil Nadu	6,21,10,839
7.	Andhra Pradesh	7,57,27,541
8.	Orissa	3,67,06,920
9.	West Bengal	8,02,21,171
10.	Pondicherry	9,73,829
11.	Andaman and Nicobar Island	3,56,265
12.	Lakshadweep Island	60,595
	Total	48,94,22,974

Source: http://cyberjournalist.org.in/census_cenindia.html

Table 7.4 Climate Change Projections for India based on an ensemble of four A-O GCM outputs

Year	Temperature change (°C)			Precipitation change (%)			SLR (cm)
	Annual	Winter	Monsoon	Annual	Winter	Monsoon	
2020s	1.36±0.19	1.61±0.16	1.13±0.43	2.9±3.7	2.7±17.7	2.9±3.7	4–8
	(1.06±0.14)	(1.19±0.44)	(0.97±0.27)	(1.05±3.7)	(−0.1±10.0)	(1.05±3.7)	
2050s	2.69±0.41	3.25±0.36	2.19±0.88	6.7±8.9	−2.9±26.3	6.7±8.9	15–38
	(1.92±0.20)	(2.08±0.85)	(1.81±0.57)	(−2.36±7.1)	(−4.8±18.9)	(−2.36±7.1)	
2080s	3.84±0.76	4.52±0.49	3.19±1.42	11.0±12.3	5.3±34.4	11.0±12.3	46–59
	(2.98±0.42)	(3.25±0.53)	(2.67±1.49)	(−0.13±15.2)	(−1.2±21.2)	(−0.13±15.2)	

Numbers in bracket are for the GHG + aerosol forcing experiments while those outside are for GHG only forcing experiments

Table 7.5 Potential effects on India's coastal area and population due to 1-m rise in sea level

State/union territory	Coastal area (million hectares)			Population (millions)		
	Total	Likely to be inundated	Percentage	Total	Likely to be affected	Percentage
Andhra Pradesh	27.504	0.055	0.19	66.36	0.617	0.93
Goa	0.370	0.016	4.34	1.17	0.085	7.25
Gujarat	19.602	0.181	0.92	41.17	0.441	1.07
Karnataka	19.179	0.029	0.15	44.81	0.25	0.56
Kerala	3.886	0.012	0.30	29.08	0.454	1.56
Maharashtra	30.771	0.041	0.13	78.75	1.376	1.75
Orissa	15.571	0.048	0.31	31.51	0.555	1.76
Tamil Nadu	13.006	0.067	0.52	55.64	1.621	2.91
West Bengal	8.875	0.122	1.38	67.98	1.6	2.35
Andaman and Nicobar Islands	0.825	0.006	0.72	0	0	0
India	139.594	0.571	0.41	416.74	7.1	1.68

Table 7.6 Fraction of land likely to be affected in case of 1-m sea-level rise along coasts of various Indian States

State	Cultivated land	Cultivable[a] land	Forest land	Land not available for agriculture
Gujarat	0.03	0.08	0.00	0.89
Maharashtra	0.39	0.21	0.09	0.31
Goa	0.65	0.03	0.00	0.31
Karnataka	0.51	0.13	0.13	0.23
Tamil Nadu	0.39	0.39	0.00	0.21
Orissa	0.68	0.15	0.05	0.12
West Bengal	0.74	0.04	0.00	0.22

Source: State sea level rise report
[a]Cultivable land is the land that can be brought under cultivation, while land not available for agriculture is the land under human settlements, commerce, trade, etc.

7.2.1 Effect on Agriculture

In India, tidal ingress and horizontal intrusion of saline waters from the direction of offshore to inshore region may extend by 35–50 km beyond the present limit (Mohanty and Ray 1987), and during storm conditions, the spread of salinity in the low-lying agricultural lands may ruin the prospect of crops. The potential impacts of 1-m sea-level rise on the land-use pattern in the coastal states in terms of the share of total land affected are shown in Table 7.6 (TERI 1996). The cultivable land would be worst affected by both inundation and intrusion due to rising sea level, and subsequently the crop production would be seriously hampered.

The coincidence of cyclonic depression and spring tide is probably the worst natural hammering

to the livelihoods of the island dwellers. Such impact was witnessed during 19 October 2005 in several islands of eastern Indian Sundarbans. The tidal amplitude of about 7 m devastated the agricultural land and freshwater ponds of the island, making thousands of people homeless (Mitra and Banerjee 2005).

7.2.2 Effect on Fishery

Projected climate change could halve or double average harvests of any given species; some fisheries may disappear, and other new ones may develop. More warm-water species will migrate poleward and compete for existing niches, and some existing populations may take on a new dominance. These factors may change

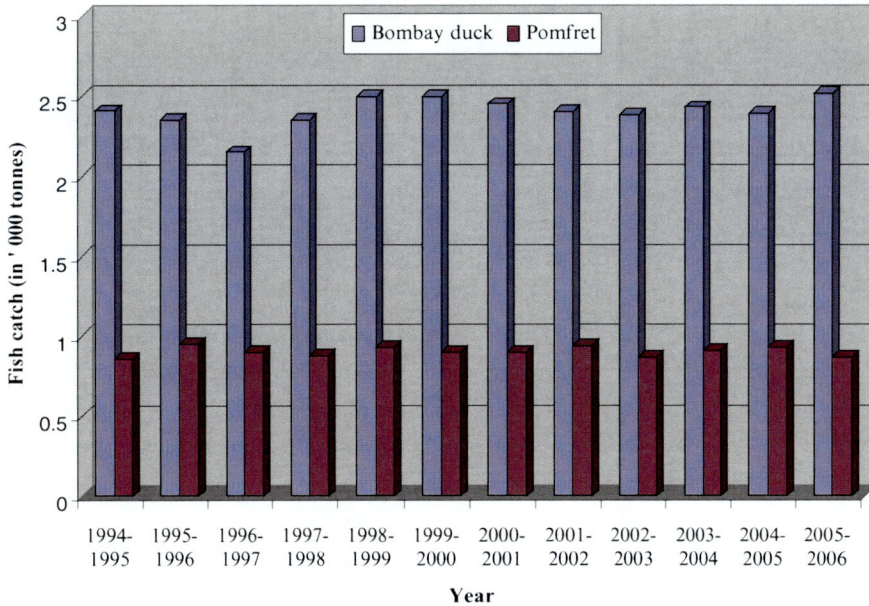

Fig. 7.1 Consistent pattern of fish catch at Kakdwip in western Indian Sundarbans

the population distribution and value of the catch. This has profound influence on economies particularly on stakeholders whose livelihoods are directly related to fishing and fishery. Apart from this direct impact of climate change on compositional change in fish spectrum, several indirect impacts like change of water quality, shelter ground and breeding ground of fishes and alteration of plankton (food for fishes) community will also pose adverse impact on fish resources. Ecosystems such as coral reefs, mangroves, estuaries and deltas are rich in biodiversity which play a crucial role in fishery production besides protecting the coastal zones from erosion by wave action. During 1996 India ranked 5th in fishery productivity in Asia (Grainger and Garcia 1996).

The rich mangrove diversity of Indian Sundarbans (West Bengal), Orissa, etc. has made the east coast rich in fishery, but recent surveys are pointing towards replacement of economically important fish species with low-priced trash variety. The example of pomfret (*Pampus* spp.) and Bombay duck (*Harpodon nehereus*) catch pattern in western and eastern Indian Sundarbans is very relevant in this context. The catch of these species is almost consistent in the western part due to more flow of freshwater through Farakka

barrage that prevented the aquatic salinity from being rising. In the eastern sector, the gradual increase of low-priced Bombay duck is a representation of increase of salinity due to obstruction of freshwater from Ganga–Bhagirathi system as a result of dying of Bidyadhari channel due to heavy siltation (Figs. 7.1 and 7.2).

The increase of highly demanding commercial fish species Hilsa (*Tenualosa ilisha*) during monsoon in the western part is also a boon for the local economy, unlike eastern part, where the estuarine waters have become unfavourable for migration of *Tenualosa ilisha* for spawning in the upstream zone. This favourable situation is attributed due to increased flow from Farakka during monsoon, which is purely a man-made cause and has no string tied with the natural forces induced by climate change. Five-year surveys (1999–2003) on water discharge from Farakka barrage revealed an average discharge of $(3.4 \pm 1.2) \times 10^3$ m^3/s. Higher discharge values were observed during the monsoon with an average of $(3.2 \pm 1.2) \times 10^3$ m^3/s and the maximum of the order 4,200 m^3/s during freshet (September). Considerably lower discharge values were recorded during premonsoon with an average of $(1.2 \pm 0.09) \times 10^3$ m^3/s and the minimum

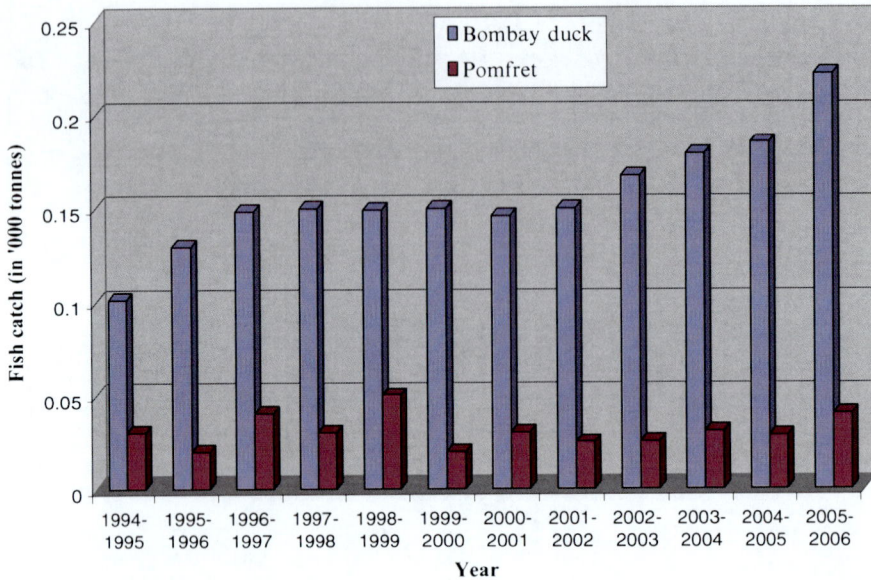

Fig. 7.2 Fish catch at Sonakhali in eastern Indian Sundarbans showing increasing trend of Bombay duck catch

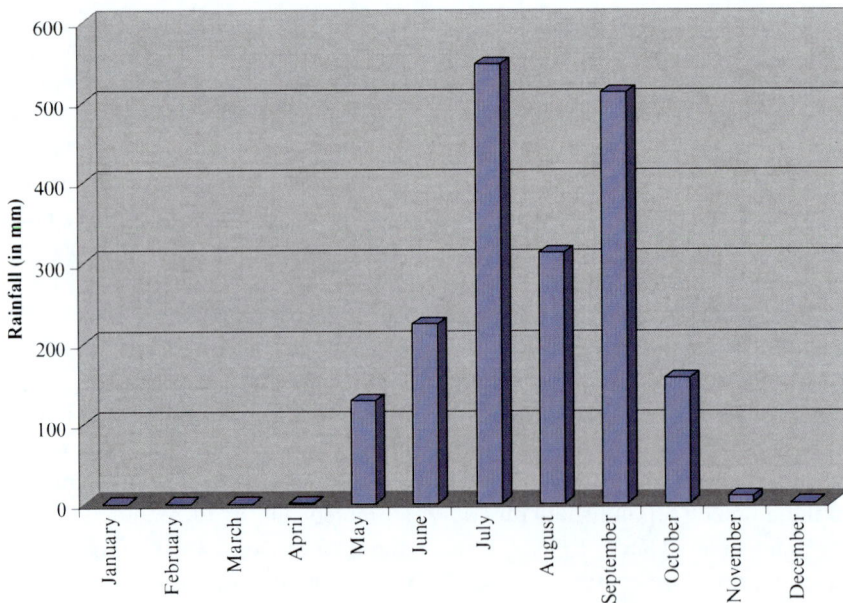

Fig. 7.3 Monthly variation of rainfall (in mm) in and around Indian Sundarbans (Data averaged over a period of 10 years; 1998–2007; *Source*: IMD, Kolkata)

of the order 860 m³/s during May. During post-monsoon discharge, values were moderate with an average of $(2.1 \pm 0.98) \times 10^3$ m³/s. The lower Gangetic deltaic lobe also experiences maximum rainfall during September (Fig. 7.3). This causes a considerable volume of surface run-off from the 60,000 km² catchment areas of Ganga–Bhagirathi–Hugli system and their tributaries. All these factors (discharge + precipitation + run-off) increase the dilution factor of the Hugli estuary in the western part of Indian Sundarbans, resulting in more population of Hilsa during monsoon.

Fig. 7.4 Impact of
sea-level rise on livelihood
via groundwater system

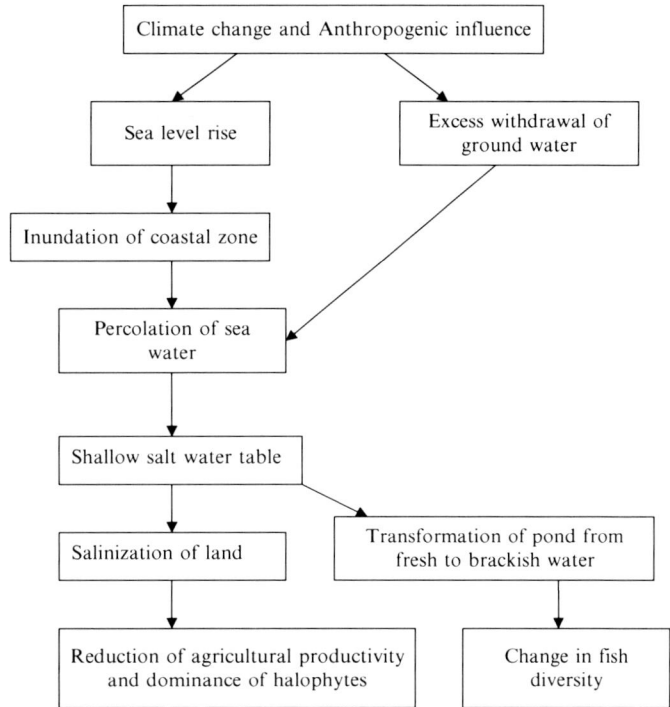

7.2.3   Effect on Groundwater Resource

In the coastal regions of Tamil Nadu, salinity of groundwater due to the intrusion of seawater into the subsurface aquifer is a major problem (Subramanian 2000). Due to excess withdrawal of groundwater, the water table has fallen too far below, thereby allowing seawater to percolate. Similarly, in Gujarat, due to uncontrolled withdrawal of groundwater, the groundwater has become highly saline, apart from the fact that depth of the water table reaching at places beyond 200 m (Subramanian 2000). Coastal aquifer system will be more contaminated with salinity, bringing greater complicacy to the problem of tapping usable groundwater (Mohanty 1990). In coastal regions of West Bengal and Orissa, the problem of freshwater is fairly acute because of the depth of water table and high cost of lifting the same from the depth of 700–1,000 m. The shallow saltwater table often makes the stored freshwater in ditches and ponds brackish and surface soil saline in nature. This reduces the agricultural productivity

and consequently the local economy gets affected. It is also interesting to note that contamination of groundwater (shallow water table) is not restricted at a particular depth, but its pulse is felt even on the above-ground floral and faunal diversity particularly in coastal zone and islands. The halophytic vegetations gradually dominate the floral community, which has a far-reaching impact on the ecological assets of the area. The impact of climate change on coastal groundwater system and subsequently on associated livelihood sectors is schematically represented in Fig. 7.4.

7.2.4   Effect on Coastal Natural Resources

Coral reefs and mangroves provide benefits under the four categories of ecosystem services defined by the 2005 Millennium Ecosystem Assessment. These are:

- Regulating – *e.g.* protection of shores from storm surges and waves and prevention of erosion

- Provisioning – *e.g.* fisheries, building materials, honey, wax and bioactive substances (of pharmaceutical interests)
- Cultural – *e.g.* tourism and spiritual appreciation
- Supporting – *e.g.* cycling of nutrients, fish nursery habitats and fish breeding grounds

The mangrove and coral reef ecosystems are among the most valuable ecosystems in terms of their benefits to humankind (particularly for the livelihood sector) as stated in the report of UNEP-WCMC (2006). Some of these benefits linked with the economic profile of the area are listed here.

- Economic valuation of ecosystems needs to be treated with caution, but annual values per km² have been calculated at US$100,000–600,000 for reefs and US$200,000–900,000 for mangroves.
- The total area of coral reefs and mangroves belies their importance in terms of fisheries, other extractive uses, shoreline protection and, in the case of reefs, tourism and recreation.
- Both mangrove and coral reef ecosystems contribute significantly to national economies, particularly those of small island developing states (SIDS), 90 % of which have coral reefs and over 75 % of which have mangroves.
- Ecosystems that can no longer provide their full ecological services have a social and economic 'cost' that can be felt locally and many miles away.

Degradation of coral reefs (on account of lowering of aquatic pH) and mangroves (through shrinkage of area and landward migration) due to climate change may cause negative impacts of high magnitude on the livelihood sectors of coastal population. Some of the possible impacts are pointed here:

- Reduced fish catches
- Decrease of tourism revenue in coastal communities
- Decrease of food security
- Malnutrition due to lack of protein
- Loss in export earnings
- Decline of the tourism industry
- Increased coastal erosion and destruction from storms and catastrophic natural events, which affects coastal residents, tourism operations and many other economic sectors

Case Study of Bangladesh

Bangladesh has a very high population density (more than 1,000 people per km²), with a literacy rate of less than 50 %, an infant mortality rate of 48/1,000 live births and 50 % of the population below the poverty line (World Bank 2004). The case study zone is located in the southwest region of Bangladesh, an area heavily influenced by the tributaries of the Ganges that flow into the Bay of Bengal. Most of the region lies less than 3 m above sea level and is subject to extreme events such as flooding, cyclones and storm surges (Ali 1999; Mirza 2002; UNEP 2002). Originally a region dominated by farming, over the last 30 years, the shrimp has grown considerably driven by a flourishing export market, international donor support and aided by successive Bangladeshi Governments since the 1980s to liberalize and diversify the economy (Deb 1998; McLachlan 2003).

The village of Subarnabad is located in the southwest region near the Indian border, with a population of approximately 2,440 occupying roughly 3.2 km². Water flow within the area is regulated largely by the Ichhamati River, which is a Ganges-dependent river. The main industry in the village and occupation for the residents is shrimp farming, resulting in the conversion of large portions of the land from production to saltwater ponds for shrimp cultivation. Many landless and poor individuals work in these farms for wage incomes, stocking and catching shrimp, building and repairing embankments and removing weeds from the farm. Subarnabad was a suitable case study because of its high proportions of poor villagers, its exposure to climate-related conditions, the importance of aquaculture in the area and the cooperation of an NGO with local credibility and contracts. The research in Subarnabad

(continued)

(continued)

was undertaken in collaboration with the project Reducing Vulnerability to Climate Change (RVCC), funded by CIDA and implemented by CARE Canada via CARE Bangladesh. The RVCC project operated in six districts in southwest Bangladesh through partnerships with 16 local partner organizations representing local and national NGOs, a community-based organization (CBO) and two research organizations, with the overall goal to increase the capacity of Bangladeshi communities in the southwest to adapt to the adverse effects of climate change (CARE 2006). Within the study area, the local partner organization was the Institute of Development Education for Advancement of Landless (IDEAL). IDEAL has a long record of participatory development initiatives in the region and has worked with the community beneficiaries in Subarnabad village to implement adaptations to climate change.

The poorest villagers of Subarnabad identified a variety of problematic conditions or exposures related principally to income generation and overall poverty. The bulk of villagers' concerns were directly or indirectly related to (a) environmental changes associated with saltwater intrusion and (b) the changes in production systems from rice and other land-based crops to saltwater shrimp farming and subsequent shifts in livelihoods.

The history of saltwater intrusion in the village can be traced back 30 years to the construction of several megaprojects and their related environmental changes. Over the last three decades, the water resource system in the southwest of Bangladesh has experienced many changes as a result of the construction and poor maintenance of the Coastal Embankment Project (CEP) along the coast of Bangladesh: the construction of the Farakka Barrage (dam) in India, local water diversions, sea-level rise

and storm surges. The result of these has been the siltation, sedimentation and rising waters of Ganges-dependent rivers, which has complicated drainage and essentially turned lands protected by polders (structures designed to protect shoreline communities from flooding, surges and saline water intrusion) into lakes, particularly during the monsoon season. This has occurred in the village of Subarnabad, where the base level of the neighbouring Ichhamati River has risen considerably, preventing effective drainage and creating severely water-logged conditions.

Water logging, increasing salinity, an absence of any effective government water management programmes and the attractive market for shrimp together have prompted an increase in shrimp farming and other types of aquaculture by larger land owners, replacing the land-based crops within the polders. Though few villagers were initially interested in shrimp farming, the difficulty of sustaining subsistence and commercial crops under the increasing salinity as neighbouring areas was converted to saline ponds for shrimp gave farmers with clear land title little choice but to lease or sell their land to shrimp producers or start shrimp farming themselves. Subsistence farmers without clear title had little means to resist the conversion of land to shrimp ponds managed by the wealthy. They were forced off the land. Shrimp farms now surround the village of Subarnabad, with a few small homestead gardens remaining in which attempts to grow crops in increasingly saline soils are still made, with predictably poor outputs. While the economic gain through the proliferation of the shrimp industry is very important for the Bangladesh national economy and a major benefit for the large shrimp operators, the associated environmental and social changes are becoming issues of growing

(continued)

(continued)

concern, particularly for poorer residents (Deb 1998). The environmental changes due to saltwater intrusion in Subarnabad directly influence the exposure of poor villagers to a number of problematic conditions, including a decrease in freshwater supply, the loss of crop production and common resources, an increase in health problems and an increase in the fragility of mud homes. Villagers of Subarnabad experienced decreased soil and water quality, as the surrounding waters were no longer considered fit for domestic use and consumption and rendered the soil unsuitable for crop cultivation.

7.3 Adaptation: A Way to Get Rid

Adaptation is the ability to respond and adjust to actual or potential impacts of changing conditions of the climate in ways that moderate harm or take advantage of any positive opportunities that the climate may afford. It is therefore a mode of coping with the surrounding environment through development and alteration/modification of some attitude, habit, habitat, etc. Even changes in the policy level or planning processes or decision-making processes are also a part of adaptation. Adaptation can be anticipatory or reactive. In the former case, the systems adjust before the initial impacts take place, but in the latter case, change is introduced in response to the onset of impacts.

Although there has been progress in studies of climate change impacts and adaptations, as well as an increased recognition of the need for programmes and policies to implement and facilitate adaptations, to date, there are still few practical on-the-ground examples of planned adaptation to climate change that reduce people's vulnerabilities (Huq et al. 2003; Smit and Wandel 2006). Studies of practical adaptation initiatives have revealed that climate and climate variability are experienced in the context

of other changing conditions (environmental, socio-economic, political), and vulnerabilities are rarely to climate change stimuli alone. It is also increasingly recognized that an effective means of ensuring that climate adaptations are undertaken is to mainstream them into development initiatives and existing priorities, such as livelihood enhancement, poverty alleviation, environmental management and sustainable development (Huq et al. 2003; Khan 2003; Klein and Smith 2003; Schipper and Pelling 2006; Smit and Wandel 2006). The utility of promoting adaptations to climate change by incorporating them into development projects has been recognized and acted upon by several development agencies, including the Canadian Development Agency (CIDA), the UK Department for International Development (DFID), the Red Cross/Red Crescent and the World Bank the Asian Development Bank. In recent times, IUCN has also taken an important role in Indian subcontinent through projects of livelihood upgradation in vulnerable sites like Indian Sundarbans.

7.3.1 Major Components of Adaptation

The process of adaptation may be 100 % successful if it is accomplished through actions that target and reduce the vulnerabilities poor people now face, since they are likely to become more prevalent as the climate changes. This approach encompasses the convergence of four distinct components who have long been tackling the issue of vulnerability reduction through their respective activities. These components are disaster risk reduction, climate and climate change, environmental management and poverty reduction. Bringing these components together and offering a common platform – and a shared vocabulary – from which to develop an integrated approach to climate change adaptation can provide an opportunity to revisit some of the intractable problems of environment and development. The starting point for this convergence is a common understanding of the concepts of adaptation, vulnerability, resilience, security,

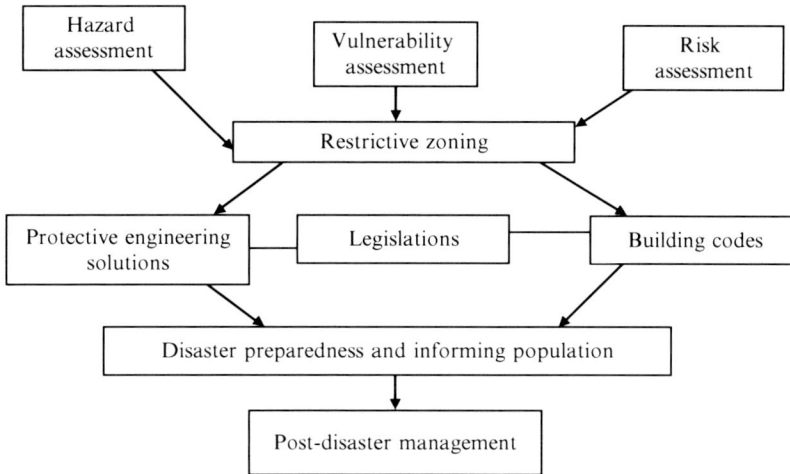

Fig. 7.5 A generalized flowchart of disaster management

poverty and livelihoods as well as an understanding of the gaps in current adaptation approaches. Taken together, they indicate a need – and an opening – for adaptation measures based on the livelihood activities of poor and vulnerable communities. This places the goal of poverty reduction at the centre of adaptation, as the capabilities and assets that comprise people's livelihoods often shape poverty as well as the ability to move out of poverty. The four major components to strengthen the adaptive capabilities to oscillating climatic conditions are discussed here in separate points.

7.3.1.1 Disaster Risk Reduction

It is a vital component to reduce the negative impacts of climate change, and it may be either anticipatory or reactive or both. Disaster may be defined as sudden or abrupt changes of the surrounding environment or processes that lead to severe damages like loss of properties and lives. Thus, disaster may disrupt the normal functioning of a society (community). Disaster cannot be prevented absolutely; rather, the adverse impacts of disaster can be minimized through proper disaster management plan (Fig. 7.5).

The vulnerable and susceptible communities (for a particular type of calamities) need to learn to cope with the disasters. For this regular training through highly efficient manpower and proper institutional mechanisms are extremely important.

The type of manpower and disaster management policy is a function of disaster type (Table 7.7), *e.g.* experts of earthquake or tsunami cannot fit into the domain of mitigating the adverse impact of drought and crop disease, which are also disasters with different origin (cause), duration and effect (Table 7.8).

The risk of disaster (both natural and human originated) may be reduced through effective disaster preparedness. The term 'disaster preparedness' refers to a specialized type of management strategy, for unavoidable circumstances, such as the natural disasters, where the communities or systems are equipped with proper forecast and warning systems or are made prepared not to succumb at the time of disastrous consequences. Such a management strategy involves several steps like:

- Preparation of a disaster plan based on the geographical features of the area, risk and hazard assessment studies
- Anticipating damage to critical facilities
- Establishment of communications and control centre
- Disaster training exercises on regular basis
- Evacuation plans
- Informing/training population
- Forecasts/warning/prediction of disasters
- Monitoring the responses of the society or community to pseudo-disaster (that can be created artificially for a certain duration)

Table 7.7 Classification of coastal disaster on the basis of origin

Natural disaster	Natural disaster with human influence	Mixed: natural/ human influence	Human-originated disaster with natural influence	Human-originated disaster
(a) Earthquake	(a) Flood	(a) Landslides	(a) Crop disease	(a) Armed conflict
(b) Tsunami	(b) Dust storm	(b) Subsidence	(b) Insect infestation	(b) Land mine explosion
(c) Volcanic eruption	(c) Drought	(c) Coastal erosion	(c) Forest fire	(c) Major (air, sea, land) traffic accidents
(d) Lightning and torrential rain		(d) Greenhouse effect	(d) Mangrove decline	(d) Nuclear/chemical accidents
(e) Cyclone/hurricane		(e) Sea-level rise	(e) Coral reef decline	(e) Oil spill
(f) Meteorite impact			(f) Acid rain	(f) Pollution
(g) El Nino			(g) Ozone layer depletion	(g) Groundwater pollution[a]
				(h) Electrical power breakdown
				(i) Pesticide pollution

[a]Groundwater pollution in coastal areas (e.g. contamination with seawater) often results due to inundation of coastal zone on account of sea-level rise. Hence, natural cause of groundwater pollution cannot be ignored

Table 7.8 Classification of coastal disasters on the basis of duration of impact, length of forewarning and frequency of type of occurrence

Disaster type	Duration of impact	Length of forewarning	Frequency or type of occurrence
Lightning	Instant	Second/hours	Random
Earthquake	Second/minutes	Second/hours	Random
Tornado	Second/hours	Minutes	Log/normal
Landslide	Second/decades	Seconds/years	Seasonal/irregular
Intense rainstorm	Minutes	Second/hours	Seasonal/diurnal; Poisson
Tsunami	Minutes/hours	Minutes/hours	Random
Flood	Minutes–day	Minutes–day	Seasonal; Markovian, gamma, log normal
Subsidence	Minutes–decades	Second–years	Sudden or progressive; irregular
Volcanic eruption	Minutes–months	Minutes–weeks	Random
Cyclone	Minutes–years	Hours–days	Seasonal/irregular
Hurricane	Hours	Second–days	Seasonal/random
Forest fire	Hours–days	Hours	Seasonal/irregular; exponential, gamma
Coastal erosion	Hours–decades	Days–weeks	Seasonal/irregular; binomial gamma
Drought	Days–months	Days–months	Seasonal/irregular
Crop/fish disease	Weeks–months	Hours–days	Seasonal

Modified after Alexander (1993)

In coastal areas that are vulnerable to multiple disasters like cyclone, flood, earthquakes, tsunami or oil spills, disaster preparedness should form the basic foundation of disaster management practice. Although, efforts are going on to minimize the disastrous consequences through proper management action plan and preparedness, in India, we still put more emphasis on post-disaster relief, recovery and rehabilitation than on pre-disaster prediction, prevention and preparedness. This approach not only increases the financial load but at the same time poses a shocking effect on the victims/community. The authors in this context feel the need of a 'prevention better than cure' policy, where the vulnerable zone (for different categories of disasters) along with rooms (space) for shelters and medical units should be clearly marked along the coastal stretch. Major stakeholders of coastal zone like ports and harbour authority, fisheries department, forest department, coastguard authority, coastal industries and local communities should be taken into confidence for such anticipatory approach. It is therefore the mind-set of the scientific and policy makers that needs to be changed fast to cope with the disasters more efficiently.

Cost-effective and well-balanced disaster preparedness scheme is needed, considering the expanse and extent of marine, estuarine and coastal zone of India.

Indian coastline extends to about 5,700 km on mainland and to about 7,500 km including two groups of islands. Western coastline has a wide continental shelf having an area of about 0.31 million km^2, which is marked by backwaters and mud flats. East coast consists of Tamil Nadu coast, Andhra coast, Orissa coast and West Bengal coast, which is flat and deltaic and is characterized by the presence of mangrove forests. Mangroves are located all along estuarine areas, deltas, tidal creeks, mud flats and salt marshes and extend to about 6,740 km^2 (about 7 % of world's mangrove areas). Major estuarine areas located along the Indian coasts extend to about 2.6 million hectares (Gouda and Panigrahy 1999). Coral reefs are predominant on small islands in Gulf of Kutch, Gulf of Mannar in Tamil Nadu and on Lakshadweep and Andaman and Nicobar groups of islands. Ecosystems such as coral reefs, mangroves, estuaries and deltas are rich in biodiversity which play a crucial role in fishery production besides protecting the coastal zones from erosion by wave action. Saving these delicate systems of nature spun over long evolutionary period of time from natural or man-made disasters is not an easy task and demands a critical network analysis on the subject. Network analysis is a methodology to study objects as part of a larger system. It starts with the assumption that a system can be represented as a network of nodes (vertices, compartments, components, etc.) and the connections between them. Realistic systems have many such interacting components. When there is a flow of matter or energy between any two objects in that system, we say there is a direct transaction between them. These direct transactions give rise to both direct and indirect relations between all the objects in the system. Network analysis, by design, provides a system-oriented perspective because it is based on uncovering patterns and influence among all the objects in a system. Therefore, it gives a view on how system components are tied to a larger web of interactions. For coastal areas, the disaster preparedness master plan should include the inter- and intra-relationships between ecological assets (like mangroves and coral reefs) and all stakeholders like River Research Institute, water management authority, coastguard authority, forest department, ports and harbour authority, coastal industries, dam management cells, fisheries department, agriculture department, irrigation department, groundwater authority, policy makers and local communities. Past data on sea level is also important to establish the trend or degree of vulnerability.

In the Indian coast, past observations on the mean sea level indicate a long-term rising trend of about 1.0 mm/year on an annual mean basis. However, the recent data suggests a rising trend of 2.5 mm/year in sea-level rise along Indian coastline. The east coast of India is more vulnerable to sea-level rise in comparison to that of the west coast. In the island system of Sundarbans delta, some islands are fast vanishing from the map rendering thousands of people permanently homeless and displaced from their original habitat. The rate of relative sea-level rise is presently approaching 3.14 mm per year near Sagar Island, and this could increase to 3.5 mm per year over the next few decades due to global warming, including the other global and local factors (Hazra et al. 2002). At Bangladesh the rate is more than double due to higher rate of deltaic subsidence. In a comprehensive study on the impact of climate change on deltaic complex (Hazra et al. 2002), it has been estimated that the Sundarbans will lose further 14 % of its land area by 2020 due to relative sea-level rise, storm surges and coastal flooding, making more than 70,000 people homeless. Climate-change-related processes coupled with anthropogenic causes may induce further damages causing salinization, species reduction, loss of forest cover or loss of

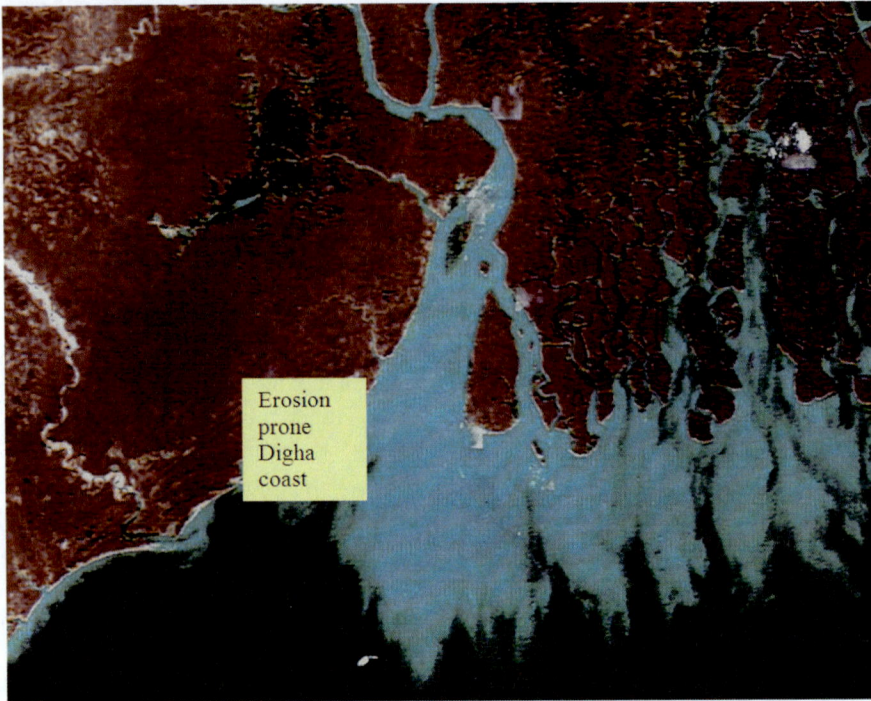

Fig. 7.6 Erosion-prone coastal site (Digha) in the northeast coast of India

food security. Whether such disasters are directly related to climate change or geological processes or human factors or a combination of all are still matters of debate, but the adverse impacts on the fragile ecosystem of Sundarbans are 100 % visible under the lens of reality.

In India there are no laws, policies, insurance or funds to protect, compensate and rehabilitate the vulnerable communities. Not only in Sundarbans but also along Digha coast (in the north east coast of India), several mouzas, which are still shown on the land-use and land-holding maps, have been engulfed by the sea (Fig. 7.6) during the last 30 years (Hazra and Mukherjee 2007). Being on a mainland coast, the inhabitants have been forced to migrate further inland without being compensated for their loss of property. This creates a situation where the properties and life along the coastline come under further risk.

Building of cyclone shelters and cyclone warning system along with a well-orchestrated disaster management plan can ameliorate the adversity of impact to a certain extent. The authors also feel the necessity of introducing some kind of climate insurance policies (like health insurance and car insurance) to cover the risk of climate-change-related disasters.

7.3.1.2 Climate and Climate Change Component

In the initial stage, this was constituted by the world's meteorological community. However, considering the need of interdisciplinary approach in mitigating the climate-change-related hazards, this component has now expanded to include experts from a wide range of disciplines like biologists, geophysical scientists, social scientists and economists. The community now includes people concerned with current weather variability and extremes as well as the projected changes in long-term climate. The policy makers of the government have also become a part of this component.

The climate and climate change component may take an effective role to combat the adverse impact of global warming, sea-level rise and associated disasters through important steps as stated here:

1. Coastal nations of the world should implement comprehensive coastal zone management plans, which should encompass issues like sea-level rise, temperature rise, acidification of ocean water and other impacts of global climate change. They should ensure that risks to coastal populations are minimized (particularly in the livelihood sector) while recognizing the need to protect and maintain important coastal ecosystems (like mangroves, coral reefs and seagrass ecosystems).

2. Coastal areas at risk should be demarcated with proper mapping. National efforts should be undertaken to (a) identify functions and resources at risk from a 1-m rise in sea level and (b) assess the implications of adaptive response measures on them. Improved mapping indicating the magnitude of risk will be a vital component for completing this task.

3. Nations should ensure that development in the coastal zone does not increase vulnerability to sea-level rise. Structural measures to prepare for sea-level rise may not yet be warranted. Nevertheless, the design and location of coastal infrastructure and coastal defences should include consideration of sea-level rise and other impacts of climate change. It is sometimes less expensive to incorporate these factors into the initial design of a structure than to rebuild it later. Thus, in all Environmental Impact Assessment (EIA) for any type of establishment in coastal zone (like tourism units, aquacultural farms, ports, harbours or fish landing centres), the Environmental Management Plan (EMP) should ensure proper design of structures, keeping the concept of sea-level rise in to consideration.

4. Actions like construction of dams; conversions of mangroves and other wetlands for aquaculture, agriculture and human habitation; harvesting of coral; and increased settlement in low-lying areas need to be reviewed critically in the light of climate change and sea-level rise.

5. Efforts should be undertaken to develop effective local-level emergency preparedness plans for reducing vulnerability to coastal storms, tidal surges, rapid erosion, tsunamis, etc. through better evacuation planning and the development of coastal defence mechanisms that recognize the impact of sea-level rise.

6. A continuing international focus on the impacts of sea-level rise needs to be maintained. Existing international organizations should be augmented with new mechanisms to focus awareness and attention on sea-level change and to encourage nations of the world to develop appropriate responses. This requires the development of a strong network between the national stakeholders (of coastal zone) and international bodies.

7. Technical assistance to developing nations should be provided. Institutions offering financial support should recognize the need for technical assistance in developing coastal management plans, assessing coastal resources (biotic and abiotic) at risk and increasing a nation's ability, through education, training and technology transfer, to address sea-level rise.

8. International organizations should support national efforts to limit population growth in coastal areas. This is mainly because the rapid population growth is the prime underlying problem with greatest impact on both the efficiency of coastal zone management and the success of adaptive response options.

9. International funding agencies should support researches and pilot projects on brackish water-oriented alternative livelihood programmes (like oyster culture, seaweed culture and brackish water aquaculture) backed up with proper eco-friendly technology. This should be carried out not at the cost of valuable coastal ecosystems (like mangroves or marshy wetlands) but aligning the coastal biotic community with these livelihood projects in a symbiotic relationship.

10. Research on the impacts of global climate change on sea-level rise should be strengthened. International and national climate

research programmes need to be directed at understanding and predicting changes in sea level, extreme events, precipitation and other impacts of global climate change on coastal areas. In this context, exchange of information and data between researchers is extremely worthy.

11. A global ocean observing network should be developed and implemented. Member nations are strongly encouraged to support the efforts of the IOC, WMO and UNEP to establish a coordinated international ocean observing network that will allow for accurate assessments and continuous monitoring of changes in the world's oceans and coastal areas, particularly sea-level change.

12. Data and information on sea-level change and adaptive options should be made widely available. An international mechanism should be identified with the participation of the parties concerned for collecting and exchanging data and information on climate change and its impact on sea level and the coastal zone and on various adaptive options. Sharing this information with developing countries is critically important for preparation of coastal zone management plans.

7.3.1.3 Environmental Management

This community includes a wide spectrum of people and institutions that deal with overall environmental issues and different aspects of environmental management such as water resources, conservation of forests and pollution monitoring and management. A major drawback of this set is the lack of coordination within the group. Foresters do not communicate sufficiently with water managers, and even within a sector such as water, many individuals and institutions (within and out of government) often have little contact with each other. Social sciences and economics also come under the banner of environmental management, but very rare instances can be cited where the social scientists collaborate with natural scientists in the process of exchange of information, idea and data. Even within the same groups, collaborative works are rare. In this connection, an examination of team structure,

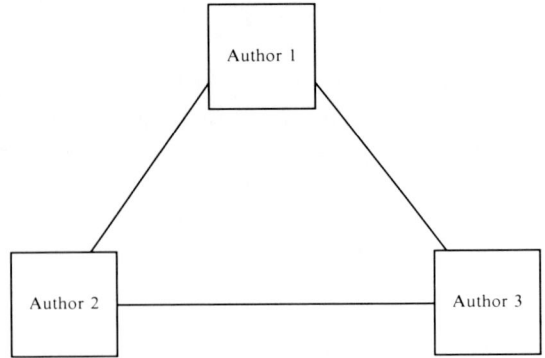

Fig. 7.7 A triangle network graph showing the paper written by three authors

network, architecture and other major influences of the IPCC's Third Assessment Report was carried on by Travis Franck, Robert Nicol and Jaemin Song in 2006. The IPCC network was analyzed similar to other collaboration network research. Authors like Newman (2001a, b, 2003, 2004), Li and Guanrong (2003), and Moody (2004) were designated as the vertices (or nodes) and papers were the edges. If three authors published a paper together, the network graph would be a triangle, since there are three vertices each connected by a common paper (Fig. 7.7).

Moody (2004) showed that social scientists tend to collaborate less than scientists due to the nature of the field, even though collaborations in social science are increasing over time. This is because social scientists are rarely dependent on large-scale lab or resources compared to pure science or environmental science research. In the present network analysis of the IPCC's Third Assessment Report, it was evaluated that natural/environmental scientists have the highest average number of collaborators per author, while, on average, social scientists have the smallest number of collaborators (Table 7.9). The average number of collaborators refers to the scope of their close researchers whom they indeed work with. It has been documented after a critical study that social scientists not only work with small numbers of people but also have a propensity to work alone. The result is that almost 20 % of papers in Volume 3 are written by a single author, while only 3 % are by a single author in Volume 1.

In terms of the number of authors per paper, Volume 1 showed the largest average number of authors per paper, while smaller numbers of authors worked together per paper in Volume 3. This leads to the conclusion that environmental scientists collaborate more often than social scientists do and, furthermore, environmental scientists also work with a more diverse range of researchers. This is the basic essence of environmental management where the task of mitigating the impact of climate change needs to be handled from a multidisciplinary platform.

The reason why environmental scientists have different collaboration patterns from social scientists and vice versa is mainly due to the nature of the field as well as exposure to diverse subjects. Most scientific researches related to climate change require large-scale experiments or data collecting procedures, which inevitably lead to collaborations among researchers. On the other hand, social science studies tend to be theoretical or qualitative, which does not necessarily require large-scale collaboration.

7.3.1.4 Poverty Reduction

Poverty-stricken people are more susceptible to adverse impact of climate change. They have no proper shelter, no fund for resettlement and no insurances against their lives, health and properties. Institutional help hardly reaches them. Hence, reduction of poverty is an important component in fighting against temperature rise, seawater intrusion and disease outbreaks, which are the clutches of climate change potential to scratch the economics of the region. This component engages a wide and diverse spectrum of specialists to utilize the available resources for boosting up the living standard of the community. The local level economic profile is also upgraded by seeking expertise in the field of agriculture, poultry, animal husbandry, pisciculture, etc. In Indian Sundarbans region, few anticipatory actions have already been initiated considering the seawater intrusion into the creeks and inlets criss-crossing the islands. These include training the local population with the technology of oyster and seaweed culture, which are widely distributed in the area. The island dwellers, however, have no idea of their edible values and economic benefits (Table 7.10). Hence, awareness programmes, trainings and workshops on regular basis are being carried on to induce these new

Table 7.9 Image of collaboration in the IPCC Third Assessment Report

Volume no.	Number of collaborators per author	Number of papers with a single author	Number of authors per paper
Volume 1	11.52	209 (2.97 %)	3.78
Volume 2	6.06	1,069 (10.61 %)	2.83
Volume 3	4.14	669 (19.22 %)	2.23

Table 7.10 Few untapped living resources in Indian Sundarbans

Brackish water resource	Taxonomic position	Economic importance
Enteromorpha intestinalis	Division – Chlorophyta	1. Used as cattle feed
	Class – Chlorophyceae	2. Used as poultry feed after mixing with trash fish dust
	Order – Ulvales	3. Used as agent of bioremediation
	Family – Ulvaceae	
	Genus – *Enteromorpha*	
	Species – *intestinalis*	

(continued)

Table 7.10 (continued)

Brackish water resource	Taxonomic position	Economic importance
 Ulva lactuca	Division – Chlorophyta Class – Chlorophyceae Order – Ulvales Family – Ulvaceae Genus – *Ulva* Species – *lactuca*	1. Consumed as food 2. Used in salad, soup, etc. 3. Used as fodder and manure
 Catenella repens	Division – Rhodophyta Class – Rhodophyceae Order – Gigartinales Family – Rhabdoniaceae Genus – *Catenella* Species – *repens*	1. Rich source of astaxanthin and therefore used as an ingredient of fish feed 2. Used as agent of bioremediation
 Saccostrea cucullata	Phylum – Mollusca Class – Bivalvia Order – Pterioida Family – Ostreidae Genus – *Saccostrea* Species – *cucullata*	1. Edible with high demand in South Asian countries 2. Shell is a source of lime 3. Shell dust is used in poultry feed as source of calcium
 Crassostrea gryphoides	Phylum – Mollusca Class – Bivalvia Order – Pterioida Family – Ostreidae Genus – *Crassostrea* Species – *gryphoides*	1. Edible with high demand in South Asian countries 2. Shell is a source of lime 3. Shell dust is used in poultry feed as source of calcium

(continued)

Table 7.10 (continued)

Brackish water resource	Taxonomic position	Economic importance
	Phylum – Mollusca	1. Edible with high demand in South Asian countries
	Class – Bivalvia	2. Shell is a source of lime
	Order – Pterioida	3. Shell dust is used in poultry feed as source of calcium
	Family – Ostreidae	
	Genus – *Crassostrea*	
	Species – *madrasensis*	

Crassostrea madrasensis

brackish water livelihood programmes to the local people.

In Bangladesh the Subarnabad, villagers' ability to access IDEAL and become an NGO beneficiary was found to be an important factor in mediating a person's ability to access adaptation measures and adopt new livelihood activities. Through the NGO, villagers were provided with training and technical support to embark upon new livelihood strategies as well as access to loans and a savings bank. The initiatives promoted in Subarnabad for income generation and food production included livelihood activities such as goat and fowl rearing, crab fattening, tree planting, halophytic (saltwater-tolerant) vegetable gardens and handicraft production (Table 7.11).

The most widely adopted strategy was household gardening with saline-tolerant crops. Goat raising and chicken and egg production were also popular, and duck raising and crab fattening were slightly less widely adopted. The success of these livelihood strategies lies in the minimal need for land requirements, as in animal rearing or crab farming, or in the use of salt-tolerant plants for both consumption and marketing for food products and handicrafts. Practices were adopted with technical and financial assistance from IDEAL and were maintained when they proved to be successful and self-sustaining. The independent adoption of innovations by those who were not direct beneficiaries of a project indicated the applicability or utility of adaptive development initiatives.

Table 7.11 NGO (IDEAL)-aided adaptation strategies

Adaptation strategy	Description
Goat rearing	Goats are purchased and used for milk, meat and as a type of savings bank where, in times of need, they can be sold
Chicken farming	These consisted of small chicken farms (70–100 chickens). Chicks are purchased at the local market, raised and sold for a profit when fully grown. In these farms, chickens are rarely used as a source of food for the farmer
Crab fattening	Crab farming involves the collection, rearing and feeding of crabs for 15 days to increase their market value. They are sold for profit in the local market
Tree planting	Homestead planting of saline-tolerant fruit and timber trees (e.g. kewra, guava) for longer-term income and fuel generation as well as a source of food
Saline-tolerant vegetable gardens	The promotion of saline-tolerant crops such as chilli and potato for consumption or for sale
Duck rearing	Ducks are raised for meat and eggs, for consumption or for sale
Hen egg production	Hens are raised for eggs, for consumption or for sale
Handicrafts	The production and marketing of mele mats and stools made out of grasses

IDEAL's strategies were found to meet immediate needs for food and income as well as enhance householders' capacity to address other stresses by improving their financial assets.

7.3.2   Levels of Adaptation

Adaptation takes place at all levels, from international to national or regional levels and even the orientation made by local communities and individuals. The development of adaptation strategies needs to recognize this and define the appropriate mix of actions at these different levels. It can be planned, where pre-meditated decisions that reflect an awareness of impacts are made, or it can be autonomous, where people or natural systems adjust to climate impacts without conscious planning decisions.

Understanding the autonomous responses is particularly important in defining the best approach to adaptation, as in many cases, they will significantly change our expectations of what will happen in the future. They also represent major policy opportunities that must not be neglected, as policies such as stimuli to markets or the dissemination of technology opportunities can be more effective, less expensive and far less demanding on limited institutional capabilities than approaches that solely rely upon planned interventions.

Holling (2001) introduced the idea of the adaptive cycle, which links different time and spatial frameworks within which adaptation should take place. Holling identified three core characteristics that shape the cycle and can therefore shape the responses of ecosystems and people to crisis. These properties are:

- The *inherent potential* of a system that is available for change. This defines the range of possible options for the future and can be thought of as the inherent 'wealth' of the system.
- The *internal controllability* of the system, which reflects the degree of connectedness between internal controlling variables and processes, along with the degree of rigidity or flexibility of these controls. According to Holling, this property determines the degree to which a system can control its own destiny.
- The *adaptive capacity*, which is the resilience of the system to unpredictable shocks. Holling pointed this property as the opposite of the vulnerability of the system.

Adaptation strategies should be based on these three general properties – wealth, controllability and adaptive capacity – as they relate to different scales and contexts. They should include local actions taken by the poor themselves in response to changing market or environmental conditions supported by larger-scale, planned responses by government or other institutions that provided adaptation measures that are beyond the control or capabilities of local communities. The need for and scale of adaptation reflects the vulnerability of people and natural systems to disruption from changes that reflect the impacts of climate conditions. *Vulnerability* is a term that is used in many different ways, usually describing a condition of susceptibility shaped by exposure, sensitivity and resilience (Kasperson et al. 1996). For poor people, vulnerability is both a condition and a determinant of poverty and refers to the (in) ability of people to avoid, cope with or recover from the harmful impacts of factors that disrupt their lives and that are beyond their immediate control. This includes the impacts of shocks (sudden changes such as natural hazards, war or collapsing market prices) and trends (e.g. gradual environmental degradation, oppressive political systems or deteriorating terms of trade). In relation to climate change, vulnerability relates to direct effects such as more storms, lower rainfall or sea-level rises that lead to displacement, and to indirect effects such as lower productivity from changing ecosystems or disruption to economic systems. With the poor being more directly dependent on ecosystem services and products for their livelihoods, the vulnerability of natural systems has profound implications. Any consideration of the need for adaptation to help poor communities to adjust to the effects of climate change must take account of all of these different forms of vulnerability. Of course, exactly how climate change impacts will affect different people in different places is largely unknown – one of the many uncertainties that surround the climate change debate. This is because of the uncertainties inherent in specifying these impacts and because the vulnerability of people will be affected by many things beyond climate change.

Central to the understanding of vulnerability is the concept of *resilience*. The resilience of poor people represents their ability to withstand the impact of the trends and shocks (due to change of environment), absorbing them while maintaining function (Folke et al. 2002). Resilience varies greatly from household to household even in one locality. It is determined by two characteristics of peoples' livelihoods: the assets they possess and the services provided by external infrastructure and institutions. Both the assets and the services are extremely broad in their scope.

Assets include the amount and quality of knowledge and labour available to the household, the physical and financial capital they possess, their social relations and their access to natural resources. External services encompass services like flood control, coastal protection and other infrastructure, transport and communications, access to credit and financial systems, access to markets, emergency relief systems and others. For many poor people in developing countries, access to these external services is extremely limited, so that their resilience is in large part a reflection of the local asset base. Strategies to strengthen the resilience of communities, and especially poor communities, should be based on the most effective combination of measures to secure and enhance the community's asset base and measures to provide improved external services. What is the best balance in any one place needs to be determined through effective assessments of local needs and capabilities. Girot (2002) quotes Folke et al. (2002) to identify three defining features of resilience in integrated human-ecological systems:

1. The amount of disturbance a system can absorb and still remain within the state of domain of attraction
2. The degree to which the system is capable of self-organization versus the lack of organization or organization forces by external factors
3. The degree to which the system can build and increase the capacity for learning and adaptation

Taken together, the reduction of vulnerabilities and the improvement of resilience of poor people to withstand the impacts of climate change will improve their *security*: that is, the extent to which they can spend their lives and conduct their livelihoods free from threats. These threats have many dimensions and characters. In countries like India, climate-related disasters are a major threat occurring almost in every year (Table 7.12).

These natural calamities are direct threats to the security of the poor people, which not only result in the displacement of people from their root but also pose adverse impact on their socio-economic conditions and livelihoods. The restoration phase can be initiated by strengthening the call for an adaptive management style that focuses on transparency and learning. Such an approach needs to involve all stakeholders in decision making and implementation at the level of landscapes and seascapes. Coalitions, including governments and their agencies, NGOs, local communities and research institutions, can support immediate actions, plan for the medium term and establish key priorities for the longer term. Whether constituted at the regional, national or international level, these coalitions should aim to bring about change in environmental management strategy to accelerate the process of adaptation of ecosystems and their components to oscillating climate of the planet Earth.

Fact Below the Carpet

For poor people and underdeveloped countries dealing with many urgent needs and many immediate problems that demand attention and investment, it is necessary to offer a process for identifying those 'win-win' options that address current realities and assist with long-term adaptation to climate change. This process can be based on three general steps: (1) understanding vulnerability – livelihood interactions; (2) establishing the legal, policy and institutional framework through which adaptation measures can be implemented; and (3) developing a national climate change adaptation strategy, including reform measures and investment options. Since livelihood is embedded in the political matrix of the nation, politicians should be sensitized with the issues, effects, mitigation and adaptation to climate change.

Table 7.12 Chronological report of natural calamities (extreme weather events) in India in recent times

Year	Natural calamity	Adverse impact
2001	Earthquake in Gujarat	Death of more than 15,000 people occurred
2001	Floods in Orissa	About 200 people died
2001	Drought in most part of North India and South India	–
2002	Floods, torrential rains and landslides in Assam, Bihar, Meghalaya, Tripura and Arunachal Pradesh	More than 10 million people were affected; death of about 300 people occurred; more than 20,000 houses were damaged with huge loss of public properties (estimated loss was around INR 1.5 billion)
2002	Severe drought and heat wave in Rajasthan, Gujarat and Madhya Pradesh	About 250 people died
2003	Heat wave in Uttar Pradesh, Haryana, Punjab, Rajasthan, Gujarat, Bihar, Orissa and Andhra Pradesh from 14 May 2003 to 06 June 2003	Approximately 1,392 people died
2003	Floods and landslides caused by monsoon rains in Assam, Bihar, West Bengal and Orissa from 13 June 2003 to 09 October 2003	Death of nearly 650 people occurred and 10,00,000 people became homeless
2003	Floods and landslides caused by monsoon rains in Uttar Pradesh, Himachal Pradesh, Rajasthan and Gujarat from 03 July 2003 to 11 September 2003	Approximately 373 people died and 9,00,000 people became homeless
2003	Severe floods in Orissa from 27 August 2003 to 10 October 2003	1,400 villages were flooded; 360 people died; 12,00,000 people became homeless; estimated damage was around USD 169 million
2003	Severe floods in North India and Bihar from 27 August 2003 to 10 September 2003	Death of about 367 people occurred; 4,00,000 people became homeless
2003	Cyclonic storms occurred in coastal Orissa from 15 December 2003 to 16 December 2003 with wind speed around 120 km/h	4,403 houses were destroyed, 13,000 houses were severely damaged; death of 45 people occurred; total estimated damage was around USD 28 million
2003	Severe drought conditions in Northwest, major parts of North India, Northeast India and parts of Andhra Pradesh (Telangana and Rayalaseema region), parts of Tamil Nadu	Crops of estimated amount of USD 25 million were destroyed; many starvation deaths were reported
2004	Severe floods in Bihar, Assam, parts of Haryana, and Punjab and parts of Gujarat	Death of 1,000 persons were reported
2004	Severe drought in Gujarat, Rajasthan, parts of Haryana, Punjab, Uttar Pradesh, Andhra Pradesh with minor heat waves in parts of Rajasthan	Death of about 50 persons occurred
2005	Heavy monsoon in Mumbai	Around 1,000 persons died
2008	Floods around Kosi River, Bihar (purely anthropogenic cause)	2.3 million people were affected in the northern Bihar; 1,000 people died

Important References

Aggarwal D, Lal M (2001) Vulnerability of Indian coastline to sea level rise. Published by Centre for Atmospheric Sciences, Indian Institute of Technology, New Delhi. www.survus.mdx.ae.uk/pdfs/3dikshas.pdf

Alexander D (1993) Natural disasters. UCL Press, London, 631pp

Ali A (1999) Climate change impacts and adaptation assessment in Bangladesh. Climate Res 12:109–116

Boesch cited in Bette Hileman (2000) Consequences of climate change. Chemical and Engineering News, 27 Mar 2000, pp 18–19

CARE Bangladesh (2006) The reducing vulnerability to climate change (RVCC) project. Report, Final September

Deb AK (1998) Fake blue revolution: environmental and socio-economic impacts of shrimp culture in the coastal areas of Bangladesh. Ocean Coast Manag 41(1):63–88

Engelman R (2009) State of the world 2009: into a warming world. W. W. Norton & Company, New York, www.worldwatch.org

Folke C, Carpenter SR, Elmqvist T (2002) Resilience and sustainable development: building adaptive capacity in a world of transformation. Scientific background paper on resilience for the process of the World Summit on Sustainable Development on behalf of The Environmental Advisory Council to the Swedish Government. Also in Ambio 31:437–440

Franck T, Nicol R, Jaemin, S (2006) Network analysis of the intergovernmental panel on climate change: an examination of team structure, network architecture, and other major influences of the IPCC's Third Assessment Report, Massachusetts Institute of Technology, Cambridge, MA

Girot P (2002) Scaling up: resilience to hazards and the importance of cross-scale linkages. Paper presented at the UNDP expert group meeting on integrating disaster reduction and adaptation to climate change, Havana, 17–19 June 2002

Gouda R, Panigrahy RC (1999) Phytoplankton in Indian estuaries. J Indian Ocean Stud 7(1):74–82

Grainger RJR, Garcia SM (1996) Chronicles of marine fishery landings (1950–1994)-Trend analysis and fisheries potential. FAO fisheries technical paper, no. 359, FAO, Rome, 51pp

Hazra S, Mukherjee M (2007) Coastal disaster management and fisher folk. In: Dr. Madhumita M (ed) Sunderban wetlands. Department of Fisheries, Aquaculture, Aquatic resources and Fishing Harbour, Govt. of West Bengal, Kolkata

Hazra S, Ghosh T, Dasgupta R, Sen G (2002) Sea level and associated changes in Sundarbans. Sci Cult 68:309–321

Holling CS (2001) Understanding the complexity of economic, ecological and social systems. Ecosystems 4:390–405

Huq S, Rahman A, Konate M, Sokona Y, Reid H (2003) Mainstreaming adaptation to climate change in least developed countries. International Institute for Environment and Development Climate Change Programme, London

IPCC (2007) Climate change: impact, adaptation and vulnerability. Contribution of working group II to the fourth assessment report of the Intergovernmental Panel on Climate Change. Cambridge University Press, Cambridge

James EN, Yohe G, Nicholls R, Manion M (2000) Sea level rise and global climate change: a review of impacts to U.S. coasts. Pew Center on Global Climate Change, Arlington; Gaffin, op. cit. note 58

Kasperson JX, Kasperson RE, Turner BL (1996) Regions at risk: comparisons of threatened environments. United Nations University Press, New York

Khan SR (2003) Adaptation, sustainable development and equity: the case of Pakistan. In: Smith JB, Klein RJT, Huq S (eds) Climate change, adaptive capacity and development. Imperial College Press, London, pp 285–316

Klein RJT, Smith JB (2003) Enhancing the capacity of developing countries to adapt to climate change: a policy relevant research agenda. In: Smith JB, Klein RJT, Huq

S (eds) Climate change, adaptive capacity and development. Imperial College Press, London, pp 317–334

Lal M, Aggarwal D (2000) Climate change and its impacts in India. Asia-Pacific Jr. Environment & Development (Communicated)

Li C, Guanrong C (2003) Network connection strengths: another power-law? Paper, City University of Hong Kong. http://arxiv.org/abs/cond-mat/0311333 on 24 Apr 2006

McLachlan SM (2003) Export-oriented shrimping, rural people, and the environment in Bangladesh: good, bad or simply ugly? In: Rahman M (ed) Globalization, environmental crisis and social change in Bangladesh. The University Press Limited, Dhaka, pp 309–329

Mirza MMQ (2002) Global warming and changes in the probability of occurrence of floods in Bangladesh and implications. Glob Environ Chang 12(2):127–138

Mitra A, Banerjee K (2005) In: Banerjee CSR (ed) Living resources of the seas: focus Sundarbans. WWF-India, Canning field office, 24 Parganas(S), W.B. 96 pp

Mohanty M (1990) In: Victor Rajamanickam G (ed) Sea level rise: background, global concern and implications for Orissa coast, India in sea level variation and its impact on coastal environment. Tamil University Press, Thanjavur. Heat wave in Orissa: a study based on heat indices and synoptic features

Mohanty M, Ray SB (1987) Some aspects of Geology and management of the Mahanadi river deltaic – complex, East coast of India. In: Abstract international symposium on coastal lowlands: geology and geotechnology, The Hague, 126pp

Moody J (2004) The structure of a social science collaboration network: disciplinary cohesion from 1963 to 1999. Am Sociol Rev 69:213–238

Newman MEJ (2001a) Scientific collaboration networks: I. Network construction and fundamental results. Phys Rev E 64:016131

Newman MEJ (2001b) The structure of scientific collaboration networks. Proc Nat Acad Sci 98:404–409

Newman MEJ (2003) The structure and function of complex networks. SIREV 45(2):167–256

Newman MEJ (2004) Co-authorship networks and patterns of scientific collaboration. Proc Nat Acad Sci 101:5200–5205

Schipper L, Pelling M (2006) Disaster risk, climate change and international development: scope for, and challenges to, integration. Disasters 30(1):19–38

Smit B, Wandel J (2006) Vulnerability, adaptation and adaptive capacity. Glob Environ Chang 16(3):282–292

Subramanian V (2000) Water: quantity-quality perspective in South Asia. Kingston International Publishers, Surrey, p 49

TERI (1996) The economic impact of one metre sea level rise on Indian coastline-methods and case studies, Report submitted to the Ford Foundation

UNEP (1989) Criteria for assessing vulnerability to sea level rise: a global inventory to high risk area. Delft Hydraulics, Delft, 51pp

UNEP (United Nations Environment Programme) (2002) Bangladesh state of the environment report 2001. United Nations Development Programme, Dhaka

UNEP-WCMC (2006) In the front line: shoreline protection and other ecosystem services from mangroves and coral reefs. UNEP-WCMC, Cambridge, 33pp

World Bank, World Development Report (1999/2000) (New York: Oxford University Press, 2000) pp 100; population from United Nations, op. cit. note 26; Sanghai population from United Nations, World urbanization prospects: The 2003 revision (New York: 2004); Sanghai from Stuart R. Gaffin, High water blues: impact of sea level rise on selected coasts and islands (Washington, DC: Environmental Defense Fund, 1997), 27pp

World Bank 2004 Data by Country. www.worldbank.org/data/countrydata/countrydata.html

Internet References

http://www.the_world_factbook.html
http://country_profiles.pdf
http://www.coastalpopulation.html
www.survus.mdx.ae.uk/pdfs/3dikshas.pdf
http://cyberjournalist.org.in/census_cenindia.html

Brackish-Water Aquaculture: A New Horizon in Climate Change Matrix

8

8.1 Brackish-Water Aquaculture: An Overview

Brackish-water aquaculture has become an important source of seaweed, shellfish and fin-fish, especially for human food and production, which is likely to expand well in the next century if sea-level rise maintains its present pace. It has both direct and indirect impacts on biodiversity through the consumption of natural resources and the production of wastes. Most of the brackish-water aquaculture (particularly the shrimp farms) has developed in the mangrove ecosystem as the water has congenial parameters and tidal actions.

There is no doubt that world population is increasing faster than global food supply. While 99 % of food comes from terrestrial agriculture (Pimentel and Giampietro 1994), this disguises the fact that in many, especially in developing countries, the bulk of animal protein comes from fish and other aquatic products. Aquatic foods have until recently been derived almost exclusively from capture fisheries sources. In recent years, however, aquaculture has been playing an increasingly important role. Brackish-water aquaculture can be defined as the farming in the estuarine, coastal, delta mouth and similar brackish-water environment. A large and increasing range of plants and animals is being cultured in diluted seawater. Unlike inland water aquaculture, brackish-water aquaculture involves the culture of plants, invertebrates not only for food but also for decoration (shells and pearls) and chemicals (alginates). Systems and methods for the most

commonly grown tropical and temperate species are summarized in Table 8.1. The terms 'intensive', 'semi-intensive' and 'extensive' are used here with regard to inputs of foods. In extensive aquaculture, the farmed organism is reliant on the environment for food or nutrients while in semi-intensive farming natural food is supplemented with additions of fertilizer and/or food, the latter usually being derived from agricultural by-products such as animal manures and rice bran. In intensive aquaculture, all, or almost all, of the nutritional requirements are supplied by the farmer and diets are largely fish meal based. There is also a correlation between intensity of production, as defined here, and energy consumption.

The brackish-water aquaculture is an economic activity that transforms natural resources through inputs of capital and labour into products valued by society. In so doing, wastes are inevitably produced. The impact of aquaculture on the environment and on biodiversity thus arises from these three processes: the consumption of resources, the aquaculture process itself and the production of wastes (Beveridge et al. 1994).

Statistics produced by the FAO show that world aquaculture production is currently around 25 million tonnes (FAO 1996), equivalent to 20 % of world fisheries (capture + culture) production by weight and around twice this by value. Production from the marine environment accounts for around 51 % of aquaculture production by weight (53 % by value) and is growing by some 5 % per annum. While only 4 % of farmed fish production comes from the sea, all farmed

A. Mitra, *Sensitivity of Mangrove Ecosystem to Changing Climate*,
DOI 10.1007/978-81-322-1509-7_8, © Springer India 2013

Table 8.1 Summary of the principle rearing systems and methods employed in tropical and temperate mariculture

Group	Species	System	Method
Tropical			
Macroalgae	*Laminaria japonica*	Beds	Extensive
	Undaria pinnatifida	Stake-and-line	
	Porphyra tenera	Rafts	
	Eucheuma spp.		
Molluscs	*Crassostrea* spp.	Suspended (rafts, longlines)	Extensive
	Mytilus spp.		
	Pecten yessoensis		
	Venerupis japonica		
	Solen spp.		
Crustaceans	*Scylla serrata*	Ponds	Semi-intensive
	Penaeus spp.		Intensive
Finfish	*Chanos chanos*	Land-based (ponds)	Intensive
	Mugil spp.	and water-based (cages)	
	Epinephelus spp.		
	Serranidae		
	Pagrus major		
	Seriola quinqueradiata		
Temperate			
Macroalgae	*Gracilaria*	Beds, rafts	Extensive
Molluscs	*Ostrea edulis*	Bottom (tressels, trays) and	Extensive
	Crassostrea spp.	suspended (rafts, longlines)	
	Mytilus spp.		
	Tapes spp.		
Finfish	*Salmo salar*	Water-based (cages) and	Intensive
	Oncorhynchus spp.	land-based (tanks)	
	Dicentrarchus labrax		

macroalgae, almost all farmed molluscs and more than 90 % of farmed crustaceans are produced in the marine and brackish-water environment. The fastest-growing sectors of mariculture are in high market value products such as shrimp and fish, production of the former having doubled over the past 5 years. By contrast, farmed production of aquatic plants and molluscs has exhibited a slow and gradual increase.

Mariculture and brackish-water aquaculture practices, which are essentially cage or pond cultures, have an impact on the carrying capacity of the coastal marine environment. The impacts are severalfold and not fully appreciated. Eutrophication is a major issue. In general from the fish feed, 85 % of the phosphorus (P), 80–88 % carbon (C) and 52–95 % of nitrogen (N) are lost to the environment (Wu 1995). Nearly 53 % of P, 23 % C and 21 % N end up in the sediment.

Specifically in the shrimp culture, such as those on the east coast of India, which are run on Thailand design, 24 % N and 13 % P were incorporated and the rest is exported to the environment (Briggs and Funge-Smith 1994). Increasing the stocking densities for a bigger harvest increased their nutrient level but not their relative ratios (Briggs and Funge-Smith 1994). The effluents from the farm contribute 35 % N and 10 % P. Additionally the wastage, fish faeces and excretion and vitamins contribute to the nutrient loading which will induce a typical algal bloom. Although not all the blooms are toxigenic, their sheer mass and degradation of organic matter make excessive demands on oxygen demands leading to anoxic conditions and fish mortalities.

Washings from the cages, chemicals such as the therapeutants, antifoulants and oil leakages pollute the environment with serious impacts.

Recently induced Chinese carp is taking over and has become a pest because it eats anything, everything and any quantity. Some of the desirable species may be suppressed and replaced by undesirable weed species. Being a monoculture, the shrimp ponds are most vulnerable to viral and fungal diseases that could cross contaminate other shrimp farms. Bacterial and fungal diseases specifically the white spot disease (WSD) and brown spot disease (BSD) are the most common and resulted in a heavy loss. Efforts to combat the shrimp diseases include naturopathy, chemotherapy and usage of antibiotics. Only 10 % of the production costs were spent on disease treatment. Both the European Economic Council and the US Food and Drug Administration have banned the import of farmed shrimp that have been exposed to antibiotics. The shrimp farmers lost business and new opportunities from countries like India, Bangladesh and Pakistan.

8.1.1 Reasons for Collapse in India

Lack of a holistic scientific approach, i.e. site-specific evaluation of the carrying capacity without risking the environment equilibrium, led to the collapse of shrimp farming in Indian subcontinent. The various factors that contributed to this are as follows: (1) diseases, (2) no holistic scientific approach, (3) no constructional management – cross-contamination of ponds, (4) nonfunctional drainage ponds, (5) overstocking, (6) crowding too many farms on a small water source, (7) did not dry the ponds before filling with water, (8) development of obnoxious algal blooms, (9) lack of cost–benefit analysis in environmental crises and (10) lack of treatment ponds preferably biotreatment. There were no extension services or arrangements for maintaining brackish-water aquaculture systems. When the shrimp farming was modest around 1980, during the first 5 years, there were no diseases. Incidence of the shrimp diseases was acute since 1993. The maxim 'Share the water share the disease' (Stewart 1998) is also valid in several Indian shrimp farms. About twenty viral shrimp diseases exist, of which six or seven cause catastrophic outbreaks.

8.1.2 Indian Supreme Court Judgement: Consequences

Agitation by affected shrimp fishermen, taken up by environmentalists in Indian maritime states like Tamil Nadu, Andhra Pradesh and Orissa, culminated in a few interim orders by the Indian Supreme Court on 11 December 1996. These include the following: (a) no shrimp farm within 500 m from high tide mark, (b) traditional and improved farmers are exempted, (c) demolish any farm within 500-m to 1,000-m zone, (d) a government authority will enforce a preparatory principles and the polluter pays principle, (e) fix compensation for environmental damage and (f) retrenched workers shall be compensated. Consequences of this judgement are far reaching: (a) loss of revenue, (b) bankruptcies, (c) unemployment, (d) loss of consumer confidence, (e) environmental deterioration and (f) loss of traditional rice farms.

Long-term observations and researches on the failure of brackish-water aquaculture in the Indian subcontinent demonstrate that although climate-changed-induced sea-level rise could result in rise in aquatic salinity in the adjacent estuaries and bays, but without scientific foundation, the blue revolution can never see the path of success.

8.1.3 Remedial Measures

In the scenario of changing climate, it is extremely necessary to give more thrust on brackish-water aquaculture as it can generate several saline-water-based livelihood options for the coastal population. The success of brackish-water culture, however, is dependent on scientific approaches to manage water quality, nutrition of the culturable species and feed management and preserve the surrounding mangrove ecosystem as the vegetation serves as the natural feed factory for the finfish, shellfish and seaweed species that may be the candidate organisms for generating alternative livelihood. Mangrove trees give protection to the aquaculture farm by way of reducing the intensity of storms, coastal surges and

Table 8.2 Value of erosion control for the present cover of mangroves

Regions	Total present area under mangroves (ha)[a]	Value of coastline protection per ha per annum (in Rs.)[b]	Value (in $US)	Total value of coastline protection per ha (in Rs.)[b]	Value (in $US)
Kachh and Jamnagar	84,900	137,606	2,867.791667	11,682,749,400	243,390,612.5
Sourashtra	1,700	137,606	2,867.791667	233,930,200	4,873,545.833
Gulf of Khambhat	3,000	137,606	2,867.791667	412,818,000	8,600,375
South Gujarat	1,500	137,606	2,867.791667	206,409,000	4,300,187.5
Total	91,100	137,606	2,867.791667	12,535,906,600	261,164,720.8

[a]*Source*: Singh (2000)
[b]After Ong et al. (1995)

Table 8.3 Value of erosion control for the potential area of mangroves

Regions	Total potential area for mangroves in the region (ha)[a]	Value of coastline protection per ha per annum (in Rs.)[b]	Value (in $US)	Total value of coastline protection per ha (in Rs.)[b]	Value (in $US)
Kachh and Jamnagar	46,707	137,606	2,867	6,427,163,442	133,899,238
Sourashtra	2,039	137,606	2,867	280,578,634	5,845,388
Gulf of Khambhat	10,606	137,606	2,867	1,459,449,236	30,405,192
South Gujarat	4,363	137,606	2,867	600,374,978	12,507,812
Total	63,715	137,606	2,867	8,767,566,290	182,657,631

[a]*Source*: Singh (2000)
[b]After Ong et al. (1995)

tidal actions that are common features of coastal zone. The vegetation also acts as powerful agent of shoreline stabilization and erosion control. A study conducted by Gujarat Ecological Commission (2004) reveals the value of erosion control by mangroves, which is not considered under the banner of direct benefit of mangroves (Tables 8.2 and 8.3).

Mangal ecotourism is another roadmap to revive the lost brackish-water aquaculture glory of India as it develops among the local population the sense biodiversity, the value of species and their conservation. Basically mangal ecotourism can be an instrument for elimination of poverty, ending unemployment, creating new skills and encouraging tribal and local crafts and cultures that are related to the endemic plants and animals. This will magnify the sense of awareness among the local mangrove-based population, who can be the part of symbiotic movement of brackish-water aquaculture and mangrove conservation.

Tobias and Mendelsohn (1991) used the travel costs method (TCM) to estimate the value of Monteverde Cloud Forest Reserve in Costa Rica for ecotourism. Using the TCM, Tobias and Mendelsohn (1991) estimated a $35 per visitor value for recreation at a 10,000 ha Costa Rican Tropical Forest Reserve. They included only Costa Rican visitors for their study. Costanza (1989) used two methods to calculate the value of coastal wetlands recreation in the United States. Using the TCM, they estimated the value at $70.67 per visitor. Using the CVM (Contingency Valuation Method), they estimated a value of $47.11 per visitor. Some studies with estimated value of mangrove forests in terms of ecotourism are presented in Table 8.4.

Apart from the mangal ecotourism, most of the common people (preferably the island dwellers and coastal population, who are the primary stakeholders of mangrove forest) are not aware of the cost of carbon sequestration by mangrove trees (the readers may look at Annexure 2A.1 of

Table 8.4 Economic value of ecotourism in mangrove forests as per previous studies

Study	Year	Area	Value (Rs. per hectare per annum)	Value ($US per hectare per annum)
Bennet and Reynolds	1993	Sarawak	19,652	409.41
Costanza	1997	Global	22,990	478.95
Leong	1999	Kuala Selangor, Malaysia	42,874	893.20

Chap. 2 of this book for the technical know-how of estimating carbon stored in mangrove trees), although this can also bring financial and ecological benefit to the stakeholders, which may develop their sense of conservation leading to a successful brackish-water aquaculture. Gujarat Ecological Foundation (2004) presented some figures on the valuation of stored carbon in mangroves (Tables 8.5 and 8.6).

8.2 Shrimp Culture

Shrimp aquaculture is an important economic activity in the coastal areas of many tropical and subtropical countries and offers opportunities to contribute to poverty alleviation, employment, community development and foreign exchange income generation. The global production of farmed finfish, crustaceans and molluscs reached over 35 million metric tonnes in 2000, while production of the three main shrimp species (*Penaeus monodon*, *P. chinensis* and *P. vannamei*) was only 2.6 % of the total quantity; they contributed to over 12 % of the value (Table 8.7), with *P. monodon* ranked highest.

Development of coastal aquaculture and shrimp farming in particular has generated considerable international debate in recent years over the actual environmental and social costs and benefits. The major environmental issues have been summarized by the 'Shrimp Farming and Environment' Consortium (World Bank/NACA/WWF/FAO 2002) as the following:

• Ecological consequences of conversion and changes in natural habitats, such as mangroves, associated with construction of shrimp ponds and other infrastructures

• Discharge of pond effluent leading to water pollution in farming and coastal areas

• Seepage and discharge of saline pond water that may cause salinity changes in ground water and surrounding agricultural land

• Use of fish meal and fish oil in shrimp diets

• Improper use of chemicals raising health and environmental concerns

• Spread of shrimp diseases

• Transboundary movements concerning the spread of genetic materials, exotic species and diseases

• Biodiversity issues primarily arising from the collection of wild seed

The degree of interaction between shrimp aquaculture and the environment depends on a wide variety of interrelated factors such as the species being cultured, the type, scale and intensity of culture practices used and the location of the farm. There has been considerable interest in this subject over the last 10 years, and at a general level, there has been substantial progress in both understanding these interactions and implementing measures to reduce their impacts. The most effective approaches have included the following:

1. Introduction, adoption and implementation of codes of practice by the industry
2. Use of environmental impact assessment techniques, both at site-specific and cumulative levels
3. Increasing use of certification to standardize sustainable farming methods and products
4. National management strategies for controlling the import and use of exotic organisms
5. Integration of shrimp farm planning into overall coastal land use zoning and management

The above discussion leads to the fact that although shrimp industry has several dark chapters, it is one of the most profitable and fastest-growing segments of the aquaculture industry. Global farmed shrimp production has grown hundred-fold (by weight) in less than two decades, from under 10,000 metric tons (MT) produced by fewer than a dozen countries in the early 1970s to over one million MT (MMT) by the late 1990s.

Table 8.5 Value of carbon sequestration for the present cover of mangroves

Regions	Total present area under mangroves (ha)[a]	Value of carbon sequestration in biomass of mangroves per ha per annum (in Rs.)[b]	Value (in $US)	Value of carbon sequestration in mangrove sediment per ha per annum (in Rs.)[c]	Value (in $US)	Total value of carbon sequestration per ha per annum (in Rs.)[b]	Value (in $US)
Kachh and Jamnagar	84,900	4,214,508,165	87,802,253.43	885,400,875	18,445,851.56	5,099,909,040	106,248,105
Sourashtra	1,700	84,389,445	1,758,113.43	17,728,875	369,351.56	102,118,320	2,127,465
Gulf of Khambhat	3,000	148,922,550	3,102,553.12	31,286,250	651,796.87	180,208,800	3,754,350
South Gujarat	1,500	74,461,275	1,551,276.56	15,643,125	325,898.43	90,104,400	1,877,175
Total	91,100	4,522,281,435	94,214,196.56	950,059,125	19,792,898.43	5,472,340,560	114,007,095

[a]*Source*: Singh (2000)
[b]After Ong et al. (1995)
[c]*Source*: After Ong (1993)

Table 8.6 Value of carbon sequestration for the potential area of mangroves

Regions	Total potential area under mangroves (ha)[a]	Value of carbon sequestration in biomass of mangroves per ha per annum (in Rs.)[b]	Value (in $US)	Value of carbon sequestration in mangrove sediment per ha per annum (in Rs.)[c]	Value (in $US)	Total value of carbon sequestration per ha per annum (in Rs.)[b]	Value (in $US)
Kachh and Jamnagar	46,707	2,318,575,181	48,303,649	487,095,626	10,147,825	2,805,670,807	58,451,475
Sourashtra	2,039	101,217,693	2,108,701	21,264,221	443,004	122,481,914	2,551,706
Gulf of Khambhat	10,606	526,490,855	10,968,559	110,607,323	2,304,319	637,098,177	13,272,878
South Gujarat	4,363	216,583,028	4,512,146	45,500,636	947,929	262,083,664	5,460,076
Total	63,715	3,162,867,758	65,893,057	664,467,806	13,843,079	3,827,334,564	79,736,136

[a]*Source*: Singh (2000)
[b]After Ong et al. (1995)
[c]*Source*: After Ong (1995)

Table 8.7 Status of shrimp in world aquaculture production (2000)

Species	Quantity		Value	
	MT	%	US $ ('000)	%
Penaeus monodon	571,497	1.6	$4,046,751	8.0
Penaeus chinensis	219,152	0.6	$1,324,969	2.6
Penaeus vannamei	143,737	0.4	$878,324	1.7
Subtotal	**934,386**	**2.6**	***$6,250,044***	***12.3***
Other species	34,650,725	97.4	$44,609,103	87.7
Total	**35,585,111**	**100**	**$50,859,147**	**100**

Source: FAO (2002)

Shrimp farming is currently practiced in over 50 countries worldwide, and the sector has grown at an annual average of over 18.8 % since 1970 (FAO 2001). By contrast, the total catch of shrimp from capture fisheries has grown at a relatively modest rate 3.8 % per year, from just over one MMT to just under three MMT over the same period. Moreover, although shrimp accounted for only 2.6 % of total global aquaculture production by weight in 1999, it represented 12.4 % of total aquaculture production by value, at US$ 6.7 billion (FAO 2001). It is perhaps not surprising therefore that shrimps currently contribute over a quarter of total global shrimp landings and constitute the single most valuable internationally traded aquaculture commodity worldwide (FAO 2000). Tables 8.8 and 8.9 present statistical information concerning global shrimp aquaculture production by species and by country for 1999 according to FAO (2001).

In developing countries like India, the demand for protein is accelerating at a rapid rate. The annual per capita fish consumption in India is only 4 kg against the recommended 31 kg by the Nutritional Advisory Committee on human nutrition (Santhanam et al. 1990). This shows that our protein demand is so great that it is imperative to increase fish production by 'aquaculture' which is the only alternative to capture fisheries. On the basic matrix of Indian scenario and climate-change-induced saline water intrusion in the upstream zone of the estuaries, the importance of aquaculture is highlighted below:

(a) Brackish-water aquaculture can be a dependable year-round source of animal protein unlike agriculture or animal husbandry.

(b) It is certainly a profitable venture than maintaining livestock as the food conversion

Table 8.8 Total world production of farmed shrimp in 1999, by weight

Shrimp species	Production (MT)	Change 1998–1999 (%)
Giant tiger prawn *Penaeus monodon*	575,842	+3.9
Whiteleg shrimp *Penaeus vannamei*	187,224	−5.6
Fleshy prawn *Penaeus chinensis*	171,972	+19.5
Penaeid shrimp *Penaeus* spp. (spp. not given)	95,634	+20.2
Banana prawn *Penaeus merguiensis*	53,109	+7.5
Metapenaeid shrimp *Metapenaeus* spp.	22,421	+1.0
Blue shrimp *Penaeus stylirostris*	12,390	−22.1
Indian white prawn *Penaeus indicus*	7,043	+13.7
Kuruma prawn *Penaeus japonicus*	2,359	−6.6
Southern white shrimp *Penaeus schmitti*	1,364	−21.3
Natantian decapods *Natantia*	904	+175.0
Akiami paste shrimp *Acetes japonicus*	270	+2.3
Redtail prawn *Penaeus penicillatus*	107	−21.9
Palaemonid shrimp, spp. not given	98	−39.9
Total	*1,130,737*	*+5.2*

Source: FAO (2001)

ratios in the fish are 1:2 against 1:3 and 1:4 in pig/chick and cattle, respectively.

(c) Brackish-water aquaculture yield is quite high compared to that of agriculture or livestock. It is estimated that a yield as much as 350 tonnes shell-on edible oysters (17.5 tonnes oyster meat) is possible per ha through coastal aquaculture.

8.2.1   Culturable Shrimp Species

The common prawn species that are abundant in the creeks, inlets and estuarine waters of Indian Sundarbans mangrove ecosystem may be grouped into penaeid and non-penaeid varieties. The penaeid prawn species are selected for aquacultural purposes owing to their bigger size and faster growth rates (Santhanam et al. 1990).

Table 8.9 Total world production of farmed shrimp in 1999 (by country)

Country/territory	Production (MT)	Change 1998–1999 (%)
Thailand	230,000	−9.0
China	170,830	+19.4
Vietnam	131,800	+13.6
Ecuador	119,700	−16.9
Indonesia	119,120	+0.8
India	114,670	+41.4
Bangladesh	81,068	+22.7
Philippines	35,898	−5.0
Mexico	29,120	+22.6
Brazil	16,750	+131.0
Malaysia	12,188	+23.9
Colombo	9,227	+23.6
Honduras	8,000	0
Taiwan Province of China	6,065	+9.3
Venezuela	6,000	0
Nicaragua	4,198	−12.2
Peru	4,005	−12.5
Sri Lanka	3,820	−41.4
Madagascar	3,486	+39.9
Belize	3,163	+92.6
Panama	2,585	−74.3
Costa Rica	2,465	+5.0
Australia	2,444	+75.4
USA	2,098	+4.9
New Caledonia	1,906	+21.5
Iran	1,800	+107.1
Japan	1,726	−13.4
Guatemala	1,403	+1.6
Cuba	1,364	−21.3
Saudi Arabia	1,300	−23.0
S. Korea	1,142	+35.0
Seychelles	227	−65.0
Guyana	162	+35.0
El Salvador	149	−33.2
Spain	138	−25.4
South Africa	120	+34.8
Suriname	105	0
Singapore	82	+24.2
Pakistan	76	+10.1
Cambodia	62	−68.5
Brunei	45	−34.8
Cyprus	43	+72.0
French Polynesia	43	−10.1
Fuji Islands	39	+11.4
Turkey	30	−88.9

(continued)

Table 8.9 (continued)

Country/territory	Production (MT)	Change 1998–1999 (%)
Guam	25	+13.6
Italy	18	−28.0
Solomon Islands	13	0
Myanmar	8	0
Albania	5	−37.5
St. Kitts and Nevis	5	+25.0
Bahamas	1	0
Total	**1,130,737**	**+5.2**

Source: FAO (2001)

The main penaeid prawns of aquacultural interest that occur in the nearshore areas and brackish waters of east and west coasts of India are *Penaeus indicus*, *Penaeus monodon*, *Penaeus semisulcatus*, *Penaeus merguiensis*, *Metapenaeus brevicornis*, *Metapenaeus dobsoni*, *Metapenaeus monoceros* and *Meta-penaeus affinis*. In addition to this, species like *Exopalaemon styliferus*, *Penaeus penicillatus*, *Acetes* sp., *Macrobrachium rude*, *Metapenaeus ensis*, *Parapenaeopsis sculptilis* and *Nemato-palaemon tenuipes* are also available in the aquatic subsystem of Indian Sundarbans. The rich species diversity of prawn and shrimp may be related to the presence of mangrove vegetation throughout the entire sector. It has also been recorded that the maximum diversity of prawn juveniles in the mangrove-dominated zone can be correlated with life cycle pattern of penaeid species. The shrimp juveniles arise in large numbers in the estuaries and mangrove waters at late mysis or post-larval stages. If they escape unhurt from various gear types employed in estuaries, they live for 3–4 months in mangrove waters (Chong 1980). Leh and Sasekumar (1984) reported that 11–62 % plant matters in the diet of penaeid prawns are derived from mangrove origin. In addition to supplying food and acting as nursery ground, mangroves provide physical shelter and protection to juveniles of prawns by reducing the wave energy through obstruction created by pneumatophores, stilt roots, woody stems and branches of mangroves trees (Niyogi et al. 1999).

A brief description on the commercially important penaeid prawn species found in and around the mangroves ecosystem of Indian Sundarbans is given here.

1. Name: *Penaeus indicus*
   Common name: White prawn
   Maximum size: 175 mm
   Identifying features:
   (a) Body creamy white with no bands over the body
   (b) Presence of numerous small brownish, greyish or greenish chromatophores scattered over the carapace and abdomen
   (c) Rostrum slender with distinct double curve
   (d) Rostral teeth formula 6–9 by 4–7
2. Name: *Penaeus monodon*
   Common name: Giant tiger prawn
   Maximum size: 330 mm
   Identifying features:
   (a) Presence of a glossy appearance
   (b) Presence of conspicuous transverse bands on carapace and abdomen, which are brown, blue, red or black
   (c) Rostrum strongly sigmoidal with distinctive double curve
   (d) Rostral teeth formula 7–9 by 2–4
3. Name: *Penaeus semisulcatus*
   Common name: Green prawn
   Maximum size: 250 mm
   Identifying features:
   (a) Presence of dark green with indistinct dark bands on abdomen
   (b) Presence of straight rostrum, short or uniformly curved
   (c) Presence of 5th periopod with exopodite
   (d) Rostral teeth formula is 6–8 by 1–4
4. Name: *Penaeus merguiensis*
   Common name: Banana prawn
   Maximum size: 200 mm
   Identifying features:
   (a) Milky white to yellowish in colour with black or dark brown speckles
   (b) Rostrum very high and assumes a triangular shape
   (c) Rostrum becomes shorter with increase in size
   (d) Rostral teeth formula 6–9 by 4–6
5. Name: *Metapenaeus brevicornis*
   Common name: Yellow prawn
   Maximum size: 130 mm

Identifying features:

(a) Pale yellowish in colour with sparsely distributed brown dots

(b) Rostrum posteriorly elevated

(c) Rostrum straight with a dorsal naked portion

(d) Rostral teeth 5–7

6. Name: *Metapenaeus dobsoni*
Common name: Yellow prawn
Maximum size: 130 mm
Identifying features:

(a) Presence of light cream or deep yellowish body with orange chromatophores

(b) Rostrum not posteriorly elevated

(c) Rostrum naked dorsally and has distinct double curve and crest

(d) Rostral teeth 7–10

7. Name: *Metapenaeus monoceros*
Common name: Ginger prawn
Maximum size: 180 mm
Identifying features:

(a) Body colour grey to green and speckled with dark brown pigments

(b) Rostrum straight, slightly uplifted and knife-like

(c) Periopod without exopodite

(d) Rostral teeth 8–12

8. Name: *Metapenaeus affinis*
Common name: Jinga shrimp
Maximum size: 170 mm
Identifying features:

(a) Body cream coloured or translucent, bluish green often with chromatophores

(b) Rostrum curved

(c) Anterior thecal plate wider posteriorly than anteriorly

(d) Rostral teeth 9–11

Although the scientific names of commercially important penaeid prawn species are stated here, recently some confusion with generic names of shrimp as highlighted in Table 8.10 has become an issue in the aquacultural sector. In the book *Penaeid and Sergestoid Shrimps and Prawns of the World (Keys and Diagnoses for the Families and Genera)*, Dr. Isabel Perez Farfante and Dr. Brian Kensley (1997) proposed some changes in the way scientists refer to the popular farmed shrimp species.

Table 8.10 Suggested name changes of some penaeid prawn species

Major farm-raised species	
Old name	New name
Penaeus vannamei	*Litopenaeus vannamei*
Penaeus stylirostris	*Litopenaeus stylirostris*
Penaeus chinensis	*Fenneropenaeus chinensis*
Penaeus indicus	*Fenneropenaeus indicus*
Penaeus japonicus	*Marsupenaeus japonicus*
Minor farm-raised species	
Old name	New name
Penaeus schmitti	*Litopenaeus schmitti*
Penaeus setiferus	*Litopenaeus setiferus*
Penaeus occidentalis	*Litopenaeus occidentalis*
Penaeus brasiliensis	*Farfantepenaeus brasiliensis*
Penaeus aztecus	*Farfantepenaeus aztecus*
Penaeus californiensis	*Farfantepenaeus californiensis*
Penaeus duorarum	*Farfantepenaeus duorarum*
Penaeus notialis	*Farfantepenaeus notialis*
Penaeus subtilis	*Farfantepenaeus subtilis*
Penaeus paulensis	*Farfantepenaeus paulensis*
Penaeus merguiensis	*Fenneropenaeus merguiensis*
Penaeus penicillatus	*Fenneropenaeus penicillatus*
No name change	
Penaeus monodon, *Penaeus esculentus* and *Penaeus semisulcatus*	

Source: Farfante and Kensley (1997)

A summary of the comparative features among the three main types of culture systems is shown in Table 8.11.

West Bengal is endowed with the highest potential resources of brackish-water aquaculture (27 % of the country's potential) among all the maritime state. The state's share of the saline soil is about 0.08 million hectare (mha) out of 2.10 mha in the country. The total potential culturable area of 48,000 ha of water area has been presently brought under brackish-water farming (Government of West Bengal, April, 2001). The brackish-water fisheries development is high in West Bengal particularly because of the extensive saline soil–water resource, human resource, favourable agro-climatic conditions, productive estuarine ecosystem including the Sundarbans and also abundance of prawn and other brackish-water finfishes.

Table 8.11 Comparison of different methods of coastal aquaculture

Sl. No.	Parameters	Traditional/extensive culture	Semi-intensive culture	Intensive culture
1.	Culture area	Bheries, Pokkali rice fields, mangrove water ways	Ponds	Ponds, cages, pens, raft-rope, etc.
2.	Engineering design and layout	May or may not be well laid-out	With provisions for effective water management	Very well-engineered system with pumps and aerators to control water quality and quantity
		Very big ponds	Manageable-sized units (up to 2 ha each)	Small ponds, usually 0.5–1 ha each
		Ponds may or may not be fully cleaned	Fully cleaned ponds	Fully cleaned ponds
3.	Special installation	No	Vegetation in prawn pond for shelter and food; aerators are often used depending on the stocking density	Paddle splashers or conventional aerators for oxygenation
4.	Seed source	Autostocking by tidal waters	Collected from wild waters, selected and stocked. Hatchery is also another source	Induced breeding for stocking uniform-sized seeds of desired species
5.	Species used	Monoculture or polyculture	Monoculture	Monoculture
6.	Stocking density	Depending on richness in tidal waters	Controlled (usually 5–15/m^2)	Controlled (max. 100/m^2)
7.	Fertilization	No	Medium	High
8.	Food and feeding regimen	Natural food of tidal waters	Natural food of tidal waters + supplementary conventional food	Natural food of tidal waters + supplementary protein–vitamin–mineral-rich pelleted feed
9.	Water quality analysis	No	Analyzed weekly or monthly	Analyzed daily or weekly
10	Pests/diseases	High	Medium	Low
11.	Loss due to predators or poaching	High	Medium	Low
12.	Survival rate	20–30 %	60–70 %	90 %
13.	Yield	1 t/ha/year	5 t/ha/year	20 t/ha/year
14.	Cost of production	Low	Medium	High
15.	Net income	Low	Medium	High
16.	Management	Less	Fair	Intensive

Based on the management practices, namely, traditional and improved traditional/extensive methods of culture are conducted in the state. The traditional shrimp farming practices are also commonly known as '*Bhasa Badha*' fishery. About 80 % of the fisheries in West Bengal belong to the traditional type of farming where large areas are enclosed and tidal water is allowed to enter along with the natural seed of shrimp and fish. No supplementary feed is given. After a considerable period, the fishes and shrimps are harvested. The average productivity of such systems lies between 500 and 900 kg/ha/year of which about 30 % is constituted by prawns/shrimps and 70 % by mullets. These fisheries are mainly located in the districts of North 24-Parganas, South 24-Parganas and Midnapore. With the advent of scientific farming in the early 1970s, the traditional farmers have also resorted to selective stocking with improvement in production levels.

The extensive farming is commonly known as improved traditional farming in the state. This system involves construction of peripheral canals/ponds of size ranging from 1 to 5 ha. Shrimp seeds at the rate of 15,000–20,000/ha are stocked. Water management is done by tidal effect. The average yield is 1,500–1,700 kg/ha, including mullets.

Presently, the average production from the state is about 436 kg/ha (MPEDA 2001). Availability of hatchery-raised seed is still a constraint in West Bengal, which is related to requirement of high aquatic salinity during the hatching process of tiger prawn. Although the state government has restricted harvesting of wild seed, some collection in remote estuarine areas of the state still continues. To meet the growing demand of the shrimp farms, there is an urgent need to set up shrimp hatcheries and facilities for testing of the seed in the state.

8.2.2 Water Quality Management: A Vital Component of Successful Shrimp Culture

Water quality management forms an integral aspect of aquaculture operations. An understanding of the complex interactions continuously taking place between the ecosystem and the stocked organism is essential to enhance the survival and production by appropriate manipulation of the aquatic environment. Where shrimp culture is concerned, any characteristic of water that affects the survival, reproduction, growth, production or management of shrimp in any way is a water quality variable. It is also of critical importance in enhancing seed production rates in hatcheries and nursery systems. It has considerable significance in controlling water pollution problems as well as environmental contamination from metabolites and oxygen depletion in culture systems. A pond with 'good' water quality will produce more and healthier organism than a pond with 'poor' water

quality. Though there are many water quality variables in pond shrimp culture, only a few of these normally play an important role. The main components of water quality that shrimp culturists should concentrate on are water temperature, salinity pH, dissolved oxygen, transparency or turbidity, nutrient concentration and concentration of pollutants.

8.2.2.1 Temperature

Temperature has a pronounced effect on chemical and biological processes. In general, biological processes such as growth and respiration double for every 10 °C increase in temperature. This means that shrimps often will grow twice as fast at 30 °C as at 20 °C, and they will require twice as much oxygen in order to meet the increasing metabolic demand at the greater temperature.

Significance

- *Regulates moulting*: Active moulting takes place as temperature approach 21–22 °C.
- *Controls metabolic rate of prawn*: In general, temperature increases the rate of excretion of waste products by prawn. Beyond 30 °C, the enzyme of tiger prawn is denatured which poses an adverse impact on the metabolic process of the prawn.
- *Controls bacterial growth*: Higher temperature augments bacterial population growth.
- *Influences respiration*: Rate of respiration increases as temperature increases; if temperature exceeds 30 °C, respiratory problem may appear.
- *Influences moulting and hardening time*: Higher temperature decreases moult time and the time required for hardening and expansion of the new shell.

Management Guidelines

Desirable temperature: 20–25 °C; shading with trees is recommended to avoid the scorchy heat of the sun.

8.2.2.2 Salinity

It is the most important factor influencing the metabolism, growth and osmotic behaviour in prawns. The optimum level of salinity is 15–25‰.

Significance
- *Causes mortality*: Shrimps are highly sensitive to sudden changes in salinity. It may cause stress or actual mortality of cultured species. So, shrimp living in water at one concentration of salinity should not suddenly be placed in water with a much higher or lower salinity.

Management Guidelines
In order to maintain the level of salinity, the shrimp culture farm is often connected to saline water source on one hand and freshwater reservoir on the other hand. Again, water may evaporate from the system leaving the sea salt behind. So, water must be periodically added to the system to offset evaporation losses, and the optimum salinity in the culture pond is also maintained by manipulating the sluice gate, which has linkage with both saline and freshwater reservoirs.

8.2.2.3 Transparency or Turbidity
The term turbid indicates that water contains suspended material, which interferes with the passage of light. In shrimp ponds, turbidity, which results from planktonic organism, is a desirable trait, whereas that caused by suspended clay particles is undesirable. If the pond receives run-off, which carries heavy loads of silt and clay, the silt settles over the pond bottom. The clay particles, which remain in suspension, restrict light penetration and limit the growth of plants. A persistent clay turbidity, which restricts visibility into the water to 30 cm or less, may prevent development of plankton blooms.

Significance
Transparent water allows solar radiation to penetrate in the water of the culture pond.

Management Guidelines
The water turbidity is normally maintained by alum, hydrated lime and gyPSUm. These chemicals help to remove the clay turbidity from the pond water.

Remark
Application of alum and gyPSUm lowers the pH of the aquatic medium; therefore, simultaneous application of lime is also done to maintain the optimum pH range.

Table 8.12 Effects of dissolved oxygen concentrations on shrimp

Dissolved oxygen concentration	Effect
Less than 1 or 2 mg/l	Lethal if exposure lasts more than a few hours
2–5 mg/l	Growth will be slow if exposure to low dissolved oxygen is continuous
5 mg/l – saturation	Best condition for good growth
Above saturation	Can be harmful if supersaturated conditions exist throughout pond volume. Normally, there is no problem

8.2.2.4 Dissolved Oxygen
Dissolved oxygen is the most critical water quality variable in shrimp culture. Shrimp farmers need to thoroughly understand factors affecting the concentration of dissolved oxygen in pond water. They also should be around of the influence of low dissolved oxygen concentrations on shrimp.

Significance
- Controls the moulting process of prawn
- Controls filter efficiency
- Controls microaerophilic pathogens

The adverse effects of low dissolved oxygen more often are expressed as reduced growth and greater susceptibility to disease. In ponds with chronically low dissolved oxygen concentration, shrimp will eat less and they will not convert food to flesh as efficiently as in ponds with normal dissolved oxygen concentrations. The influence of dissolved oxygen concentration on shrimp is summarized in Table 8.12.

As light passing through pond water is rapidly reduced with depth and photosynthesis occurs in the surface layer of water, dissolved oxygen concentration declines with depth. In ponds with a lot of plankton, dissolved oxygen concentration may fall to 0 mg/l at depth of 1.5 or 2 m. For this reason, it is best to use relatively shallow ponds (1.0–1.5 m) for shrimp culture. Concentrations are particularly low during periods of cloudy weather. The production of oxygen on a cloudy day is less than on clear or partly cloudy day, so dissolved oxygen concentration does not increase to usual afternoon levels. This results in lower than usual dissolved oxygen concentrations the following morning. Extended periods of cloudy weather may result in dangerously low dissolved oxygen concentration even in ponds with moderately heavy plankton blooms.

In ponds with heavy plankton blooms, scum of algae often forms at the surface. Occasionally, the algae in this scum will suddenly die and their decomposition will result in depletion of dissolved oxygen.

There is also a marked fluctuation in dissolved oxygen concentration during a 24-h period in ponds. Concentrations of DO are lowest in the early morning just after sunrise, increase during daylight hours to a maximum in late afternoon and decrease again during the night. This daily fluctuation of DO in shrimp ponds apparently has little effect on feeding and growth as long as the minimum DO concentration for the day does not drop below 1 or 2 mg/l in the early morning and then rises near saturation within a few hours after sunrise. If DO concentration remains at less than 3 or 4 mg/l for prolonged periods, shrimps cease to feed or grow well.

Management Guidelines
- Application of up to 6 or 8 mg/l of potassium permanganate has been recommended as it is supposed to oxidize organic matter and lower the demand for dissolved oxygen in the pond.
- Application of calcium hydroxide has been recommended to destroy organic matter in ponds with low dissolved oxygen concentration and thereby reduce rates of oxygen consumption by bacteria.
- The only really effective procedure for preventing fish mortality during periods of extremely low dissolved oxygen involves the use of mechanical device for aeration.

8.2.2.5 pH
The optimum pH level in an ideal aqua cultural farm is between 7.5 and 8.5. The acid and alkaline death points for pond shrimp are approximately pH 4 and pH 11, respectively. Even though shrimp may survive, production will be poor in pond with early morning pH values between 4 and 6 and between 9 and 10.

Significance
- The water pH has great impact on the metabolic and physiological processes in prawn.
- As pH drops below 7.0, bacterial (beneficial bacteria also) activity is inhibited. As the pH approaches 8.0, ammonia toxicity increases and bacterial activity (pathogenic) also increases.

The pH of natural water is greatly influenced by the concentration of carbon dioxide, an acidic substance. Phytoplankton and other aquatic vegetations remove carbon dioxide from the water during photosynthesis, so the pH of a body of water rises during the day and decreases during the night, because of accumulation of CO_2 in the pond water as a result of respiration. The daily fluctuation of pH should not exceed 0.45. The pH of the aqua cultural farm also decreases due to abundant accumulation of sediments on the pond bottom, which accelerates the decomposition process by aerobic and anaerobic bacteria producing NH_3 and H_2S.

Management Guidelines
- By using calcareous material, pH of the water will be stabilized below 7.5. Use of shell and dolomite usually maintains a pH of 7.0–7.2.
- As the number of prawn in a system increases, waste increases and water tends to become acidic. Some acid producing compounds are removed by aeration.

8.2.2.6 Carbon Dioxide
Shrimp can survive in waters containing up to 60 mg/l of CO_2, provided dissolved O_2 concentration is high. When dissolved O_2 is low, photosynthesis is not proceeding rapidly and carbon dioxide concentration rises because CO_2 released by respiration is not absorbed by phytoplankton for use in photosynthesis. Because of the relationship of carbon dioxide to respiration and photosynthesis, carbon dioxide concentration usually increases during the night and decreases during the day. Particularly, high concentration of CO_2 occurs in ponds after phytoplankton die off, after destruction of thermal stratification and during cloudy weather.

8.2.2.7 Nitrogen Metabolites

In semi-intensive and intensive prawn culture units, nitrogen is produced in different forms in the culture pond. The main causes of production of nitrogenous compounds are listed below:

- Decomposition of feed load by bacteria
- Excreta of shrimp
- Algal crash (specially sudden crash of cyanobacteria or blue-green algae)
- Reduction of nitrate to nitrite in anaerobic mud of pond

Ammonia

In water, ammonia nitrogen occurs in two forms, unionized ammonia and ammonium ion, in a pH- and temperature-dependent equilibrium:

$$NH_3 + H_2O = NH_4^+ + OH^-$$

Shrimp pond waters usually have pH values around 8, and at this pH, total ammonia nitrogen concentration as high as 10 mg/l probably would not kill shrimp. However, to avoid stress, concentration above 2 mg/l should be avoided. Unionized ammonia may be quite high in the afternoon and negligible during the night.

Significance

- Ammonia is toxic to prawn. If ammonia concentrations increase in the water, ammonia excretion by shrimp diminishes, and levels of ammonia in blood and other tissues increase. The result is an elevation in blood pH and adverse effects on enzyme-catalyzed reactions and membrane stability.
- Ammonia increases oxygen consumption by tissues, damages gills and reduces the ability of blood to transport oxygen.
- Disease susceptibility also increases in fish exposed to sublethal concentration of ammonia.

Management Guidelines

Total ammonia should not exceed 1 mg/l. Digging and subsequent drying of the pond or exposing the pond to air can remove the ammonia. Aeration of pond bottom enhances nitrification, and the depletion of ammonia pool in the sediment is compensated by the application of hydrated lime and CaO.

Nitrite

Nitrite is an important hydrological parameter, which not only regulates the growth of prawn but also plays a major role in the survival of the culturable species.

Significance

When fish absorbs nitrite, it reacts with haemoglobin to form methaemoglobin that is not an effective oxygen carrier. So, continued absorption of nitrite can lead to hypoxia and cyanosis.

Management Guidelines

- Nitrite levels should not exceed 0.5 mg/l.
- Dead prawn/fish and other debris should be removed as soon as possible to keep the system clean.

8.2.2.8 Hydrogen Sulphide

Under anaerobic conditions in the shrimp culture pond, certain heterotrophic bacteria can use sulphate and other oxidized sulphur compounds as terminal electron acceptors in metabolism and excrete sulphide as given below:

$$SO_4^{2-} + 8H^+ = S^{2-} + 4H_2O$$

The sulphide excreted is an ionization and participates in the following equilibria:

$$H_2S = HS^- + H^+$$

$$HS = S^{2-} + H^+$$

Concentration of hydrogen sulphide of 0.01–0.05 mg/l is lethal to shrimp, and any detectable concentration of H_2S in pond water is considered undesirable.

Significance

Unionized hydrogen sulphide is toxic to shrimp, but the ions resulting from its dissociation are not appreciably toxic. Its presence beyond a level of 0.033 ppm is found to affect the growth of shrimp.

Management Guidelines

The normal remedial measure is the application of lime in acid dominated soil (preferably during pond preparation period).

Table 8.13 Required water quality for better growth of culturable species

Variables	Method of measurement	Guidelines for protecting aquatic ecosystems
1. Temperature	Mercury thermometer	Between 28 and 30 °C; should be less than 2 °C change
2. Salinity	Argentometric method/refractometer	5–25 psu
3. Transparency	Secchi disk visibility method	Between 35 and 60 cm; should not change by more than 10 % of seasonal mean in coastal waters
4. Dissolved oxygen	Standard DO meter	Between 5 and 10 ppm
5. pH	Standard pH meter	Between 7.5 and 8.0
6. Ammoniacal nitrogen	Phenate method	<1 ppm
7. Nitrate nitrogen	Diazonium salt method (spectrophotometric method)	Between 0.2 and 0.5 ppm
8. Phosphate	Ammonium molybdate method (spectrophotometric method)	Between 0.2 and 0.5 ppm
9. Total alkalinity	Titration method	100–150 ppm

8.2.2.9 Phytoplankton Level

Since a stable phytoplankton development can effectively maintain the health of aquaculture pond, therefore, phytoplankton have great ecological significance. Further, a proper concentration of phytoplankton eliminates development of filamentous algae and reduces predation of shrimp by birds and cannibalism among the shrimp.

A stable watercolour is an indication of optimum growth of phytoplankton. Colour changes are always associated with a succession of species. Development of heavy phytoplankton blooms leads to super saturation with O_2 up to 175 % at 1400 h reducing to even up to 42 % at 0500 h. The species composition of phytoplankton can be altered by:

- Lowering the salinity to reduce dinoflagellates
- Increasing the salinity to encourage diatoms
- Inoculating the desired species from other ponds or from cultures

Phytoplankton density can also be altered by a water exchange, increasing the pH through liming and reducing the pH by tea seed cake, acetic acid, ground trash fish, etc. These changes in plankton communities may be a matter of concern to the aquaculturists. However, unless plankton becomes so dense that dissolved oxygen problems occur or so thin as to encourage underwater weeds, the changes do not affect fish production appreciably. By

monitoring Secchi disk visibility on a regular schedule (once or twice weekly) and observing the appearance of the water, the information on phytoplankton in the culture pond can be obtained.

8.2.2.10 Concentration of Pollutants

Water discharged from shrimp ponds during routine water exchange and for harvest contains nutrients, organic matter and suspended solids. These substances represent potential pollutants because they can cause water quality deterioration in receiving waters. Thus, effluents are considered to be a major environmental problem in shrimp farming.

Unionized NH_3, Hg, Cu, Cd, Zn, malathion, parathion and endosulphur are the major pollutants often witnessed in shrimp culture farms. The heavy metals accumulate inside the prawn tissues and affect the growth of prawns. Malathion, parathion and endosulphur induce chronic soft-shell syndrome in shrimp. The optimum levels of few parameters related to growth of shrimp are highlighted in Table 8.13.

These pollutants can be removed by implementing plants in shrimp culture ponds. The treatment plants in the recent time utilize macroalgae to accumulate the metals from the ambient media. The nitrate, phosphate and ammonia increment can be checked by employing filter-feeding bivalves like *Saccostrea cucullata* and *Crassostrea* sp. in the treatment pond. The high concentration of nutrients in aquacultural wastewater triggers the algal growth in the treatment pond, which is controlled by employing filter-feeding bivalves, as they are voracious eaters of phytoplankton.

8.2.3 Shrimp Diseases

Like any living organisms, shrimp have specific physiological functions for growth and development, which are greatly influenced by various environmental factors. Any impairment in the function may lead to abnormal condition of an organism, which is known as disease. Many experts describe the phenomenon, disease, as an expression of complex interaction of three factors: the host (shrimp), the environment and the disease-causing organism (pathogen). A decline in host's immunity is the main cause of disease. A lot of factors will impair shrimp health and the most important predisposing factors leading to diseases in shrimp culture are:
1. Adverse environment
2. High stocking density with limited water exchange facilities
3. Nutritional deficiency/poor nourishment
4. Accumulation of unused feed
5. Inadequate aeration
6. Suboptimal or heavy blooms in the pond
7. Physical injury
8. Presence of virulent pathogens in high count

8.2.4 Host

Like any other crustaceans, shrimp host's body is covered by exoskeleton, which is regularly replaced by a new one during moulting. The moulting process exerts requirement of energy on the shrimp and renders the organism susceptible to disease agents or cannibalism. In addition, the shrimp's nutritional well-being, size and immune response determine its degree of resistance to disease agents. Behavioural characteristics such as burrowing at the pond bottom also expose the shrimp to adverse condition in the pond.

8.2.5 Environment

The environment in aquaculture comprises the pond soil, rearing water and the various living organisms in it. The living organisms include not only shrimp but also other aquatic fauna and flora including pathogenic organisms. The survival and growth of the organisms is largely influenced by various physico-chemical parameters such as pH, dissolved oxygen, temperature and light. Any abnormal change in these factors will adversely affect shrimp in the culture system. For example, high ammonia level and low dissolved oxygen are stressful and may affect the survival of shrimp. High ammonia level in brackish-water shrimp ponds arises from uneaten (residual) feeds that are mostly deposited in the pond bottom. In this context, few mangrove-based floral feeds were tested to monitor their effect on the growth and survival of shrimp (*Penaeus monodon*) and freshwater prawn (*Macrobrachium rosenbergii*) in and around the mangrove ecosystem of Sundarbans. The water quality of the culture pond was also considered as one of the parameters for this innovative experimental study as the output of such effort can generate alternative livelihood, upgrade the culture pond environment and protect the culture species from diseases. Technical reports of few such experimental approaches are presented as Annexures 8A.1 and 8A.2.

8.2.6 Pathogen

Various pathogenic organisms may be present in the aquaculture system. They may be the part of the natural flora and fauna of the rearing water or pond soil. Mere presence of these organisms may not cause any disease condition. However, when present in large numbers, these may invade the injured tissues, get established, and multiply resulting in disease and death. Nevertheless, the quantitative level of pathogen is influenced largely by prevailing culture condition such as availability of food source, temperature, dissolved oxygen and pH.

Table 8.14 Viruses affecting cultured and wild penaeid shrimp

Name	Approx. virion size	Nucleic acid	Probable classification	References
Baculovirus penaei (BP)	50–75 × 300 nm	ds DNA	Baculovirus	Couch (1974)
Baculoviral midgut gland necrosis (BMN)	75 × 300 nm	ds DNA	Baculovirus	Sano et al. (1981)
Infectious hypodermal and hematopoietic necrosis virus (HHNV)	22 nm	ss DNA	Parvovirus	Lightner et al. (1983a)
Monodon baculovirus (MBV)	75 × 300 nm	ds DNA	Baculovirus	Lightner et al. (1983b)
Hepatopancreatic parvovirus (HPV)	22–24 nm	ss DNA	Parvovirus	Lightner et al. (1985)
Type C baculovirus (TCBV)	75 × 300 nm	ds DNA	Baculovirus	Brock and Lightner (1990)
Lymphoid parvo-like virus	25–30 nm	ss DNA	Parvo-like virus	Owens et al. (1991)
Iridovirus (IRIDO)	136 nm	ds DNA	Iridovirus	Lightner and Redman (1993)
Haemocyte-infecting nonoccluded baculovirus	90 × 640 nm	ds DNA	Baculo-like virus	Owens (1991)
White spot syndrome virus (WSSV)	80–330 nm	ds DNA	Baculovirus	Wongteerasupaya et al. (1995); Lightner and Redman (1998)
Reo-like virus	8 × 70 nm	ds DNA	Reo-like virus	Tsing and Bonami (1987)
Lymphoid organ vacuolization virus (LOVV)	30 × 88 nm	ss DNA	Toga-like virus	Bonami et al. (1992)
Rhabdovirus of penaeid shrimp (RPS)	75 × 125 nm	ss DNA	Rhabdovirus	Nadala et al. (1992)
Yellow-head virus (YHV)	44 × 173 nm	ss DNA	Rhabdovirus	Flegel et al. (1995)
Taura syndrome virus (TSV)	30–32 nm	ss DNA	Rhabdovirus	Lightner et al. (1995)

8.2.6.1 Viral Diseases of Shrimp

Shrimps encounter diseases from all bio-aggressors such as viruses, bacteria, parasites and fungi, both in hatcheries and culture ponds. Viral diseases constitute the most serious problems of shrimp culture due to high infectivity, pathogenicity and total lack of curative measures. So far, 15 viruses infecting cultured shrimps have been recorded across the shrimp-farming countries of the world (Table 8.14), and viral diseases recorded from Indian shrimp farms are explained in Table 8.15.

8.2.6.2 Bacterial Diseases of Shrimp

The bacteria-causing diseases of penaeid shrimp constitute part of the natural microbial flora of seawater. Accumulation of unutilized feed and metabolites of shrimp in the culture tanks/ponds enriches the water with organic matter that supports the growth and multiplication of bacteria and other microorganisms.

Bacterial infections of shrimp are primarily stress related. Adverse environmental conditions or mechanical injuries are important predisposing factors of bacterial infections and disease. The most common shrimp pathogenic bacteria belong to the genus *Vibrio*. Other gram-negative bacteria such as *Aeromonas* spp., *Pseudomonas* spp. and *Flavobacterium* spp. are also occasionally implicated in shrimp diseases. Various bacterial diseases of shrimp in India are given in Table 8.16.

8.2.6.3 Fungal Diseases of Shrimp

The shrimps are susceptible to fungal attack and particularly stages are highly vulnerable (Table 8.17). Other fungi such as *Fusarium* spp. cause infections in nauplii, protozoea, juveniles and adults. Oomycetous fungi such as *Saprolegnia* spp. and *Leptolegnia* spp. are also known to affect shell of shrimp and produce dark necrotic lesions causing gradual mortality.

Table 8.15 Viral diseases of shrimp in India

Name of the disease	Causative agent	Signs and symptoms	Prevention and control
Monodon Baculovirus (MBV) disease	Monodon Baculo virus, a single enveloped, rod-shaped, occluded double-stranded DNA virus. It occurs freely or within proteinaceous polyhedral occlusion bodies in the virions measuring 75–300 nm	Shrimp lethargic with surface and gill fouling	MBV infection can be prevented through avoidance by quarantine methods, destruction of contaminated stocks and disinfection of contaminated facilities. Good farm management can minimize this disease
		Hepatopancreatic cells of all life stages of the prawn except egg, nauplius and protozoa 1 and 2 stages are affected	
		High cumulative mortalities	
Hepatopancreatic Parvo Virus (HPV) Disease	Hepatopancreatic Parvo virus (HPV), a single-stranded DNA virus of 22–24 nm size	Gross signs of HPV infection may not be specific	Infection can be prevented through avoidance by quarantine method and destruction of infected stocks
		In severe infection, there are signs of an atrophied hepatopancreas, poor growth rate and anorexia and secondary infection by pathogenic *Vibrios*	
White spot disease	White spot baculovirus (WSBV)	Rapid reduction in food consumption	Avoidance of the disease by quarantine methods, destruction of contaminated stocks and disinfection of the culture facility.
		Become lethargic	Use of physical disinfectants like UV radiation and ozone and chemical disinfectants, like sodium hypochlorite, benzalkonium chloride and povidone iodine at proper doses has found useful in inactivating the WSV from the rearing systems
		Have a loose cuticle with white spots of 0.5–2.0 mm in diameter, which are more apparent on the inside surface of the carapace. These spots represent abnormal deposits of calcium salts by the cuticular epidermis	
		High mortality rates with cumulative mortalities reaching 100 % within 3–10 days of the onset of clinical signs	
Yellow-head virus disease	Yellow-head virus, a rhabdovirus or paramyxovirus	Sudden decline in feed consumption during grow-out, followed by fasting and then death	Exclusive use of virus-free quality seed and broodstock
		Yellowish head	Maintenance of water transparency between 25 and 35 cm
		Change of colour in hepatopancreas	Chlorination and the use of iodine are very effective

8.2.6.4 Parasitic Diseases of Shrimp

Among the disease-causing organisms of shrimp, parasites, especially protozoan parasites, form an important group (Table 8.18). Following are the major disease problems caused by the protozoa:

- Protozoan fouling
- Cotton shrimp disease
- Enterozoic cephaline gregarine infection
- Invasive protozoan infection

8.2.6.5 Non-infectious Diseases of Shrimp

Non-infectious disease are common in the grow-out farms (Table 8.19), as influence of

Table 8.16 Bacterial diseases of shrimp in India

Name of the disease	Causative agent	Signs and symptoms	Prevention and control
Bacterial Septicaemia (Vibrio disease)	Bacteria such as *Vibrio alginolyticus*, *Vibrio anguillarum*, *Vibrio parahaemolyticus*, *Vibrio* spp.	Shrimps are lethargic and show abnormal swimming behaviour	Maintenance of good water quality and reduction of the organic load by increased water exchange
		Periopods and pleopods may appear reddish due to expansion of chromatophores	Fed shrimp with antibiotic fortified feeds (only after ascertaining in vitro sensitivity of the pathogen), e.g. feeds containing oxytetracycline along with proper water and pond management
		Shrimps may show slight flexure of the abdominal musculature	
		In severely affected shrimps, the gill covers appear flared up and eroded	
		In more severe cases, extensively melanized black blisters can be seen on the carapace and abdomen	
Luminescent bacterial disease (a serious problem in hatcheries)	Luminescent bacteria, namely, *Vibrio harveyi*	Infected larvae appear luminescent in darkness	Use of ultraviolet irradiated and chlorinated (calcium hypochlorite 200 ppm for 1 h) water
		Suffer heavy mortality	Exchange of 80 % water daily with UV-sterilized/sand-filtered seawater
Brown spot disease (shell disease or rust disease)	*Vibrio* spp., *Aeromonas* spp. and *Flavobacterium* spp.	Animals show presence of brownish to black eroded areas on the body surface and appendages	Reduction of organic load in water by increased water exchange
			Avoidance of unnecessary handling and overcrowding to minimize chances of injury and infection
			Induction of moulting by applying tea seed cake may be useful
Necrosis of appendages	Epibiotic bacteria such as *Vibrio* spp., *Pseudomonas* spp., *Aeromonas* spp. and *Flavobacterium* spp.	Tips of walking legs, swimmerets and uropods of shrimp undergo necrosis and become brownish and black	Maintenance of good water quality
			Induction of moulting by applying tea seed cake may be useful
Vibriosis in larvae	Bacteria, namely, *Vibrio alginolyticus*, *Vibrio anguillarum* and *Vibrio parahaemolyticus*	Larvae show necrosis of appendages	Maintenance of good water quality and reduction of the organic load by increased water exchange
		Expanded chromatophores	10–15 ppm EDTA to the rearing water
		Empty gut	
		Absence of faecal strands and poor feeding	
Filamentous bacterial disease	Filamentous bacteria such as *Leucothrix mucor*	Larvae show fouling of gills, appendages and body surface	Maintenance of good water quality with optimal physico-chemical parameters
		Moulting of affected shrimp is impaired and may die due to hypoxia	0.25–1 ppm copper sulphate bath treatment for 4–6 h

Table 8.17   Fungal disease of shrimp in India

Name of the disease	Causative agent	Signs and symptoms	Prevention and control
Larval mycoses (one of the most devastating diseases in shrimp hatcheries)	Oomycetous fungi, *Lagenidium* spp., *Sirolpidium* spp. and *Haliphthoros* spp. These fungi are filamentous, non-septate and coenocytic	Larvae appear opaque followed by sudden mortality	Removal of bottom sediments and dead larvae periodically
		The protozoeal and mysis stages are highly susceptible	Disinfection of the tanks and other equipments in the hatchery from time to time
			Treatment with Treflan (Trifluralin) 0.1–0.2 ppm bath for 1 day

Table 8.18   Parasitic diseases of shrimp in India

Name of the disease	Causative agent	Signs and symptoms	Prevention and control
Protozoan fouling	Peritrichous ciliates such as *Zoothamnium*, *Epistylis*, *Vorticella* and *Acinata*	Shrimps are restless and their locomotion and respiratory functions are hampered	Maintenance of good water quality
		Heavily infected, larger shrimp often have fuzzy mat-like appearance on the body surface, appendages and gills	Reduction of organic substance, silt and sediment on the pond bottom
		Animals show brownish discolouration due to algal filaments or debris entangled with the epibiont	Maintenance of optimum dissolved oxygen level (5–6 ppm) and frequent exchange of water
			Treatment with formalin 15–25 ppm (single treatment) for ponds or dip treatment of affected animals in 50–100 ppm for 30 min
Cotton shrimp disease or milk shrimp disease	Microsporeans such as *Agmasoma*, *Ameson* and *Pleistophora*	Striated muscle of shrimp becomes opaque and white	Affected animals should be destroyed and buried away from the farm. Before stocking, the possible conditioning host/intermediate host should be eliminated
		Muscle appears crooked	No treatment has been reported for penaeids
		In severely affected shrimps, the exoskeleton appears bluish black, and white tumour-like swelling may be found on the gills and subcuticle	
		Infection in gonads, heart, haemolymph vessel, hepatopancreas	
Enterozoic cephaline gregarine infection	Cephaline gregarines such as *Nematopsis* and *Cephalolobus*	Shrimps show loss of appetite, lethargy and weakness	Avoidance of wild seeds
		Low levels of mortalities	Elimination of intermediate hosts from the culture system can prevent the disease occurrence
Invasive protozoan infection	Ciliate protozoa, *Paranophrys*- and *Paraoronema*-, *Leptomonad*-like organisms	Noticed in a few cases in hatcheries and often heavy mortalities have been recorded	Control and preventive measures are not reported

Table 8.19 Non-infectious diseases of shrimp in India

Name of the disease	Causative agent	Signs and symptoms	Prevention and control
Soft-shell syndrome	Sudden fluctuation in water salinity and temperature	Drastic reduction in feed consumption and daily growth rate during the middle of culture	Proper water exchange is an effective remedial measure; 20–50 % daily water exchange is useful
	High soil pH	Shrimp losses weight and turns into a soft leathery shell	Pond flushing should be done regularly
	Highly reducing conditions in soil	Shrimp has a paper-like carapace with a gap between the muscle tissue and exoskeleton	
	Low organic matter in soil	Gills are reddish to brown in colour and shrimp become sky-blue in colour instead of the normal brown-black	
	Low phosphate content and pesticide pollution in water	Severely affected *P. indicus* often show undulating gut in the first three abdominal segments	
	Nutritional deficiency		
	Insufficient water exchange		
Black gill disease	Presence of excessive levels of toxic substances such as nitrite, ammonia, heavy metals and crude oils in the culture water	Shrimps have gills with black to brown discolouration, which is due to the deposition of melanin at sites of massive haemocyte accumulation, followed by dysfunction and destruction of whole gill processes	Control of phytoplankton bloom
	High organic load, heavy siltation and reducing conditions in rearing pond	In acute cases necrosis and atrophy of the gill lamellae may be apparent	Control of high rate of sedimentation
	Attack of certain bacterial, fungal and protozoan pathogens		Low DO concentration is prevented by using aerators
			Proper water exchange is necessary
			The rate of feeding is made low
Red disease	Definite causative agent is not known. Microbial toxin in rancid or spoiled diets or in detritus of ponds rich in organic matter may lead to disease	Juveniles and adult shrimp have reddish discolouration in body, pleopods and gills	Water transparency should be maintained between 25 and 35 cm
		In the cephalothorax, excessive fluid with foul odour may be found	Water exchange is necessary
			Use of oxytetracycline mixed with 3–5 g of pelleted food is necessary
			Excessive feeding checked

(continued)

Table 8.19 (continued)

Name of the disease	Causative agent	Signs and symptoms	Prevention and control
Cramped tail disease	Exact cause not known, but environmental and nutritional causes have been suggested	Shrimps have rigid dorsal flexure of the abdomen, which cannot be straightened	Maintenance of healthy conditions in the pond with proper feeding with balanced diet may be helpful in the prevention/control of this disease
		Shrimps lie on their sides at the bottom of the pond and are susceptible to cannibalism	
Gas-bubble disease	Super saturation of atmospheric gases and oxygen in pond	Presence of gas bubble in the gills or under the cuticle	Super saturation of the gases must be avoided to prevent the disease
	The threshold saturation level causing gas bubble in case of nitrogen is 118 % and for oxygen is 250 % of normal saturation	Severely affected or dead shrimp may float near the water surface	
Muscle necrosis	Poor environmental conditions such as low oxygen levels and salinity or temperature shock	Shrimps are characterized by the presence of white opaque areas in body musculature, usually in the lower abdomen or sometimes in the appendages	Avoidance of overcrowding, proper handling and maintenance of favourable environmental factors may help to control the disease
	Overcrowding and poor handling	In severe cases, sloughing of the affected areas occurs due to secondary bacterial infection leading to death	

nutritional factors, environmental factors such as temperature extremes and oxygen depletion and toxicity from biotic and abiotic origins becomes critical during the lengthy culture period.

Proper and accurate diagnosis forms the important step in any disease control and prevention programme. To diagnose a shrimp disease problem, some general information on the farming activity and condition of cultured organisms is required. These are the following:

1. *Background information about the farming practices*
   (a) Pond condition
   (b) Stocking parameters
   (c) Management practices
   (d) Environmental parameters
2. *Field observation for signs and symptoms*
   (a) *Observation of behaviour of shrimp*:

   (i)   *Escape reflex*
   (ii) *Swimming at the surface*
   (iii) *Moulting behaviour*
   (iv) *Feeding behaviour*
   (b) *Observation of external signs and symptoms*:
   (i)   *Colour and nature of exoskeleton*
   (ii) *Eye colour*
   (iii) *Appendages*
   (iv) *Musculature*
   (c) *Examination of internal organs for pathological signs*:
   (i)   *Gills*
   (ii) *Hepatopancreas*
   (iii) *Haemolymph*
3. *Collection of samples for laboratory investigation*
   For accurate diagnosis of the disease, typical and representative sample of infected animals should be collected.

8.3   Oyster Culture

8.3.1   Major Types of Edible Oysters

The edible oysters of the world comprise about 100 species and are widely distributed in all temperate and tropical coasts. In India, the common species are *Saccostrea cucullata, Crassostrea madrasensis, Crassostrea gryphoides, Crassostrea rivularis* and *Crassostrea discoidea*. Out of these five dominant species, the first three species are very common in the Indian Sundarbans. Patchy settlement of oysters covers many square miles of bottom in the intertidal zone of high saline areas. They are found attached to rocks, boulders and several underwater structures, submerged branches and trunks of mangroves, concrete embankments and piles and even on lighthouse bases.

8.3.1.1   Edible Oysters Found in Indian Sundarbans

1. *Saccostrea cucullata* (Born)	
Phylum	Mollusca
Class	Bivalvia
Order	Pterioida
Family	Ostreidae
Genus	*Saccostrea*
Species	*cucullata*

Identifying features:

1. Shell variable in shape, valves unequal, left valve thicker and larger than the right valve
2. Sometimes valves are flat, sometimes shallow cup-like; outer margin with a series of folds interlocking with each other
3. Shell whitish or greyish white, marked with deep purple towards the margin
   Distribution in India: West Bengal (North and South 24 Parganas), Andhra Pradesh, Andaman and Nicobar Islands, Gujarat, Kerala, Maharashtra, Orissa and Tamil Nadu.
   Habit and Habitat: The species is highly variable in shape, growing in clusters on rocks, bricks, wooden piles or jetties throughout the Indian Sundarbans. Sometimes they are

also observed on the stem of mangrove plants and also on molluscan shells, when no other suitable substratum is available. A salinity range of 15 to 28‰ is optimum for their growth, and survival and high dilution factor witnessed during monsoon season causes considerable mortality.

2. *Crassostrea madrasensis* (Newton and Smith)	
Phylum	Mollusca
Class	Bivalvia
Order	Pterioida
Family	Ostreidae
Genus	*Crassostrea*
Species	*madrasensis*

Identifying features:

1. Shell very heavy and bulky, variable in shape.
2. The shape, size and even the impression of muscle scars change due to the environmental conditions, and this creates a lot of problem in proper identification.
   Distribution in India: West Bengal (North and South 24 Parganas), Andhra Pradesh, Karnataka, Kerala, Gujarat, Maharashtra, Orissa and Tamil Nadu.
   Habit and Habitat: The species is observed in the lower stretch of Indian Sundarbans, attached to the sluice gates, bricks and dykes and also to the lighthouse and jetties. The species prefers a salinity range of 15–32‰ and cannot withstand high turbidity.

3. *Crassostrea gryphoides* (Schlotheim)	
Phylum	Mollusca
Class	Bivalvia
Order	Pterioida
Family	Ostreidae
Genus	*Crassostrea*
Species	*gryphoides*

Identifying features:

1. Shell stout, bulky, elongated and irregular in shape
2. Inner margin pearly white; cavity beneath the hinge well marked
3. Muscle scar broad with more or less oblong striations on it

Distribution in India: West Bengal (North and South 24 Parganas)

Habit and Habitat: The species is mainly found on the lower height of sluice gates and sometimes on hard substrata like lighthouse and boulders. Mass mortality of the species is observed during the period of monsoon as it prefers an aquatic salinity around 20–32‰.

8.3.2   Importance of Oysters

8.3.2.1   Edible Value

Oysters have traditionally been appreciated as one of the popular seafood in Malaysia, Korea, Japan as well as in European coastal region. This bivalve after proper purification is often consumed raw with pepper and lime. Alternatively, it may be prickled or brined or canned. Steaming of meat followed by smoking is popular in western countries. Oyster omelette or sandwich is an attractive dish in tropical restaurants. Oyster sauce is yet another product, especially in Thailand and Malaysia. It is prepared by boiling oyster meat until a thick broth is produced. The concentrate is then mixed with water and bottled. The resulting meats are sun dried and sold for use as flavour enhancers in various dishes (Santhanam 1990). However in India, the importance of oyster as edible item has not at all penetrated in the north Indian states. It is very much popular in Goa and south Indian states. The culture of edible oysters has been initiated in the south Indian states, but yet to get a concrete marketability momentum.

In Indian Sundarbans area, where conservation of marine biotic resources sometime becomes at stake due to illegal fishing, poaching in the reserve forests, cutting of mangroves, etc., culture of edible oysters may be an alternative source of income for the local people. This alternative livelihood scheme may bind significant portion of the coastal population in the economic matrix, which may ultimately lead to reduction of pressure on the adjacent forest zone.

Researchers have collected detailed information on mangrove oysters, largely because they can be valuable food (Tack et al. 1992; Ruwa and

Polk 1994). The oysters grow fast and attain an average size of 80–90 mm weighing 100–200 g with meat forming 8–10 % at the end of 1 year. From culture operations in a 3-year period in 0.25 ha area, the estimated production of oyster is about 125 tonnes with a meat yield of about 10 tonnes. At the end of each year, approximately 42 tonnes of oyster can be harvested.

The meat of oyster contains essential food components like protein (10 %), fat (4 %), carbohydrate (6 %), vitamins (like A, B, B_{12}) and essential minerals (Na, K, Ca, P, Fe, etc.).

8.3.2.2   Different Methods of Oyster Culture

The culture of edible oysters dates back to first century B.C. Although several methods of culture have been standardized centuries ago, very few countries like France, Japan, Korea, the USA and the Netherlands have taken up this venture on a commercial scale. In India the Central Marine Fisheries Research Institute (CMFRI) has done extensive work on the rack and tray culture of oysters using spats collected on lime-coated tiles or oyster shell rens. Experimental works on oyster culture carried out at Athankarai, Bhcemunipatnam backwaters, north of Visakhapatnam, Goa and Mulki estuary have also been proved to be encouraging. In order to initiate oyster culture in any brackish-water environment, it is very important to judge the aquatic salinity range and extinction coefficient (turbidity) as these parameters often cause mass mortality of oysters. In addition to these, tidal amplitude, water temperature, nutrient load, phytopigment concentration and phytoplankton density are also important parameters to be considered. Hence, site selection for edible oyster culture should be done on the basis of the following criteria:

- Protection against excessive wind and wave action (particularly applicable for on-bottom culture)
- Water quality in terms of sewage and industrial wastes (being sedentary in nature oysters tend to accumulate conservative substances in the body tissues)
- Water quality in terms of salinity, pH and temperature (salinity requirement for culture is

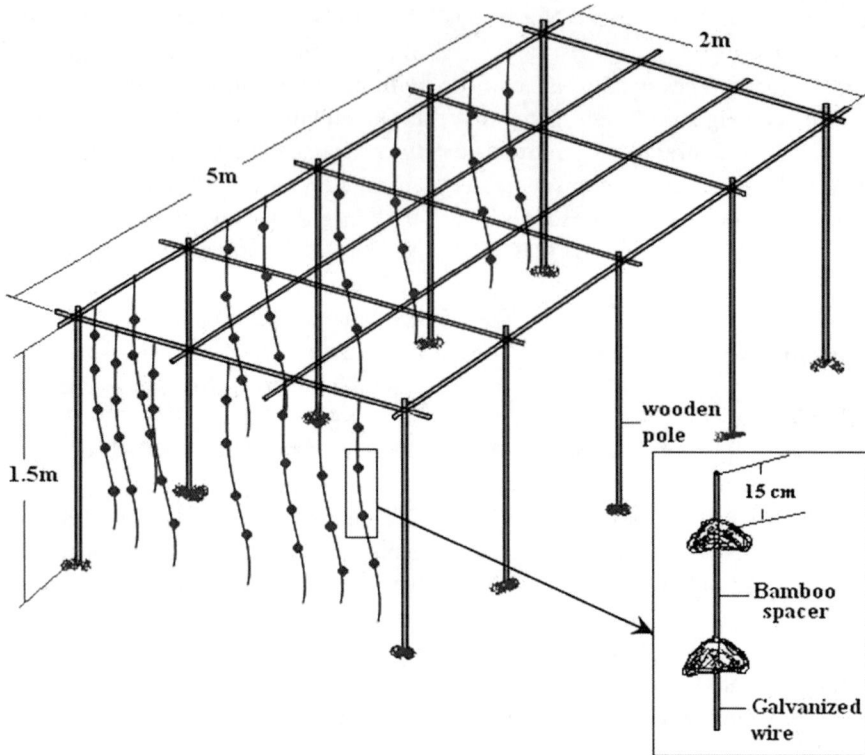

Fig. 8.1 Rack and string culture

20–35‰, pH requirement is around 8.3 and temperature requirement is 25–30 °C)

- Tidal characteristics and current pattern (as frequent water exchange is extremely important for oyster culture)
- Optimum nutrient level and adequate phytoplankton density (as nutrition of oyster is obtained through consumption of phytoplankton)
- Relative abundance of predators, like boring gastropod *Cymatium*, starfish and crabs (sites with high population and diversity of predators are always avoided as they cause massive damage to oyster population)
- Availability and abundance of adult oyster stock in the vicinity of the culture site (this is for ensuring oyster seed supply on regular basis)

There are two common methods for culturing oysters and these are:

1. *On-bottom culture*
2. *Off-bottom culture*

In case of Indian Sundarbans, considering the background of siltation rate and turbidity of the aquatic phase, off-bottom culture is most preferable and among various types of off-bottom culture most economic methods are:

1. *Rack and string culture*
2. *Rack and tray culture*

Both these methods are very much suited in protected and deeper areas (>5 m) having soft bottom particularly inside the mangrove-sheltered creeks. Brief description of each of these methods is given here.

1. *Rack and String Culture*

Two rows of tar-coated bamboo poles are planted vertically at the bottom (substratum) and a rack is prepared by fixing bamboo poles or thin rods crosswise on top. Galvanized wires or tar-painted units are hung vertically from the crossbars, to which seedlings adhere. About 5–10 pieces of empty shells are hung at intervals of about 15–20 cm on 1.5 m string, and 20 such strings are hung vertically for every 3.3 m² area (Fig. 8.1).

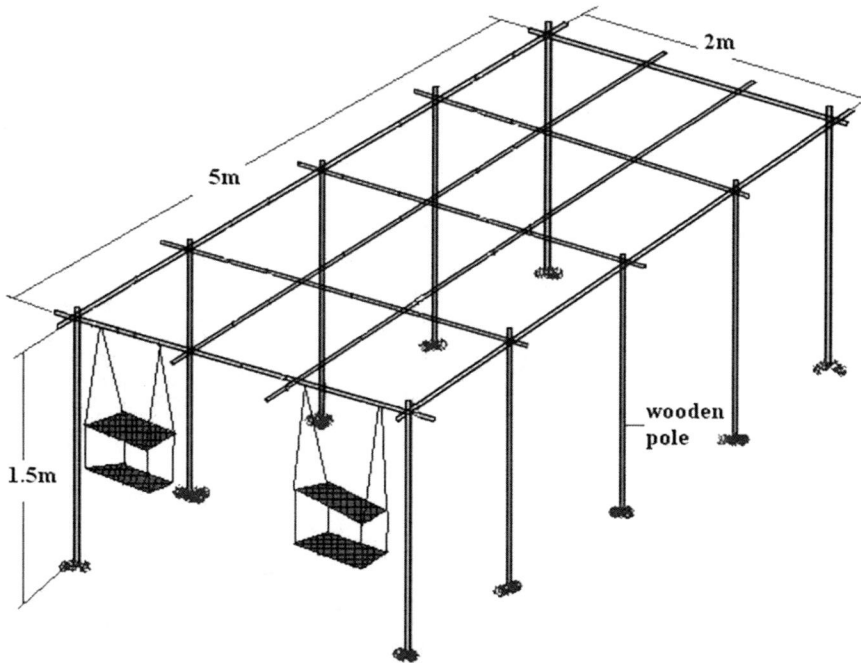

Fig. 8.2 Rack and tray culture

2. *Rack and Tray Culture*

In this culture method, racks are prepared similarly as in previous method and culture trays are hung vertically instead of strings (Fig. 8.2).

Culture trays are of several types (Fig. 8.3) as discussed below:

• *Tyre Tray*

Discarded motorcycle tyres can be used to make durable and cheap trays. The tyres are turned inside out to form the frame for the tray. A leather punch is used to make holes along the bottom edge of the inverted tyres. The spacing of the holes depends on the mesh size of Netlon to be used. The Netlon is cut to the diameter of the inverted tyres and tied in place with HDPE twine. A cover is commonly used to keep predators out. These types of trays can carry about 100 oysters of marketable size.

• *Netlon Tray*

A frame is made of 1½ PVC pipe cut to the desired length. Netlon is cut to fit the frame and tied in place. A tray measuring 60 m × 100 m is a convenient size. Suspension ropes are attached to the corners. The tray can carry 100 oysters. This type of tray is more expensive than tyre trays, but this is comparatively long lasting.

• *Plastic Basket Tray*

Plastic baskets are also suitable for nursing and rearing oyster spats. Holes are drilled in the upper corners for tying the suspension ropes. Best-quality baskets are chosen; otherwise, they cannot last for long period.

Note: Instead of rack and tray and rack and string culture, concrete cubical pillars can also be constructed which is again a suitable substratum for the setting of oyster spats.

Oyster Culture Technology with Special Reference to Indian Sundarbans

The Indian Sundarbans is a vibrating ecosystem from the point of view of physico-chemical variables, and significant spatial and temporal variations of water salinity, pH, nutrient level and turbidity exist in this dynamic geographical locale (Mitra et al. 2004). The lower stretch of the estuarine complex and high saline zones (*e.g.*

Fig. 8.3 Different types of oyster culture tray

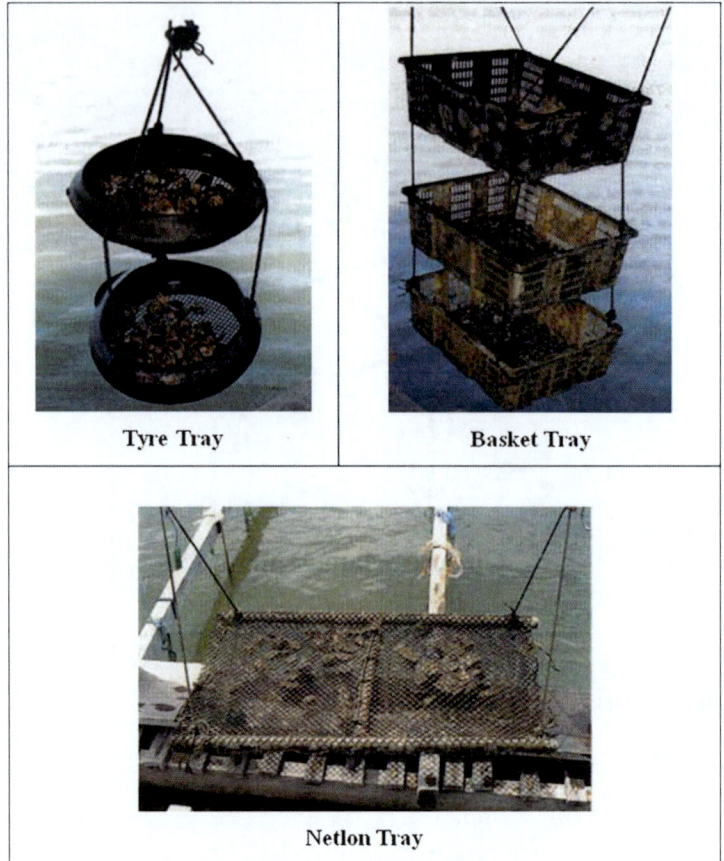

Tyre Tray

Basket Tray

Netlon Tray

Namkhana block, Sagar block, Patharpratima block) is extremely favourable for edible oyster culture. *Saccostrea cucullata* is the dominant oyster species in Indian Sundarbans and mean population density is around 385/m² (Table 8.20).

In context of the lower stretch of Indian Sundarbans, the relevant parameters, which have major role in regulating the oyster populations, are surface water temperature, salinity, turbidity (expressed as extinction coefficient), wave action, nutrient (nitrate, phosphate and silicate) and phytoplankton population. A brief account of each of these parameters is necessary to justify the commercial viability of oyster culture in this mangrove-dominated deltaic lobe at the apex of Bay of Bengal.

Surface Water Temperature

Surface water temperature has profound influence on biotic community (Mitra et al. 2004). It regulates the composition of phytoplankton community (the major food of oysters) by increasing the dominancy of flagellates in warm stratified waters of coastal regions (Strickland 1983; Maestrini and Graneli 1991). It has been observed that low temperature and low dissolved oxygen often limit the germination of phytoplankton cyst (Ishikawa and Taniguchi 1994; Blanco 1995). In this context, it is to be mentioned that the environment of Indian Sundarbans is ideal for oyster culture as temperature lower than 20 °C seldom reaches in this deltaic lobe, which is extremely congenial for phytoplankton stock acceleration. The water temperature also influences the process of gametogenesis and breeding cycle in marine and estuarine organisms. Most of the members of nekton and benthos in coastal West Bengal breed during the onset of summer, when the temperature of the ambient water is in a state of gradual increase.

Table 8.20 Monthly variation of oyster population density (No. m^{-2}) at three stations in Indian Sundarbans Biosphere Reserve

Month	Station I (Sagar light house)	Station II (Henry's island – first sluice gate)	Station III (Henry's island – second sluice gate)
June 1987	629.6	154.9	200.8
July	601.5	152.7	184.1
August	498.7	131.7	178.3
Sept	493.8	130.9	178.2
Oct	493.0	130.3	175.3
Nov	486.9	128.7	171.7
Dec	486.4	124.3	170.8
Jan 1988	465.9	125.0	168.9
Feb	465.3	124.8	165.8
March	564.6	123.4	169.2
Apr	617.8	169.0	190.6
May	624.7	164.5	189.4
June	623.9	161.1	179.2
July	611.8	144.9	158.0
August	586.8	152.0	153.6
Sept	501.2	129.1	133.8
Oct	500.9	129.8	130.7
Nov	500.2	126.4	129.0
Dec	496.5	126.1	128.9
Jan 1989	488.8	124.8	128.3
Feb	486.1	124.1	128.1
March	485.7	123.6	139.5
Apr	511.2	161.0	156.8
May	519.9	164.2	161.3

In coastal West Bengal and adjacent estuarine systems, the seasonal variation of surface water temperature is not drastic between premonsoon and monsoon seasons. The premonsoon period (March–mid-June) is characterized by a mean surface water temperature around 34 °C. The monsoon period (mid-June–October) shows a surface water temperature around 32 °C (mean), and the postmonsoon period (November to February) is characterized by cold weather with a mean surface water temperature around 23 °C (Mitra 2000).

Surface Water Salinity

In context to Indian Sundarbans, the aquatic salinity is the principal regulating component influencing spawning, growth and survival of oyster species. The spatial variation of salinity in different loca-

tions of Indian Sundarbans may be attributed to the proximity of the station to the coastal Bay of Bengal where the average salinity is 32.5‰ during premonsoon, 16.0‰ during monsoon and 28.0‰ during postmonsoon (Mitra 2000). A significant lowering of surface water salinity values is observed in all the areas of Indian Sundarbans during monsoon period (July to October). In addition to direct precipitation, the run-off from the adjoining landmasses also plays a significant role in reducing the salinity to very low values. The freshwater brought into the estuarine and coastal zone through a large number of existing riverine systems also increases the dilution factor and subsequently decreases the salinity. The Hugli and Mooriganga rivers are the major contributors of freshwater in the Bay of Bengal zone as they are connected to Himalayan glaciers (Fig. 8A.1.1). La Fond (1954) explained that the decline in salinity of the surface waters in coastal zone is mainly due to the riverine contribution during monsoon. The subsequent 8 months from November to June is a period of restoration. This upward trend of salinity continues till the summer maximum is reached during June. Considering the salinity profile of the entire deltaic lobe of Indian Sundarbans, the lower stretch of the Hugli–Matla estuarine complex is favourable for the culture of edible oyster as salinity below 10‰ hardly reaches in this marine-dominated estuarine stretch.

Turbidity

Turbidity in the estuarine and coastal waters is the effect of suspended particulate matters, which are basically contributed by land drainage and turbulence within the aquatic ecosystem that churns the bottom sediments and transfer them in suspension. As a result of turbidity, the reduction of light intensity in the vertical column of seawater occurs due to which the phytoplankton growth is hindered. Thus, turbidity often limits the food reservoir of oysters. Loosanoff and Tommers (1948) studied the effect of suspended solids on feeding rate of oysters and concluded that oysters are very sensitive to suspended silt and there is an inverse relationship between concentration of suspended load and rate of pumping. Dead and dying oysters found in turbid waters invariably contain large amount of silt in

Fig. 8.4 Idea site for oyster culture within the mangrove creek

their gills (Loosanoff 1962). The aquatic phase of Indian Sundarbans is highly turbid owing to run-off from the adjacent landmasses and nature of underlying substratum, and hence, off-bottom culture is preferred to on-bottom culture technology. Vertical cubical pillars (around 3 m height) in mangrove-sheltered creeks are also suitable cultch materials for oyster spat setting in deltaic Sundarbans region.

Wave Action

Waves and currents affect the process of setting oyster larvae on cultch materials or substratum. In general, oyster spats are found at the bottom where the current action is weak than the surface. In Indian Sundarbans region, the waves and current patterns are not extreme except during Norwesters or equinoxes conditions. During Norwesters, the wind speed rises above 100 km/h and is usually accompanied by huge tidal waves. When cyclonic incidences coincide with the spring tides, wave height can rise over 5 m above the mean sea level. Ripple waves appear in the month of October, November and December when wind-generated wave height varies approximately from 0.20 to 0.35 m. In the month of April to August, large wavelets are formed in the

shelf region, which starts breaking when they approach towards the coastal margin. Wave height rises up to 2 m during this period, which causes maximum scouring of landmasses. Wave actions, micro- and macro-tidal cycles and long shore currents are recorded in most of the islands in this ecosystem. With these wave and current characteristics, open fronts of estuarine and coastal zone are not advisable to initiate oyster culture in the river mouth (facing open sea); rather cultch materials need to be placed in sheltered inlets, creeks and mangrove-dominated delta distributaries (Fig. 8.4).

The author of the book along with his group of scholars successfully carried out an oyster culture programme in the creeks of mangrove forests of Indian Sundarbans during 2010. The species survived even at salinity 24–28 psu and exhibited considerable protein content (Fig. 8.5). Interested readers may go through Annexure 8A.3 at the end of the chapter to get a detailed analysis of biochemical composition of the cultured oyster.

Surface Water Nutrients

In Indian Sundarbans region, nutrients are basically contributed through land run-off (that

Fig. 8.5 Lime-coated boulder to trap and grow the oyster spats in mangrove creeks of Indian Sundarbans

increases during monsoon) and litter and detritus of mangrove vegetation. All the nutrients (nitrate, phosphate and silicate) exhibit high values during monsoon and lowest during the premonsoon periods. The large discharges of freshwater from land run-off during the monsoon may contribute to the observed increase in nutrient content in the aquatic subsystem. Nutrient entrapped in the bottom sediments may get released to the overlying water column due to turbulence during the monsoon. All these sources saturate the water in and around Indian Sundarbans with considerable concentrations of nutrients due which phytoplankton growth never shows a decreasing trend.

Phytoplankton Population

These are free-floating minute plants that constitute the drifting community of shallow seas, estuaries and bays, where light can penetrate. They are the major food for oyster and other filter-feeding bivalves and form the primary biological component of any aquatic system, sweet or saline, which play the chief role of energy transference to higher trophic level. The phytoplankton diet of oysters is composed of diatoms such as *Chaetoceros affinis*, *Skeletonema costa-tum*, *Thalassiosira subtilis* and *Nitzschia closterium*, the phytoflagellates *Isochrysis galbana* and *Pavlova* sp. and the microgreen alga *Chlorella salina*. According to Nagappan Nayar (1987), the optimum algal cell concentration for hatchery rearing of edible oyster is around 1.0 million cells/ml. The aquatic phase of Indian Sundarbans sustains 102 species of phytoplankton (Mitra et al. 2004), with a dominancy of *Coscinodiscus* population. The mean population density of these free-floating producers is around 106 cells/m$^3 \times 10^3$, although seasonal trend and oscillation are marked features. It has been confirmed from a number of studies that phytoplankton biomass is a direct function of nutrient load in ambient aquatic phase, and with respect to Indian Sundarbans, the sources of phytoplankton nutrients are litters and detritus contributed by adjacent mangrove vegetation, anthropogenic inputs from the city of Kolkata and the newly developing Haldia port-cum-industrial complex and sometimes drainage wastes from shrimp culture farms. Hence in conclusion, it can be advocated that the brackish-water stretch in the lower sector of Indian Sundarbans is an ideal nursery ground for the culture and rearing of edible oyster species.

Fact Below the Carpet

Sea-level rise does not necessarily lead to loss of mangrove areas, especially where there are significant tides, because these intertidal mudflats accrete vertically and maintain their elevation relative to sea level where the supply of sediment is sufficient, but what is a matter of concern is the loss of livelihood (like paddy and carp culture), which are dependent on freshwater. Considering sea-level rise inevitable and saline water intrusion as the consequence, thrust should be given to brackish-water centric livelihood. Apart from shrimp, oyster and seaweed culture, the pharmaceutical industries may take drive on conservation of horseshoe crabs. The amoebocytes in the blue blood of horseshoe crabs are sources of Carcinoscorpius amoebocyte lysate (CAL), Tachypleus amoebocyte lysate (TAL) and Limulus amoebocyte lysate (LAL) that are used for bacterial endotoxin test (BET). Multiplication of these species by creating suitable niche in the mangrove intertidal mudflats and using them as blood donor can open up another livelihood option for the mangrove-based population.

Important References

Beveridge CA, Ross JJ, Murfet IC (1994) Branching mutant *rms-2* in *Pisum sativum*: grafting studies and endogenous indole-3-acetic acid levels. Plant Physiol 104:953–959

Blanco J (1995) Cyst production in four species of neritic dinoflagellates. J Plankton Res 17:165–182

Bonami JR, Lightner DV, Redman RM, Poulos BT (1992) Partial characterization of a togavirus (LOVV) associated with histopathological changes of the lymphoid organ of penaeid shrimps. Dis Aquat Organ 14:145–152

Briggs MRP, Funge-Smith SJ (1994) A nutrient budget of some intensive marine shrimp ponds in Thailand. Aquacult Fish Manage 25:789–811

Brock JA, Lightner DV (1990) Diseases of crustacea. Diseases caused by microorganisms. In: Kinne O (ed) Diseases of marine animals, vol 3. Biologische Anstalt Helgoland, Hamburg, pp 245–349

Chong VC (1980) Maturation and breeding ecology of the white prawn Penaeus merguiensis de Man. In: Furtado JI (ed) Tropical ecology and development. Proceedings of the Vth international symposium of tropical ecology. International Society of Tropical Ecology, Kuala Lumpur, pp 1175–1184

Costanza R, Sklar FH, White ML, Day JW Jr (1989) A dynamic spatial simulation model of land loss and marsh succession in coastal Louisiana. In: Day JW Jr, Conner WH (eds) Physical processes ecological dynamics and management implications: results of research in the Atchafalaya delta region. Louisiana Sea Grant College Program, Baton Rouge, pp 190–200

Couch JA (1974) Free and occluded virus similar to *Baculovirus* in hepatopancreas of pink shrimp. Nature 247(5438):229–231

FAO (1996) State of the world agriculture. FAO, Rome

FAO (2000) Land resources and potential constraints at regional and country levels. Food and Agriculture Organization of the United Nations (FAO), Rome

FAO (2001) Assessment of the World food security situation. Committee on Food Security, twenty-seventh session, document no CFS: 2001/2. FAO, Rome

FAO (2002) The state of food insecurity in the world 2001. Published by the Food and Agriculture Organization of the United Nations Viale delle Terme di Caracalla, 00100 Rome, Italy, pp 4–7

Farfante I, Kensley B (1997) Penaeoid and sergestoid shrimps and prawns of the world. Keys and diagnoses for the families and genera, vol 175, Memoires du Museum National d' Histoire Naturelle. Editions du Muséum, Paris, pp 1–233

Flegel TW, Sriurairtana S, Wongteerasupaya C, Boonsaeng V, Panyim S, Withyachumnarnkul B (1995) Progress in characterization and control of yellow-head virus of *Penaeus monodon*. In: Browdy CL, Hopkins JS (eds) Swimming through troubled water, proceedings of the special session on shrimp farming, aquaculture '95. World Aquaculture Society, Baton Rouge, pp 76–83

Gong WK, Ong JE (1990) Plant biomass and nutrient flux in a managed mangrove forest in Malaysia Estuarine. Coast Shelf Sci 31:519–530

Ishikawa A, Taniguchi A (1994) The role of cysts on population dynamics of *Scrippsiella* spp. (Dinophyceae) in Onagawa Bay, northeast Japan. Mar Biol 119:39–44

La Fond LC (1954) On upwelling and sinking off the East Coast of India. Andhra Univ Waltair, Mem Oceanogr 49(1):117–121

Leh MUC, Sasekumar A (1984) Feeding ecology of prawns in shallow waters adjoining mangrove shores. In: Proceedings of the Asian symposium on mangrove environment – research and management, Kuala Lumpur, 1984, pp 331–353

Lightner DV, Redman RM (1993) A putative iridovirus from the penaeid shrimp *Protrachypene precipua* Burkenroad (Crustacea: Decapoda). J Invertebr Pathol 62:107–109

Lightner DV, Redman RM (1998) Shrimp diseases and current diagnostic methods. Aquaculture 164: 201–220

Lightner DV, Redman RM, Bell TA, Brock JA (1983a) Detection of IHHN virus in *Penaeus stylirostris* and *P. vannamei* imported into Hawaii. J World Maricult Soc 14:212–225

Lightner DV, Redman RM, Bell TA (1983b) Observations on the geographic distribution, pathogenesis and morphology of the baculovirus from *Penaeus monodon* Fabricius. Aquaculture 32:209–233

Lightner DV, Redman RM, Hasson KW, Pantoja CR (1995) Taura syndrome in *Penaeus vannamei*: histopathology and ultrastructure. Dis Aquat Organ 21:53–59

Lightner DV, Redman RM, Williams RR, Mohney LL, Clerx JPM, Bell TA, Brock JA (1985) Recent advances in penaeid virus disease investigations. Infectious hypodermal and hematopoietic necrosis a newly recognized virus disease of penaeid shrimp. J World Aquacult Soc 16:267–274

Loosanoff VL (1962) Effects of turbidity on some larval and adult bivalves. In: Proceedings of the Gulf and Caribbean Fisheries Institute. Springer 14:80–95

Loosanoff VL, Tommers FD (1948) Effect of suspended silt and other substances on the rate of feeding oysters. Science 107:69–70

Maestrini SY, Graneli E (1991) Environmental conditions and ecophysiological mechanisms which led to the 1988. Chrysochromulina polylepis bloom: an hypothesis. Oceanol Acta 14:397–413

Marine Products Export Development Authority (MPEDA) (2001) Fishery status: Departmental report, vol 3, pp 77–87

Mitra A, Banerjee K, Bhattacharyya DP (2004) The other face of mangroves. Published by Department of Environment, Govt. of West Bengal, Kolkata

Nadala ECB Jr, Lu Y, Loh PC, Brock JA (1992) Infection of *P. stylirostris* (Boone) with a rhabdovirus isolated from *Penaeus* spp. Fish Pathol (Gyobyo Kenkyu) 27:143–147

Nayar KN (1987) Technology of oyster farming. Oyster culture-status and prospects. CMFRI Bull 38:59–62

Niyogi S, Mitra A, Aich A, Choudhury A (1997) *Sonneratia apetala* – an indicator of heavy metal pollution in the coastal zone of West Bengal (India). In: Iyer CSP (ed) Advances in environmental science.

Educational Publishers and Distributors, New Delhi, pp 283–287

Niyogi S, Mitra A, Saha Shyamalendu B, Amalesh C (1999) Interrelationship between diversity of prawn juveniles and mangroves in Sager Island, West Bengal, India. Proc Zool Soc 52(2):85–88

Ong JE (1982) Mangroves a carbon source and sink. Chemosphere 27:1097–1107

Ong JE (1993) Mangroves – a carbon source and sink. Chemosphere 27:1097–1107

Ong OC, Yamane HK, Phan KB, Fong HKW, Bok D, Lee RH, Fung BKK (1995) Molecular cloning and characterization of the G protein γ subunit of cone photoreceptors. J Biol Chem 270:8495–8500

Owens L (1993) Description of the first haemocytic rod-shaped virus from a penaeid prawn. Dis Aquat Organ 16:217–221

Owens L, DeBeer S, Smith J (1991) Lymphoidal parvovirus-like particles in Australian penaeid prawns. Dis Aquat Organ 11:129–134

Pimentel D, Giampietro M (1994) Global population, food, and the environment. Trends Ecol Evol 9(6):239

Ruwa RK, Polk P (1994) Patterns of spat settlement recorded for the tropical oyster *Crassostrea cucullata* and the barnacle, *Balanus Amphitrite* in a mangrove creek. Trop Zool 7:121–130

Santhanam R (1990) Influence of thermal effluents on the phytoplankton ecology of mangrove estuaries of Tuticorin, South East-coast of India. Int Symp Ecol Mangrove Ecosyst 24–30

Santhanam R, Ramanathan N, Jegatheesan G (1990) Coastal aquaculture in India. CBS Publishers and Distributors, New Delhi

Sano T, Okamoto N, Nishimura T (1981) A new viral epizootic of Anguilla japonica Temminck and Schlegel. J Fish Dis 4:127–139

Singh HS (2000) Mangroves in Gujarat: current status and strategy for conservation. Gujarat Ecological Education and research (GEER) Foundation, Gandhinagar

Stewart JE (1998) Sharing the waters: an evaluation of site fallowing, year class separation and distances between sites for fish health purposes on Atlantic salmon farms. Canadian technical report of fisheries and aquatic sciences, vol 2218. Department of Fisheries and Oceans, Dartmouth Nova Scotia, Canada, p 56

Strickland R (1983) The fertile Fjord. Plankton in Puget sound. Washington Sea Grant, Seattle

Tack JF, Van den-Berghe E, Polk P (1992) Ecomorphology of *Crassostrea cucullata* in mangrove creek. Hydrobiolopension feeding ascidians and jellyfish. Ophelia 41:329–344

Tsing A, Bonami JR (1987) A new virus disease of the tiger shrimp *Penaeus japonicus* Bate. J Fish Dis 10:139–141

Tobias D, Mendelsohn R (1991) Valuing ecotourism in a tropical rain-forest reserve. Ambio 29(2):91–93

Wongteerasupaya C, Vickers JE, Sriurairatana S, Nash GL, Akarajamorn A, Boonsaeng V, Panyim S,

Tassanakajon A, Withyachumnarnkul B, Flegel TW
(1995) A non-occluded, systemic baculovirus that
occurs in cells of ectodermal and mesodermal origin
and causes high mortality in the black tiger prawn,
Penaeus monodon. Dis Aquat Organ 21:69–77

World Bank, NACA, WWF and FAO (2002) Shrimp farming and the environment. A World Bank, NACA, WWF
and FAO Consortium Program "To analyze and share

experiences on the better management of shrimp analyze and share experiences on the better management
of shrimp aquaculture in coastal areas". Synthesis
report. Published by the Consortium, WWF-
International p 126

Wu RSS (1995) The environmental impact of marine fish
culture: towards a sustainable future. Mar Pollut Bull
31(4–12):159–166

Annexure 8A.1: Sustainable Freshwater Aquaculture in Mangrove-Dominated Indian Sundarbans Using Floral-Based Feed

Abhijit Mitra, Rajrupa Ghosh, Ajoy Mallik, Kunal Mondal, Sufia Zaman and Kakoli Banerjee

Abstract

Indian aquaculture is gradually evolving from the level of subsistence activity to that of an industry. This transformation has been made possible with the development and standardization of many new technologies. In the early 1990s, aquaculture created great enthusiasm and interest among entrepreneurs especially for shrimp farming in coastal areas. However, shrimp farming is a capital-intensive activity and uncontrolled mushrooming growth of it has led to outbreak of diseases and attributed environmental issues calling for closure of shrimp farms.

Although India has vast freshwater resources, they are not fully exploited except for carp culture in limited scale. Freshwater fish culture employing composite fish culture technology has become popular for use in large number of tanks and ponds in the country. To meet the raw material required by the processing units for export demand, there is urgent need to expand production base of India. In addition, it is always stressed that there is a need to utilize our natural resources productively to ensure the much needed food security.

Considering the high export potential, the giant freshwater prawn, *Macrobrachium rosenbergii*, the scampi, has immense potential for culture in India. About four million ha of impounded freshwater bodies in the various states of India offer great potential for freshwater prawn culture. Scampi can be cultivated for export through monoculture in existing as well as new ponds or with compatible freshwater fishes in existing ponds. Since the world market for scampi is expanding with attractive prices,

A. Mitra (✉) • S. Zaman
Department of Marine Science, University
of Calcutta and Department of Oceanography,
Techno India University, Salt Lake Campus,
Kolkata, West Bengal 700 091, India
e-mail: abhijit_mitra@hotmail.com

R. Ghosh • A. Mallik • K. Mondal • K. Banerjee
Department of Marine Science, University
of Calcutta, 35, B.C. Road, Kolkata,
West Bengal 700 019, India

there is great scope for scampi production and export. However, the quality of the product must be of high grade to keep the export chain sustainable. To fulfil this objective, the present pilot programme was initiated at P.S.-Swarupnagar, Sub-Div:Basirhat, Dist-North 24 Parganas in West Bengal. In order to maintain the quality of the product as well as of the culture environment (pond), attempt was undertaken to replace the animal (trash fish) ingredient of prawn feed with floral components, and for this the salt-marsh grass of Sundarbans, *Porteresia coarctata*, was used as protein source to the culture species. The technology of feed preparation has been highlighted in the report. The formulated feed from the salt-marsh grass not only improved the aquatic health of the pond but at the same time increased the growth and protein level of the prawns. In the present venture, the overall production picture in the experimental ponds speaks in favour of the efficacy of the herbal-based prawn feed. The investigation also shows that prawn farming in the freshwater system of Indian Sundarbans is an economically feasible project and the return can be enhanced if specially formulated herbal feed is provided to the culture species instead of the traditional one.

List of Acronyms

S. No.	Abbreviation	Full form
1.	DOC	Days of culture
2.	BOD	Biological oxygen demand
3.	COD	Chemical oxygen demand
4.	OD	Optical density
5.	CI	Condition index
6.	FCR	Feed conversion ratio
7.	Δf	Change in feed mass
8.	Δb	Change in biomass
9.	APHA	American Public Health Association
10.	ANOVA	Analysis of variance
11.	MPN	Most probable number
12.	DO	Dissolved oxygen

8A.1.1 Introduction

In the Indian subcontinent, aquaculture contributed over one-third of the country's total fish production of 6.1 million tonnes in 2003. The total aquaculture production of 2.2 million tonnes was valued at US$ 2.5 billion of which carp alone was responsible for as much as 1.87 million tonnes. On the other hand, the production of high-valued shellfish species, namely, giant river prawn, produced 30,000 tonnes, while shrimps from brackish water, mainly *P. monodon*, produced 1,15,000 tonnes (FAO 2005). Aquaculture plays an important role in Indian economy and is greatly related with the socio-economic condition of the fish farmer. Apart from supplying quality protein to the consumer, this sector plays a major role in providing employment opportunities and earning foreign exchange. After the high risk of disease outbreak and environmental pollution, the aquaculturists have shifted their interest from shrimp (*Penaeus monodon*) culture towards scampi culture. The giant freshwater prawn (*Macrobrachium rosenbergii*) known as scampi in commercial parlance is a highly valued delicious food and commands very good demand in both domestic and export market. *Macrobrachium rosenbergii* culture is gradually gaining momentum in the present *era* owing to its price, taste, fast growth rate, less susceptibility to diseases and its compatibility to grow with carps. It is now farmed in many countries; the major producers (>200 million tonnes) are Bangladesh, Brazil, China, Ecuador, India, Malaysia, Taiwan Province of China and Thailand (FAO 2002). About four million ha of impounded freshwater bodies in the various states of India offer great potential for freshwater prawn culture. Scampi can be cultivated for export through monoculture in existing as well as new ponds or with compatible freshwater fishes in existing ponds. Today freshwater aquaculture has witnessed diversification through the incorporation of high-valued species like scampi and has increased its production from 455 tonnes in 1992 to over 30,000 tonnes in 2003. Freshwater aquaculture activity is prominent in the eastern part of the country, particularly the states of West Bengal, Orissa and Andhra Pradesh with new areas coming under culture in the states of Punjab, Haryana, Assam and Tripura. The state of Andhra Pradesh dominates the sector with over 86 % of the total production in India with approximately 60 % of the total water area dedicated to prawn farming, followed by West Bengal. In the state of West Bengal, the existing culture system includes both monoculture and polyculture with Indian major carps in ponds. Grow-out stocking densities range from 0.5 to 2.5 scampi/m^2 in polyculture and 1–5/m^2 in monoculture. The culture period is 6–8 months at the beginning of southwest monsoon (June/July with temperature around 27–30 °C), and the scampi are fed with farm-made or commercial feeds (Mitra et al. 2005). In Indian Sundarbans, freshwater aquaculture, in general, is practised with the utilization of low to moderate levels of inputs, especially organic-based fertilizers and feed. The concept of using balanced diet or feed rich in protein for boosting the production is still in an embryo stage. In many pockets of North 24 Parganas district, meat/flesh of live mussels and several gastropod species (mainly *Telescopium telescopium*) is used with the aim to increase the quantum of final harvest.

In Indian Sundarbans, standard practices in freshwater prawn culture include:

- Pond fertilization with organic manures from cattle or poultry as well as inorganic fertilizers like urea and single super phosphate.
- Provision of supplementary feeds mainly in the form of a mixture of rice bran/wheat bran and groundnut/mustard oilcake in equal ratio.
- Meat/flesh of molluscs is often used as supplementary feed.

It is however documented from several literatures that in a successful *Macrobrachium* culture,

the main thrust is generally given to the dietary protein content. The retention of dietary protein for growth is the goal of nutritionists for the development of cost-effective diets. It has been observed from a series of pilot projects that successful and sustainable aquaculture is a direct function of proper feed which should be nutritionally balanced, eco-friendly and economically viable (Kaushik 1990). For producing cost-effective diet and maintaining an ecologically sound culture system environment, nowadays, artificial feeds particularly of floral origin are extensively used.

Studies conducted on the proximate analysis of green seaweeds, salt-marsh grass and mangrove litter revealed considerable percentage of protein in these flora (Chakraborty and Santra 2008; Banerjee et al. 2009a, b; Hoq et al. 2002). The biomass of these floral species, as documented from the field study, also indicates the viability of using these species as ingredients for mass production of prawn feed. On this background matrix, the present programme was initiated to develop a sustainable scampi culture on the foundation of mangrove floral resources. Another important objective of the present study is to maintain the stability of aquatic health, which deteriorates mostly due to use of traditional feed (having trash fishes as the main ingredients) and flesh of molluscs.

8A.1.2 Objectives

Mangroves and their associate halophytes are unique vegetations with multiple ecological benefits, but linking these endemic vegetations directly with the livelihood of the Sundarban people is the primary aim of this programme. To reach this goal, we designed our proposal to undertake the following objectives:

- To screen the candidate flora (salt-marsh grass) for protein content (quantitative estimation) with the aim to develop eco-friendly nutritive feed for freshwater prawn (*Macrobrachium rosenbergii*)
- To develop location-specific fish feed preparation technology (through incorporation of floral pulp rich in protein as the major feed ingredient)

- To investigate the impact of formulated feed on the prawn biomass, survival rate, condition index (a function of length and weight) and FCR
- To investigate the interrelationship between protein-rich fish feed and protein level in prawn tissues, through regular monitoring of protein content of culture species
- To investigate the impact of formulated feed on water quality in terms of salinity (as salt water often penetrates the pond from adjacent brackish-water canal), pH, dissolved oxygen (DO), BOD, COD, nutrient concentrations, phytopigment level and organic carbon of bottom soil of ponds

This experimental venture was designed not only to upgrade the ecological health of the culture pond but at the same time to understand the responses of the culture species towards the feed prepared from mangrove associate floral species. It is expected that such venture may lead to economic upliftment of the poverty-stricken island dwellers and open window of alternative livelihood for the local community.

8A.1.3 Physiography of the Study Area

The study area is located at P.S.-Swarupnagar, Sub-Div:Basirhat, Dist-North 24 Parganas in West Bengal. The geographic location of the study area is 22°48′19″ N latitudes and 88°54′21.5″ E longitudes. Physically it is a wetland (Bilballi) along both sides of Sonai canal. The area of the Bilballi is around 15 km². The Sonai canal is connected to the Ichhamati River (Fig. 8A.1.1). The wetland gets submerged under water during monsoon (July–October) and postmonsoon (November–February) and remains exposed during premonsoon (March–June).

The area around the Bilballi wetland is dominated by the fishermen community, who are mostly engaged in wild catch from the canal. However, a section of this fishing community practises polyculture in ponds.

8A.1.4 Materials and Methods

8A.1.4.1 Selection of Pond

The selection of ponds was done on the basis of water availability and water quality. The study area Bilballi is a low-lying land, and adequate water is available throughout the culture period (February 2010–September 2010) which is one of the prime requisites for scampi culture. The salinity of the culture area is also low (3–6 psu) and the pH ranges from 6.80 to 8.10. Water is drawn from the adjacent Sonai canal to fill the culture ponds. The pond availability throughout the entire culture period was ensured after negotiating with the pond owner (also the beneficiary) on the basis of the following terms and conditions:

- Feed has to be supplied (as per the requirement) by the University of Calcutta.
- The university will have no claim on the final harvest (except for research related data).
- Day-to-day monitoring of the cultured species and feeding will be done by the beneficiary as per the instruction and training of the researchers of Calcutta University.
- The university research group will be monitoring the water quality and cultured species in every fortnightly interval, and the beneficiary will be required to assist the team during this monitoring phase.

8A.1.4.2 Pond Preparation and Water Filling

Two ponds were selected, of which one was experimental and the other was treated as control. The area of the experimental pond is 7,500 m², and the control pond is 10,000 m². At the very initial stage of the experiment, attention was given on scientific pond preparation. For this purpose, ponds were dried sufficiently in order to decompose all organic matters, to oxidize different toxic compounds present in the soil of pond bottom and also to eliminate undesirable filamentous algal mat and eggs of different predatory fishes, crab, etc. Then lime was applied accordingly to maintain

Fig. 8A.1.1 Map showing the culture site in the Bilballi wetland (*red mark*)

soil pH and neutralize the organic acid, pyrite, etc. present in the pond bottom. After the preparation of the pond, water was filled which was mainly done by storing the rainwater and by pumping the water from the adjacent Sonai canal.

8A.1.4.3   Stocking of Seeds

Prawn seed collection is a common practice in coastal West Bengal, which is presently discouraged by all sections of the society due to its linkage with several environmental issues like ecological crop loss (mass destruction of several fish juveniles), uprooting of mangrove seedlings and health problems of seed collectors. To step aside all these dark environmental issues, seeds

were procured from a commercial hatchery of Nellore district of Andhra Pradesh and stocked on 5 February 2010 at a rate of two individual/m^2 with initial size 0.80 cm and 0.01 g body weight in each pond. The mean stocking weight was determined from a sample of 100 prawn seeds that were blotted to free from water. Before stocking all the prawn seeds are well acclimatized to avoid temperature and pH shocks (Sarver et al. 1982).

8A.1.4.4   Feed Preparation and Feeding Rate

The feed preparation was initiated since the beginning of the programme from the first week of February 2010. The feed composition of the

Table 8A.1.1 Feed ingredients

Ingredients	% by weight
Soybean dust	10
Mustard oil cake (MOC)	34
Rice bran	17
Wheat bran	4
Porteresia powder	30
Vitamin + mineral (mix)	5

Table 8A.1.2 Feed ingredients of locally used commercial feed

Ingredients
Soybean dust (10 %)
Mustard oil cake (34 %)
Rice bran (17 %)
Wheat bran (4 %)
Fish/prawn head dust (30 %)
Vitamin + mineral (mix) (5 %)

experimental pond (Table 8A.1.1) shows the presence of 30 % dust of salt-marsh grass *Porteresia coarctata*, which is a mangrove associate species of deltaic Sundarbans.

The *Porteresia* powder was prepared by drying the pulp taken out from the plant material. The pulp was dried in hot air oven at 45–50 °C and then it was powdered. All the ingredients were weighed and mixed well. Dough was prepared with warm water which was passed through a pelletizer to obtain the desired feed pellets. All of these were carried out in the laboratory of Department of Marine Science, University of Calcutta.

The commercial feed (mostly available from the local market) was provided to the control pond whose ingredient composition is given below in Table 8A.1.2.

As a part of scientific culture, feed chart (Table 8A.1.3) was maintained on the basis of DOC during the culture period in the experimental pond.

8A.1.4.5 Analysis of Physico-chemical Parameters

The success of prawn culture and production of prawn is a direct function of water quality, and the most relevant aquatic parameters in context to

Table 8A.1.3 Feed chart

First month	25 % of biomass
Second month	20 % of biomass
Third month	15 % of biomass
Fourth month	10 % of biomass
Fifth month	5 % of biomass
Sixth month	2 % of biomass
Seventh month	No feed
Eighth month	No feed

aquaculture are listed below along with detailed methodology:

(a) Surface water temperature
(b) Surface water salinity
(c) Surface water pH
(d) Soil pH
(e) Organic carbon content of the soil
(f) Dissolved oxygen
(g) Nutrient (nitrate, phosphate, silicate) concentration
(h) BOD
(i) COD
(j) Total coliform
(k) Phytopigment concentration

All these parameters were regularly monitored at an interval of 15 days (*fortnightly interval*) following the standard protocols (Strickland and Parsons 1972; American Public Health Association (APHA), American Water Works Association, Water Pollution Control Federation 1998) as discussed here.

(a) *Surface water temperature*
    Surface water temperature was recorded in both experimental and control ponds by a Celsius thermometer at an interval of every 15 days.

(b) *Surface water salinity (psu)*
    Surface water salinity was checked in the field by refractometer and cross-checked in the laboratory by argentometric method.

(c) *Surface water pH*
    Surface water pH was recorded in the field by a portable pH meter (sensitivity = ±0.1) during each field trip.

(d) *Soil pH*
    Sediment was collected from pond bottom during each field trip. 10 g of the collected sediment sample was dried in hot air oven. The dried sample was stirred with 100 ml of distilled water, and it was kept for few hours.

Then the pH of the supernatant water was measured by using portable pH meter (sensitivity = ±0.1)

(e) *Organic carbon content of the soil*
The organic carbon content of the collected pond sediment was estimated using the procedure of Walkey and Black (Walkley and Black 1934). 1 g of dried soil sample was taken in a conical flask, and 6 ml of distilled water and 1 ml of phosphoric acid were added. It is kept in hot air oven for 10 min. Then 10 ml $K_2Cr_2O_7$ and Ag_2SO_4 was added and it is kept in dark for 30 min. After that 200 ml distilled water, 10 ml phosphoric acid and 1 ml DPA indicator were added. Then it was titrated against Mohr salt which was taken in the burette. Finally organic carbon is calculated as % of carbon using the following formula:

$$\% \text{ of carbon} = (3.95 \div g) \times (1 - T / S)$$

where G = weight of sample in g
S = Mohr salt solution for blank
T = Mohr salt solution for sample

(f) *Dissolved oxygen*
The sample was taken from the surface water of the pond in the Winkler bottle of capacity to approximately 150 ml. The water was allowed to flow at a moderate speed avoiding air bubbles. 1 ml of Winkler-I solution was added followed by 1 ml of Winkler-II solution. The bottle was carefully closed with a stopper avoiding the trapping of air bubbles and the bottle was vigorously shaken; after the precipitation settled, 1 ml (1:1) H_2SO_4 solution was added. The bottle was kept in a dark place and titrated in a maximum time of 1 h. The dissolved sample was quantitatively washed down into the conical flask and was titrated with standard thiosulphate solution till the very pale straw colour remained. Then 1 ml of starch solution was added, and the titration was continued to get colourless solution. The volume of thiosulphate was noted and DO of the pond water was calculated using the following formula:

$$\text{DO}(\text{mg} / 1) = (V_1 \times N \times 32{,}000) / 4\left(V_2 - \frac{2}{125}\right)$$

where N = strength of sodium thiosulphate
V_1 = volume of sodium thiosulphate
V_2 = volume of the sample taken

(g) *Nutrient (nitrate, phosphate, silicate) concentration*
Surface water for nutrient analysis was collected in clean TARSON bottles and transported to the laboratory in ice-freezed condition. Triplicate samples were collected from the same collection site to maintain the quality of the data. The standard spectrophotometric method of Strickland and Parson (1972) was adopted to determine the nutrient concentration in surface water. Nitrate was analyzed by reducing it to nitrite by means of passing the sample with ammonium chloride buffer through a glass column packed with amalgamated cadmium fillings and finally treating the solution with sulphanilamide. The resultant diazonium ion was coupled with *N*-(1-napthyl)-ethylene diamine to give an intensely pink azo dye. Determination of the phosphate was carried out by treatment of an aliquote of the sample with an acidic molybdate reagent containing ascorbic acid and a small proportion of potassium antimony tartrate. Dissolved silicate was determined by treating the sample with acidic molybdate reagent. The resultant silico-molybdic acid was reduced to molybdenum blue complex by ascorbic acid and incorporating oxalic acid prevented formation of similar blue complex by phosphate. SYSTRONIC UV–VIS spectrophotometer (Type117, Sr.No.690) was used for nutrient (NO_3, PO_4 and SiO_3) analysis at their respective wavelengths.

(h) *BOD (Biochemical oxygen demand)*
1. Two 300 ml BOD bottles were half-filled with dilution water. With a large-tipped pipette, the pre-calculated amount of sample was dispensed into each of the two 300 ml of BOD bottles. Then each bottle was filled with dilution water and the stopper was inserted and all air bubbles were excluded.
2. An additional two 300-ml BOD bottles with only dilution water were filled, and the stopper was inserted as in step 1.

3. At 20 °C, one bottle containing diluted samples and one containing only dilution water were incubated.

4. A DO determination on the remaining BOD bottles from step1 and step 2 was run and the initial DO content was recorded.

5. After 5 days, DO determination tests were done with the incubated bottles. The DO content of the incubated bottles was recorded. There should not be an increase or decrease of more than 0.2 mg/l of DO between initial dilution water and final dilution water. Large changes may be caused by improper techniques or contaminated dilution water.

$$\text{BOD} = (D_1 - D_2) - (B_1 - B_2) f / P \text{ mgl}^{-1}$$

where D_1 is the dissolved oxygen of diluted sample immediately after preparation, mg/l; D_2 is dissolved oxygen of diluted sample after 5 days incubation at 20 °C, mg/l; P is decimal volumetric fraction of sample used; B_1 is dissolved oxygen of seed control before incubation, mg/l; B_2 is dissolved oxygen of seed control after incubation, mg/l; and f is ratio of seed in sample to seed in control = (% seed in D_1)/(% seed in B_1).

(i) *COD (chemical oxygen demand)*
Organic substances in the sample were oxidized by potassium dichromate in 50 % sulphuric acid solution at reflux temperature. Silver sulphate was used as a catalyst, and mercuric sulphate was added to check chloride ion interference.

(j) *Total coliform*
For the microbial analysis of surface water in terms of *total coliform* load, the *most probable number (MPN)* procedure by *Multiple Fermentation Technique (5 test tube method)* was followed as stated in APHA (1998). The technique involves inoculating the sample and/or its several dilution in a liquid medium of lauryl tryptose broth. After expiry of the incubation period, the tubes were examined for gas and acid production by the coliform organisms. This test is known as *presumptive test*. Since the organisms other than the coliforms may also produce this reaction, the positive tubes from the presumptive test were subjected to a *confirmatory test* using *Brilliant Green Lactose Bile Broth*. The density of bacteria was calculated on the basis of positive and negative combination of the tubes using MPN table. In the case of *water samples*, the results were expressed in MPN/100 ml.

(k) *Phytopigment (chlorophyll a) concentration*
Phytopigment concentration of the ambient aquatic phase was analyzed in order to monitor the productivity of the water. For pigment analysis, 1 l of surface water, collected from each of the pond, was filtered through a 0.45-μm Millipore membrane fitted with a vacuum pump. The residue along with the filter paper was dissolved on 90 % acetone. And it was kept in a refrigerator for about 24 h in order to facilitate the complete extractions of the pigment. The solution was centrifuged for about 20 min under 5,000 rpm, and the supernatant solution was considered for the determination of the chlorophyll pigment by recording the optical density at 750, 664, 647 and 630 nm with the help of SYSTRONICS UV–VIS spectrophotometer (Type177, Sr. No. 690). All the extinction values were corrected for a small turbidity blank by subtracting the 750 nm signal from all the optical densities, and finally phytoplankton pigment was estimated as per the following expressions of Jeffrey and Humprey (1975):

$$\text{Chl } a = 11.85 \text{ OD}_{664} - 1.54 \text{ OD}_{647} - 0.08 \text{ OD}_{630}$$

The values obtained from the equation were multiplied by the volume of the extract (in ml) and divided by the volume of the water (in litre) filtered to express the chlorophyll content in mg/m^3.

8A.1.4.6 Monitoring of Zootechnical Parameters

Individual weights and lengths of prawns ($n = 100$) were taken at fortnightly interval during the entire culture period, and the relevant response variables were determined for each control and experimental ponds.

The length–weight relationship of the cultured species was determined to evaluate the proportionality in growth for both control and experimental ponds. Length–weight relationships have been extensively used for estimation of weight from length due to technical difficulties and the amount of time required to record weight in the field, conversion of growth in length equations to growth in weight for use in stock assessment models, estimation of the biomass from length observations and estimation of the condition factors of the aquatic species. In addition to the above, length–weight relationships are useful for understanding spatial and temporal variations of life histories of cultured species in response to environmental variables, feed type, etc.

Condition index (C.I.) was analyzed at fortnightly interval during the culture period as per the expression: $C.I. = W/L^3 \times 100$, where W = weight of the cultured species (in g) and L = length of the cultured species (in cm).

Percentage weight gain was calculated as the difference in weight from the average final weight with respect to the initial weight; weight gain = [(average individual final weight – average individual initial weight)/average individual initial weight] × 100.

Feed consumption was estimated on the basis of the total amount of feed provided to the cultured species during the culture tenure (6 months). Feed conversion ratio (FCR) was analyzed after the harvesting of shrimps as per the expression: $FCR = \Delta f/\Delta b$, where Δf = change in feed biomass and Δb = change in body biomass of the cultured species.

The survival rate was measured as percentage of the difference of stocking number and production volume (No.) at the end of the culture period.

Body pigmentation was assessed for each treatment on prawn cooked for 5 min in boiling water and comparing the orange-red colouration with Roche SalmoFan™ colour score. The astaxanthin content of the harvested prawn was also analyzed by the standard chemical method.

8A.1.4.7 Protein Estimation

Protein of feed and prawns were estimated by Lowry's method as per the following procedure:

Protein sample was mixed with 10 ml phosphate buffer in tissue homogenizer and was homogenized and centrifuged. The supernatant was used to estimate the proteins. 0.2, 0.4, 0.6, 0.8 and 1 ml of working standard of bovine serum albumin were pipetted out in a series of previously cleaned test tubes and volume was made out to 1 ml. 1 ml supernatant was also pipetted out into one another test tube. A tube containing 1 ml distilled water was taken as a blank. 5 ml of reagent C was added to each tube and after well mixing kept for 10 min. 0.5 ml of reagent D was added to all test tubes and mixed well. The mixture was incubated in room temperature at dark condition for 30 min. Blue colour was developed and OD values were measured at 600 nm. Then a graph was drawn and factor was calculated. The concentration of protein was calculated using following relation:

$$\text{Protein concentration (\%)} = \text{Factor} \times \text{OD value of the sample} \times 100$$

8A.1.4.8 Statistical Analysis

Analysis of variance (ANOVA) was computed between all the selected parameters (indicators of our experiment) considering both control and experimental ponds to evaluate the differences caused by inclusion of *Porteresia coarctata* dust in the feed. All statistical calculations were performed with SPSS 9.0 for Windows.

8A.1.5 Results

8A.1.5.1 Physico-chemical Parameters

The results of physico-chemical parameters of the two ponds (control and experiment) are shown in Table 8A.1.4 and Fig. 8A.1.2.

The surface water temperature during the study period ranged from 28.3 to 36.1 °C with a mean value of 32.4 ± 2.87 °C in the control pond, while in the experimental pond the value ranged from 28.3 to 36.2 °C with a mean value of 32.4 ± 2.93 °C.

Table 8A.1.4 Monthly variations of environmental parameters in experimental and control ponds during 8 months culture period of freshwater prawn (*Macrobrachium rosenbergii*)

	February		March		April		May		June		July		August		September	
	E	C	E	C	E	C	E	C	E	C	E	C	E	C	E	C
Surface water temperature (°C)	28.3 (28.0–28.6)	28.3 (28.1–28.6)	29.1 (29.0–29.4)	29.2 (29.0–29.6)	30.3 (30.0–30.7)	30.5 (30.0–30.6)	33.8 (33.5–33.9)	33.7 (33.5–33.9)	35.4 (35.0–35.7)	35.4 (35.0–35.7)	36.2 (36.0–36.6)	36.1 (35.9–36.4)	34.0 (33.8–34.2)	34.1 (33.8–34.3)	32.2 (32.0–32.4)	32.1 (32.0–32.5)
Dissolved oxygen (ppm)	5.85 (5.80–5.91)	5.93 (5.90–5.97)	5.41 (5.35–5.42)	5.38 (5.33–5.40)	5.46 (5.42–5.48)	5.13 (5.10–5.16)	5.09 (5.05–5.15)	4.52 (4.49–4.55)	4.87 (4.84–4.91)	4.11 (4.09–4.16)	4.58 (4.53–4.60)	3.35 (3.31–3.39)	4.89 (4.87–4.92)	3.89 (3.85–3.91)	5.00 (4.98–5.04)	3.91 (3.80–3.95)
pH	8.02 (8.00–8.03)	8.1 (8.09–8.11)	8.00 (7.99–8.02)	7.98 (7.96–8.00)	8.00 (7.98–8.01)	7.73 (7.71–7.74)	7.95 (7.93–7.97)	7.68 (7.67–7.69)	7.91 (7.89–7.93)	7.57 (7.56–7.59)	7.9 (7.88–7.91)	7.4 (7.38–7.41)	7.82 (7.81–7.83)	6.93 (6.92–6.95)	7.8 (7.78–7.81)	6.88 (6.86–6.89)
Salinity (‰)	4.89 (4.85–4.90)	5.12 (5.10–5.15)	5.67 (5.65–5.69)	6.05 (6.03–6.07)	4.88 (4.85–5.16)	5.12 (5.09–5.17)	5.25 (5.23–5.29)	5.83 (5.80–5.87)	4.69 (4.65–4.72)	4.9 (4.88–4.95)	3.9 (3.85–3.91)	3.85 (3.80–3.88)	3.00 (2.98–3.04)	3.4 (3.29–3.45)	4.66 (4.64–4.70)	5.12 (5.02–5.14)
Nitrate (µgat/l)	10.67 (10.11–12.02)	9.81 (9.60–10.99)	11.66 (11.05–12.01)	13.17 (11.05–13.35)	13.89 (12.00–13.99)	16.06 (15.99–16.08)	15.22 (15.18–15.25)	19.76 (19.70–19.82)	14.87 (13.10–14.99)	21.33 (20.05–21.76)	12.44 (11.43–13.02)	23.75 (22.00–24.80)	13.69 (12.90–14.17)	23.8 (21.95–24.00)	15.05 (14.52–16.11)	25.01 (25.37–25.98)
Phosphate (µgat/l)	1.06 (0.99–1.12)	0.95 (0.82–1.03)	1.02 (0.97–1.08)	0.98 (0.86–1.03)	1.33 (1.22–1.41)	1.78 (1.75–1.82)	1.89 (1.85–1.90)	2.06 (1.83–1.91)	2.01 (1.97–2.06)	2.33 (2.29–2.38)	1.98 (1.96–1.99)	3.30 (2.95–3.08)	1.74 (1.68–1.82)	3.68 (3.42–3.90)	1.65 (1.60–1.75)	3.44 (3.38–3.50)
Silicate (µgat/l)	26.45 (25.00–27.01)	25.32 (24.60–26.10)	23.41 (22.89–24.00)	27.35 (26.10–28.33)	34.87 (33.10–35.32)	31.09 (29.17–32.11)	29.67 (28.55–30.10)	27.9 (26.29–28.14)	28.88 (27.35–29.16)	28.7 (26.28–29.13)	27.45 (26.30–29.00)	28.6 (27.43–29.82)	25.40 (23.90–25.80)	26.90 (25.15–27.21)	33.23 (31.10–34.69)	29.18 (28.73–30.27)
BOD	3.2 (3.0–3.3)	3.5 (3.3–3.4)	5.1 (5.0–5.2)	6.4 (6.0–6.5)	4.8 (4.0–5.0)	7.5 (6.9–7.7)	5.3 (4.9–5.6)	8.0 (7.7–8.2)	6.8 (6.5–7.0)	10.0 (9.8–10.3)	5.9 (5.5–6.4)	12.1 (12.0–12.5)	4.7 (4.5–4.9)	11.5 (11.2–11.7)	4.0 (3.8–4.3)	13.6 (13.2–13.8)
COD	73 (70–76)	95 (90–98)	81 (76–83)	113 (110–115)	79 (75–80)	123 (120–125)	85 (80–90)	126 (120–135)	90 (85–95)	119 (110–125)	86 (80–90)	120 (115–125)	81 (75–85)	115 (110–120)	72 (70–75)	131 (128–134)
Total coliform (MPN/100 ml)	365 (360–373)	359 (350–365)	423 (415–430)	528 (518–540)	567 (550–570)	617 (610–620)	426 (418–430)	702 (698–715)	418 (402–425)	756 (750–760)	394 (390–412)	788 (780–795)	380 (370–386)	895 (890–905)	384 (378–390)	1136 (1129–1140)

Chlorophyll a (mg/m³)	2.9 (2.88–2.91)	2.93 (2.92–2.94)	2.98 (2.95–2.99)	2.95 (2.93–2.99)	3.15 (3.12–3.17)	3.28 (3.25–3.30)	3.26 (3.20–3.29)	3.66 (3.65–3.69)	3.71 (3.68–3.73)	4.02 (3.99–4.05)	3.58 (3.55–3.60)	3.92 (3.90–3.98)	3.44 (3.42–3.46)	4.33 (4.30–4.35)	3.10 (3.07–3.12)	4.57 (4.55–4.58)
Soil pH	7.10 (7.08–7.12)	7.05 (7.03–7.06)	7.20 (7.18–7.22)	8.02 (8.00–8.03)	7.25 (7.24–7.27)	7.60 (7.59–7.62)	7.25 (7.23–7.27)	7.95 (7.93–7.97)	7.10 (7.08–7.12)	7.70 (7.68–7.71)	7.12 (7.11–7.13)	7.30 (7.28–7.31)	7.05 (7.04–7.08)	6.2 (6.19–6.22)	7.02 (7.00–7.03)	5.25 (5.23–5.27)
Organic carbon (%)	1.08 (1.06–1.10)	1.03 (1.00–1.04)	1.11 (1.08–1.15)	1.42 (1.10–1.15)	1.15 (1.12–1.16)	1.56 (1.16–1.19)	1.09 (1.17–1.22)	2.77 (1.35–1.40)	1.33 (1.50–1.55)	2.98 (1.96–1.99)	2.00 (2.46–2.48)	3.11 (2.80–2.87)	2.09 (1.56–1.62)	3.63 (2.94–2.97)	2.14 (1.45–1.49)	3.89 (3.07–3.17)

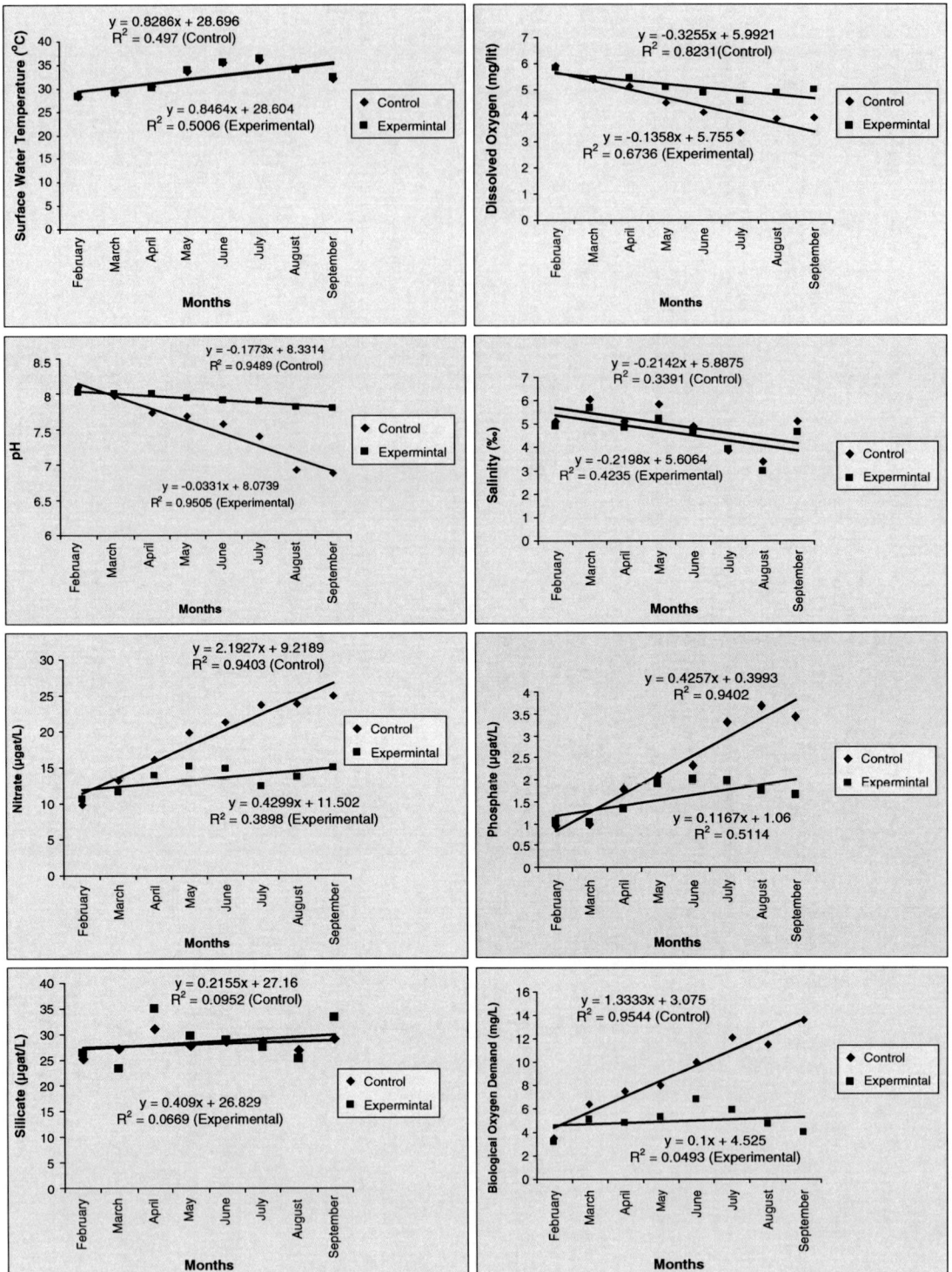

Fig. 8A.1.2 Hydrological parameters and soil conditions of control and experimental ponds with trend line equations

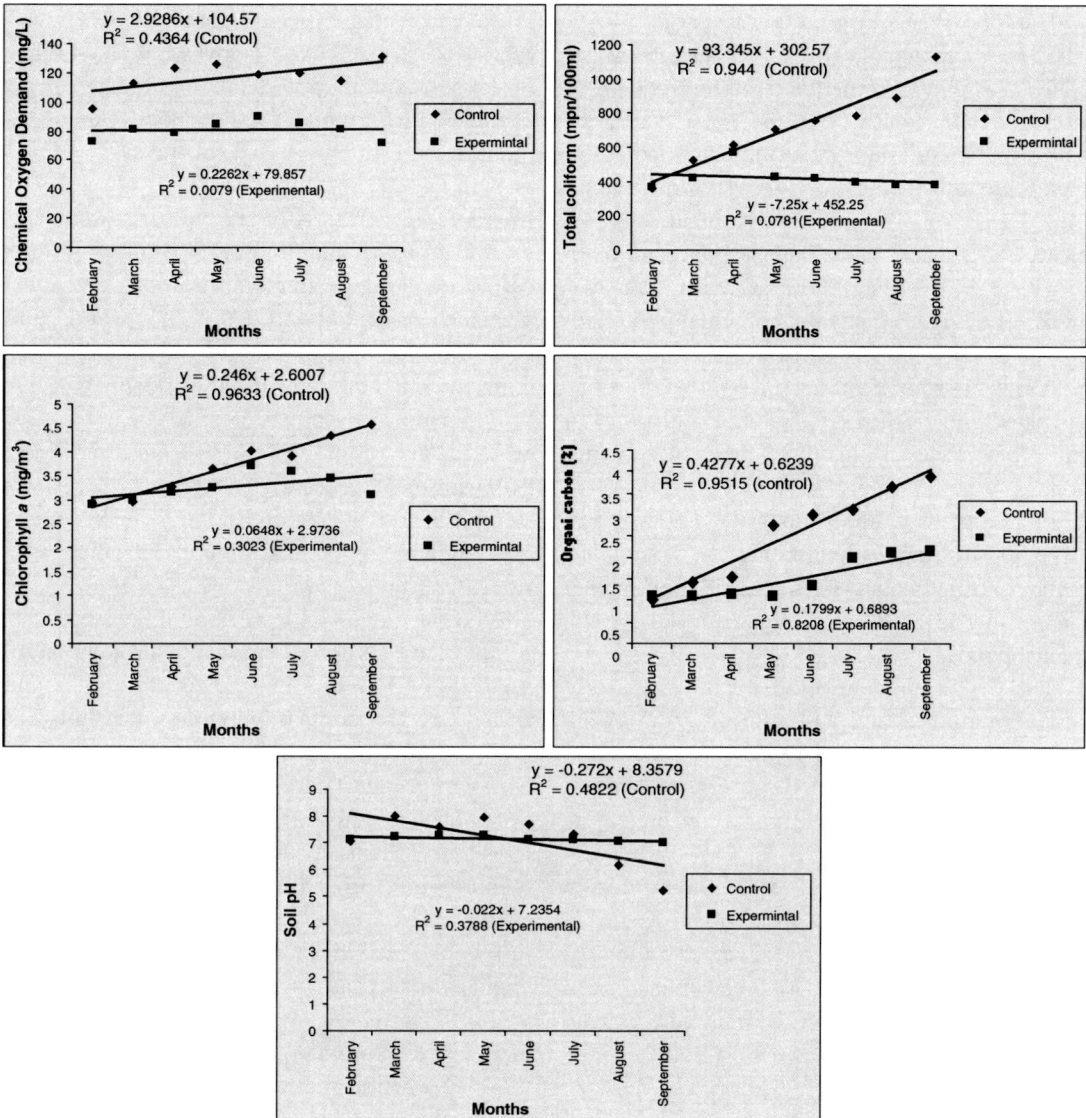

Chemical Oxygen Demand (mg/L) vs Months:
$y = 2.9286x + 104.57$
$R^2 = 0.4364$ (Control)
$y = 0.2262x + 79.857$
$R^2 = 0.0079$ (Experimental)

Total coliform (mpn/100ml) vs Months:
$y = 93.345x + 302.57$
$R^2 = 0.944$ (Control)
$y = -7.25x + 452.25$
$R^2 = 0.0781$ (Experimental)

Chlorophyll a (mg/m^3) vs Months:
$y = 0.246x + 2.6007$
$R^2 = 0.9633$ (Control)
$y = 0.0648x + 2.9736$
$R^2 = 0.3023$ (Experimental)

Organi carbon (%) vs Months:
$y = 0.4277x + 0.6239$
$R^2 = 0.9515$ (control)
$y = 0.1799x + 0.6893$
$R^2 = 0.8208$ (Experimental)

Soil pH vs Months:
$y = -0.272x + 8.3579$
$R^2 = 0.4822$ (Control)
$y = -0.022x + 7.2354$
$R^2 = 0.3788$ (Experimental)

Fig. 8A.1.2 (continued)

The surface water salinity ranged from 3.40 to 6.05 psu with a mean value of 4.92 ± 0.90 psu in the control pond and 3.00 to 5.67 psu with a mean value of 4.61 ± 0.82 psu in the experimental pond.

The surface water pH ranged from 6.88 to 8.10 with a mean value of 7.53 ± 0.44 in the control pond and from 7.80 to 8.02 with a mean value of 7.92 ± 0.08 in the experimental pond.

The DO values ranged from 3.35 to 5.93 mg/l with a mean value of 4.52 ± 0.87 mg/l in the control pond and from 4.58 to 5.85 mg/l with a mean value of 5.14 ± 0.4 mg/l in the experimental pond.

The nutrient (nitrate) ranged from 9.81 to 25.01 µgat/l with a mean value of 19.08 ± 5.53 µgat/l in the control pond and from 10.67 to 15.22 µgat/l with a mean value of 13.43 ± 1.68 µgat/l in the experimental pond.

The phosphate concentration ranged from 0.95 to 3.68 µgat/l with a mean value of 2.25 ± 1.04 µgat/l in the control pond and from 1.02 to 2.01 µgat/l with a mean value of 1.58 ± 0.39 µgat/l in the experimental pond.

In case of silicate the concentration ranged from 25.32 to 31.09 µgat/l with a mean value of 28.13 ± 1.71 µgat/l in the control pond and from 23.41 to 34.87 µgat/l with a mean value of 28.67 ± 3.87 µgat/l in the experimental pond.

The BOD value ranged from 3.5 to 13.6 mg/l with a mean value of 9.07 ± 3.34 µgat/l in the control pond and from 3.2 to 6.8 mg/l with a mean value of 4.97 ± 1.1 µgat/l in the experimental pond.

In case of COD the value ranged from 95 to 131 mg/l with a mean value of 117.75 ± 10.85 µgat/l in the control pond and from 72 to 90 mg/l with a mean value of 80.87 ± 6.22 µgat/l in the experimental pond.

The Chl a concentration during the study period ranged from 2.93 to 4.57 mg/m^3 with a mean value of 3.7 ± 0.61 mg/m^3 in the control pond and from 2.90 to 3.71 mg/m^3 with a mean value of 3.26 ± 0.28 mg/m^3 in the experimental pond.

The total coliform load (MPN value) during the study period ranged from 359/100 ml to 1,136/100 ml with a mean value of 722.62 ± 235.33 MPN/100 ml in the control pond and from 365/100 ml to 567/100 ml with a mean value of 419.62 ± 63.53 MPN/100 ml in the experimental pond.

The organic carbon of the pond bottom soil during the study period ranged from 1.03 to 3.15 % with a mean value of 1.956 ± 0.9 % in the control pond and from 1.08 to 2.47 % with a mean value of 1.45 ± 0.45 % in the experimental pond.

The pH of the pond bottom soil ranged from 5.25 to 8.02 with a mean value of 7.13 ± 0.95 in the control pond and from 7.02 to 7.25 with a mean value of 7.13 ± 0.08 in the experimental pond.

8A.1.5.2 Zootechnical Parameters

Prawns fed with *Porteresia*-based diet exhibited higher final weights and better weight gain at the end of the experiment (Figs. 8A.1.3 and 8A.1.4). C.I. values of prawns were also higher in experimental ponds than control pond (Table 8A.1.5), which implies a better environment in experimental pond for the survival and growth of the species. The survival rate was found to be 71.2 % in the control pond and 76.5 % in experimental pond. The biotic indicators of the experimental approach are summarized in Table 8A.1.6. The trend line equations of length–weight relationship for experimental and control ponds during the culture period are shown in Figs. 8A.1.2 and 8A.1.3. The allometric equations reveal the proportionate increase of weight with respect to length in experimental pond. The R^2 values also indicate good significance of the trend lines. On contrary, in case of control pond the allometric equations reveal completely a different picture where there is a disproportionate increase of weight with respect to length. The R^2 values for control pond are insignificant showing minimum goodness of fit.

An important factor governing the consumer acceptance and market value of many cultivated fish and prawn species is the pink or red colouration of their flesh or boiled exoskeleton (Brun and Vidal 2006). In the wild, this colouration is achieved through the ingestion of carotenoid pigments particularly astaxanthin contained within invertebrate food organisms (Johnson et al. 1977; Ibrahim et al. 1984). The *Porteresia*-based feed in the present study resulted in higher astaxanthin values in harvested prawns of experimental pond (115.56 ppm) as reflected through darker orange-red colouration of shrimp exoskeleton in comparison to control pond (84.38 ppm). Roche SalmoFan™ colour score showed the value of 24 in control pond, much less than experimental pond with a colour score of 29. The protein content was also higher in the prawns fed with *Porteresia*-based feed (Table 8A.1.7).

The present pilot-scale study speaks in favour of healthy pond environment, better and proportionate growth, higher survival rate and low FCR values through use of *Porteresia*-based feed.

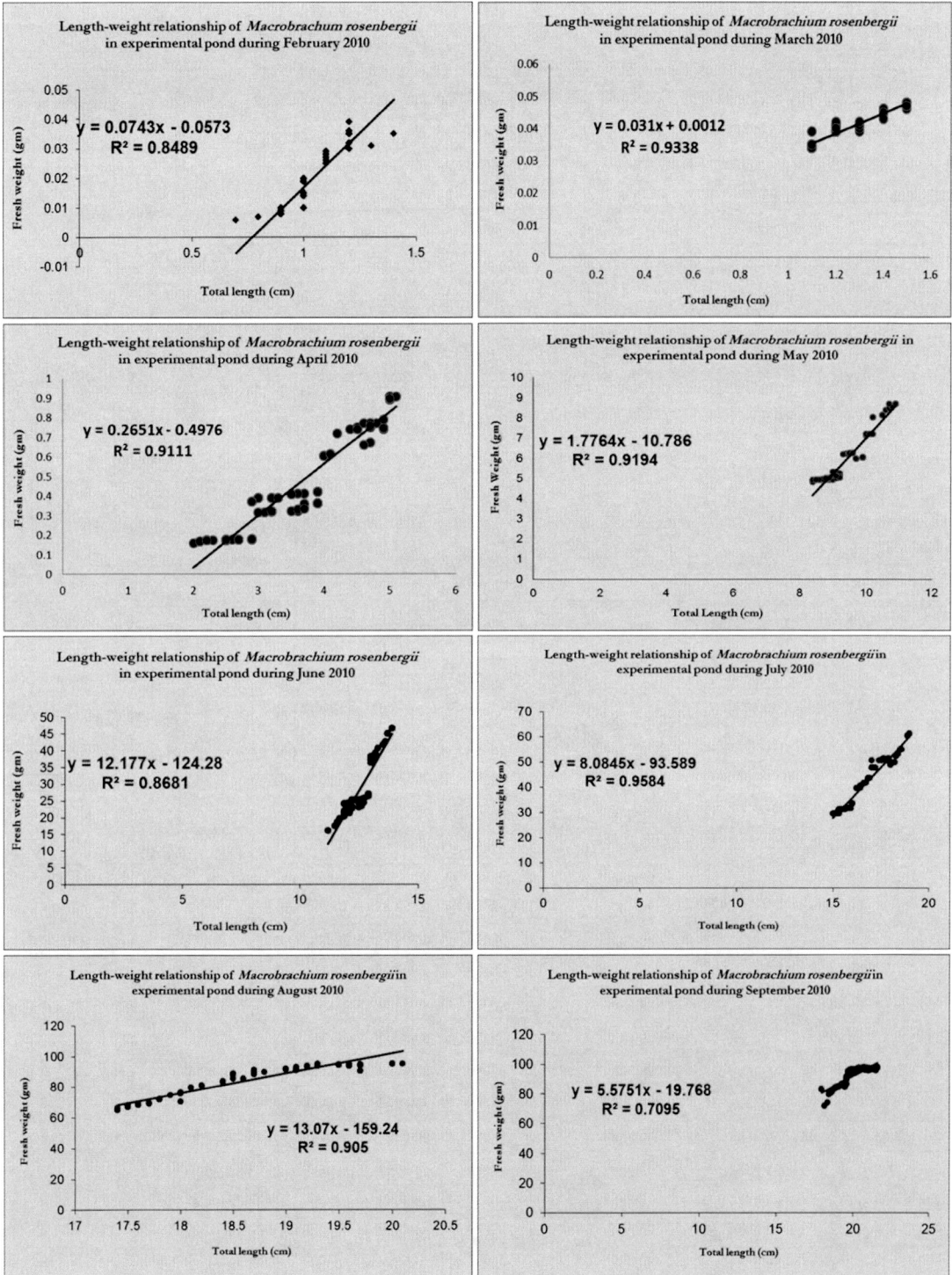

Fig. 8A.1.3 Length–weight relationship of *Macrobrachium rosenbergii* in experimental pond

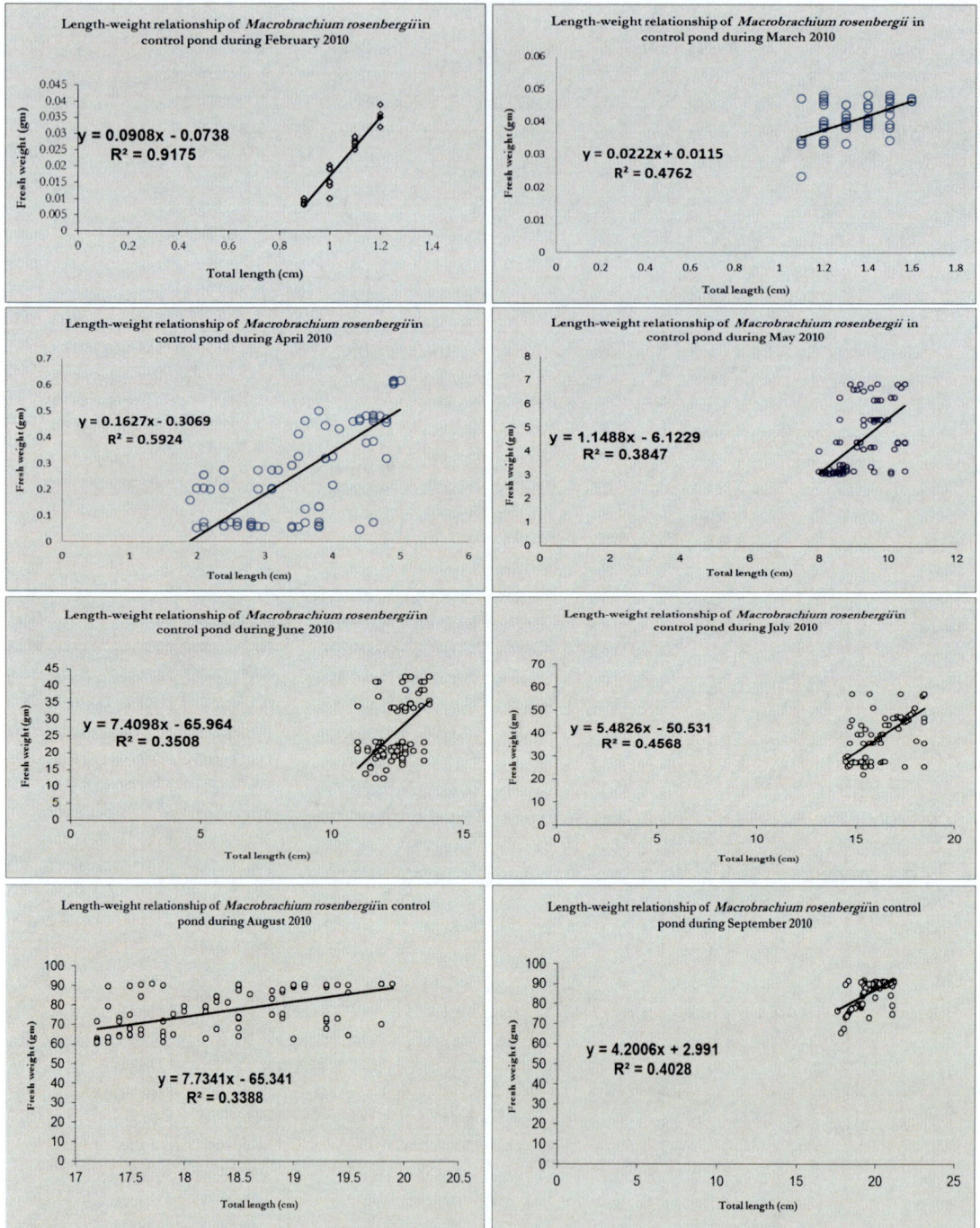

Fig. 8A.1.4 Length–weight relationship of *Macrobrachium rosenbergii* in control pond

8A.1.6 Discussion

ANOVA results indicate no significant differences between surface water temperature, salinity and pH between two ponds which may be attributed to location of both the ponds in the same area. The linear distance between experimental and control pond is approximately 50 m. Significant differences with respect of organic

Table 8A.1.5 Monthly variation of condition index in experimental and control ponds

Months	Experimental	Control
February	1.694	1.624
March	1.898	1.75
April	0.954	0.595
May	0.708	0.561
June	1.474	1.348
July	0.892	0.887
August	1.303	1.243
September	1.129	1.113

Table 8A.1.6 A comparative account of zoo-technical parameters (associated with *M. rosenbergii* culture) in control and experimental ponds

Parameters	Control pond	Experimental pond
Prawn biomass (n = 100 for each month)	Initial biomass = 0.016 g	Initial biomass = 0.017 g
	Final biomass = 86.003 g	Final biomass = 92.676 g
Survival rate	71.2 %	76.5 %
Specific growth rate (%/day)	3.578	3.584
Length–weight relationship	Mostly non-uniform (weight has not proportionately increased with length) vide Fig. 8A.1.3	Uniform; vide Fig. 8A.1.2
FCR	1.428	1.213
Condition index (average of 8 months)	1.14	1.25

carbon and pH of pond bottom soil, dissolved oxygen, nutrient load, BOD, COD and phytopigment concentrations of water were observed ($p<0.01$) which clearly indicates the difference in water quality due to application of different types of feed. The formulated feed prepared from salt-marsh grass not only upgraded the water but also reduced the total coliform count which was higher in case of control pond (where commercial feed available in local market was applied). This commercial feed contains trash fish and shrimp dust as a source of protein. The residual commercial feed deteriorated the water quality by increasing the organic carbon, nutrient load (except silicate), BOD, COD and total coliform.

The trend lines of these variables along with their respective equations and R^2 values confirm the variation of water quality through application of mangrove-based feed.

Critical analysis of the zootechnical parameters reveals better growth of the species cultured in experimental pond. In addition to increase of survival rate of the cultured species, the specific growth rate has also increased in the experimental pond (Tables 8A.1.6 and 8A.1.8).

Analysis of the length–weight relationship reveals some interesting features like proportionate increase of length and weight of the prawns in the experimental pond throughout the culture tenure. On the contrary, such proportionality has decreased with the increase of age of the stocked individuals in the control pond (Figs. 8A.1.2 and

Table 8A.1.7 Protein level in formulated feed and prawn

Month	Experimental pond		Control pond	
	Feed protein%	Prawn protein%	Feed protein%	Prawn protein%
February	35.16 ± 1.2	25.58 ± 1.3	15.23 ± 1.5	25.58 ± 1.4
March	34.16 ± 1.7	28.34 ± 1.6	15.23 ± 1.4	27.98 ± 1.8
April	34.16 ± 1.5	30.26 ± 1.4	15.23 ± 1.6	29.34 ± 1.6
May	33.16 ± 1.5	34.89 ± 1.4	15.23 ± 1.5	31.77 ± 1.8
June	34.16 ± 1.4	35.78 ± 1.3	16.23 ± 1.4	34.13 ± 1.9
July	36.16 ± 1.3	36.12 ± 1.5	17.23 ± 1.4	34.55 ± 1.7
August	No feed	36.65 ± 1.6	No feed	34.90 ± 1.6
September	No feed	37.90 ± 1.5	No feed	35.04 ± 1.7

Table 8A.1.8 A comparative picture of growth performance, survival rate, condition index and FCR (Feed Conversion Ratio) related to *M. rosenbergii* culture in control and experimental ponds

Parameters	Experimental pond	Control pond
Pond dimension (m²)	7,500	10,000
Stocking date	05.02.2010	05.02.2010
Stocking density	2/m²	2/m²
Initial average weight (g)	0.017	0.016
Total no. of seed stocked = (area × stocking density)	15,000	20,000
Date of harvest	30.09.2010	30.09.2010
Final average weight (g)	92.676	86.003
Final harvest (in tonnes)	1.063	1.224
Specific growth rate (%/day)	3.584	3.578
Survival rate (%)	76.5	71.2
Condition index	1.25	1.14
Feed given (kg)	1,289.42	1,747.87
FCR	1.213	1.428

8A.1.3). This feature indicates that the formulated feed from *Porteresia coarctata* regulates the length–weight relationship throughout the culture period in a uniform pattern. The deviation from uniformity in control pond may be attributed to use of commercially available feed from local market which was fed without any regularity as stated in the feed chart (Table 8A.1.3). This may increase the load of residual feed in the pond bottom leading to less biomass production in the control pond. The FCR value is a litmus test of the situation.

Low FCR value in experimental pond indicates that majority of the feed has been converted into biomass which is also an indication acceptability of *Porteresia coarctata*-based feed by the species. In case of control pond, the FCR value (relatively higher than the experimental pond) reflects a major quantum of wastage with low input in the biomass sector. The maximum percentage of feed wasted in case of control pond remained as residual feed. The residual feed degrades the water quality and pond environment which is again indicated by the condition index value. The index is the reflection of the

health of ambient environment of cultured species, and its lower value (as seen in case of control pond) is a reflection of degraded environment. ANOVA performed with the monthly condition index values (Table 8A.1.5) indicates significant difference ($p < 0.01$) between the two ponds. The *Porteresia coarctata*-based prawn feed seems to be the major player for such variation.

Plant proteins have been found to be relatively poorly utilized in crustaceans in terms of growth in comparison to protein of animal origin. However, digestibility studies in freshwater prawn have indicated that the species can efficiently digest both plant and animal protein sources (Ashmore et al. 1985). The omnivory of freshwater prawn permits the use of a wide variety of locally available feedstuffs including commercial by-products as ingredients in formulated diets. To create a balanced diet, it is necessary to establish the minimum protein level to provide essential amino acids (Guillaume 1997; Tacon and Akiyama 1997). However, in the present study, the amino acids for *P. coarctata* diet have not been determined. Hence, the protein level of the salt-marsh grass may be the factor responsible for acceleration of growth and survival percentage. Millikin et al. (1980) indicated that *M. rosenbergii* species attains best growth at 40 % protein level in feed. Castell et al. (1989) have concluded that protein level ranging between 30 and 38 % resulted in the best growth of the species.

The results of the present studies partially differ with the findings of Du and Niu (2003), who conducted a study in tank water on *M. rosenbergii* fed with diets where 0, 20, 50, 75 and 100 % of fish meal is replaced by soybean meal. They concluded that soybean meal, without supplementation of amino acids or other additives, is not suitable as a major protein source in freshwater prawn diets. However, Weidenbach (1980) reported that prawns are able to adjust to the absence of feed pellets by increasing consumption of available vegetation. Tidwell et al. (1993) indicated that animal ingredients like fish could be partially or totally replaced by soybean meal and distiller by-products in diets for the pond production of freshwater prawns. According to Tidwell et al. (1995), prawns may be able to adjust to reductions in the nutritional value of prepared diets (i.e. protein source and vitamin and mineral content) by increasing predation on natural fauna (i.e. macro invertebrates) in the pond. All these studies reflect the wide adaptability of the species to different categories of feed. With this wide range of metabolic adaptability, inclusion of salt-marsh grass in the feed ingredient list of freshwater prawn will not only upgrade the ecological environment but will also ensure a better quality product and a livelihood option for the local inhabitants.

8A.1.7   Summary

Aquaculture has become a peak industry in the present millennium, which involves farming with prawn, cuttlefish, squid, lobster and other such culinary delights actually 'cultivated' in aquatic enclosures under scientifically controlled conditions (Rajkhowa 2005). The use of nutrient-rich feed continues to gain wide acceptance in the aquaculture industry in order to boost up the quality of the aquacultural products. Such feed results in substantial reduction in the overall variable cost of an operation through improved animal performance (indicators are length–weight relationship, specific growth rate, etc.), better FCR and improved water quality due to reduction in the amount of nutrients and suspended solids (*i.e.* faecal matter and uneaten residual food particles) in the culture system. *Porteresia*-based formulated feed showed better growth performance of the cultured species with respect to condition index values and survival rate (Table 8A.1.8). Body pigmentation improved in the cultured species of experimental pond and showed significantly higher astaxanthin level than the controlled pond. A series of experiments are still needed for time testing the results and make the programme sustainable for the poor island dwellers of lower Gangetic delta.

8A.1.8   A Way Forward

The farming of giant river prawn (*Macrobrachium rosenbergii*) has gained increased interest in recent years, due to its high economic value and an annual production of over 30,000 tonnes through the use of monoculture practice. In addition, the sector has been witnessing increased interest in diversification with the inclusion of high-valued species, including medium and minor carps, catfishes and murrels. While carp and other finfishes are grown for the domestic market, a large proportion of freshwater prawn production is exported. Aquaculture in India is usually practised with the utilization of low to moderate levels of inputs, especially organic-based fertilizers and feed. India utilizes only about 40 % of the available 2.36 million hectares of ponds and tanks for freshwater aquaculture, and there is still enough room for expansion. In areas like the one where this pilot programme was carried out, there is a great potential for aquaculture. However, this is practised in a non-scientific way without any water and feed management. The inevitable results are disease outbreak and mass mortality of the cultured species particularly shrimps and prawns. Under this circumstance, the present programme and its success motivated the local aquaculturists. Suggestions came from their end to scale-up the venture, which is beyond the capacity of the academic institute like the University of Calcutta. Hence, the Forest Department, Govt. of West Bengal, being the major stakeholder of Sundarbans, was approached with the proposal of training the local people the art of prawn feed preparation from salt-marsh grass, *Porteresia coarctata*. The department agreed to this proposal. It is expected that this untapped service of mangroves may be used for the betterment of local people and island dwellers of Indian Sundarbans. The coordinator feels that a publication from the end of MFF, IUCN, may spread the message of such nonconventional use of mangroves to stakeholders engaged in the pisciculture sector.

The Sundarbans mangrove region is a threatened ecosystem due to a multitude of factors like prolonged overexploitation of its natural resources, its use as a sink of anthropogenic wastes, industrial and maritime wastes generated in the upstream of the rivers flowing through the region, high population pressure around the region and the resultant shrinkage of the area brought about by clearing of forest land for agriculture and tiger prawn culture and lack of proper ecological management. The present programme has immense ecological and economic relevance in connection to these threats, due to its connection to the following lanes:

- Upgradation of the freshwater system (canals, ponds, ditches, etc.) and therefore clearance of mangrove areas for the culture of *Penaeus monodon* may be totally avoided.
- Involvement of the local people in three livelihood tiers: preparation of fish feed, eco-friendly culture practice and nursery development of *Porteresia coarctata* for raw material backup to sustain the floral-based fish feed industry. It is expected that such involvement will restrict a sizable fraction of the people from intruding into the forest.
- Scientific utilization of mangrove floral resources for sustainable pisciculture practice in the area.
- Improvement of aquatic health in terms of physico-chemical parameters due to replacement of animal ingredients (like trash fish dust, shrimp dust) in the traditional fish feed with floral components.
- Introduction of a new technology be fitted to the area.
- Economic upliftment of the local people.

References of Annexure 8A.1

American Public Health Association (APHA), American Water Works Association, Water Pollution Control Federation (1998) Standard methods for the examination of water and wastewater, 20th edn. American Public Health Association, Washington, DC

Ashmore SB, Standby RW, Moore LB, Malecha SR (1985) Effect on growth and apparent digestibility of diets varying in gram source and protein level in *Macrobrachium rosenbergii*. J World Maricult Soc 16:205–216

Banerjee K, Ghosh R, Homechaudhuri S, Mitra A (2009a) Biochemical composition of marine macroalgae from Gangetic delta at the apex of Bay of Bengal. Afr J Basic Appl Sci 1(5–6):96–104

Banerjee K, Ray D, Basu S, Chakraborty B, Mitra A (2009b) A comparative study of Astaxanthin level in

mangrove species. Proc Natl Acad Sci India, Sec B 79(Pt.II):135–142

Behanan L, Mukandan MK, Korath A, Sherif PM (1992) Formulation and evaluation of a particulated feed for *Macrobrachium rosenbergii* post-larvae. In: Silas EG (ed) Freshwater prawn. Kerala Agriculture University, Thrissure, pp 238–240

Boonyaratpalin M, Suraneiranat O, Tunpibal T (1998) Replacement of fish-meal with various types of soybean products in diets for the Asian sea bass, *Lates calcarifer*. Aquaculture 161:67–78

Boyd CE (1990) Water quality in ponds for aquaculture. Alabama Agricultural Experiment Station, Auburn University, Alabama, 482pp

Castell JD, Kean JC, D'Abramo LR, Conklin DE (1989) A standard reference diet for crustacean nutrition. I. Evaluation of two formulations. J World Aquacult Soc 20(3):93–99

Chakraborty S, Santra SC (2008) Biochemical composition of eight benthic algae collected from Sunderban. Indian J Mar Sci 37:329–332

D'Abramo LR, Brunson MW (1996) Production of freshwater prawns in ponds. SRAC (Southern Regional Aquaculture Center) Publication No. 484, Stoneville

Daniel S (1981) Introducing *Macrobrachium rosenbergii*. Freshw Mar Aquarium 4(7):32–34

Du L, Niu CJ (2003) Effects of dietary substitution of soybean meal for fish meal on consumption, growth, and metabolism of juvenile giant freshwater prawn, *Macrobrachium rosenbergii*. Aquacult Nutr 9(2):139–143

FAO (2002) Fishstat Plus (v. 2.30), 15.03.2002. FAO, Rome

FAO (2005) Aquaculture production, 2003. Yearbook of fishery statistics, vol 96/2. Food and Agriculture Organization of the United Nations, Rome

Gitte MJ, Indulkar ST (2005) Evaluation of marine fish meat incorporated diets on growth and survival of post-larvae of *Macrobrachium rosenbergii* (de Man). Asian Fish Sci 18:323–334

Guillaume J (1997) Protein and amino acid. In: D'Abramo RL, Concknil ED, Akiyama DM (eds) Crustacean nutrition, vol 6, Advances in World Aquaculture Society. World Aquaculture Society, Louisiana, pp 26–41

Hall MR, Van Hamm EH (1998) The effect of different types of stress on blood glucose in giant tiger prawn *Penaeus monodon*. J World Aquacult Soc 29:290–299

Hari B, Kurup BM (2003) Vitamin C (Ascorbyl 2 Polyphosphate) requirement of freshwater prawn *Macrobrachium rosenbergii* (de Man). Asian Fish Sci 15:145–154

Hilton JW, Harrison KE, Silnger SJ (1984) A semipurified test diet for *Macrobrachium rosenbergii* and the lack of need for supplemental lecithin. Aquaculture 37(3):209–215

Hoq ME, Islam ML, Paul HK, Ahmed SU, Islam MN (2002) Decomposition and seasonal changes in nutrient constituents in mangrove litter of Sundarbans mangrove, Bangladesh. Ind J Mar Sci 31(2):130–135

Ibrahim A, Shimizu C, Kono M (1984) Pigmentation of cultured red sea bream, *Chrysophrys major*, using astaxanthin from antarctic krill, *Euphausia superba*, and a mysid, *Neomysis* sp. Aquaculture 38:45–57

Jeffrey SW, Humphrey GF (1975) New spectrophotometric equations for determining chlorophylls *a*, *b*, *c*, and *c2* in higher plants, algae and natural phytoplankton. Biochem Physiol Pflanzen 167:191–194

Jintasataporn O, Tabthipwon P, Yenmark S (2004) Substitution of golden apple snail meal for fishmeal in giant freshwater prawn, *Macrobrachium rosenbergii* (de Man) diets. Kasetsart J (Nat Sci) 38:1–7

Johnson EA, Conklin DE, Lewis MJ (1977) The yeast *Phaffia rhodozyma* as a dietary pigment source for salmonids and crustaceans. J Fish Res Board (Canada) 34:2417–2421

Kaushik SJ (1990). Use of alternative protein sources for the intensive rearing of carnivorous fishes. In Flos R, Tort L, Torres P (eds) Mediterranean aquaculture. Ellis Horwood, Chichester, pp 125–138

Kaushik SJ, Cravedi JP, Lalles JP, Sumpter J, Fauconneau B, Laroche M (1995) Partial or total replacement of fish meal by soybean protein on growth, protein utilization, potential estrogenic or antigenic effects, cholesteroemia and flesh quality in rainbow trout, *Oncorhynchus mykiss*. Aquaculture 133:257–274

Krishnankutty N (2005) Plant proteins in fish feed. Curr Sci 88:865–867

Krishnankutty N, Sujatha TR (2003) Role of plant proteins in formulated fish feeds. Curr Sci 85:247–249

Md Hasanuzzaman AF, Md Siddiqui N, Md Chisty AH (2009) Optimum replacement of fishmeal with soybean meal in diet for *Macrobrachium rosenbergii* (De Man 1879) cultured in low saline water. Turk J Fish Aquat Sci 9:17–22

Millikin MR, Fortner AR, Fair PH, Sick LV (1980) Influence of several dietary protein concentrations on growth, feed conversion and general metabolism of the juvenile prawn (*Macrobrachium rosenbergii*). In: Proceedings of the World Mariculture Society, Elsevier Scientific Publishing, Amsterdam vol 11

Mitra G, Mukhopadhyay PK, Chattopadhyay DN (2005) Nutrition and feeding in freshwater prawn (*Macrobrachium rosenbergii*) farming. Aqua Feeds Formul Beyond 2(1):17–19

New MB (1995) Status of freshwater prawn farming: a review. Aquacult Res 26:1–54

Nyirenda J, Mwabumba M, Kaunda E, Sales J (2000) Effect of substituting animal protein sources with soybean meal in diets of *Oreochromis karogae* (Trewavas 1941). Naga, The ICLARM Q 23:13–15

Rajkhowa I (2005) Action in aquaculture-opportunities in aquatic specialization. Bus Today (May 22 Issue):131

Sandifer PA, Smith TIJ (1975) Effect of population density on growth and survival of *Macrobrachium rosenbergii* reared in re-circulating water management system. Proc World Maricult Soc 6:43–53

Sandifer PA, Smith TIJ (1977) Intensive rearing of post larval Malaysian prawn(*Macrobrachium rosenbergii*) in a closed cycle nursery system. Proc Maricult Soc 8:225–235

Sandifer PA, Smith TIJ (1985) Freshwater prawns. In: Huner JV, Brown EE (eds) Crustacean and mollusk

aquaculture in the United States. Avi Publishing, Westport, pp 63–125

Sarver D, Malecha S, Onizuka D (1982) Possible sources of variability in stocking mortality in post larval *Macrobrachium rosenbergii*. In: New MB (ed) Giant prawn farming, developments in aquaculture and fisheries science, vol 10. Elsevier Scientific Publishing, Amsterdam, pp 99–113

Smith TIJ, Sandifer PA (1979) Development and potential of nursery system in the farming of Malaysian prawns. In: Proceedings of the World Mariculture Society, Elsevier Scientific Publishing Company, Amsterdam 10:369–384

Smith TIJ, Jenkins WE, Sandifer PA (1983) Enclosed prawn nursery systems and effect of stocking juveniles *Macrobrachium rosenbergii* in ponds. J World Maricult Soc 14:111–125

Strickland JDH, Parsons TR (1972) A practical handbook of sea-water analysis, 2nd edn. J Fish Res Bd Can 167:311pp

Tacon AG, Akiyama DM (1997) Feed ingredients. In: D' Abramo RL, Concknil ED, Akiyama DM (eds) Crustacean nutrition, vol 6, Advances in World Aquaculture Society. World Aquaculture Society, Louisiana, pp 411–472

Tidwell JH, Webster CD, Yancey DH, Abramo LRD (1993) Partial and total replacement of fish meal with soybean meal and distiller's by-products in diets for pond culture of the freshwater prawn (*Macrobrachium rosenbergii*). Aquaculture 118(1–2):119–130

Tidwell JH, Webster CD, Sedlacek JD, Weston PA, Knight WL, Hill SJ Jr, D'Abramo LR, Daniels WH, Fuller MJ, Montafiez JL (1995) Effects of complete and supplemental diets and organic pond fertilization on production of *Macrobrachium rosenbergii* and associated benthic macroinvertebrate populations. Aquaculture 138:169–180

Walkley A, Black IA (1934) An examination of Degtjareff method for determining organic carbon in soils: effect of variations in digestion conditions and of inorganic soil constituents. Soil Sci 63:251–263

Watanabe T, Pongmaneerat J, Sato S, Takeuchi T (1993) Replacement of fish meal by alternative protein sources in rainbow trout diets. Nippon Suissan Gakkaishi 59:1573–1579

Weidenbach RP (1980) Dietary components of prawns reared in Hawaiian ponds. In: Proceedings of the giant prawn conference international foundation for science report 9, Bangkok

Weindenbach RP (1982) Dietary components of freshwater prawns reared in Hawaiian ponds. In: New MB (ed) Giant prawn farming, vol 10, Developments in aquaculture and fisheries science., pp 257–267

Willis SA, Hagood RW, Eliason GT (1976) Effects of four stocking densities and three diets on growth and survival of post-larval *Macrobrachium rosenbergii* and *M. acanthurus*. In: Proceedings of the World Mariculture Society, vol 7, pp 655–665

Annexure 8A.2: Study on the Role of Mangrove-Based Astaxanthin in Shrimp Nutrition

Abhijit Mitra, Rajrupa Ghosh, Ajoy Mallik,
Kunal Mondal, Sufia Zaman and Kakoli Banerjee

8A.2.1 Preface

In the Indian subcontinent, tiger prawn (*Penaeus monodon*) is the single dominant item in the export basket of marine food, which accounts for almost two thirds of the total export earnings. Although several species of crabs, lobsters, oysters, mussels, sea cucumbers and fin fishes are in the list of exportable items, it is shrimp farming which has opened the avenue of large-scale livelihood in coastal villages of our country. The backbone of this aquaculture sector, however, depends largely on proper feed and pond management, which have been very poorly progressed and neglected in terms of research, development and technology transfer.

The rearing of large number of shrimps in relatively confined conditions necessitates a detailed understanding of their nutrition in order to provide a diet that is adequate for their optimum growth and well-being. Adequate and balanced diet for cultured shrimp is the foundation on which the success of commercial shrimp farming stands. Shrimp farming, on a global scale, till a

A. Mitra (✉) • S. Zaman
Department of Marine Science, University
of Calcutta and Department of Oceanography,
Techno India University, Salt Lake Campus,
Kolkata, West Bengal 700 091, India
e-mail: abhijit_mitra@hotmail.com

R. Ghosh • A. Mallik • K. Mondal • K. Banerjee
Department of Marine Science, University
of Calcutta, 35, B.C. Road, Kolkata,
West Bengal 700 019, India

few years back, was largely dependent upon 'natural' food with some supplementation of the by-products of agriculture, fishing and industry, such as slaughterhouse wastes; tiny shrimps, snail meat, clam meat, mussel meat in raw conditions; grain wastes; and silkworm pupae. At very low stocking densities, these diets were adequate as most of the nutrient requirements for shrimp/prawn were satisfied from natural sources. However, at high stocking densities, shrimps are dependent on artificial feed, benefiting slightly from natural feed. Thus, at high stocking densities, inadequate supplementary feeding leads to malnutrition with the resultant consequences like poor growth, increased disease susceptibility and parasitic and bacterial infestation. This was the primary cause behind the failure of blue revolution in the mid-1990s, when disease problems totally devastated the shrimp industry.

Nutritionally complete and balanced diet helps to promote faster growth of the cultured species and resistant against diseases. Nowadays special substances are incorporated in fish feed in low dose to enhance feed intake, growth and feed conversion to biomass and resistance against diseases. These are called *additives* and are obtained from various natural sources. The market value of shrimp and consumer acceptability is determined on the basis of colour, which is basically the reflection of carotenoid content in the shrimp tissue. The colouration of shrimp has been observed to be a function of astaxanthin content, which is an important constituent of additives used in shrimp feed. It is a naturally occurring carotenoid pigment possessing strong biological antioxidant property. Apart from imparting attractive colour, astaxanthin exhibits strong free radical scavenging activity and protects against lipid peroxidation and oxidative damage of LDL-cholesterol, cell membranes, cells and tissues (http://www.astaxanthin.org/). Many types of fish and crustaceans including salmon, red sea bream, shrimp and lobsters accumulate astaxanthin in their body tissues. In most cases, astaxanthin has a red-orange to pinkish tinge, and in some live crustaceans, the astaxanthin molecule is bound to a protein. During the process of cooking, the bond between the protein and astaxanthin is disrupted allowing us to see the red-orange colour of astaxanthin (http://www. astafactor.com/in-nature.htm). Considering the marine source of astaxanthin, the proposed project is aimed at formulation of shrimp feed by using *astaxanthin derived from coastal floral sources* (*Avicennia marina*, *Avicennia alba*, *Avicennia officinalis*, *Sonneratia apetala*, *Porteresia coarctata*, *Enteromorpha intestinalis*, *Ulva lactuca*, *Catenella repens* and *Sueda* sp.).

8A.2.2   Project Task

The main tasks of the project encompass the introduction of healthy seeds from hatchery (wild catch will be avoided), their culture through scientific methods by way of water management and soil management, feed preparation technology (from floral extract rich in astaxanthin), feed management and finally the cost–benefit analysis of the beneficiaries.

8A.2.3   Objectives

The prime objectives of the present programme are:
- To screen the candidate flora for astaxanthin content (quantitative estimation) with the aim to develop eco-friendly nutritive feed for shrimp (*Penaeus monodon*)
- To develop location-specific shrimp feed preparation technology (through incorporation of floral astaxanthin) and impart the same to the local people through workshops,

awareness programmes and meeting at panchayat level
- To investigate the impact of formulated feed (with astaxanthin based additive) on the shrimp biomass, survival rate, condition index and FCR
- To investigate the interrelationship between astaxanthin-rich shrimp feed and astaxanthin level in shrimp tissues, through weekly monitoring of cultured shrimp's astaxanthin content
- To investigate the impact of formulated feed (with astaxanthin based additive) on water quality in terms of salinity, DO, nutrient load and phytopigment level

8A.2.4   Introduction

Aquaculture-related feed technology for the last two decades has advanced a lot, and as a result, different types of artificially made compounded feed are being used for the culture of shrimps in different countries (SEAFDEC 1981; Liu and Mancebo 1983; Shigueno 1984). However, in this context, it is very much pertinent to note that knowledge of nutrient requirements for shrimp gathered by different successful feed manufacturers throughout the world are not well documented at this time. Every company keeps such information as a *trade secret*. Multinational companies have greater advantage in the ability to improve its feed quality rapidly and efficiently, since sufficient funds are provided for R & D (Chaudhuri 1995). Feed cost is the major limitation to profitable aquaculture in most areas. Traditionally, formulated shrimp feed contains high levels of marine protein sources. While used mainly as a source of protein, marine meals are also valued for their content of essential fatty acids, cholesterol, attractants and other unidentified growth factors. As more information becomes available on shrimp nutrition, synthetic components such as amino acids and fatty acids were used with successful results as a means to lower feed costs. More plant protein such as soybean meal is being used to replace the expensive marine proteins. Soybean meal also has the advantage of being resistant to oxidation and spoilage and is naturally clean from organisms such as fungi, viruses and bacteria that are harmful to shrimp.

To date, limited research has been conducted to evaluate the nutritive value of soybean meal as a partial or complete replacement for marine protein sources in shrimp feeds. Levels as high as 20–40 % soybean meal give acceptable performance. Results are variable within the range, however, and depend on several factors such as water quality, salinity, age of shrimp, diet composition and inclusion of synthetic amino acids. Soybean meal has been projected and estimated to remain an abundant and economical source of feed protein relative to fish meal well into the next century. Long-term success of shrimp industry will depend on increased use of soybean meal in combination with supplemental amino acids. The feed becomes nutritionally well balanced and disease resistive with the *incorporation of astaxanthin* from external sources. Today farmed salmon are fed with a diet containing natural astaxanthin to achieve the same astaxanthin profile as their wild counterparts. Esterified astaxanthin, found in *Haematococcus pluvialis*, is a stable form and is believed to be stored in the tissue without oxidation. Although astaxanthin level is an important criterion of fish species in terms of quality, research and development in this particular branch has not yet crystallized (Ziegler 1989; Shigeo et al. 1994; Finkelstein et al. 1995; Graves et al. 1996). The production of astaxanthin from *Phaffia* sp. has been extensively studied. *Phaffia* is yeast that naturally produces astaxanthin. *Phaffia* cells normally produce about 300 ppm per dry mass. This level of production is too low to develop a commercially viable synthesis. Considerable research has been performed to increase the productivity of astaxanthin synthesis in *Phaffia* (R. Mawson, U.S. Patent, **5453565**, 1995). The result of this effort has encouraged several companies in the world to produce a *Phaffia* product containing astaxanthin for the fish and shellfish aquaculture industry.

The single-celled alga *Haematococcus pluvialis* has also been extensively studied as a host to produce astaxanthin (Furubayashi 1991). Technology has been developed to take advantage of the physiology of this alga. Under growing conditions, this alga does not produce astaxanthin. However, when the culture is subjected to stress in which nutrients are eliminated from the growth medium, then the alga produces and accumulates astaxanthin. The levels of astaxanthin can be very high, and there are reports of astaxanthin accumulation of greater than 4 % per dry mass. Under most conditions, this level of astaxanthin synthesis and accumulation occurs only after several weeks of growth. Methods to reduce the time required to produce astaxanthin are currently under extensive study by several groups, and pilot-scale production is under way at several sites. As with *Phaffia*, cells of *Haematococcus pluvialis* have been found to be a suitable delivery vehicle for astaxanthin for aquaculture, and no further purification of the astaxanthin is required. A number of groups have investigated a variety of organisms and systems to produce various carotenoids. The plant *Adonis aestivalis* produces astaxanthin in the petals of the flower. Researchers have developed varieties of *Adonis* with an increased astaxanthin content (Mawson 1995). The present programme of screening the mangroves as a source of astaxanthin has not been reported yet from any part of the world.

8A.2.4.1   Astaxanthin: An Overview

Oxygen is an indispensable molecule for the growth and survival of aerobic organisms in the planet Earth. The entire mechanism of aerobic respiration resulting in the liberation of ATP is triggered by oxygen. However, this gaseous lifeline of the planet has an important demerit as it poses oxidative stress. Oxidative stress has been defined as a disturbance in the cell or organism related to *pro-oxidant–antioxidant* balance in favour of the former (Sies 1991). It differs from any other stresses in that its primary effectors, the reactive oxygen species (ROS), can arise largely in the course of normal cell metabolism (Marova et al. 2005). Oxidative stress is involved in several pathological problems, especially in chronic degenerative diseases as diabetes, atherosclerosis, cancer, and Alzheimer's disease. Question

Integrated Antioxidant System

```
        Preventive      Scavenging        Repair
        Antioxidants    Antioxidants      Enzymes

Enzymes                      Small molecules

                  Hydrophilic        Lipophilic
                  (glutathione,      (tocopherol,
                  ascorbate,         carotenoids
                  bilirubin uric     etc.)
                  acid,
                  flavonoids etc.)
```

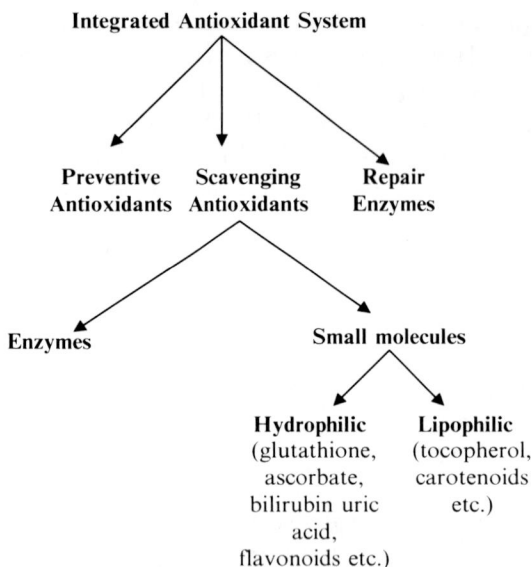

Scheme 8A.2.1 Components of integrated antioxidant system

arises how the organisms get rid of the oxidative stress. Till date the answer is related to the antioxidant defence mechanism of aerobic organisms. This defence mechanism is provided by integrated antioxidant system, which has three distinct components (Scheme 8A.2.1), each equipped to reduce oxidative stress and the resultant adverse effects.

Preventative antioxidants suppress the formation of free radicals. Radical-scavenging antioxidants, such as the flavonoid compounds and vitamin C, serve to 'mop up' excess free radicals. Thus, scavenging antioxidants remove the ROS once formed, thereby preventing radical chain reaction (Marova et al. 2005). Repair enzymes play an important role in repairing and removing ROS damaged molecules. Vitamin E and the carotenoids are very important biological antioxidants that have both preventative and radical-scavenging roles.

Astaxanthin is a carotenoid. It belongs to a larger class of phytochemicals known as terpenes. It is classified as a xanthophyll, which means 'yellow leaves'. Like many carotenoids, it is a colourful, fat/oil-soluble pigment. Astaxanthin can be found in microalgae, yeast, salmon, trout, krill, shrimp, crayfish, crustaceans and the feathers of some birds. Professor Basil

Weedon was the first to map the structures of astaxanthin (Fig. 8A.2.1).

Astaxanthin, unlike some carotenoids, does not convert to vitamin A (retinol) in the human body. Too much vitamin A is toxic for a human, but astaxanthin is not. However, it is a powerful antioxidant; it is ten times more capable than other carotenoids, which is due to its pure antioxidant nature, unlike other pigments which have pro-oxidant features (Fig. 8A.2.2).

In nature, a typical xanthophylls-producing unicellular microalgae is *Haematococcus pluvialis*, well known for its massive accumulation of ketocarotenoids, mainly astaxanthin up to 4 % of its dry mass and its acyl esters, in response to various stress conditions, *e.g.* nutrient deprivation or high irradiation. Also, the yeast *Phaffia rhodozyma* has been widely used for astaxanthin production in fed-batch fermentation processes using low-cost materials as substrates (An et al. 2001; Chociai et al. 2002; Vazquez et al. 1998). Because of antioxidative properties and the increasing amount of astaxanthin needed as a supplement in the aquaculture of salmonoids and other seafood, there is growing interest in finding out the natural reservoir of astaxanthin. The present project is a venture towards this mission in the framework of lower Gangetic region, which sustains the famous Sundarbans mangrove ecosystem.

8A.2.5 Physiography of the Study Area

The Indian Sundarbans is one of the most biologically productive and taxonomically diversified, low-lying, mangrove detritus-based, open, dynamic, heterogeneous coastal ecotones situated at the apex of the Bay of Bengal (between $21°13'$ and $22°40'$ N latitude and $88°03'$ to $89°07'$ E longitude) (Fig. 8A.2.3). The region is bordered by Bangladesh in the East, the Hugli river in the West, Dampier and Hodges line in the North and the Bay of Bengal in the South. The important rivers in this deltaic lobe from west to east are Hugli, Muriganga, Saptamukhi, Thakuran, Matla, Gosaba and Harinbhanga that finally end up at Bay of Bengal. This mangrove

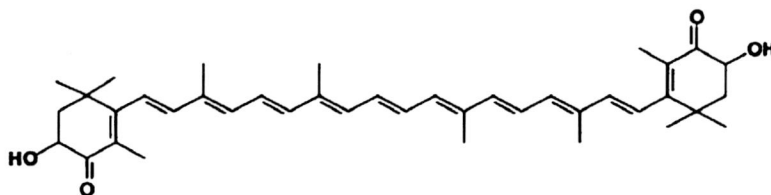

Fig. 8A.2.1 Chemical structure of astaxanthin

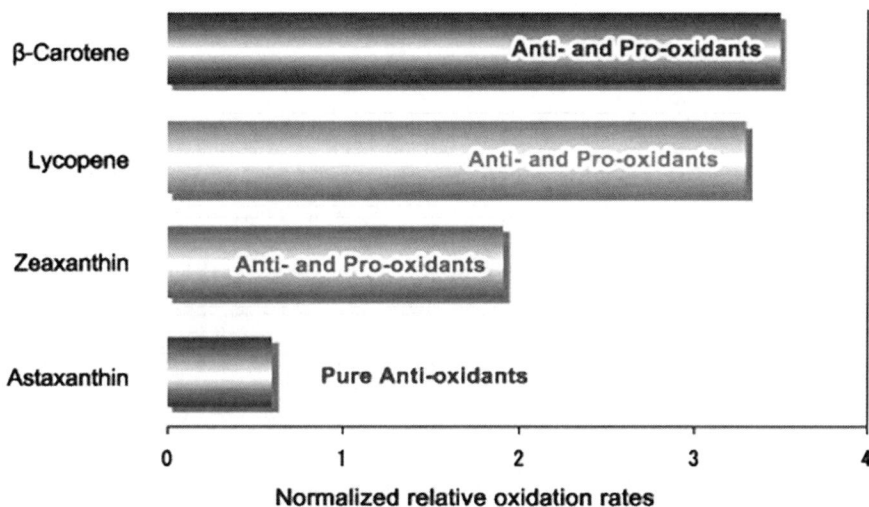

Fig. 8A.2.2 Comparative Anti- and Pro-oxidant properties of different carotenoids

forest has been declared as the world's largest mangrove forest and is declared the World Heritage Site by IUCN in 1987, Biosphere Reserve under Man and Biosphere Programme by UNESCO in 1989 and is a proposed Ramsar Site, which sustains 34 true mangrove species and about 50–55 mangrove associate species. It is the only mangrove forest in the planet Earth inhabited by the Royal Bengal Tiger (*Panthera tigris tigris*). The aquatic subsystem in and around this deltaic lobe is the cradle of several species of finfish, nursery of different variety of shell fish and reservoir of several biological resources – still untapped. The landscape of Indian Sundarbans covering an area of 9,630 km² encompasses mangrove forests, riverine, estuarine, coastal and marine habitats. On one hand, it exhibits enormous diversity based on its genesis, geographical location, hydrological regimes and substrate factors, and on the other hand, it also sustains rare endemic genetic material

which demands preservation and proper sustainable utilization for the benefit of mankind.

The climate of the area is humid (up to 96 %) and tropical with temperature ranging from 11.8 to 34.5 °C. The climate is monsoonal with an average rainfall of 1,900 mm. During monsoon months (July–October), the estuarine system becomes dominated by freshwater resulting in strong predominance of ebb tides. From November to February, the system becomes salinity gradient dominated, and during premonsoon period (March–June) due to less freshwater discharges, effects of tide are considerably accentuated resulting in the system more or less marine dominated. The low-lying tidal flats of the quaternary period have been developed from alluvial deposits of river Hugli, Saptamukhi and Matla together with tidal incursions. The soil consists of clayey loam or different black clay; there is no rock. The areas are about 1 m above the mean sea level and

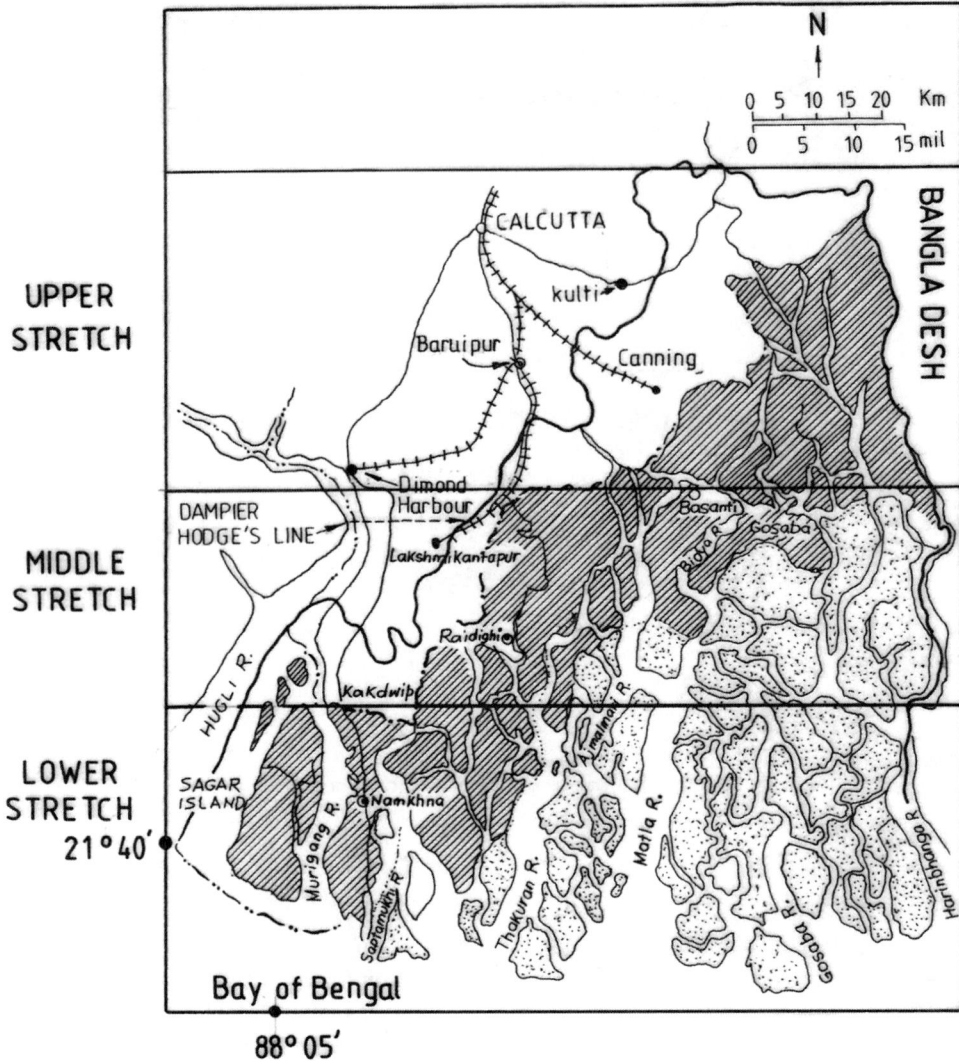

Fig. 8A.2.3 Map showing Indian Sundarbans

are submerged under saline estuarine waters for several hours in the spring tide twice a day.

There are about 102 islands in Indian Sundarbans (54 inhabited and 48 uninhabited) which supports human habitation of about 4.2 million people. This landscape has changed remarkably due to large-scale human intervention, overexploitation, demographic pressure, loss of habitats and change of ecological condition. Several of the earlier workers reported that many species have become extinct or are in a very threatened or degraded state. However, the geographical area is a rich reservoir of biotic resources for future.

8A.2.6 Materials and Methods

8A.2.6.1 First Phase: May 2007– April 2008: Screening of Mangroves for Astaxanthin

The entire network of the present programme encompassed the sampling of the leaves of ten dominant mangrove species during the low tide period *from the Jharkhali island during May 2007 to April 2008.* Leaves of the selected species were collected from two different portions (submerged lower zone and exposed upper zone) of the same plant. The lower region of the tree gets inundated during the high tide condition and the upper region of the same plant remains unexposed to tidal water. In addition to true mangrove species, the astaxanthin level of few associates like *Porteresia coarctata*, *Enteromorpha intestinalis*, *Ulva lactuca*, *Catenella repens* and *Sueda* sp. was also monitored. Salinity, pH, temperature, dissolved oxygen and nutrient load of the ambient water were analyzed simultaneously to pinpoint the hydrological parameters to which the plant species are exposed in natural condition. The collected leaves were thoroughly washed with ambient water followed with deionized water and oven dried at 110 °C overnight. The extraction of astaxanthin was done in organic solvent as per the standard method and analyzed spectrophotometrically. The mean results of all the analyses (of 12 months; May 2007 to April 2008) are shown in Table 8A.2.2.

8A.2.6.2   Second Phase: May 2008– April 2009: Shrimp Nutrition

On the basis of astaxanthin concentration, *Porteresia coarctata* was selected for feed preparation because of its high astaxanthin content. Accordingly two types of feed (for control and experimental ponds at Jharkhali) were selected. The experimental pond was provided with *Porteresia coarctata* dust to replace the fish and shrimp meal component (Table 8A.2.1) of the traditional fish feed. The second phase was thus devoted for shrimp feed preparation with the aim to replace the animal ingredients of the feed with plant matter.

8A.2.6.3   Third Phase: May 2009– April 2010: Shrimp Tissue and Water Quality Analysis

8A.2.6.3.1   Water Quality Analysis

Hydrological parameters were analyzed at fortnightly interval for a 90-day culture period (2 January–2 April 2010).

The *surface water salinity* was recorded by means of an optical refractometer (Atago, Japan)

Table 8A.2.1 Feed ingredients for shrimp culture in control and experimental ponds

Component	Control (g)	Experimental (g)	%
Rice bran	250	250	50
Wheat bran	250	250	
Fish meal	150	Salt-marsh grass	30
Shrimp meal	150	dust = 300	
Sun flower oil cake	50	50	10
Mustard oil cake	50	50	
Vitamin + mineral mixture	50	50	5
Binder	30	30	3
Shark oil	20	20	2

in the field and cross-checked in laboratory by employing Mohr–Knudsen method (Strickland and Parsons 1968). The correction factor was found out by titrating the silver nitrate solution against standard seawater (IAPO Standard Seawater Service, Charlottenlund, Slot Denmark; chlorinity = 19.376 psu). This laboratory method was applied to estimate the salinity of standard seawater procured from NIO and a standard deviation of 0.02 % was obtained for salinity. The average accuracy for salinity (in connection to the triplicate sampling) is ± 0.28 psu.

Glass bottles of 125 ml were filled to overflow from collected water samples and Winkler titration was performed for the determination of *dissolved oxygen*.

The *pH of surface water* was recorded through a portable pH meter (Hanna, USA), which has an accuracy of ± 0.1.

A Secchi disk was used to measure the *transparency* of the water column, and the data was used to calculate the euphotic depth.

Surface waters were analyzed for *nutrient concentrations* (nitrate, phosphate and silicate) following the standard spectrophotometric method (Strickland and Parsons 1972).

Organic carbon content of pond bottom soil was estimated by the standard titration method (Walkey and Black 1934).

8A.2.6.3.2   Shrimp Tissue Analysis

Individual weights and lengths of shrimps (sample size = 50 from each pond) were taken at fortnightly interval during the 90-day culture period, and the relevant biological variables

Table 8A.2.2 Mean astaxanthin content in mangrove and associate species collected from Jharkhali island of Indian Sundarbans during May 2007 to April 2008

Mangrove species	Astaxanthin content (mg/kg)		Percentage increase (%)
	Submerged	Exposed	
Avicennia officinalis	396.56	297. 75	33.19
Avicennia alba	441.33	328.02	34.54
Avicennia marina	435.59	319.10	36.51
Sonneratia apetala	162.80	103.49	57.31
Aegiceras corniculatum	120.86	98.15	23.14
Aegialitis rotundifolia	105.92	82.41	28.53
Ceriops decandra	91.09	67.44	35.07
Heritiera fomes	761.00	398.54	90.95
Rhizophora apiculata	84.90	56.22	51.01
Bruguiera gymnorrhiza	461.38	397.11	16.18
*Porteresia coarctata	607.56		
*Enteromorpha intestinalis	120.78		
*Ulva lactuca	56.43		
*Catenella repens	129.05		
*Sueda sp.	87.55		

Remark: Mangrove associates (denoted with * sign) were not monitored during 2 tidal phases. They were collected during low tide condition

were determined for each control and experimental ponds.

Condition index was analyzed as per the expression: $C.I. = W/L^3 \times 100$, where W = weight of the cultured species (in g) and L = length of the cultured species (in cm).

Percentage weight gain was documented by calculating the difference in weight from the average final weight with respect to the initial weight; weight gain = [(average individual final weight – average individual initial weight)/average individual initial weight] × 100.

The survival rate was measured as percentage of the difference of stocking number and production volume (No.) at the end of the culture period.

Feed conversion ratio (FCR) is the weight of feed consumed per unit of body weight gain and was analyzed after the harvesting of shrimps as per the expression: $FCR = \Delta f/\Delta b$, where Δf = change in feed biomass and Δb = change in body biomass of the cultured species.

Astaxanthin in the shrimp tissue was estimated as per the standard spectrophotometric method (Schuep and Schierle 1995) and body pigmentation of the cultured shrimp was assessed (for each treatment) after boiling the

shrimp for 5 min in water and comparing the orange-red colouration with Roche SalmoFan™ colour score.

8A.2.7 Results and Discussion

8A.2.7.1 Astaxanthin Level in Mangroves

The astaxanthin level in the selected mangrove species (collected from Jharkhali region) exhibits significant variations. It is of the order *Heritiera fomes* > *Bruguiera gymnorrhiza* > *Avicennia alba* > *Avicennia marina* > *Avicennia officinalis* > *Sonneratia apetala* > *Aegiceras corniculatum* > *Aegialitis rotundifolia* > *Ceriops decandra* > *Rhizophora apiculata* (Table 8A.2.2). The relatively greater astaxanthin content in the submerged leaves of mangroves confirms the synthesis of astaxanthin content under stressful condition. However, more studies are needed to confirm the influence of tidal influx and subsequent salinity fluctuation on astaxanthin level in the mangrove floral parts. The present data may serve as baseline information on the regulatory

Table 8A.2.3 Astaxanthin level in different organisms

Natural astaxanthin sources	Astaxanthin concentration (ppm)
Salmonoids	~5
Plankton	~60
Krill	~120
Arctic shrimp	~1,200
Phaffia Yeast	~8,000
Haematococcus pluvialis	~40,000

Source: http://algatech.com/astax.htm

role of tidal submergence on astaxanthin level in the estuarine and coastal vegetation. The enhancement of astaxanthin production under stressed condition of organisms is a matter of interest, and several researches are still being undertaken to pinpoint the reaction pathway of astaxanthin production by inducing stress of varied nature. Many types of yeast have been described with an increase ability to produce carotenoids when they grow under unfavourable environment (Certik et al. 2005). Several workers have reported both in the dark and light the enhancement of the accumulation of astaxanthin in cysts of *Haematococcus pluvialis* under salt stress conditions. The present study points to higher astaxanthin level in those leaves of the mangroves that are inundated for 10–12 h by tidal waters of Jharkhali station having typical estuarine water characteristics (salinity = 10–25.85 psu; pH = 7.98–8.28; temperature = 29.8–31.5 °C; dissolved oxygen = 5.93–6.10 mg/l; NO_3 = 15.09–21.04 µgat/l; PO_4 = 1.12–1.39 µgat/l and SiO_3 = 64.44–83.16 µgat/l). The steep enhancement of astaxanthin level in the inundated Sundari leaves (*Heritiera fomes*) clearly reflects the highest degree of stress posed by water salinity on this species. *Heritiera fomes*, being freshwater-loving mangrove species, cannot tolerate high salinity (Mitra and Pal 2002), and thus, acceleration of astaxanthin production may probably be a part of its adaptation to cope with the stenohaline condition of coastal and estuarine environment that becomes acute during high tide. The astaxanthin level of mangrove flora is thus a function of its physiological system, which is extremely species specific. Highest astaxanthin was recorded in mangrove associate *Porteresia*

coarctata (commonly known as salt-marsh grass) in comparison to other species and therefore considered for feed preparation. The level of astaxanthin estimated in the mangrove floral species of Indian Sundarbans is less than the existing natural mega-reservoir of astaxanthin like *Phaffia rhodozyma* and *Haematococcus pluvialis* (Table 8A.2.3).

8A.2.7.2 Shrimp Nutrition and Growth

Shrimps fed with *Porteresia* diet exhibited higher final weights and better weight gain (Table 8A.2.4) at the end of the experiment (27.2 g final weight) in comparison to control pond (20.5 g final weight). Condition index values of shrimp were also higher in experimental ponds (3.87 ± 0.67) than control ponds (2.98 ± 0.55) (Table 8A.2.4). The FCR value for control pond was 1.67 and for experimental pond was 1.27. The survival rate was found to be 58.4 % in the control pond and 69.2 % in experimental pond. The present pilot-scale study speaks in favour of healthy pond environment, better growth, higher survival rate and low FCR values through the use of *Porteresia-based* feed.

An important factor governing the consumer acceptance and market value of many cultivated fish and shrimp species is the pink or red colouration of their flesh or boiled exoskeleton (Brun and Vidal 2006). In the wild, this colouration is achieved through the ingestion of carotenoid pigments particularly astaxanthin contained within invertebrate food organisms (Johnson et al. 1977; Ibrahim et al. 1984). The *Porteresia*-based feed in the present study resulted in higher astaxanthin values in shrimps of experimental pond (15.32 ± 1.22 ppm; $n = 10$) as reflected through darker orange-red colouration of shrimp exoskeleton in comparison to control pond (8.66 ± 0.78 ppm). The Roche SalmoFan™ colour score showed the value of 23 and 29 for shrimps from control pond and experimental ponds, respectively (Table 8A.2.4), which confirms the variation of astaxanthin level due to different feed ingredients.

Table 8A.2.4 Success indicators of the research programme

Parameters	Control pond	Experimental pond
Condition index of shrimp	2.98 ± 0.55	3.87 ± 0.67
Wt. at harvest (g)	27.2	20.5
Astaxanthin in shrimp (ppm)	8.66 ± 0.78	15.32 ± 1.22
Roche SalmoFan™ colour score	23	29
Survival rate (%)	58.4	69.2
FCR	1.67	1.27

Table 8A.2.5 Variation in the physico-chemical parameters of the culture ponds

Parameters	Control pond	Experimental pond
Surface water temperature (°C)	29.4 ± 0.22	29.4 ± 0.22
Surface water salinity (psu)	18.56 ± 0.20	15.08 ± 0.13
pH	7.88 ± 0.35	8.08 ± 0.03
Transparency (cm)	17.02 ± 2.02	24.3 ± 2.58
Dissolved oxygen (mg/l)	4.66 ± 0.77	5.47 ± 0.12
Nitrate (µgat/l)	19.5 ± 1.21	16.6 ± 1.01
Phosphate (µgat/l)	2.22 ± 0.05	2.19 ± 0.29
Silicate (µgat/l)	64.01 ± 2.23	64.32 ± 2.67
Chlorophyll a (mg/m^3)	1.74 ± 0.51	1.92 ± 0.16
Soil organic carbon (%)	1.18 ± 0.16	0.97 ± 0.10

Values are the mean ± SD

8A.2.7.3 Water Quality

Hydrological parameters of the shrimp culture ponds are a reflection of the quality of feed provided to the cultured species, and the condition index values symbolize the suitability of the environment for the species (Maceina and Murphy 1998). The hydrological parameters are recorded in Table 8A.2.5.

Surface water temperature in both the culture ponds showed more or less parallel trend of variation throughout the study period. The uniformity in temperature profile is due to the location of both the ponds in the same site that experience similar weather and climate. Water temperature plays a major role in shrimp enzyme kinetics which may have a regulatory influence on their growth (Mitra et al. 2006). It also affects the

process of moulting during the post-larval stage of shrimps (WWF-India 2006).

The *salinity* of the Hugli–Matla estuarine complex is known to exhibit intensive variations (Saha et al. 1995). The difference in salinity between ponds may be attributed to the different soil salinity (as substratum of the pond) that leaches soluble salt to the pond water. The relatively higher C.I. values in the experimental pond with less salinity prove the efficiency of formulated feed in combating the stress posed by salinity.

Shrimp culture directly affects the pH of the pond bottom through deposition of excess feed, shrimp excreta, dead shrimps, etc. This shifts the soil and overlying aquatic pH towards acidic condition. In the present study, such condition was not observed owing to the traditional practice of liming at a regular interval of time.

Dissolved oxygen (DO) is a vital parameter regulating the aquatic life. The shrimp health is a direct function of dissolved oxygen and its diurnal variation. Excessive organic load in the system results in lowering the DO value during night/dawn posing threat to the survival of aquatic life. In the present study the DO level in the control pond showed lower value owing to deposition of organic carbon at the bottom of the pond. The significant variation of DO between ponds may be attributed to different growth rate of the culture species and also the use of different types of feed. Traditional feed contains dry fish dust and trash shrimp dust which lowers the DO due to their utilization for oxidizing the residual matter (BOD value increases under this situation). Floral-based feed, on the other hand, generates very limited residue due to which DO remains almost unaltered.

Transparency controls the phytoplankton standing stock in shrimp culture ponds due to their dependency on the solar radiation for photosynthesis. The experimental pond provided with formulated feed showed increased transparency due to its unique binding property. The ready acceptance of the *Porteresia*-based feed by the cultured species in the experimental pond may be the basis of reduced suspended particulate matter in the aquatic phase of the experimental pond.

Nutrients (comprising of nitrate, phosphate and silicate), budget in the aquatic phase of the culture ponds, are regulated through quantum of excretory products of the cultured species, left-over feed and also by the churning of the pond bed (due to run-off from the adjacent landmasses).

High concentration of nitrate in the control pond may be due to leaching of the feed ingredients (particularly from animal component in traditional feed) in pond water and also the faecal matter that generates ammonia (Mitra and Choudhury 1995).

The phosphate concentration during the study period showed no significant variation between the ponds owing to ban imposed on washing utensils, clothes and other daily household activities during the culture period.

The silicate level of the ponds may be attributed to substratum or pond bottom composition. In both the control and experimental ponds, no significant variation in silicate was observed.

Soil organic carbon was greater in control pond due to more generation of residual feed and excreta in the absence of any feed management. On the contrary, the lower value of organic carbon in the experimental pond is an indication of better acceptability of *Porteresia*-based feed by shrimp due to which wastage was a minimum.

8A.2.8 Looking Ahead

The mangrove ecosystem of Indian Sundarbans is one of the most biologically productive and taxonomically diverse ecotone of Indian subcontinent with a unique reservoir of bioactive substances. The detection of antioxidant astaxanthin in the floral parts of these typical estuarine and coastal vegetations adds a new dimension to these halophytes. Astaxanthin is an important feed ingredient with wide application both in pisciculture and animal husbandry sector owing to its antioxidant nature. Since the animals cannot synthesize carotenoids within their system, pigments must be supplemented to their feeds, allowing the assimilation and providing the characteristic pigmentation of the cultured aquacultural species, egg yolk, etc. for increasing the quality and consumer acceptance in the market place (Johnson and An 1991). This

Table 8A.2.6 Application mangrove antioxidant property

Application field	Avenues
Conservation of biodiversity	1. Diversion of local inhabitants from illegal intrusion into the forest through introduction of nonconventional livelihood programmes
	2. Maintaining carotenoid-enriched floral nursery or floral park
Livelihood schemes	1. Setting up of fish, poultry and animal feed units based on mangrove-originated astaxanthin
	2. Initiation of eco-friendly aquaculture practice through fish feed rich in astaxanthin or any other natural antioxidant
	3. Production of export quality aquacultural items
Ecology and environment	1. Upgradation of water and pond bottom health due to replacement of the animal ingredients with floral ingredients in the feed of the cultured species

will not only upgrade the nutrition sector of animal husbandry and aquaculture but will also increase the immunity power of the cultured fish species and domesticated livestock. An ecologically fragile ecosystem sustaining a large fraction of poverty-stricken population like Sundarbans needs an alternative livelihood programme not only to upgrade their economic profile but also to realize the utility and application of their surrounding vegetation as a part of strengthening the root of conservation. The present programme may open an avenue of preparing fish feed, poultry feed and cattle feed by utilizing the antioxidant base of mangroves through involvement of the local people. This will defray the people from illegal entry into the forest and will also improve the animal and fish nutrition sector of the area through setting up of small-scale feed units. The antioxidant reservoir of Sundarbans mangrove ecosystem has several future applications (Table 8A.2.6), and hence, proper policy is needed to blend the biotechnological approach with the livelihood components of the island dwellers.

References of Annexure 8A.2

An GH, Jang BG, Cho MH (2001) Cultivation of the carotenoid hyperproducing mutant 2A2N of the red yeast *Xanthophyllomyses dendrorhous* (*Phaffia rhodozyma*) with molasses. J Biosci Bioeng 92:121–125

Banerjee K (2009) Income generation through eco-friendly pisciculture in Indian Sundarbans: an innovative approach towards sustainable green technology. Report of WOS-B scheme, Department of Science and Technology, Government of India, pp 1–20

Banejee K, Ghosh R, Homechaudhuri S, Mitra A (2009) Seasonal variation in the biochemical composition of red seaweed (*Catenella repens*) from Gangetic delta, northeast coast of India. J Earth Syst Sci 118(5):1–10

Biswas G, Jena JK, Singh SK, Patmajhi P, Muduli HK (2006) Effect of feeding frequency on growth, survival and feed utilization in mrigal, *Cirrhinus mrigala* and rohu, *Labeo rohita* during nursery rearing. Aquaculture 254:211–218

Bjerkeng B (2000) Carotenoid pigmentation in salmonid fishes – recent progress. In: Cruz-Suarez LE, Ricque-Marie D, Tapia-Salazar M, Olvera-Novoa MA, Cerecedo-Olvera R (eds) Avances en Nutricion Acuicola V – Memorias del Quinto Simposium Internacional de Nutricion Acuícola, Merida, 19–22 Noviembre 2000, vol 5. Universidad Autonoma de Nuevo Leon, Monterrey; ISBN 970-694-52-9, pp 71–89

Boussiba S (2000) Carotenogenesis in the green algae *Haematococcus pluvialis*: cellular physiology and stress response. Plant Physiol 108:111–117

Bratova K, Ganovski KH (1982) Effect of Black Sea algae on chicken egg production and on chick embryo development. Vet Med 19:99–105

Brun HL Vidal F (2006) Shrimp pigmentation with natural carotenoids. Feed Technol Aquacult Asia-Pac Mag 3(1):34–35

Buttle LG, Crampton VO, Williams PD (2001) The effect of feed pigment type on flesh pigment deposition and colour in farmed Atlantic salmon, *Salmo salar* L. Aquacult Res 32:103–111

Certik M, Masrnova S, Sitkey V, Minarik M, Emilia B (2005) Biotechnological production of astaxanthin. Chem Listy 99:s237–s240

Chakraborty S, Santra SC (2008) Biochemical composition of eight benthic algae collected from Sunderban. Indian J Mar Sci 37(3):329–332

Chaudhuri A (1995) Brand equity or double jeopardy? J Prod Brand Manag 4(1):26–32

Chociai MB, Machado IMP, Fontana JD, Chociai JG, Busta SB, Bonfim TMB (2002) Cultivo da levedura *Phaffia rhodozyma* (*Xanthophyllomyses dendrorhous*) em processo descontinuo alimentado para producao astaxantina. Brazillian J Pharm Sci 38:457–462

Cruz-Suarez LE, Ricque-Marie D, Tapia-Salazar M, Guajardo-Barbosa C (2000) Uso de harina de kelp (*Macrocystis pyrifera*) en alimentos para camaron. In: Cruz-Suarez LE, Ricque-Marie D, Tapia-Salazar M, Olvera-Novoa MA, Cerecedo-Olvera R (eds) Avances en Nutricion Acuicola V – Memorias del Quinto Simposium Internacional de Nutricion Acuícola, Merida, 19–22 Noviembre 2000, vol 5. Universidad Autonoma de Nuevo Leon, Monterrey; ISBN 970-694-52-9, pp 227–266

Cruz-Suarez LE, Tapia-Salazar M, Nieto-Lopez MG, Guajardo-Barbosa C, Ricque-Marie D (2009) Comparison of *Ulva clathrata* and the kelps *Macrocystis pyrifera* and *Ascophyllum nodosum* as ingredients in shrimp feeds. Aquacult Nutr 15:421–430

Daniels WH, Robinson EH (1986) Protein and energy requirements of juvenile red drum (*Sciaenops ocellatus*). Aquaculture 53:243–252

Darachai J, Piyatiratitivorakul S, Kittakoop P, Nitithamyong C, Menasveta P (1998) Effect of astaxanthin on larval growth and survival of giant tiger prawn, *Penaeus monodon*. In: Flegel TW (ed) Advances in shrimp biotechnology. National Center for Genetic Engineering and Biotechnology, Bangkok, pp 117–121

Finkelstein LM, Burke MJ, Raju NS (1995) Age-discrimination in simulated employment contexts: an integrative analysis. J Appl Psychol 80:652–663

Gomes E, Dias J, Silva P, Valente L, Empis J, Gouveia L, Bowen J, Young A (2002) Utilization of natural and synthetic sources of carotenoids in the skin pigmentation of gilthead seabream (*Sparus aurata*). J Eur Food Res Technol 214:287–293

Hansen HR, Hector BL, Feldmann J (2003) A qualitative and quantitative evaluation of the seaweed diet of North Ronaldsay sheep. Anim Feed Sci Technol 105:21–28

Harker M, Tsavalos AJ, Young AJ (1996) Factors responsible for astaxanthin formation in the chlorophyte *Haematococcus pluvialis*. Bioresour Technol 55:207–214

Hasan MR (2001) Nutrition and feeding for sustainable aquaculture development in the third millennium. In: Aquaculture in the third millennium. Technical proceedings of the conference on aquaculture in the third millennium, Bangkok, 20–25 Feb 2000. NACA, Bangkok and FAO, Rome, pp 193–219

Hashim R, Mat-Saat NA (1992) The utilization of seaweed meals as binding agents in pellet feeds for snakehead (*Channa striatus*) fry and their effects on growth. Aquaculture 108:299–308

Ibrahim A, Shimizu C, Kono M (1984) Pigmentation of culture red sea bream, *Chrysophrys major*, using astaxanthin from Antarctic krill, *Euphausia superba*, and a mysid, *Neomysis* sp. Aquaculture 38:45–57

Jamu DM, Ayinla OA (2003) Potential for the development of aquaculture in Africa. NAGA 26:9–13

Johnson EA, An GH (1991) Astaxanthin from microbial sources. Crit Rev Biotechnol 11:297–326

Johnson EA, Conklin DE, Lewis MJ (1977) The yeast *Phaffia rhodoyma* as a dietary pigment source for salmonids and crustaceans. J Fish Res Board Canada 34:2417–2421

Kader MA, Hossain MA, Hasan MR (2005) A survey of the nutrient composition of some commercial fish feeds available in Bangladesh. Asian Fish Sci 18:59–69

Kaushik S (1990) Use of alternative protein resources for the intensive rearing of carnivorous fish. In: Flos R, Tort L, Torres P (eds) Mediterranean aquaculture. Hellis Horwood Ltd., Chichester, pp 125–138

Khan MA, Jafri AK, Chadha NK (2004) Growth and body composition of rohu (*L. rohita*) fed compound diet: winter feeding and rearing to marketable size. J Appl Ichthyol 20:265–270

Kobayashi M, Kakizono T, Nishio N, Nagai S (1992) Effects of light intensity, light quality and illumination cycle on astaxanthin formation in a green alga *Haematococcus pluvialis*. J Ferment Bioeng 74:61–63

Leupp JL, Caton JS, Soto-Navarro SA, Lardy GP (2005) Effects of cooked molasses blocks and fermentation extract or brown seaweed meal inclusion on intake, digestion, and microbial efficiency in steers fed low-quality hay. J Anim Sci 83:2938–2945

Liu MS, Mancebo VJ (1983) Pond culture of *Penaeus monodon* in the Philippines: survival, growth and yield using commercially formulated feed. J World Maricult Soc 14:75–85

Lovell RT (1989) Nutrition and feeding of fish. Van Nostrand Reinhold, New York

Lowry OH, Rosebrough NJ, Farr AL, Randell RJ (1951) Protein measurements with folin-phenol reagent. J Biol Chem 193:1433–1437

Maciena MJ, Murphey BR (1988) Variation in the weight to length relationship among Florida and Northern largemouth bass and their inter-specific F1 hybrid. Trans Am Fish Soc 117:232–237

Marova I, Macuchova S, Kotrla R, Hiemer J (2005) How antioxidant intake influences lipid metabolism and antioxidant status in patients with type 2 diabetes and/or hypelipidamia. Chem Listy 99:s251–s253

Mawson PR (1995) Publication number US5453565 A, Publication type – Grant, Application number – US 08/321,100. Publication date 26 Sept 1995, Filling date 11 Oct 1994

Meyers SP, Latscha T (1997) Carotenoids. In: DAbramo LRD, Conclin DE, Akiyama DM (eds) Crustacean nutrition, vol 6, Advances in world aquaculture. The World Aquaculture Society, Baton Rouge, pp 164–193

Mitra A (1993) Metallic pollution in the Hooghly estuary: effects and proposed control measures. J Ind Ocean Stud 1(1):49–53

Mitra A (2000) The northwest coast of the Bay of Bengal and deltaic Sunderbans. In: Sheppard CRC (ed) Seas at the millennium: an environmental evaluation. Pergamon, Amsterdam, pp 145–160

Mitra A, Banerjee K (2005a) Living resources of the sea: focus Indian Sundarbans. Published by WWF-India Secretariat, Sunderbans Landscape Project, New Delhi, 120pp

Mitra A, Banerjee K (2005b) Mangroves and associated flora. In: Lt. Col. Banerjee SR (ed) Living resources of the sea: focus Indian Sundarbans, 1st edn, vol 1. World Wide Fund for Nature-India, Canning Town, WWF-India, Kolkata pp 55–70

Mitra A, Bhattacharyya DP (2003) Environmental issues of shrimp farming in mangrove ecosystem. J Ind Ocean Stud 11(1):120–129

Mitra A, Choudhury A (1993) Heavy metal concentrations in oyster *Crassostrea cucullata* of Sagar Island, India. Ind J Environ Health NEERI 35(2):139–141

Mitra A, Choudhury A (1995) Causes of water pollution in prawn culture farms. J Ind Ocean Stud 2(3):230–235

Mitra A, Pal S (2002) The oscillating mangrove ecosystem and the Indian Sundarbans. In: Shakti B, Farida T (eds) Published by WWF-India, WBSO, Kolkata

Mitra A, Choudhury A, Zamadar YA (1992) Seasonal variations in metal content in the gastropod *Cerithedia (Cerithedeopsis) cingulata*. Proc Zool Soc 45:497–500

Mitra A, Trivedi S, Choudhury A (1994) Inter-relationship between trace metal pollution and physico chemical variables in the framework of Hooghly estuary. J Ind Ports 10:27–35

Mitra A, Mandal T, Bhattacharya DP (1999) Concentration of heavy metals in *Penaeus* sp. of brackish water wetland ecosystem of West Bengal, India. Ind J Environ 2(2):97–106

Mitra A, Chakraborty R, Banerjee K, Banerjee A, Mehta N, Berg H (2006) Study on the water quality of the shrimp culture ponds in Indian Sundarbans. Ind Sci Cruiser 20(1):34–43

Mohanty SN, Das KM, Sarkar S (1995) Effect of feeding varying dietary formulations on body composition of rohu fry. J Aquacult 3:23–28

Moretti VM, Mentasti T, Bellagamba F, Luzzana U, Caprino F, Turchini GM, Giani I, Valfre F (2006) Determination of astaxanthin stereoisomers and colour attributes in flesh of rainbow trout (*Oncorhynchus mykiss*) as a tool to distinguish the dietary pigmentation source. J Food Addit Contam 23:1056–1063

Moss SM (1994) Growth rates, nucleic acid concentrations, and RNA/DNA ratios of juvenile white shrimp, *Penaeus vannamei* Boone, fed different algal diets. J Exp Mar Biol Ecol 182:193–204

Mukhopadhyay A, Pal R (2002) A report on biodiversity of algae from coastal West Bengal (South and North 24 Parganas) and their cultural behaviour in relation to mass cultivation programme. Ind Hydrobiol 5:97–107

Mukhopadhyay N, Ray AK (1999) Improvement of quality of sesame (*Seasamum indicum*) seed meal protein with supplemental amino acids in feeds for rohu, *Labeo rohita*, (Hamilton), fingerlings. Aquacult Res 30:549–557

Mukhopadhyay N, Ray AK (2001) Effect of amino acids supplementation on the nutritive quality of fermented linseed meal protein in the diets for rohu, *Labeo rohita* (Hamilton), fingerlings. J Appl Ichthyol 17:220–226

Nickell DC, Bromage NR (1998) The effect of timing duration of feeding astaxanthin on the development and variation of fillet color and efficiency of pigmentation in rainbow trout (*Oncorhyncus mykiss*). Aquaculture 169:233–246

NRC (1983) Nutrients requirements of warm water fishes and shell-fishes. National Academy Press, Washington, DC, 102pp

Okuzumi J, Takahashi T, Yamane T, Kitao Y, Inagake M, Ohya K, Nishino H, Tanaka Y (1993) Inhibitory effects of fucoxanthin, a natural carotenoid, on N-ethyl-N'-nitro-N-nitrosoguanidine-induced mouse duodenal carcinogenesis. Cancer Lett 68:159–168

Penaflorida VD, Golez NV (1996) Use of seaweed meals from *Kappaphycus alvarezii* and *Gracilaria heteroclada* as binders in diets of juvenile shrimp *Penaeus monodon*. Aquaculture 143:393–401

Phillips AM (1972) Calorie and energy requirements. In: Halver JE (ed) Fish nutrition. Academic, New York, pp 2–29

Prather EE, Lovell RT (1973) Response of intensively fed channel catfish to diets containing various protein energy ratio. In: Proceedings of 27th South-Eastern Association of game and fish commissioner, vol 27, pp 455–459

Rajkhowa I (2005) Action in aquaculture – opportunities in a quaint specialization. Bus Today (May 22 Issue):131

Saha SB, Ghosh BB, Gopalakrishnan V (1975) Plankton of Hooghly estuary with special reference to salinity and temperature. J Mar Biol Assoc India 17(1):107–120

Sarada R, Tripathi U, Ravishankar GA (2002) Influence on stress on astaxanthin production in *Haematococcus pluvialis* grown under different culture conditions. Proc Biochem 37:623–627

Satoh S (2000) Common carp, *Cyprinus carpio*. In: Wilson RP (ed) Handbook of nutrient requirement of finfish. CRC Press, Boca Raton/Ann Arbor/Boston/London, pp 55–68

Schuep W, Schierle J (1995) Astaxanthin determination of stabilized, added astaxanthin in fish feeds and premixes. In: Carotenoids isolation and analysis, vol 1. Birkhauser Verlag, Basel, pp 273–276

SEAFDEC: Southeast Asian Fisheries Development Center (1981) Fishery statistical bulletin for The South China Sea Area, 1979. SEAFDEC Secretariat, Bangkok

Sen S (1998) Environment, law of sea and coastal zone. WWF – India, Eastern Region, Kolkata

Shigueno K (1984) Intensive culture and feed development in Penaeus japonicus. In: Proceedings of first international conference on the culture of penaeid prawns/shrimps Iloilo City Philippines. SEAFDEC Aquaculture Department, Iloilo City, pp 123–130

Shigeo et al (1994) http://www.lib.kobe-u.ac.jp/repository/90000288.pdf. Accessed 6 Oct 2011

Shyong WJ, Huang CH, Chen HC (1998) Effects of dietary protein concentration on growth and muscle composition of juvenile. Aquaculture 167:35–42

Sies H (1991) Oxidative stress II. Oxidants and antioxidants. Academic, London

Simpson KL, Chichester CO (1981) Metabolism and nutritional significance of carotenoids. Annu Rev Nutr 1:351–374

Storebakken T, No HK (1992) Pigmentation in rainbow trout. Aquaculture 100:209–229

Strand A, Herstad O, Liaaen-Jensen S (1998) Fucoxanthin metabolites in egg yolks of laying hens. Comp Biochem Physiol 119:963–974

Strickland JDH, Parsons TR (1968) A manual for seawater analysis. Bull Fish Res Board Can 311pp

Strickland JDH, Parsons TR (1972) A practical handbook of seawater analysis. J Fish Res Board Canada (Ottawa) 167:311pp

Suarez-Garcia HA (2006) Efecto de la inclusion de alginate y harina de algas Sargassum sp y Macrocystis pyrifera sobre la estabilidad en agua, digestibilidad del alimento y sobre el crecimiento del camaron blanco *Litopenaeus vannamei*. Undergraduate thesis, Universidad Autonoma de Nuevo Leon, Nuevo León

Tacon AGJ, De Silva SS (1997) Feed preparation and feed management strategies within semi-intensive fish farming systems in the tropics. Aquaculture 151:379–404

Tecator (1983) Fat extraction on feeds with the Soxtec System HT – the influence of sample preparation and extraction media. Application note AN 67/83 (1983.06.13). Soxtec system HT manual, Tecator AB, Hoganas

Tjahjono AE, Hayama Y, Kakizono T, Terada Y, Nishio N, Nagai S (1994) Hyper-accumulation of astaxanthin in a green alga *Haematococcus pluvialis* at elevated temperatures. Biotechnol Lett 16:133–138

Torrisen OJ, Hardy RW, Shearer KD (1989) Pigmentation of salmonids – carotenoid deposition and metabolism. CRC Crit Rev Aquat Sci 1:209–225

Trevelyan WE, Harrison JS (1952) Studies on yeast metabolism. 1. Fractionation and micro-determination of cell carbohydrates. Biochem J 50:298–303

Turner JL, Dritz SS, Higgins JJ, Minton JE (2002) Effects of *Ascophyllum nodosum* extract on growth performance and immune function of young pigs challenged with *Salmonella typhimurium*. J Anim Sci 80:1947–1953

Valente LMP, Gouveia A, Rema P, Matos J, Gomes EF, Pinto IS (2006) Evaluation of three seaweeds *Gracilaria bursapatoris*, *Ulva rigida* and *Gracilaria cornea* as dietary ingredients in European seabass (*Dicentrarchus labrax*) juveniles. Aquaculture 252:85–91

Van der Meer MB, Zamora JE, Verdegem MCJ (1997) Effect of dietary lipid level on protein utilization and the size and proximate composition of body compartments of *Colossoma macropomum* (Cuvier). Aquacult Res 28:405–417

Vazquez M, Santos V, Parajo JC (1998) Fed-batch cultures of *Phaffia rhodozyma* in xylose-containing media made from wood hydrolysates. Food Biotechnol 12:43–55

Walkey A, Black IA (1934) An examination of the effect of the digestive method for determining soil organic matter and a proposed modification of the chronic and titration method. Soil Sci 37:29–38

White DA, Page GI, Swaile J, Moody AJ, Davies SJ (2002) Effect of esterification on the absorption of astaxanthin in rainbow trout, *Oncorhynchus mykiss* (Walbaum). Aquacult Res 33:343–350

Wilson RP (2000) Channel catfish, *Ictalurus punctatus*. In: Wilson RP (ed) Handbook of nutrient requirement of finfish. CRC Press, Boca Raton/Ann Arbor/Boston/London, pp 35–53

WWF-India (2006) Training manual on ecofriendly and sustainable fishery, New Delhi. WWF-India, Canning Office, Sundarbans, 130pp

Yan X, Chuda Y, Suzuki M, Nagata T (1999) Fucoxanthin as the major antioxidant in *Hijikia fusiformis*, a common edible seaweed. Biosci Biotechnol Biochem 63:605–607

Ziegler RG (1989) Importance of α-carotene, β-carotene and other phytochemicals in the etiology of lung cancer. J Natl Cancer Inst 88:612–615

Annexure 8A.3: Seasonal Variation of Biochemical Composition in Edible Oyster (*Saccostrea cucullata*) of Indian Sundarbans

Abhijit Mitra, Sufia Zaman and Kakoli Banerjee

Abstract

Protein, lipid, glycogen, moisture and ash content of the edible oyster species were analyzed on a monthly basis during 2004 and 2005 from the cultured site at Chotomollakhali in eastern sector of Indian Sundarbans. The culture of edible oyster was initiated in this part of the country in order to provide alternative livelihood to the poverty-stricken population in the Indian Sundarbans, which may otherwise lead to destruction of natural resources of this mangrove ecosystem. Simultaneous monitoring of hydrological parameters (surface water temperature, salinity, pH, nitrate, phosphate and silicate) and phytopigment level of the ambient water was also carried out in this cultured site to investigate the interrelationship between the hydrological parameters and biochemical composition of the oyster tissue. The 2-year study indicates significant seasonal oscillation of the major biochemical constituents of the oyster.

Keywords

Indian Sundarbans • Mangrove • Phytopigment • *Saccostrea cucullata*

8A.3.1 Introduction

Indian Sundarbans is one of the most biologically productive and taxonomically diversified, low-lying, mangrove detritus-based, open, dynamic,

A. Mitra (✉) • S. Zaman
Department of Marine Science, University
of Calcutta and Department of Oceanography,
Techno India University, Salt Lake Campus,
Kolkata, West Bengal 700 019, India
e-mail: abhijit_mitra@hotmail.com

K. Banerjee
Department of Marine Science, University
of Calcutta, 35, B.C. Road, Kolkata,
West Bengal 700 019, India

heterogeneous coastal ecotone situated at the apex of the Bay of Bengal (between 21°13′ and 22°40′ N latitude and 88°03′ to 89°07′ E longitude). The entire forest of this unique ecosystem acts as a potential reservoir of marine biotic resources. The lower stretch of the estuarine complex and high saline zones are extremely favourable for the survival and growth of edible oyster (Mitra and Banerjee 2005). *Saccostrea cucullata* is the dominant oyster species in Indian Sundarbans although *Crassostrea gryphoides* and *Crassostrea madrasensis* are also reported in the basal part of hard substrata (like sluice gates, jetties, pillars of fish landing stations, light house). Oyster is a good source of protein, vitamins, minerals and trace elements. Researchers have collected detailed information on edible oysters, largely because they can be valuable food (Tack et al. 1992; Ruwa and Polk 1994). However, many internal and environmental factors including pollutants can affect the growth and reproductive success of marine bivalves (Mac MacDonald and Thompson 1985; Steele and Mulcahy 1999). This research programme highlights the seasonal variation of biochemical composition (% lipid, % protein, % glycogen, % moisture and % ash) in the edible oyster (*Saccostrea cucullata*) sampled from the eastern sector of Indian Sundarbans with respect to hydrobiological parameters.

8A.3.2   Materials and Methods

The entire network of the present programme comprised of the monthly sampling and collection of oysters (*Saccostrea cucullata*) from Chotomollakhali island in the eastern sectors of Indian Sundarbans for a period of 2 years (2004–2005) along with simultaneous monitoring of hydrobiological parameters (surface water temperature, salinity, pH, nitrate, phosphate, silicate and phytopigment concentration, *i.e.* Chl *a*, Chl *b* and Chl *c*). The different phases of the programme are discussed separately.

8A.3.2.1   Phase 1: Collection of Oysters

Edible oyster species, *Saccostrea cucullata* (Born) were collected from the cultured site of Chotomollakhali island in the eastern sector of Indian Sundarbans. These species are highly variable in shape, growing in clusters on rocks, bricks, wooden piles or jetties, and also settle on the stems of mangrove plants and on molluscan shells. Twenty samples of almost uniform size (mean length 8.0 cm) were collected at monthly interval from the cultured site during January 2004 to December 2005. They were transported live to the laboratory after proper washing and removal of

fouling organisms from the outer surface of the shells for further biochemical analysis.

8A.3.2.2   Phase 2: Analysis of Hydrobiological Parameters

Hydrological parameters around the oyster culture site like surface water temperature, salinity, pH, nitrate, phosphate, silicate and phytopigment concentration of the ambient aquatic phase were analyzed on a monthly basis as per the standard methodology outlined in Strickland and Parsons (1968), APHA (1998). Surface water temperature of the aquatic medium in the sampling station was measured by a Celsius thermometer (scale ranging from 0 to 100 °C). The salinity of the surface water was measured by means of refractometer and cross-checked in the laboratory by employing 'Mohr–Knudsen' method as outlined by Strickland and Parsons (1968). The correction factor was found out by titrating the silver nitrate solution against standard seawater (IAPO Standard Seawater Service, Charlottenlund, Slot Denmark, chlorinity = 19.376 ppt.). pH of the ambient water was determined by a portable pH meter (sensitivity = ±0.02).

Surface water for nutrient analysis was collected in clean TARSON bottles and transported to the laboratory in ice-freezed condition. Triplicate samples were collected from the same culture site to maintain the quality of the data. The standard spectrophotometric method of Strickland and Parsons (1968) was adopted to determine the nutrient concentrations in the surface water. Nitrate was analyzed by reducing it to nitrite which was determined by treating the samples with a solution of sulphanilamide, and the resultant diazonium ion was coupled with *N*-(1-napthyl)-ethylene diamine to give an intensely pink azo dye. The reduction was then carried out by treating the sample with ammonium chloride and passing it through a glass column packed with amalgamated cadmium fillings. The determination of phosphate was carried out by treatment of an aliquot of the sample with acidic molybdate reagent containing ascorbic acid and a small quantity of potassium antimony tartrate. Dissolved silicate was determined by treating the sample with acidic molybdate reagent. The resultant silico-molybdic acid was reduced to molybdenum blue complex by ascorbic acid and incorporating oxalic acid prevented formation of similar blue complex by phosphate. Systronics Digital Spectrophotometer (Type-16S) was used for nutrient (nitrate, phosphate and silicate) analysis at their respective wavelengths.

Phytopigment concentration of the ambient aquatic phase was analyzed in order to monitor the food reservoir of the cultured species. For pigment analysis, 1 l of surface water collected in black bottles from sampling station was filtered through a 0.45-μm Millipore membrane fitted with a vacuum pump. The filter paper was transferred to a homogenizer containing acetone. The contents were grounded thoroughly and placed in refrigerator for 24 h in order to facilitate the complete extraction of phytopigment. Finally the chlorophyll density was estimated as per Jeffrey and Humphrey (1975) with the help of 'SHIMADZU UV 2100' spectrophotometer.

8A.3.2.3 Phase 3: Biochemical Analysis

The biochemical analyses were done on tissue samples pooled from ten individual oysters. The samples (average length of 8.0 cm) were collected from the cultured site at monthly intervals. They were washed with double-distilled water and processed for biochemical analysis.

Total protein was estimated using Lowry's et al. (1951) method. The assay used 50 mg of the dried sample homogenized in 10 ml phosphate buffer followed by collection of supernatant after centrifuge. The supernatant was treated with complex forming reagent (2 % Na_2CO_3: 1 % $CuSO_4$, $5H_2O$: 2 % sodium potassium tartrate = 100:1:1) followed by addition of Folin's reagent. The optical density was determined at 750 nm using a spectrophotometer (Systronics Digital Spectrophotometer; Type-16S). BSA (Bovine Serum Albumin) was used as standard for the preparation of calibration curve.

Gravimetric method of Barnes and Blackstock (1973) was used for the estimation of lipid concentration. A sample of 0.5 g was homogenized in 5 ml of double-distilled water and allowed to stand for overnight in the refrigerator. For the gravimetric determination of lipid, aliquots of the homogenate were extracted in 5 ml of 1:2 (v/v) methanol:chloroform (Folch et al. 1957). Lipid residues were weighed using a Mettler AB 204-S microbalance after evaporation of the chloroform using liquid nitrogen.

Methodology of Hewitt (1958) was used for the determination of glycogen content. 100 mg sample was homogenized in 5 ml sulphuric acid reagent and extracted overnight at 5 °C. Then the homogenate solution was centrifuged and subdivided into two portions. One portion was incubated in a water bath at 95 °C for 4 h, and the other portion was stored at 5 °C. The optical density of both portions was determined at 340 nm using a spectrophotometer (Systronics Digital Spectrophotometer; Type-16S). D-glucose was used as standard solution for the preparation of calibration curve.

The excess water of the oyster tissue was soaked using a Whatman filter paper (No. 1). The meat was then homogenized by a tissue homogenizer. A part of the homogenized tissue was oven dried at 100 °C to determine the moisture content. The ash content in oyster tissue was determined by placing the oyster sample in the Muffle Furnace overnight at 400 °C. The ash was then weighed and expressed in percentage.

8A.3.3 Results and Discussion

In Indian Sundarbans region, edible oyster (*Saccostrea cucullata*) exhibits a unique seasonal cycle with respect to their biology and biochemical composition.

The physico-chemical variables showed significant seasonal variations in the sampling station during the study period. Surface water temperature was high during premonsoon (March–June) and monsoon (July–October) and low during postmonsoon (November–February). The surface water salinity and pH were highest in

Table 8A.3.1 Hydrobiological parameters of the selected sampling station during January 2004 to December 2005

Month	Salinity (‰)	pH	Water temp. (°C)	Chl a (mg/m^3)	Chl b (mg/m^3)	Chl c (mg/m^3)	NO$_3$ (µg at/l)	PO$_4$ (µg at/l)	SiO$_3$ (µg at/l)
Jan (2004)	21.44	7.68	28.6	2.22	1.35	1.42	21.14	2.12	63.18
Feb	21.69	7.80	29.8	2.53	1.47	1.55	18.63	2.08	49.22
Mar	22.37	7.82	29.9	2.65	1.66	1.70	16.82	1.54	51.49
April	22.99	7.84	32.8	2.88	1.71	1.84	15.03	1.38	46.15
May	25.86	7.86	33.3	2.91	1.66	1.73	12.44	1.21	46.30
June	27.04	8.00	33.7	3.02	1.78	1.85	10.52	1.08	38.66
July	19.45	7.79	34.1	2.37	1.55	1.62	19.65	2.14	52.20
Aug	16.98	7.40	33.6	2.29	1.46	1.51	23.71	2.48	57.15
Sept	13.72	7.36	33.0	1.79	1.27	1.34	25.82	2.50	60.30
Oct	17.88	7.60	32.1	2.72	1.33	1.49	20.99	2.36	52.37
Nov	19.59	7.62	30.8	2.61	1.69	1.80	17.16	2.22	46.14
Dec (2004)	20.36	7.65	27.2	2.89	1.48	1.65	22.05	2.09	40.24
Jan (2005)	22.30	7.67	28.7	2.09	1.23	1.37	19.65	2.19	55.72
Feb	23.18	7.76	29.7	2.42	1.36	1.48	18.70	2.10	53.00
Mar	23.17	7.79	29.9	2.55	1.53	1.61	18.56	1.67	48.06
April	24.46	7.80	32.9	2.79	1.68	1.77	17.11	1.56	43.21
May	24.80	7.87	33.4	2.80	1.70	1.80	14.09	1.33	37.89
June	26.17	8.01	34.0	2.92	1.77	1.82	13.46	1.27	35.01
July	20.19	7.77	34.2	2.29	1.44	1.50	19.05	2.29	57.99
Aug	18.76	7.40	33.8	2.20	1.30	1.42	24.15	2.50	60.03
Sept	14.50	7.36	33.3	1.50	1.16	1.30	27.38	2.61	71.47
Oct	18.71	7.55	32.9	2.64	1.20	1.29	23.39	2.45	62.83
Nov	21.60	7.60	30.9	2.50	1.38	1.41	19.42	2.39	58.20
Dec (2005)	21.42	7.64	27.4	2.33	1.40	1.48	17.89	2.31	52.76

the season of premonsoon and lowest in monsoon that might be due to excessive evaporation in premonsoon and heavy precipitation and subsequent discharge of freshwater run-off from the adjacent city of Kolkata in monsoon. The quantum of discharge in the form of sewage also increases the nutrient load in the aquatic phase during monsoon (Table 8A.3.1). Report states that 1,125 million litres of wastewater is discharged per day through Hugli estuary. The lower stretch receives waste and wastewater load of 396×10^8 km^3/h along with the annual run-off 493 km^3. The total volume of sewage discharge from the environment of Kolkata has been estimated 350 m (Mukherjee 2003). The minimum nutrient load in the water of Indian Sundarbans during premonsoon may be attributed to their uptake by the phytoplankton community that propagates in March/April in the present geographical locale. The average composition of phytoplankton is

$(CH_2O)_{108}(NH_3)_{16}H_3PO_4$, and in case of siliceous diatom, it is slightly modified as $(CH_2O)_{108}$ $(NH_3)_{16}H_3PO_4(SiO_4)_{40}$ as stated in the standard literature (Riley and Chester 1971), which justifies the incorporation of nutrients in cell system of phytoplankton leading to the reduction in nutrient concentration of ambient water. Apparently this fact has been highlighted through an inverse relationship between nutrient level and phytopigment concentration (Table 8A.3.3).

The biochemical composition of the oyster tissue showed significant seasonal variation of protein (5.49–11.87 %), lipid (5.51–10.76 %), glycogen (1.02–6.95 %), moisture (74.60–77.10 %) and ash (0.20–3.09 %) content during the 2-year study period (Table 8A.3.2).

Analysis of biochemical composition of oysters showed significant seasonal variation. It was seen that reproductive cycle greatly influences the protein–lipid–glycogen content of the tissue.

Table 8A.3.2 Biochemical composition (in %) of edible oyster (*Saccostrea cucullata*) collected from the selected sampling station during January 2004 to December 2005

Month	Protein	Lipid	Glycogen	Moisture	Ash
Jan (2004)	11.60	8.49	2.29	74.60	3.02
Feb	11.87	9.92	1.90	74.70	1.61
Mar	10.31	9.75	2.63	75.10	2.21
April	8.12	8.93	4.59	75.80	2.56
May	8.47	9.15	6.28	75.90	0.20
June	5.73	10.09	7.30	76.20	0.68
July	10.58	7.95	2.15	76.80	2.52
Aug	10.16	8.10	1.82	77.10	2.82
Sept	10.39	6.43	3.20	77.00	2.98
Oct	10.09	6.92	4.15	76.50	2.34
Nov	10.29	6.33	5.42	76.30	1.66
Dec (2004)	9.49	6.51	5.95	75.40	2.65
Jan (2005)	11.15	10.11	2.42	75.10	1.22
Feb	11.01	10.01	1.86	74.80	2.32
Mar	10.36	10.32	2.59	75.10	1.63
April	8.00	9.12	4.72	75.30	2.86
May	8.33	8.80	6.95	74.83	1.09
June	5.49	10.76	7.23	76.10	0.42
July	10.46	9.03	1.02	76.40	3.09
Aug	10.21	8.98	1.99	76.80	2.02
Sept	10.43	7.12	3.01	76.70	2.74
Oct	10.46	6.57	4.56	76.40	2.01
Nov	10.13	6.11	5.55	76.20	2.01
Dec (2005)	9.65	6.11	5.92	76.30	2.02

Protein value reached maximum in the months of January–February and September–October, *i.e.* in the pre-spawning period. In the months of April–June and December, it showed minimum value, *i.e.* immediately after spawning, the protein value sharply decreased. During other times of the year, protein showed average value of 9.70 %.

Considerable seasonal variations of glycogen and lipid content in oyster tissues were also observed during the study period. There were significant positive correlations of glycogen level in the oyster tissue with seawater salinity, temperature and pH, which confirms that glycogen content in oyster hiked up before spawning during the period of high salinity, temperature and pH in late premonsoon (Table 8A.3.3). Add correlation value of moisture and ash.

In the pre-spawning phase when protein and glycogen content reached maximum value, lipid content exhibited minimum value indicating possible interconversion between them. As carbohydrates are recognized as the major energy source in bivalves, (Gabbott 1975) lower value of lipid concentration during prespawning period indicates possible mobilization of lipid towards glycogen to provide energy necessary for the spawning process. The present study reveals that biochemical composition of edible oyster (*Saccostrea cucullata*) in Indian Sundarbans is controlled by the seasonal influence of hydrobiological parameters, which also governs the reproductive cycle in oysters. A long-term monitoring of relevant hydrobiological parameters is therefore needed to explore the environmental potential of the region to develop large-scale oyster culture practice.

Sundarbans, being the only mangrove-dominated tiger land in the planet Earth, is presently

Table 8A.3.3 Interrelationship between the relevant hydrobiological parameters and biochemical composition of edible oyster (*Saccostrea cucullata*) in the selected sampling station during January 2004 to December 2005

Combination	*r*-value	*p*-value
Salinity × protein	−0.5935	<0.01
pH × protein	−0.5626	<0.01
Water temp. × protein	−0.4287	<0.05
Chl *a* × protein	−0.6153	<0.01
Chl *b* × protein	−0.7034	<0.01
Chl *c* × protein	−0.7051	<0.01
NO$_3$ × protein	0.6620	<0.01
PO$_4$ × protein	0.7362	<0.01
SiO$_3$ × protein	0.7051	<0.01
Salinity × glycogen	0.4942	<0.05
pH × glycogen	0.4024	IS
Water temp. × glycogen	0.0570	IS
Chl *a* × glycogen	0.6220	<0.01
Chl *b* × glycogen	0.5287	<0.05
Chl *c* × glycogen	0.5521	<0.01
NO$_3$ × glycogen	−0.5791	<0.01
PO$_4$ × glycogen	−0.5460	<0.05
SiO$_3$ × glycogen	−0.6455	<0.01
Salinity × lipid	0.6450	<0.01
pH × lipid	0.6429	<0.01
Water temp. × lipid	0.1333	IS
Chl *a* × lipid	0.2707	IS
Chl *b* × lipid	0.4137	IS
Chl *c* × lipid	0.3952	IS
NO$_3$ × lipid	−0.5134	<0.05
PO$_4$ × lipid	−0.6249	<0.01
SiO$_3$ × lipid	−0.3842	IS
Salinity × moisture	−0.5798	<0.01
pH × moisture	−0.5259	<0.05
Water temp. × moisture	0.5700	<0.01
Chl *a* × moisture	−0.3043	IS
Chl *b* × moisture	−0.2151	IS
Chl *c* × moisture	−0.2434	IS
NO$_3$ × moisture	−0.3768	IS
PO$_4$ × moisture	0.4374	IS
SiO$_3$ × moisture	0.3320	IS
Salinity × ash	−0.6636	<0.01
pH × ash	−0.5524	<0.01
Water temp. × ash	−0.0936	IS
Chl *a* × ash	−0.4968	<0.05
Chl *b* × ash	−0.4528	<0.05
Chl *c* × ash	−0.4441	IS
NO$_3$ × ash	0.6832	<0.01
PO$_4$ × ash	0.6085	<0.01
SiO$_3$ × ash	−0.5619	<0.01

IS insignificant

under severe stress due to natural calamities, erosion, shrimp culture-related problems and lack of proper planning in resource management. Although the region is flooded with several species of seaweeds, edible molluscs and several organisms with biomedical values, but hardly any initiative has been taken to link these untapped biological resources with the economics of the state. Under such circumstances, promotion of edible oyster culture may be an alternative livelihood scheme for the local population.

References of Annexure 8A.3

APHA (American Public Health Association) (1998) Standard methods for examination of water and waste water, 20th edn. APHA, Washington DC

Barnes H, Blackstock J (1973) Estimation of lipids in marine animals and tissues: detailed investigation of the sulpho-phosphovanillin method for 'total' lipids. J Exp Mar Biol Ecol 12:103–118

Folch JM, Lees M, Sloane-Stanley GH (1957) A simple method for the isolation and purification of total lipids from animal tissues. J Biol Chem 226:497–509

Gabbott PA (1975) Strong cycles in marine bivalve mollusks: a hypothesis concerning the relationship between glycogen metabolism and gametogenesis. In: Barnes H (ed) Proceedings of the ninth European marine biology symposium. Aberdeen University Press, Aberdeen, pp 191–211

Giese AC, Hart MA, Smith AM, Cheung AM (1976) Seasonal changes in body component indices and chemical composition in the Pisno clam *Tivela stultorum*. Comp Biochem Physiol 22:549–561

Graves et al (1996) United states patent, Patent Number 5,510,551, 23 Apr 1996. Site Reference: www.google.com/patents/?vid=USPAT5510551&output=pdf. Accessed 06 July 2013

Hewitt BR (1958) Spectrophotometric determination of total carbohydrate. Nature 182:246–247

Jeffrey SW, Humphrey GR (1975) New spectrophotometric equations for determining chlorophylls *a*, *b*, *c$_1$* and *c$_2$* in higher plants, algae and natural phytoplankton, Biochem Physiol Pflanzen 167:191–194

Lowry OH, Rosebrough NJ, Farr AL, Randall RJ (1951) Protein measurement with the folin phenol reagent. J Biol Chem 193:265–275

MacDonald BA, Thompson RJ (1985) Influence of temperature and food availability on the ecological energetics of the giant scallop *Placopecten magellanicus* II. Reproductive output and total production. Mar Ecol Prog Ser 25:295–330

Mitra A, Banerjee K (2005) In: Banerjee Lt. Col. SR (ed) Living resources of the sea: focus Indian Sundarbans. Conservation approaches and few success stories of

WWF - India. Published by WWF-India, Sundarbans Landscape Project, Canning Field Office, West Bengal

Mukherjee M (Compiled and Edited) (2003) A draft report on Sundarban wetlands. Department of Fisheries, Government of West Bengal, Kolkata

Riley JP, Chester R (1971) Introduction to marine chemistry. Academic Press, Inc, London

Ruwa RK, Polk P (1994) Patterns of spat settlement recorded for tropical oyster *Crassostrea cucullata* (Born 1778) and the barnacle, *Balanus amphitrite* (Darwin 1854) in a mangrove creek. Trop Zool 1:121–131

Steele S, Mulcahy MF (1999) Gametogenesis of the oyster *Crassostrea gigas* in southern Ireland. J Mar Biol Assoc UK 79:673–686

Strickland JDH, Parsons TR (1968) A practical handbook of seawater analysis. Fish Res Bd Can Bull 167:311pp

Tack JF, Van den-Berghe E, Polk P (1992) Ecomorphology of *Crassostrea cucullata* (Born 1778) (Ostreidae) in mangrove creek (Gazi Kenya). Hydrobiolopension feeding ascidians and jellyfish. Ophelia 41:329–344